Opportunistic Mobile Social Networks

Opportunistic Mobile Social Networks

Edited by
Jie Wu and Yunsheng Wang

CRC Press is an imprint of the
Taylor & Francis Group, an **informa** business

CRC Press
Taylor & Francis Group
6000 Broken Sound Parkway NW, Suite 300
Boca Raton, FL 33487-2742

First issued in paperback 2019

ISBN-13: 978-1-4665-9494-4 (hbk)
ISBN-13: 978-0-367-37843-1 (pbk)

Library of Congress Cataloging-in-Publication Data

Opportunistic mobile social networks / edited by Jie Wu, Yunsheng Wang.
pages cm
Includes bibliographical references and index.
ISBN 978-1-4665-9494-4 (hardback)
1. Ad hoc networks (Computer networks) 2. Mobile computing. 3. Online social networks. I. Wu, Jie, 1961- II. Wang, Yunsheng, 1984-

TK5105.77.O67 2014
006.7'54--dc23 2014013288

Visit the Taylor & Francis Web site at
http://www.taylorandfrancis.com

and the CRC Press Web site at
http://www.crcpress.com

Dedication

To our parents, Zengchang Wu, Yeyi Shao, Keming Wang, and Yingyan Wang

Contents

List of Figures

List of Tables

Preface

Over the past few decades, social networks have attracted massive interest from scholars in fields as diverse as sociology, biology, physics, business, politics, and computer science. From these diverse fields, researchers have found that many systems can be represented as networks, and that there is much to be learned by studying those networks. With the rapid growth of the Internet and the web, large-scale social network analysis has become possible for researchers. The most important difference between the traditional and new social networks is that the traditional theories of social networks have not been very concerned with the structure of naturally occurring networks. Traditional social network analysis is deep and elegant, but it is not especially relevant to networks arising in the real world. The emergence of recent mobile devices and their applications have brought about a new landscape in studying social networks.

The recent availability of mobile devices coupled with recent advancements in networking capabilities make opportunistic networks one of the most promising technologies for next-generation mobile applications. Opportunistic networks are commonly defined as a type of network where communication is challenged by sporadic and intermittent contacts, as well as frequent disconnections and reconnections, and where the assumption of the existence of an end-to-end path between the source and the destination is relinquished. Connectivity disruptions, limited network capacity, energy and storage constraints of those participating, mobile devices, and the arbitrary movement of nodes are only a few of the challenges that must be dealt with by the protocol stack. Clearly, current Internet protocols (i.e., the TCP/IP protocol stack) suffer and can fail under such conditions. Opportunities can be useful for building both ad hoc and delay-tolerant networks for data, but they can also be mined for information about mobility and social structures. However, to do either of these, users need to be persuaded to share resources, either at the information level, which impacts privacy, or at the communications level, which impacts their own network performance.

With new challenges brought up by the aforementioned emerging mobile technology in social networks and opportunistic networks, we have recently witnessed the rise of an emerging cross-disciplinary field called *opportunistic mobile social networks*, which has started to receive much attention from practitioners, scholars, and the general public. An opportunistic mobile social network can be described as a platform that provides services via hand held and wireless devices, mainly for the purpose of fostering and maintaining social interactions and connections. Therefore, an opportunistic mobile social network can be considered a form of social network where services are provided with mobility as an added value. Thanks to mobility, many emergent applications related to social networking are now available for individuals, business enterprises, and governments.

The convergence of social networks and opportunistic networks has its own implications for theory and practice. From a theoretical perspective, new research domains have emerged to tackle opportunistic mobile social networks from technological, social, behavioral, legal, and ethical standpoints. From a practical perspective, it is a topic that can notice new forms of collaboration, such as the one between mobile network operators and social networking sites, to offer new innovative services. Moreover, new opportunities are now available to individuals, business organizations, and governments, such as location-based services, content distribution systems, early warning systems in crisis management, and business cooperation monitoring. Indeed, these implications call for urgent attention to further investigate all related and significant issues of opportunistic mobile social networks, so as to advance our understanding and knowledge in this context.

The main goal of this book is to collect the recent development on theoretical, algorithmic, and application-based aspects of opportunistic mobile social networks. This book will be of particular value to academics, researchers, practitioners, government officials, business organizations (e.g., executives, marketing professionals, and resource managers), and even customers—those working in, participating in, or even those interested in fields related to social networks. The content of the book will be especially useful for students in areas like social networks, informatics, wireless networks, data mining, and administrative sciences and management, but also applies to students of education, economy, or law, who would benefit from the information, cases, and examples therein.

This book is expected to serve as a reference book for developers in the telecommunications industry, and for a graduate course in computer science and engineering. Our focus is to expose readers the technical challenges of opportunistic mobile social networking, and to offer some ideas on how we might overcome them. This book is organized in four areas with a total of 16 chapters. Each area corresponds to an important snapshot, according to what we believe, in this fast-growing field. Although several books have emerged recently in this area, none of them address all four areas in terms of critical issues and possible solutions.

- Fundamental concepts and models in opportunistic mobile social networks (Chapters 1–5)
- Routing and forwarding schemes in opportunistic mobile social networks (Chapters 6–9)

- Privacy, security, and economics in opportunistic mobile social networks (Chapters 10–12)
- Applications and testbeds in opportunistic mobile social networks (Chapters 13–16)

Introducing fundamental concepts and models in opportunistic mobile social networks, Chapter 1 presents a systematic analytical study of the constrained information flow problem, which models a pair of networks (social and communication) as a composite graph. Chapter 2 reviews the recent literature of social influence in complex social networks. Chapter 3 provides a comprehensive overview of the fundamental characteristics of link-level connectivity in opportunistic networks, which is crucial in understanding and evaluating network performance. Chapter 4 uses WiFi interactive to discover and predict temporal networks and human population dynamics. Chapter 5 shows how mobility and dynamic network structure impact the processing capacity of opportunistic mobile networks for cloud applications.

In a discussion of routing and forwarding schemes, which spans Chapters 6 to 9, Chapter 6 provides a comprehensive overview of the routing schemes proposed in opportunistic mobile social networks, with a focus on encounter-based unicasting and social-based unicasting. A brief overview of several multicast approaches is also given. Chapter 7 takes an in-depth look into multicast protocols, which are classified based on the number of copies of the multicast message for opportunistic mobile social networks. Chapter 8 focuses on providing pervasive data access to mobile users without the support of cellular or Internet infrastructures. Chapter 9 adopts a data-driven approach, which is based on multiple mobility traces collected from conferences, university campuses, and metropolitan cities, to address four challenges: efficiency, utilization, scalability, and trust.

Issues of privacy, security, and economics in opportunistic mobile social networks are examined in chapters 10 through 12. Chapter 10 applies privacy-preserving techniques with packet forwarding to enhance communication performance and protect users' sensitive information from disclosure. Chapter 11 surveys a collection of approaches that have been recently proposed in the literature to address the need for minimizing privacy leakage during opportunistic user profile exchange. Chapter 12 introduces economics concepts to help formalize the idea of incentives for rewarding long-term participation.

In the final area of applications and testbeds, Chapter 13 deals with a P2P search framework for intelligent crowdsourcing in opportunistic mobile social networks. Chapter 14 introduces a framework for mobile peer rating using a multi-dimensional metric scheme, based on encounter and location sensing. Chapter 15 investigates Vehicular Ad hoc NETworks (VANETs), as a particular class of opportunistic mobile social networks, under the assumption of social networking for vehicular applications (i.e., safety and entertainment applications). Chapter 16 develops a network emulation testbed called QOMB, that can be used to validate the efficient operation of opportunistic network applications and protocols in scenarios that involve both node mobility and wireless communication.

We would like to express our gratitude to all authors. This book would not be possible without their generous contributions. Our special thanks are given to CRC senior editor, Richard O'Hanley, for his encouragement and guidance. Finally, we want to thank our families for their support and patience during this project. Readers are welcome to send their comments and suggestions to jiewu@temple.edu and ywang@kettering.edu.

Jie Wu
Temple University

Yunsheng Wang
Kettering University

About the Editors

Jie Wu is chair and Laura H. Carnell Professor in the Department of Computer and Information Sciences at Temple University. He is also an Intellectual Ventures endowed visiting chair professor at the National Laboratory for Information Science and Technology, Tsinghua University. His current research interests include mobile computing and wireless networks, routing protocols, cloud and green computing, network trust and security, and social network applications. Dr. Wu regularly publishes in scholarly journals, conference proceedings, and books. He serves on several editorial boards and on organization committees of ACM and IEEE conferences. Dr. Wu is the recipient of the 2011 China Computer Federation (CCF) Overseas Outstanding Achievement Award.

Yungsheng Wang is an assistant professor in the Department of Computer Science in Kettering University, Flint, Michigan. He received a B.Eng. in electronic and information engineering from Dalian University of Technology, Dalian, China, in 2007; a M.Res. in telecommunication from University College London, London, in 2008; and a Ph.D. from the Department of Computer and Information Sciences, Temple University, Philadelphia, Pennsylvania, in 2013. His research interests include various topics in the application and protocols of wireless networks. Currently, his research focuses on the efficient communication in delay tolerant networks and opportunistic mobile social networks. Dr. Wang serves as track co-chair of the International Workshop on Mobile Sensing, Computing and Communication (MSCC) conducted in conjunction with ACM Mobihoc, 2014. He also serves on technical program committees of several IEEE conferences.

Contributors

Amotz Bar-Noy
Department of Computer Science
The Graduate Center of the City University of New York
New York, New York

Prithwish Basu
Advanced Networking Technologies
Raytheon BBN Technologies
Cambridge, Massachusetts

Ben Baumer
Department of Mathematics and Statistics
Smith College
Northampton, Massachusetts

Razvan Beuran
Hokuriku StarBED Technology Center
National Institute of Information and Communications Technology
Ishikawa, Japan

Eyuphan Bulut
Cisco Systems
Richardson, Texas

Guohong Cao
Department of Computer Science and Engineering
The Pennsylvania State University
University Park, Pennsylvania

Chi-Kin Chau
Department of Computing and Information Science
Masdar Institute
Masdar City, United Arab Emirates

Gianpiero Costantino
Istituto di Informatica e Telematica del CNR
Pisa, Italy

Michel Diaz
LAAS-CNRS
Toulouse, France

Do Young Eun
North Carolina State University
Raleigh, North Carolina

Wei Gao
Department of Electrical Engineering and Computer Science
The University of Tennessee
Knoxville, Tennessee

Khaled A. Harras
Computer Science Department
Carnegie Mellon University
Doha, Qatar

Ahmed Helmy
Department of Computer and
Information Science and Engineering
University of Florida
Gainesville, Florida

Buster O. Holzbauer
Rensselaer Polytechnic Institute
Troy, New York

Wenjie Hu
Department of Computer Science
and Engineering
The Pennsylvania State University
University Park, Pennsylvania

Andreas Konstantinidis
Department of Computer Science
University of Cyprus
Nicosia, Cyprus

Udayan Kumar
Department of Computer and
Information Science and Engineering
University of Florida
Gainesville, Florida

Chul-Ho Lee
North Carolina State University
Raleigh, North Carolina

Xiang Li
Adaptive Networks and Control
Laboratory
Department of Electronic Engineering
Fudan University
Shanghai, China

Xiaohui Liang
Department of Electrical & Computer
Engineering
University of Waterloo
Waterloo, Ontario, Canada

Thomas D.C. Little
Department of Electrical and Computer
Engineering
Boston University
Boston, Massachusetts

Chengyin Liu
Sun Yat-Sen University
Guangzhou, Guangdong, P.R. China

Cong Liu
Sun Yat-Sen University
Guangzhou, Guangdong, P.R. China

Rongxing Lu
Department of Electrical & Computer
Engineering
University of Waterloo
Waterloo, Ontario, Canada

Fabio Martinelli
Istituto di Informatica
e Telematica del CNR
Pisa, Italy

Subhankar Mishra
Department of Computer and
Information Science and Engineering
University of Florida
Gainesville, Florida

Shinsuke Miwa
Hokuriku StarBED Technology
Center
National Institute of Information and
Communications Technology
Ishikawa, Japan

Toshiyuki Miyachi
Hokuriku StarBED Technology Center
National Institute of Information and Communications Technology
Ishikawa, Japan

Abderrahmen Mtibaa
Electrical & Computer Engineering
Texas A & M University
Doha, Qatar

Anh-Dung Nguyen
Department of Mathematics Computer Science and Automatic Control
ISAE, University of Toulouse
Toulouse, France
LAAS-CNRS, Toulouse, France

Paolo Santi
Istituto di Informatica e Telematica del CNR
Pisa, Italy

Patrick Senac
Department of Mathematics Computer Science and Automatic Control
ISAE, University of Toulouse
Toulouse, France
LAAS-CNRS, Toulouse, France

Xuemin (Sherman) Shen
Department of Electrical & Computer Engineering
University of Waterloo
Waterloo, Ontario, Canada

Yoichi Shinoda
Research Center for Advanced Computing Infrastructure
Japan Advanced Institute of Science and Technology
Ishikawa, Japan

Boleslaw K. Szymanski
Rensselaer Polytechnic Institute
Troy, New York

My T. Thai
Department of Computer and Information Science and Engineering
University of Florida
Gainesville, Florida
and
Ton Duc Thang University
Ho Chi Minh City, Vietnam

Athanasios V. Vasilakos
Department of Electrical and Computer Engineering
National Technical University of Athens
Athens, Greece

Anna Maria Vegni
Department of Engineering
University of Roma Tre
Rome, Italy

Wei Wang
Sun Yat-Sen University
Guangzhou, Guangdong, P.R. China

Yunsheng Wang
Department of Computer Science
Kettering University
Flint, Michigan

Jie Wu
Department of Computer and Information Sciences
Temple University
Philadelphia, Pennsylvania

Demetrios Zeinalipour-Yazti
Department of Computer Science
University of Cyprus
Nicosia, Cyprus

Huiyuan Zhang
Department of Computer and Information Science and Engineering
University of Florida
Gainesville, Florida

Kuan Zhang
Department of Electrical & Computer Engineering
University of Waterloo
Waterloo, Ontario, Canada

Yi-Qing Zhang
Adaptive Networks and Control Laboratory
Department of Electronic Engineering
Fudan University
Shanghai, China

Chapter 1

Social-Communication Composite Networks

Prithwish Basu
Advanced Networking Technologies
Raytheon BBN Technologies
Cambridge, Massachusetts

Ben Baumer
Department of Mathematics and Statistics
Smith College
Northampton, Massachusetts

Amotz Bar-Noy
Department of Computer Science
The Graduate Center of the City University of New York
New York, New York

Chi-Kin Chau
Department of Computing and Information Science
Masdar Institute
Masdar City, United Arab Emirates

CONTENTS

1.1 Introduction

The recent explosive growth in online social networks has been fueled by the proliferation of high-speed and highly available communication networks such as the Internet and broadband cellular wireless networks, as well as the increasing popularity of mobile network-ready devices such as "smartphones" and tablets. People tend to share information with other people they know, who subsequently forward that information along various links in the social network—this occurs either verbatim (for example, the directives from a commander flow through the *chain of command*) or after modifications (for example, propagation of rumors, gossip, or news on Twitter). A social network's topology thus *constrains* or *guides* the flow and spread of information through it. These constraints can force the information to traverse much longer paths in the underlying communication network between its originator and its ultimate consumers. This phenomenon, known as *stretch*, is justified because the intermediaries may play a critical role in interpreting or modifying the information or they may serve as important links in the acquaintance chain, without whom the originator and the ultimate consumers would not have known each other.

When information gets *stretched*, the total time for it to spread through the entire social network is often different from the time taken to simply *multicast* the information on the underlying communication network to the set of ultimate consumers. An additional undesirable side-effect of the "stretch" phenomenon is that an information object may traverse a communication link or a node several times during the process, thus increasing resource consumption. While this is not a major issue for lightweight content such as text (e.g., 140 character Twitter messages), it can be a significant problem for multimedia content, especially in mobile ad hoc network (MANET) or disruption-tolerant network (DTN) settings where accessing multimedia content directly from a server over a flaky network may not be feasible.

In this chapter, we present a systematic analytical study of the constrained information flow problem—in particular, we model a pair of networks (social and communication) as a *composite graph*—a structure that results from embedding or mapping the social network into the communication network using *embedding / mapping functions*. A mapping function maps a node in the social network to one or more in the communication network when the former *uses* the latter as his/her communication portal(s). We consider unicast, broadcast, and multicast versions of this scenario. We introduce several "composite graph" metrics that capture the effect of constraining the flow of information in the communication network due to the social network, for example, *composite path stretch, composite broadcast time, composite betweenness centrality*, etc. We analytically study how these metrics scale with the sizes of both networks under consideration under various random graph models and mapping functions. The above modeling / analysis can be useful in an application scenario such as the following: workers or soldiers equipped with wireless communication devices have been deployed at a disaster relief site and their group leader disseminates messages to them following a specific chain of command, which is essentially a social network. These messages trace a logical path in the social network that translates to a potentially longer physical path in the underlying communication network (which is a MANET or a DTN).

Information multicast through a chain-of-command hierarchy can also be modeled in the composite graph framework. For many operations and missions in practice, mere topological proximity to certain recipients of a message does not warrant its direct delivery to the latter. Instead, certain hierarchical policies that define different roles and ranks of network nodes may constrain the message flow through the network. For example, in military networks, communications between various nodes may need to be observed and then cleared by individuals located higher in the chain-of-command hierarchy, which is nothing but a social network. It is often the case that a subset of nodes in the hierarchy are interested in participating together in a multicast session. Therefore, we are motivated to construct multicast "routes" that *connect* these nodes while being constrained by the relationships in the hierarchy.

1.1.1 Related Work

There are three classes of related work in this area: graph embedding, network science approaches to studying composite networks, and overlay networks in the Internet.

Graph embedding has received attention in the parallel computing domain where the problem is to map a *task graph* onto a multiprocessor interconnection network (also known as *host graph*) [6, 23, 15], and in the ubiquitous computing domain where the problem is to map heterogeneous task graphs on non-regular networks such as mobile ad hoc networks [4], while attempting to determine the optimal mapping (or task to processor assignment) function such that metrics such as delay-to-task-completion, edge dilation (or stretch), node/edge congestion, etc., are minimized. Instead of the aforementioned "optimization" approaches, in this chapter, we follow the "scaling law analysis" approach where both the graphs and the mapping function are given (deterministic or stochastic), and we characterize how a different set of appropriate "constrained" metrics such as composite path stretch, composite diameter, broadcast time, and composite betweenness centrality scale as a function of composite graph attributes.

There is a large body of work pertaining to the embedding of one metric space into another—in particular, normed spaces such as d-dimensional Euclidean space $\mathbb{R}^d$)—with "low-distortion." This has been summarized well in [16]. This entails establishing the necessary and sufficient conditions on the properties of the two spaces for finding such embedding functions that yield a particular distortion, and in many cases finding the best embedding function [2]. A related idea of finding embeddings is popular in geographic routing—virtual coordinates are assigned to nodes in a hyperbolic space, and such an embedding guarantees that a greedy algorithm on the virtual coordinate space yields a route between every source and destination, if one exists [19].

Various flavors of layered or composite networks have received some attention in the network science literature. Kurant and Thiran propose the Layered Complex Network model [20] for studying load in transportation networks. They considered 2-layer graphs where the physical graph corresponds to the transportation network and the logical graph corresponds to the traffic flow between various cities—they use computational methods to determine different levels of *load* on various transportation sectors in Europe. In comparison, our approach is analytical and we study metrics that have not been studied in [20]. A recent analytical line of research considers interdependent networks such as power grid and communication networks [8]—they use percolation theory to determine the fraction of nodes whose removal is likely to generate cascading failures in such networks. Leicht and D'Souza show that percolation thresholds of composite networks is lower than the individual networks, when considered separately [21]. While these approaches are all analytical, they study a different graph metric, i.e., degree of failure tolerance.

Overlay networks have received a lot of attention in the computer networking literature in the past decade [22]. Works such as CAN [25] and CHORD [28] attempt to design good distributed hash tables for P2P applications—for storing (key, value) pairs *overlaid* on top of the Internet, so that efficient insertion and retrieval of hashed content is feasible from any part of the network. While this is a good example of a composite network, its similarities with our approach are slim. While overlay networks attempt to design good overlay graphs for the purpose of optimization of insertion/lookup overhead, in our problem space, the social network graph is given, and we are interested in a different set of information flow metrics. Moreover, unlike the Internet, which is a complete graph (or clique) for the purpose of connectivity in P2P applications, our underlying network is a multi-hop network, in general.

The focus of this chapter is not *to find* the best embedding function that yields a low distortion—rather, it is *to analyze* the distortion (or stretch) of an information flow that results from a *random* embedding of the nodes of the first graph onto the second graph, in distribution or in expectation. The material in this chapter has been derived in part from two recent publications co-authored by us [5, 3].

Our contributions in this chapter can be summarized as follows:

1. Novel models and metrics for constrained information flow in composite networks.
2. Mathematical analysis of scaling laws for constrained composite path stretch when a social network path is randomly mapped onto a general graph under both one-to-one and many-to-one mappings.
3. Scaling laws for constrained composite broadcast time of a tree social network (chain of command) randomly mapped onto different communication networks.
4. A hierarchy-compliant multicast algorithms for composite network multicast.
5. Validation of a subset of these results using two historical deployments of military chain-of-command networks as well as the FOAF (friend of a friend) data set embedded on a geometric communication graph.

We show that the composite betweenness centrality metric yields significantly better insights about the structure of a communication network compared to classic betweenness centrality computed on a single network. We also demonstrate that one has to be willing to pay a 25% overhead for adhering to the social network structure in certain realistic composite network multicast deployment scenarios.

1.2 Composite Graph Models

We define the *composite graph* $\mathcal{G}$ of two graphs G_1 and G_2 to be the 3-tuple (G_1, G_2, R), where $R \subseteq V(G_1) \times V(G_2)$ is an *embedding / mapping relation* between the vertex sets $V(G_1)$ and $V(G_2)$ of the two graphs, respectively. In general, every element of R may have multiple attributes associated with it but in this preliminary study we only consider a binary relation. This relation may be time-varying when information is replicated or moves from one communication node to another over time. Time-varying relations are outside the scope of this chapter.

1.2.1 Metrics on Composite Graphs

We first define *constrained composite path stretch*, a metric that is useful for measuring how many physical communication hops are spanned by a logical information flow under a given embedding of the logical flow on a physical network.

Throughout this chapter, let $\mathcal{G} = (G_1, G_2, R)$ be a composite graph, with $V_i = V(G_i)$ the vertex set of graph i and R an embedding relation as mentioned above. Unless otherwise noted, $P_k = P_{uv} = \{u = v_0, v_1, ..., v_k = v\}$ is a path of length k in

G_1, and $d_{G_2} : V_2 \times V_2 \rightarrow \mathbb{R}$ is a shortest path distance metric in G_2. For clarity, we introduce the notion of an *itinerary* in a graph.

Definition 1.1 (Itinerary) Given a list of vertices $v_0, v_1, \ldots, v_k$ in a graph, an *itinerary* is a not necessarily simple path passing through $v_0, v_1, \ldots, v_k$ *in order*, for which the path connecting consecutive vertices (v_i, v_{i+1}) is a shortest path, for all $0 \leq i \leq k-1$.

Intuitively, an itinerary is the shortest possible path through a sequence of not necessarily neighboring vertices.

Definition 1.2 (cstretch) Given composite graph $\mathcal{G} = (G_1, G_2, R)$, the constrained composite path stretch of $P_{uv} = \{u = v_0, v_1, ..., v_k = v\}$ in $\mathcal{G}$ is defined as:

$$cstretch_{G_2}(P_{uv}) = \sum_{i=0}^{k-1} \max_{\substack{s,t \in V_2: \\ (v_i,s) \in R \wedge (v_{i+1},t) \in R}} \{d_{G_2}(s,t)\}. \tag{1.1}$$

Equivalently, $cstretch_{G_2}(P_{uv})$ is the longest itinerary through the vertices in G_2 that are images of the vertices of P_{uv} in G_1 under the mapping R. Note that in general, R is not necessarily a bijection, and so there may be multiple vertices in G_2 that correspond to a single vertex in G_1.

CStretch characterizes the scenario with a stringent requirement that the information needs to traverse the nodes in the path P_{uv} *in order*, and in the process need to traverse the appropriately mapped nodes in G_2. This is not a far-fetched scenario—in military systems, the chain-of-command (modeled by graph G_1) often mandates a piece of information to flow through the logical chain even though the ultimate recipient of the information may be in close proximity to the origin and the intermediate nodes are farther away from them. The reason behind this is that information often needs to get refined or obfuscated at each level of the logical chain before it is passed on further. Similarly, even in non-military applications (such as online social networks such as Twitter) information such as news or gossip is often routed along logical paths of friends who may be physically located all over the globe at large "Internet distances" from each other.

In the composite graph setting, the notion of diameter[1] can be extended to that of the *constrained composite diameter*, which can be defined in terms of constrained composite path stretch.

Definition 1.3 (ccd) The constrained composite diameter of $\mathcal{G}$ is defined as

$$ccd(\mathcal{G}) = \max_{u,v \in V_1} cstretch_{G_2}(P_{uv}). \tag{1.2}$$

[1]Diameter is the maximum length of the shortest path between any pair of nodes in a graph. It is an important measure for communication networks because it gives us a sense of the amount of time required (in the worst case) to traverse a network.

The *CStretch* metric captures the extra distance in G_2 that a message has to travel in order to move through a path in G_1. We need a different metric to capture the combined stretch for a message traveling through a chain-of-command *tree* in a composite graph. In this context, it is more natural to consider the *constrained composite broadcast time* metric.

Definition 1.4 (cbtime) Let T be a tree in G_1, with root u. Then the constrained composite broadcast time of T in the composite graph $\mathcal{G}$ is defined as

$$cbtime_{G_2}(T) = \max_{v \in T} cstretch_{G_2}(P_{uv}). \tag{1.3}$$

The constrained composite broadcast time represents the stretch necessary to send a message through a chain-of-command tree that is deployed in a network topology. This may be of interest, for example, in a disaster relief situation when information needs to travel from a central director to end caregivers while relief workers are deployed in the field. In other words, it measures the time at which the last worker received the message that was broadcast through the chain of command.

We are also interested in measuring the traffic load on a particular edge in G_2 as a result of the flows along the edges in G_1.

Definition 1.5 (Load Indicator) For a specific edge $e = (x, y) \in G_2$, we say the edge bears a load from $v_i, v_{i+1} \in V(G_1)$ in the composite graph (G_1, G_2, R) if and only if e lies along a shortest path from a vertex $w_i \in G_2$ to $w_j \in G_2$, where $(v_i, w_i) \in R$ and $(v_{i+1}, w_j) \in R$. Let P_{ij} be any shortest path from $w_i \in G_2$ to $w_j \in G_2$. Then,

$$\chi_e(v_i, v_{i+1}) = \begin{cases} 1 & \text{if } e \in P_{ij} \text{ and } ((v_i, w_i), (v_{i+1}, w_j)) \in R \\ 0 & \text{otherwise.} \end{cases}$$

Definition 1.6 (cload) Let $P_{uv} = \{u = v_0 \rightarrow v_1 \rightarrow v_2 \rightarrow \cdots \rightarrow v_k = v\}$ be a path in graph G_1. Then the composite load on $e \in E(G_2)$ of P_{uv} in G_2 is defined as:

$$cload_{G_2}(P_{uv}, e) = \sum_{i=0}^{k-1} \chi_e(v_i, v_{i+1}).$$

Note that $0 \leq cload_{G_2}(P_{uv}, e) \leq k$. Naturally, we want to determine the maximum and expected measures of load upon any edge in G_2.

Finally, we extend the notion of *betweenness centrality* to composite graphs in order to measure the load on certain vertices and edges in G_2.

Definition 1.7 (cvbc) If (G_1, G_2, R) is a composite graph, let σ_{st} be the number of shortest paths in G_2 between s and $t \in V(G_2)$, and $\sigma_{st}(u)$ be the number of shortest paths in G_2 between s and t which pass through vertex u. Then the *composite vertex* *betweenness centrality* of a vertex $u \in V(G_2)$ is given by

$$cvbc(u) = \sum_{\substack{s \neq u \neq t \in V(G_2) \\ (v,w) \in E(G_1) \wedge \{(v,s),(w,t)\} \subseteq R}} \frac{\sigma_{st}(u)}{\sigma_{st}}. \tag{1.4}$$

Definition 1.8 (cebc) If (G_1, G_2, R) is a composite graph, let σ_{st} be the number of shortest paths in G_2 between s and $t \in V(G_2)$, and $\sigma_{st}(e)$ be the number of shortest paths in G_2 between s and t which pass through edge e. The *composite edge betweenness centrality* of an edge $e \in V(G_2)$ is given by

$$cebc(e) = \sum_{\substack{s \neq t \in V(G_2) \\ (v,w) \in E(G_1) \wedge \{(v,s),(w,t)\} \subseteq R}} \frac{\sigma_{st}(e)}{\sigma_{st}}. \tag{1.5}$$

1.3 Composite Stretch Analysis

In this section, we focus on analyzing *random embedding relations*, where vertices in G_1 are mapped to vertices in G_2 via some random process π. In particular, we study two cases:

1. Each vertex in G_1 is mapped to a vertex in G_2 that has been sampled uniformly at random *with replacement*. This is the many-to-one scenario, where many "social network" nodes can get mapped to the same communication network node.

2. Each vertex in G_1 is mapped to a vertex in G_2 that has been sampled uniformly at random *without replacement*. This is the one-to-one scenario, where a communication network node can host at most one social network node.

Specifically, we characterize the distribution of the constrained composite path stretch of P_k over uniform random embeddings into G_2. We first prove some general results that apply to any graph G_2, and then illustrate scaling laws for a few well-known graph families.

1.3.1 Theoretical Results

For any graph $G = (V, E)$, let D_G be the geodesic graph distance matrix between all pairs of vertices $v_i, v_j \in V$. That is, each entry d_{ij} in D_G represents the shortest path distance from v_i to v_j in G. Then we note that the sum of the geodesic distances $\Delta_G = \sum_{v_i, v_j \in V} d_{ij}$, is a constant depending only on the structure of G.

Lemma 1.1
Let G be a graph with $|V| = n$, and let X be a random variable denoting the geodesic

distance between two vertices of G chosen uniformly at random. Then:

$$\mathbb{E}[X] = \begin{cases} \frac{\Delta_G}{n(n-1)}, & \text{when sampling without replacement} \\ \frac{\Delta_G}{n^2}, & \text{when sampling with replacement.} \end{cases}$$

Proof 1.1 The case where sampling is done with replacement is clear: since there are n^2 pairs of vertices from which to choose, the expression given is the average distance. If sampling is done without replacement, then Δ_G double-counts the distance for each of the $\binom{n}{2}$ unique pairs of vertices. Note that the n diagonal entries in D_G contribute nothing to Δ_G.

Corollary 1.3.1 *There is no asymptotic difference in $\mathbb{E}[X]$ between sampling vertices with or without replacement.*

Proof 1.2 From the preceding lemma, it follows that the ratio of $\mathbb{E}[X]$, when sampling without replacement, to $\mathbb{E}[X]$, when sampling with replacement, is $1 + \frac{1}{n} \to 1$ as $n \to \infty$.

Next, we show that the expected stretch of a link is independent of the choices of vertices already mapped, regardless of whether sampling is done with or without replacement.

Lemma 1.2
Let $v_1, v_2, ..., v_i$ be a sequence of vertices chosen uniformly at random from V (with or without replacement), and let X_i be the random variable giving the distance between v_i and v_{i-1}. Then $\mathbb{E}[\mathbb{E}[X_{i+1}|v_1, v_2, ..., v_i]] = \mathbb{E}[X_2]$.

Proof 1.3 While the statement may be obvious for the case of sampling with replacement, we exercise more care for the case where sampling is done without replacement, and prove the statement combinatorially. For the RHS, select one vertex uniformly at random and color it red (call it v_1). Then select another from the remaining and color it blue (v_2). The RHS counts the expected distance between these two vertices. We now argue that the LHS counts the same. To see this, first color one vertex blue (call it v_{i+1}), and another vertex red (v_i). Now color $i-1$ other vertices green ($v_{i-1}, ..., v_1$). The LHS counts the expected distance between the blue vertex and the red vertex.

This leads us to a general theorem about the expected composite stretch of a path.

Theorem 1.3.1 *For a path P_k embedded uniformly at random into any graph G_2 (with the sampling performed with or without replacement),*

$$\mathbb{E}[cstretch^{\pi}_{G_2}(P_k)] = k \cdot \mathbb{E}[X], \tag{1.6}$$

where X is the random variable giving the distance between two randomly chosen vertices in G_2.

We emphasize that the expectation is being taken over the uniform random embedding R_π. But as we saw in Lemma 1.1, for a specific G_2, if the sampling method of R_π is known, then the expected distance $\mathbb{E}[X]$ is a constant.

Composite Diameter: In addition to the average case, we also want to describe the worst-case *cstretch* for a random embedding. It is easy to see that if R_π samples vertices with replacement, then each successive link in any path can simply bounce back and forth between the furthest two vertices in G_2. Thus, $ccd(\mathcal{G}) = diam(G_1) \cdot diam(G_2)$. However, when R_π samples vertices without replacement, the problem is an instance of MAX-TSP, which is MAX SNP-hard [14]. However, a greedy approximation heuristic works well in practice.

1.3.2 Composite Stretch of Some Special Graphs

Theorem 1.3.1 shows that the expected stretch of a path is equal to the length of the path times a constant depending only on the structure of G_2 and the distribution of the random embedding. In what follows, we present examples of some well-known graph families, and illustrate how their structure affects the distribution of *cstretch*.

d-dimensional Discrete Lattice: Let $D_n^d = \{0,1,...,n-1\}^d$ be the d-dimensional discrete lattice on n^d points, and consider a composite graph with $G_2 = D_n^d$. On this graph topology, geodesic distance is equivalent to the ℓ_1-norm (Manhattan distance) between two points in D_n^d. Thus, $d_{G_2}(v,w) = \sum_{i=1}^d |v_i - w_i|$, and summing all n^{2d} of these pairs gives

$$\Delta_{G_2} = \sum_{v,w \in V} \sum_{i=1}^{d} |v_i - w_i| = \frac{dn^{2d+1}}{3}\left(1 - \frac{1}{n^2}\right). \tag{1.7}$$

It follows from Lemma 1.1 and Theorem 1.3.1 that under a random uniform embedding with replacement into the d-dimensional discrete lattice,

$$\mathbb{E}[cstretch_{G_2}^{\pi}(P_k)] = \frac{kdn}{3}\left(1 - \frac{1}{n^2}\right). \tag{1.8}$$

Note that in this case it is also straightforward to fully explicate the distribution of X. For any $1 \le i \le d$, let $X_i = |v_i - w_i|$. Then the probability mass function for X_i is

$$p_{X_i}(\delta) = \begin{cases} \frac{1}{n} & \text{if } \delta = 0 \\ \frac{2(n-\delta)}{n^2} & \text{otherwise} \end{cases}, \tag{1.9}$$

since each coordinate can take on any of n values, and there are $n - \delta$ ways to achieve each value of δ between 0 and $n-1$. Since the X_i's are independent and identically distributed, we can extract (among other things), the second moment of X:

$$\mathrm{Var}[X] = d \cdot \frac{(n^2-1)(n^2+2)}{18n^2}. \tag{1.10}$$

We can infer from this that the expected stretch is not likely to deviate significantly from its mean.

For the discrete lattice, we have that $diam(G_2) = d(n-1)$, so as mentioned above, the *ccd* for P_k is $k(n-1)$. For the non-trivial "without replacement" scenario, we implemented a greedy approximation heuristic, and verified that *ccd* for both without and with replacement scenarios are $O(n^2)$.

Cycle: Let C_n be the cycle of length n, and consider uniform discrete mappings from P_k onto C_n. Clearly, the maximum distance between two vertices in C_n is $\lfloor \frac{n}{2} \rfloor$. But, for each possible distance x between 0 and $\frac{n}{2}$, there are exactly n such pairs for $x = 0, \frac{n}{2}$, and exactly $2n$ such pairs otherwise (we assume that in the case of a tie, only one shortest path is kept). It is thus straightforward to show that

$$\Delta_{C_n} = \begin{cases} \frac{n(n^2-1)}{4} & \text{if } n \text{ is odd} \\ \frac{n^3}{4} & \text{if } n \text{ is even} \end{cases} . \tag{1.11}$$

Application of Lemma 1.1 and Theorem 1.3.1 then reveal that for random uniform embeddings onto C_n,

$$\mathbb{E}[cstretch^{\pi}_{C_n}(P_k)] = k \cdot \left(\frac{n}{4} + o(1)\right) . \tag{1.12}$$

Greedy is optimal on C_n, since if n is odd, it finds $n-1$ pairs at distance $\lfloor \frac{n}{2} \rfloor = diam(G_2)$ from each other, which is optimal by definition. On the other hand, if n is even, it picks *all* $\frac{n}{2}$ pairs at distance $\frac{n}{2} = diam(G_2)$ from each other, and another $\left(\frac{n}{2} - 1\right)$ pairs at the next greatest distance $\left(\frac{n}{2} - 1\right)$.

Balloon graph: Next, we consider a graph family with some interesting properties. Let $B_{n,m}$ be a balloon graph consisting of a string (line graph) of length m, connected to a balloon (clique) of size $n-m$, for any $0 \leq m < n$. For clarity, we specify that vertices $\{v_0, \ldots, v_m\}$ make up the string, while vertices $\{v_m, \ldots, v_{n-1}\}$ make up the balloon (see Figure 1.1). Note that for any two indices $0 \leq i < j \leq n-1$ in this graph, we have that

$$d_{B_{n,m}}(v_i, v_j) = \begin{cases} j - i & \text{if } i < j \leq m \\ m + 1 - i & \text{if } i < m \leq j \\ 1 & \text{if } m \leq i < j. \end{cases}$$

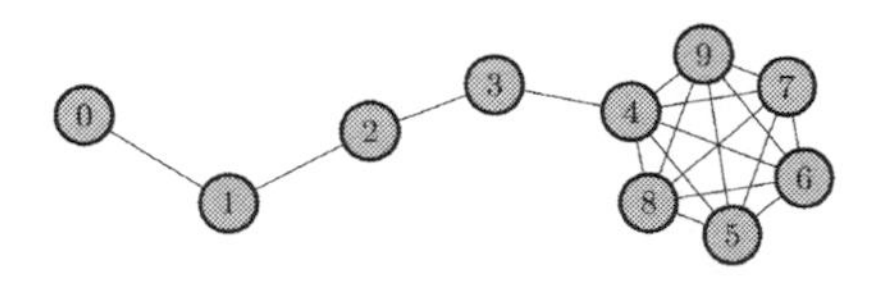

Figure 1.1: The balloon graph $B_{10,4}$.

In particular, note that $diam(B_{n,m}) = m+1$. In computing the distance matrix, we distinguish three cases based on the indices of the two vertices chosen:

1. If $i \leq j \leq m$, then both vertices lie in the string, which is D^1_{m+1}. This contributes $\Delta_{D^1_{m+1}}$ toward $\Delta_{B_{n,m}}$.

2. If $m \leq i \leq j$, then both vertices lie in the balloon, and it is clear that on the complete graph $K_n, \Delta_{K_n} = n^2 - n$, since every pair of vertices are connected by an edge, but there are n ways to choose the same vertex twice.

3. If $i < m < j$, then one vertex lies in the string, and the other lies in the balloon. Consider any vertex w_j in the balloon. Its distance from the set of vertices in the string is simply $m+1, m, m-1, ..., 2$. Thus, the contribution to $\Delta_{B_{n,m}}$ is

$$2(n-m-1)\sum_{i=2}^{m+1} i = m(m+3)(n-m-1).$$

Adding these three quantities yields

$$\Delta_{B_{n,m}} = -\frac{2}{3}m^3 + (n-2)m^2 + \left(n - \frac{4}{3}\right)m + n^2 - n. \tag{1.13}$$

The reader may verify that setting $m = 0$ corresponds to the special case where the balloon graph is itself a clique, while setting $m = n-1$ yields the special case where $B_{n,n-1} = D^1_n$.

By Theorem 1.3.1 and Lemma 1.1, the expected stretch for a path of length k onto $B_{n,m}$ is thus:

$$\mathbb{E}[cstretch^{\pi}_{B_{n,m}}(P_k)] = k \cdot \left(1 + O\left(\frac{m^2}{n}\right)\right) \tag{1.14}$$

Random Geometric Graph: Lastly, we consider the composite stretch when P_k is mapped onto a random geometric graph $G_2 = RGG(n, r(n))$, where $r(n)$ is the radius of communication. That is, G_2 consists of n vertices placed uniformly at random in $[0,1]^2$, wherein any two vertices are connected with an edge if and only if the Euclidean distance between them is at most $r(n)$. Gupta and Kumar [13] showed that a radius of connectivity of $r(n) = \sqrt{\frac{\ln n + c(n)}{\pi n}}$ ensures asymptotic connectivity in the RGG with high probability if and only if $c(n) \to +\infty$. In all of our discussions on RGG in this chapter, we assume that the radius of connectivity is at least this large, i.e., $r(n) = \Omega(\sqrt{\ln n / n})$.

As before, Theorem 1.3.1 still applies, so it remains only to characterize the distribution of the random variable X giving the geodesic distance between two vertices in $RGG(n, r(n))$ selected uniformly at random. Note that in contrast to the previous examples we have considered, we now have two sources of randomness: 1) the randomized construction of the RGG; and 2) the random uniform embedding. If the

Euclidean distance between two vertices in an RGG is δ, then recent results confirm that with high probability, the geodesic distance X differs from its minimum of δ/r by at most a constant [7].

Theorem 1.3.2 *With high probability, the expected geodesic distance in $RGG(n,r(n))$ satisfies*

$$\frac{\Delta(2)}{r(n)} \leq \mathbb{E}[X] \leq \kappa(n) \cdot \frac{\Delta(2)}{r(n)}, \tag{1.15}$$

where $\Delta(2) \approx 0.5214054331$ is a known constant, and $\kappa(n) \geq 1$ is $O(1)$.

Proof 1.4 Let v, w be two vertices in $RGG(n,r(n))$ selected uniformly at random, and set $\delta = ||v-w||_2$. Clearly, $X \geq \delta/r$. Conversely, if $\delta = \Omega(\log^{3.5} n/r^2)$, then by a result from [7], $X = O(\delta/r)$.

Taking expectation yields the result, since $\mathbb{E}[\delta] = \Delta(2)$ is a known constant [29].

Synthetic analysis suggests that $\kappa(n) < 1.3$ for $n > 1000$. Therefore, as before, we can easily bound (from above) the expected composite stretch.

Corollary 1.3.2 *For $r(n)$ sufficiently large (i.e., greater than the critical connectivity threshold), the composite stretch of a path P_k on a random geometric graph $RGG(n,r(n))$ satisfies with high probability:*

$$\mathbb{E}[cstretch^{\pi}_{RGG}(P_k)] = k \cdot \kappa(n) \cdot \frac{\Delta(2)}{r(n)} = O\left(k \cdot \sqrt{\frac{n}{\ln n}}\right). \tag{1.16}$$

1.3.3 Average vs. Worst-Case Analysis

We have so far characterized the average case (expected $cstretch$) and the worst case (ccd) for a random uniform embedding of a path onto several graph families. For both the lattice and the cycle, these quantities were of the same order of magnitude. A natural question is:

> Are there graphs for which the ratio of the maximum $cstretch$ to the average $cstretch$ of P_k is *not* $O(1)$?

Indeed, the balloon graph is one such graph. As the diameter of $B_{n,m}$ is $m+1$, the maximum stretch is $diam(G_1) \cdot (m+1)$. If we let $\phi(B_{n,m})$ be the ratio of the maximum $cstretch$ to the mean $cstretch$, we can see that:

$$\phi(B_{n,m}) = \frac{diam(G_1)(m+1)}{diam(G_1)\left(1+O\left(\frac{m^2}{n}\right)\right)} = O\left(\frac{n}{m}\right).$$

In particular then, for $m = \sqrt{n}$, the ratio of the maximum stretch to the mean stretch for the balloon graph $B_{n,m}$ is $O(\sqrt{n})$. Explicit calculations reveal that for $m = \sqrt{n}$, in fact $\mathbb{E}[X] \to 2$ as $n \to \infty$.

Table 1.1 Summary of Path Stretch Metrics for Uniform Random Embeddings of P_k

G_1	G_2	$\mathbb{E}[cstretch]$	$\max[cstretch]$
P_k	D_n^d	$\frac{kdn}{3}\left(1-n^{-2}\right)$	$kd(n-1)$
	C_n	$k\cdot\left(\frac{n}{4}+o(1)\right)$	$k\cdot\lfloor\frac{n}{2}\rfloor$
	$B_{n,m}$	$k\cdot\left(1+O\left(\frac{m^2}{n}\right)\right)$	$k(m+1)$
	$RGG(n,r(n))$	$O\left(k\sqrt{\frac{n}{\ln n}}\right)$	

More interesting is the fact that this gap appears to be mainly an artifact of the difference between sampling with and without replacement. The results of our greedy algorithm for CCD without replacement suggest that with $m=\sqrt{n}$; the CCD and expected $cstretch$ are of the same order of magnitude.

Table 1.1 summarizes our theoretical results.

1.4 Composite Broadcast Time

In this section, we analytically characterize the expected composite broadcast time for tree topologies. Social networks for information dissemination commonly have tree structures (more on this in Section 1.7), hence this analysis can be useful for specific communication network deployment scenarios. Let T_k be a k-node tree of height h and maximum (out)degree δ, for some $1\leq\delta<k$. We assume that T_k exists in some G_1, and examine the constrained composite broadcast time for sending a message from the root to each of the other nodes.

Star Topology: We begin with the special case where T_k is a k-star. First, we introduce a notation. Let

$$p_k=\frac{1}{\binom{n-1}{k}}\left(\underbrace{0,\ldots,0}_{k},1,\binom{k}{k-1},\ldots,\binom{n-2}{k-1}\right)\in\mathbb{R}^n$$

be a column vector, and note that $||p_k||_1=1$. The ith entry in p_k represents the probability that the ith largest among n values is returned, when this value is the maximum among a subset of size k chosen uniformly at random. Furthermore, let $f:\mathbb{R}^{m\times n}\to\mathbb{R}^{m\times n}$ be the function that sorts the rows of a matrix in ascending order from left to right. That is,

$$D=\begin{bmatrix}d_1\\d_2\\\vdots\\d_m\end{bmatrix}\Rightarrow f(D)=\begin{bmatrix}sort(d_1)\\sort(d_2)\\\vdots\\sort(d_m)\end{bmatrix},$$

where d_i is the ith row of D. Finally, $v_m=\frac{1}{m}(1,\ldots,1)\in\mathbb{R}^m$.

Theorem 1.4.1 *For any graph G_2, the broadcast time of a star of size k satisfies*

$$\mathbb{E}[cbtime_{G_2}(S_k)] = v_n^T \cdot f(D_{G_2}) \cdot p_k. \tag{1.17}$$

Proof 1.5 Let d_i be the ith row of D_{G_2}, and suppose that the root of the star S_k is mapped to node i in G_2. The broadcast time of S_k is the maximum $cstretch$ from among its k children. But since the jth entry of p_k is the probability that the jth largest value in d_i will be returned, the inner product $\langle sort(d_i), p_k \rangle$ gives the expected value of the maximum of the k $cstretch$es. Multiplication on the left by v_n^T simply averages these n values over all n rows.

Note that this is consistent with Theorem 1.3.1 for the special case where $k = 2$. Theorem 1.4.1 allows us to compute the broadcast time of a k-star for a variety of graph families, and we later use these as building blocks for bounds on general trees. Moreover, Theorem 1.4.1 improves on the trivial upper bound of $diam(G_2)$ for the broadcast time of a star. A better bound can be derived by considering the average eccentricity of G_2. The eccentricity ε of a vertex in a graph is defined as the maximum geodesic distance between that vertex and any other.

Corollary 1.4.1 *For any graph G_2, the broadcast time of a star of size k satisfies*

$$\mathbb{E}[cbtime_{G_2}(S_k)] \leq \frac{1}{n} \sum_{v \in V_2} \varepsilon(v). \tag{1.18}$$

Proof 1.6 Substituting p_{n-1} in place of p_{k-1} returns the average eccentricity of the vertices in G_2.

Corollary 1.4.1 provides a better bound than the diameter, but is not nearly as good as when using Theorem 1.4.1 directly. To illustrate how Theorem 1.4.1 can be used for a specific G_2, we provide an upper bound on the broadcast time of a star, when G_2 is the line lattice above.

Corollary 1.4.2 *For $G_2 = D_n^1$, the line lattice, the broadcast time of a star of size k satisfies*

$$\mathbb{E}[cbtime_{G_2}(S_k)] \leq \frac{k}{k+1} \cdot n. \tag{1.19}$$

Proof 1.7 The maximum product on the right certainly occurs at $d_1 =$

$(0,1,2,...,n-1)$, which is already sorted. Thus,

$$\begin{aligned}\langle d_1, p_k\rangle &= \frac{1}{\binom{n-1}{k}} \sum_{j=k+1}^{n} (j-1)\binom{j-2}{k-1} \\ &= \frac{k}{\binom{n-1}{k}} \sum_{j=k+1}^{n} \binom{j-1}{k} \\ &= \frac{k}{\binom{n-1}{k}} \binom{n}{k+1} = \frac{k}{k+1} \cdot n.\end{aligned}$$

Tree topology: For any tree T_k with maximum degree δ, let δ_i be the maximum out-degree among nodes at height $1 \leq i \leq h$ in T_k.

Observation 1.4.1 (Cbtime: Lower Bound)

$$\mathbb{E}[cbtime_{G_2}(T_k)] \geq \mathbb{E}[cstretch_{G_2}(P_h)]$$

Proof 1.8 The lower bound represents the expected *cstretch* of a path of length h, which is the longest in T_k. No other single path from the root to a leaf could have expectation longer than this, so the expectation for the tree must be at least this large.

Observation 1.4.2 (Cbtime: Upper Bound)

$$\mathbb{E}[cbtime_{G_2}(T_k)] \leq \sum_{\ell=1}^{h} \mathbb{E}[cbtime_{G_2}(S_{\delta_\ell})]$$

Proof 1.9 The upper bound represents the sum (over all h levels of T_k) of the expected composite broadcast time for the largest star graph S_{δ_ℓ} at each level ℓ. This the maximum expected broadcast time, since no path from the root to a leaf could take longer than this.

Combining Theorem 1.3.1 with Observations 1.4.1 and 1.4.2 yields the following bounds on the expected broadcast time of a general tree.

Corollary 1.4.3 *For any tree T_k of height h and maximum out-degree δ,*

$$h \cdot \mathbb{E}[X] \leq \mathbb{E}[cbtime_{G_2}(T_k)] \leq h \cdot \mathbb{E}[cbtime_{G_2}(S_\delta)],$$

where X is the r.v. giving the expected geodesic distance between two vertices in G_2.

Proof 1.10 The lower bound is an application of Theorem 1.3.1 to Observation 1.4.1, while the upper bound follows from Observation 1.4.2 and the fact that $\delta = \max_{1 \leq \ell \leq h} \delta_\ell$.

1.5 Composite Betweenness Centrality

Here, we characterize the betweenness centrality of composite graphs (G_1, G_2, R) as defined in Section 1.2. Since analytical characterization in closed form is difficult in general, we consider two special cases of interest. First, we consider the case where G_1 is the path graph P_k and G_2 is a line lattice D_n^1. Since all shortest paths are unique, in this simple case, we refer to the centrality of an edge as its "load."

1.5.1 *Constrained Composite Load on Path Graphs*

We consider the load on an edge for a path P_k mapped onto the discrete line D_n^1 under a random uniform embedding π (with replacement), as in Section 1.3.2. Let W_i be the random variable giving the index of vertex in D_n^1 corresponding to the vertex $v_i \in G_1$. Then any edge $e_j = (w_j, w_{j+1}) \in D_n^1$ is traversed if $W_i \leq j$ and $W_{i+1} \geq j+1$, or if $W_i \geq j+1$ and $W_{i+1} \leq j$. Let $\chi_{e_j}(v_i)$ be an indicator random variable for the event that the edge e_j is crossed by the path stretch from v_i to v_{i+1}. As W_i and W_{i+1} are independent, we can say that

$$\begin{aligned} Pr[\chi_{e_j}(v_i) = 1] &= Pr[W_i \leq j] \cdot Pr[W_{i+1} \geq j+1] + Pr[W_i \geq j] \cdot Pr[W_{i+1} \leq j+1] \\ &= 2 \cdot \left(\frac{j+1}{n}\right)\left(1 - \frac{j+1}{n}\right). \end{aligned}$$

Note that $\chi_{e_j}(v_i)$ has the Bernoulli distribution with probability $p_n(j) = 2\left(\frac{j+1}{n}\right)\left(1 - \frac{j+1}{n}\right)$, and thus the number of times that the edge e_j is traversed for a random uniform mapping π has the binomial distribution, for k trials with probability of success $p_n(j)$. The expected constrained composite load on e_j is thus $k \cdot p_n(j)$, and the expected load over entire D_n^1 is thus

$$\begin{aligned} \mathbb{E}[cload^{\pi}_{D_n^1}(P_k)] &= \frac{1}{n}\sum_{j=0}^{n-1} \mathbb{E}[cload^{\pi}_{D_n^1}(P_k, e_j)] \\ &= \frac{2k}{n}\sum_{j=0}^{n-1}\left(\frac{j+1}{n}\right)\left(1 - \frac{j+1}{n}\right) \\ &= \frac{k}{3}\left(1 - \frac{1}{n^2}\right). \end{aligned}$$

Note that by Equation 1.8, this is $\mathbb{E}[cstretch^{\pi}_{D_n^1}(P_k)]/n$.

1.5.2 *Composite Centrality in Manhattan Grid Networks*

Next, we consider a slightly more complicated scenario, where $G_1 \equiv T_n$ is a tree on n vertices and $G_2 \equiv D^2_{\sqrt{n}}$ is a $\sqrt{n} \times \sqrt{n}$ Manhattan grid, and $R : V(G_1) \to V(G_2)$ is a random uniform mapping. For simplicity of exposition, we consider the case where $\sqrt{n}$ is an odd positive integer, and hence $\sqrt{n} = 2m+1$ for some positive integer m.

Consider the contribution toward composite betweenness centrality due to a pair of vertices s and t in G_2. If we assume that R allows mapping of nodes *with replacement*, then the node with the maximum composite betweenness centrality in the worst case is the node in the center of G_2, i.e., the one with Cartesian coordinates (m,m). This is true when all neighboring nodes in G_1 get alternately mapped by R with replacement to node $s = (0,0)$ and node $t = (2m,2m)$, respectively.

Since the length of every shortest path from s to t is $4m$, and exactly half of those edges must go in each direction, the number of shortest (Manhattan) paths between s and t is given by: $\sigma_{st} = \binom{4m}{2m}$. The number of shortest paths from s to t that pass through $u = (m,m)$ is given by:

$$\sigma_{st}(u) = \sigma_{su} \times \sigma_{ut} = \binom{2m}{m} \times \binom{2m}{m}.$$

Using Stirling's approximation $n! \approx \sqrt{2\pi}\, n^{n+\frac{1}{2}}\, e^{-n}$ and algebraic simplification, we have:

$$\frac{\sigma_{st}(u)}{\sigma_{st}} = \frac{(2m)!^2}{m!^4} \times \frac{(2m)!^2}{(4m)!} = \frac{(2m)!^4}{m!^4(4m)!} \approx \sqrt{\frac{2}{\pi m}} = \frac{2}{\sqrt{\pi(\sqrt{n}-1)}}. \tag{1.20}$$

Since G_1 is a tree on n nodes, it has $n-1$ edges. In the worst-case scenario, the endpoints of each edge will get mapped by R to corner nodes $s = (0,0)$ and $t = (2m,2m)$ repeatedly, and thus node u will have equal contributions to its composite centrality from each of the above $n-1$ edges. This yields the following estimate of the worst-case maximum composite betweenness centrality of the composite graph $(T_n, D^2_{\sqrt{n}}, R)$.

$$\max\{cvbc(u)\} \approx (n-1)\frac{2}{\sqrt{\pi(\sqrt{n}-1)}} = \frac{2}{\sqrt{\pi}}(\sqrt{n}+1)\sqrt{\sqrt{n}-1} \approx \frac{2}{\sqrt{\pi}} n^{\frac{3}{4}} \tag{1.21}$$

We have verified by numerical simulations that the scaling law in Equation 1.21 is accurate. We note that it is difficult to characterize Equation 1.20 for a general G_2, thus making an analytical characterization of $\max cvbc(u)$ difficult. Furthermore, the average-case maximum composite betweenness centrality is likewise difficult to compute in closed form. Therefore, in the next section, we conduct a simulation-based study of a more realistic scenario where G_1 is a chain-of-command tree, G_2 is a geometric graph deployment scenario, and R is a given mapping.

1.6 Multicast in Composite Networks

In this section, we consider the problem of multicast in a composite network denoted by graph $\mathcal{G} = (G_1, G_2, R)$, where G_1 denotes the social network that codifies the relationships among nodes, G_2 is the physical network connecting these nodes, and R is a relation between the nodes in G_1 and G_2. In composite network multicast, the general goal is to cost-effectively connect a given subset of *terminal* nodes in G_2 while obeying certain data flow precedence constraints imposed by the structure of G_1.

1.6.1 Preliminaries

First, let us consider the multicast problem in a single network. Given a graph $G = (V,E)$, there is a set of *terminals* (or sinks) $M \subseteq V, M \neq \varnothing$, which are required to participate in multicast communication. Let $|M| = k$. We typically denote $T = (V(T),E(T))$ as a subgraph of G, where $V(T) \subseteq V$ and $E(T) \subseteq E$.

Definition 1.9 (Edge-weighted multicast problem) Each edge $e \in E$ has an associated cost $c(e)$. $V \backslash M$ are called *Steiner nodes*, which may participate in multicast communication if their involvement can enable connectivity between terminals in M. Solve $\min_T \sum_{e \in E(T)} c(e)$, subject to: (1) T is a connected subgraph of G, and (2) each pair of nodes $u, v \in M$ are connected via T.

The above problem (also known edge-weighted Steiner problem whose solution yields a tree T) has been studied extensively in the past. The problem is NP-hard not only for the general graph setting [17] but also for the special cases of Euclidean (L_2) norm [10] and Manhattan (L_1) norm [11]. It is well known that many variants of edge-weighted multicast networks (as known as Steiner network problems) have constant approximation algorithms [1, 26], based on LP relaxation and randomized rounding techniques.

Definition 1.10 (Node-weighted multicast problem) Each node $v \in V$ has an associated cost $c(v)$. Using previously defined notation, we define the following problem:[2] Solve $\min_T \sum_{v \in V(T)} c(v)$, subject to: (1) T is a connected subgraph of G, and (2) each pair of nodes $u, v \in M$ are connected via T.

Klein and Ravi gave the first approximation algorithm for the node-weighted Steiner tree problem with polynomial time complexity [18], this yields a cost within a constant factor of the best-possible approximation algorithm, and an approximation guarantee of a factor of $2(\log|M|)$ [18]. Note that it is well known that the set-cover problem can be reduced to the node-weighted Steiner tree problem in an approximation-preserving manner, thus showing that the node-weighted Steiner tree problem is $\Omega(\log|M|)$-approximable. The Klein-Ravi algorithm greedily and iteratively merges a set of subtrees to form a multicast tree. Initially, the set of subtrees consists of singleton sets of terminals $\mathcal{T}_{(0)} = \{\{v\} \mid v \in M\}$. At step t, the algorithm finds a node $v \in V$ and a subset of subtrees $F \subseteq \mathcal{T}_{(t-1)}$ such that the following dis-

[2]This is the node-weighted Steiner tree problem, which is a generalization of the edge-weighted Steiner problem since any instance of the latter can be converted to an instance of the former by a simple transformation that involves adding intermediate nodes to the graph for each edge.

tance metric is minimized: $\widetilde{\text{dist}}(v,F) \triangleq \frac{c(v)+\sum_{T\in F}\text{dist}(v,T)}{|F|}$, where $\text{dist}(v,T)$ is the total node-weighted cost of a shortest path from v to reach the nodes in T, excluding the costs of the two end nodes. Then, the subtrees in F will be merged with v using shortest paths to form one single subtree in $\mathcal{T}_{(i)}$. We summarize the Klein-Ravi algorithm below, since it will be used as a building block for composite network multicast.

The running time of the KleinRavi_Alg algorithm is $O(|V|^3 \log |V|)$, because finding the minimum $\widetilde{\text{dist}}(v,F)$ among all F can be achieved by first sorting the set of shortest paths of every node to v in a decreasing order, and then inserting each subtree incrementally into F to determine the minimum $\widetilde{\text{dist}}(v,F)$ (see [18]).

Since it is possible to transform a graph with both node and edge costs to one with only node costs, we only consider the purely node-weighted version of the problem in this chapter.[3]

1.6.2 Hierarchy-Compliant Multicast

Now we consider a useful special class of composite network multicast problems for composite graphs $\mathcal{G} = (G_1, G_2, R)$ where the social graph G_1 is a hierarchy of roles of network nodes V that can be captured by an acyclic graph $H = (V_H, E_H)$, and a mapping $h(\cdot) : V \mapsto V_H$. (Therefore, R is the inverse of the mapping h).

A path in a graph can be regarded as a sequence of nodes in the order it traverses. Given a path P in G, we write $h(P)$ as the sequence of the corresponding roles in H. A sequence Q is said to be a subsequence of Q', if removing some elements in Q' can create Q.

A subgraph T is said to be *H-compliant*, if any pair of nodes in M are connected in T, then there exist a path P in T and a path Q in H, such that Q is a subsequence of $h(P)$. Computing the optimal subgraph T is tantamount to solving the composite network multicast problem.

Definition 1.11 (H-compliant multicast problem) Solve $\min_T \sum_{v\in V(T)} c(v)$, subject to: (1) T is an H-compliant connected subgraph of G, and (2) each pair of nodes $u, v \in M$ are connected via T.

Figure 1.2 illustrates this class of constrained multicast networks with an example. An organizational hierarchy is shown on the left, and the corresponding physical deployment topology is shown on the right. Node 1 (root of the hierarchy) wants to multicast data to a set of nodes $\{4,5,7,9\}$ (circles). However, the hierarchical constraints constrain the flow of the data as the "supervisors" of these aforementioned nodes must act as intermediaries (nodes $\{2,3\}$ here). This may be necessary as the latter have better context about the subordinates than the root of the organization and they may want to even modify or embellish the message. Therefore, not only does the multicast sink set have to expand to include these intermediaries, the order in which

[3] While there are more recent algorithms that improve upon $O(\log |M|)$ approximation by a constant factor, e.g., [12], we decided to use the Klein-Ravi algorithm for our investigations, as the latter is simpler.

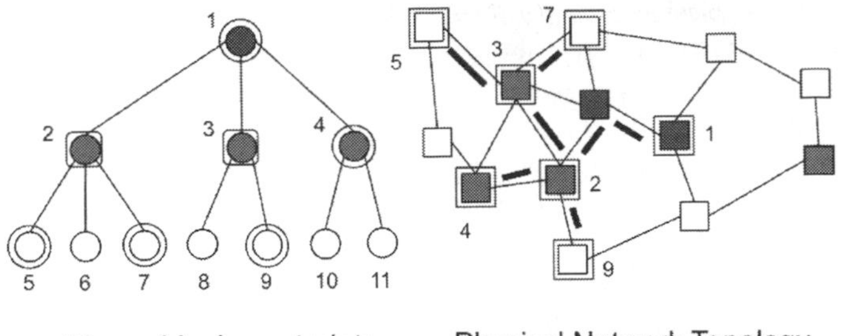

Figure 1.2: An example of composite network multicast.

the message is delivered to the sinks must obey the hierarchy. This may cause the information flow to get "stretched," and hence we need efficient algorithms for finding low-cost multicast structures that strike the balance between hierarchical constraints and exploit the properties of the physical network.

Note that if the underlying network G_2 is a wireless network, and wireless multicast advantage (WMA) can be leveraged, i.e., one transmission can reach multiple receivers simultaneously, the natural analog for H-compliant multicast is as follows:

Definition 1.12 (H-compliant wireless multicast problem) Solve $\min_T \sum_{v \in V(T)} c(v)$, subject to: (1) T is an H-compliant connected subgraph of G, and (2) $u \in M$ is either in T or is a neighbor of a node in T.

Although our formulation captures general hierarchies, most practical policy-driven hierarchies are tree based (e.g., command-and-control hierarchy). Hence, in this chapter, we will focus on the construction of hierarchy-compliant multicast networks with respect to tree hierarchy.

1.6.3 Algorithms for H-Compliant Multicast

In this section, we give efficient algorithms for composite network multicast, for the special case where G_1 is a hierarchical graph, and G_2 can be any graph. In this model, a node can *read* the message content only if it is signed by its parent node in the hierarchy. Nodes can, however, freely route messages in the network without *reading* their contents.

We consider the special case where the acyclic graph $G_1 = H = (V_H, E_H)$ mentioned in Section 1.6.2 is in fact a tree and the function $h(\cdot)$ is bijective. An edge $(u,v) \in E_H$ means that node v can receive a message from node u only after it is *signed* (or authenticated) by u. Without such authentication, the message can be neither *read* nor *modified* by v, although it can be *forwarded* along.

As mentioned in Section 1.6.2, the hierarchy-compliant multicast (HCM) problem involves augmenting the set of terminal nodes M by their ancestors in H. In particular, $M' = M \cup \{u \mid \forall s \in M, u = ancestor(s,H)\}$. We consider two models here.

We first consider the caching-friendly version of the problem (CFM), which is more straightforward. If caching the message is allowed and the message is not modifiable, it may be distributed via node-weighted Steiner multicast and cached before the *authentication/authorization* occurs. The latter can happen in hierarchical precedence order starting from the root node R toward the respective leaves in M along the edges of the tree H. The control overhead of the authentication process is typically negligible compared to that of the overhead of distribution of the actual data.

In the *strict-precedence* version of the problem, ancestor nodes cannot receive the message *before* their subordinates receive it. This phenomena could indeed occur if we simply execute the KleinRavi_Alg (Algorithm 1) on (G,M'). We also observe that in order to adhere to the strict precedence constraints, a message may have to traverse certain vertices and edges more than once. For example, if $(u-v-w)$ is a fragment of communication graph G and the message m currently resides at u and v, and the precedence constraints demand that $\{(h(v),h(u)),(h(u),h(w))\} \subseteq E_H$, then after receiving m from v, u will transmit a copy of m toward w, thus traversing the edge (u,v) again. This may be unavoidable if u desires to make modifications to m before forwarding it to w.

Algorithm 1: KlcinRavi_Alg [G, M, c $(\cdots)$]

1. Set $\mathcal{T}_{(t)} = \{\{v\} \mid v \in M\}$
2. $t = 0$
3. **While** $|\mathcal{T}_{(0)}| > 1$ **do**
4. $\quad t \leftarrow t+1$
5. $\quad$ Find $v \in V$ and $F \subseteq \mathcal{T}_{(t-1)}$ such that $\widetilde{\text{dist}}(v,F)$ is minimized
6. $\quad T' \leftarrow$ subtree by merging F with v using shortest paths from v
7. $\quad \mathcal{T}_{(t)} \leftarrow (\mathcal{T}_{(t-1)} \backslash F) \cup \{T'\}$
8. **end while**
9. Output $\mathcal{T}_{(t)}$

HCM cannot be directly mapped to a Steiner tree problem. In fact, the lowest-cost topological structure need not necessarily be a tree. The steps to compute optimal and approximate HCM structures are listed in Algorithm 2.

Note that it is possible to solve HCM optimally in a reasonable amount of time (unlike CFM) when the degree of H is constant and does not grow with $|V|$ or $|M|$. This is because the overall optimization problem can be decomposed into multiple *smaller-sized* Steiner tree problems, each of which can be solved optimally (and independently) using an integer programming formulation (Optimal_Steiner). At lines 4 and 7 in Algorithm 2, although the number of nodes in the graph is $|V|$, the number

Algorithm 2: BFSteiner_Alg$[\mathcal{G} = (H, G, h^{-1}), M, R \in V, c(\cdot)]$

1 $M' = M \cup \{u \mid \forall s \in M, u = ancestor(s, H)\}$
2 $H' = induced\_subtree(M', H)$
3 $C \leftarrow \{c \mid c \in children(R, H')\}$ ▷ breadth-first traverse H' from R
4 $S_R \leftarrow$ Steiner_Alg$[G, \{R\} \cup C, c(\cdot)]$ ▷ Solve mini-Steiner tree problem.
Steiner_Alg could be Optimal_Steiner or KleinRavi_Alg
5 **while** $C \neq \varnothing$ **do** ▷ BFS and solve mini-Steiner tree problems
6 $\quad c \leftarrow popfront(C)$
7 $\quad S_c \leftarrow$ Steiner_Alg$[G, \{c\} \cup children(c, H'), c(\cdot)]$
8 $\quad pushback(C, children(c, H'))$
9 $SubG \leftarrow \bigcup_{u \in H'} S_u$ ▷ Union of Steiner trees

of terminals is typically much smaller than $|M'|$ and is bounded by the out-degree of H', which is in turn bounded by the out-degree of H.

Theorem 1.1
BFSteiner_Alg$[G, M, H, R, c(\cdot)]$ *returns the optimal subgraph* T *of* G *when* Optimal_Steiner *is used.*

Proof 1.11 We use the notation defined in Section 1.6.2. H' is the tree rooted at R induced on H by the multicast terminal set M'. Consider an arbitrary subtree T' of H' which is rooted at node $r \in V_H$. Let $C_{H'}(r)$ be the direct children of r in H'. Any optimal subgraph *SubG* that connected all terminals in M in an H-compliant manner must contain a subgraph S_u of G that connects node $u \in V_G$ (such that $h(u) = r$) with certain nodes $v_1, v_2, \ldots, v_{|C_{H'}(r)|} \in V_G$ such that $\{h(v_1), h(v_2), \ldots, h(v_{|C_{H'}(r)|})\} = C_{H'}(r)$. Now, S_u can be computed independently of other such subgraphs that are computed during the execution of BFSteiner_Alg because none of the portions of the previously computed subgraphs can be reused during the computation of S, for if these portions are reused, there will be a violation of H-compliance. Moreover, there will be violation of H-compliance if one or more of nodes in set $\{u, v_1, v_2, \ldots, v_{|C_{H'}(r)|}\}$ do not get included in T. This follows from the fact that $h(\cdot)$ is bijective. Since S_u can be computed independently, if an optimal Steiner tree algorithm is used to compute it, then $SubG = \bigcup_{u \in M'} S_u$.

Corollary 1.1
If H *has maximum degree* k, BFSteiner_Alg$[\mathcal{G} = (\mathrm{H}, \mathrm{G}, \mathrm{h}^{-1}), \mathrm{M}, \mathrm{R}, \mathrm{c}(\cdot)]$ *returns a subgraph* T *with cost within* $2\log(k+1)$ *of that of the optimal subgraph* T, *when* KleinRavi_Alg *is used.*

Proof 1.12 Since the maximum number of terminals in each "mini" Steiner tree computation is $k+1$, according to Klein-Ravi's approximation guarantee of

$2\log(k+1)$, the cost of each mini-Steiner tree is within at most a $2\log(k+1)$ factor of the optimal. Therefore, the total cost obeys the same bound.

H-compliant wireless multicast BFSteiner_Alg can be naturally extended to the wireless network setting. Essentially, instead of repeatedly computing Steiner trees that connect H-compliant nodes at consecutive levels in H, one needs to repeatedly compute Steiner CDSes that can connect the H-compliant nodes at consecutive levels in H. In particular, Lines 4 and 7 in Algorithm 2 need to be replaced by a call to either the Optimal Steiner CDS algorithm or to Guha and Khuller's approximation algorithm for computing Steiner CDS [12].

We note that HCM is significantly different from overlay multicast [27]. In the latter, the overlay network is used to perform point-to-point unicast to achieve multicast, mainly because multicast is not widely supported in the underlay network. In contrast, in our setting, multicast is available as a primitive and the precedence constraints impose an overlay structure on top of the multicast-capable network substrate. Hence the solution space for our algorithms is likely to be richer.

1.7 Simulation-Based Evaluation

In this section, we use a variety of simulation-based approaches to study how the composite stretch, composite broadcast time, and composite betweenness centrality metrics behave under various choices of real (or at least realistic) social and communication networks. We first study the case where the social network (or G_1) is a hierarchical network like a chain of command that exists in military missions or in disaster relief operations. Our second case study is that of a social network that is richer than a tree, as in friendship relationships.

1.7.1 Chain of Command

Figures 1.3(a) and 1.4(a) present the chain-of-command hierarchy from within a representative brigade reporting structure in the US Army and UK army, respectively. Both data sets were synthesized from various sources in the public domain.[4] [5] [6] [7] [8] In each case, commands move from the highest-ranked (root, G at top) node to the lower ranked nodes in the tree until they reach the lowest-ranked (leaf, shown at bottom) nodes. Though not depicted in these figures, each node in G_1 is deployed to a physical location in space. In Figures 1.3(b) and 1.4(b), we show the geometric graph G_2 constructed from these physical locations by adding a communication edge

[4] British Army website: http://www.army.mod.uk.

[5] Ministry of Defense website: http://www.mod.uk.

[6] T. F. Mills, "Land Forces of Britain, the Empire, and the Commonwealth." http://www.regiments.org.

[7] Kenneth Macksey, *First Clash, Combat Close-Up in World War Three*, (Royal Tank Regiment), Berkley Books, First edition, March 1, 1988.

[8] Carl Schulze, *The British Army of the Rhine (Europa Militaria)*, No. 19, Crowood Press, 64 pp., June 1995.

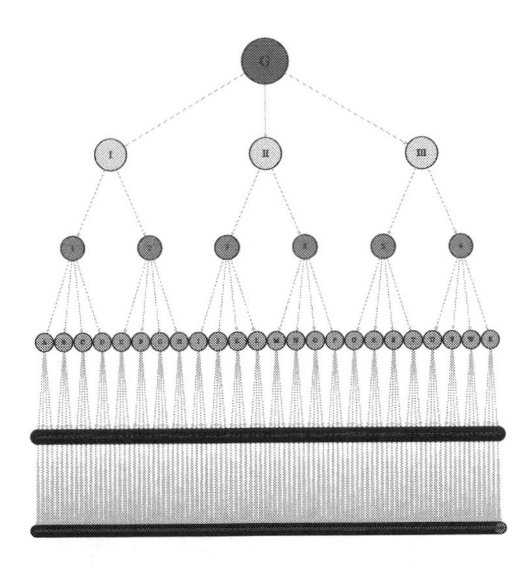

Number of Nodes = 514 , Diameter = 5

(a) G_1

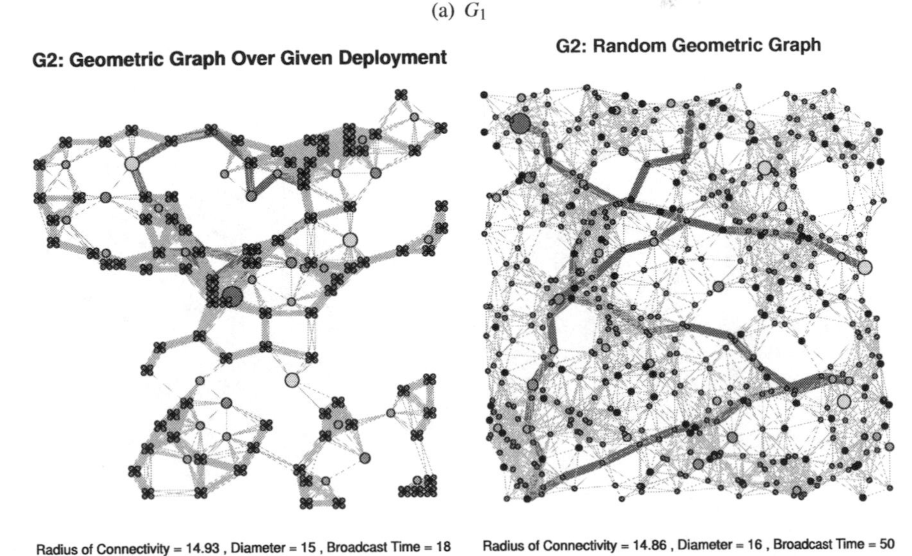

(b) G_2 (c) G_2 under S2

Figure 1.3: Realistic US army *chain-of-command* networks. (a) a chain-of-command social network (tree); (b) historical deployment of the chain of command shown in (a); (c) one random deployment of the chain of command shown in (a) under scenario S2.

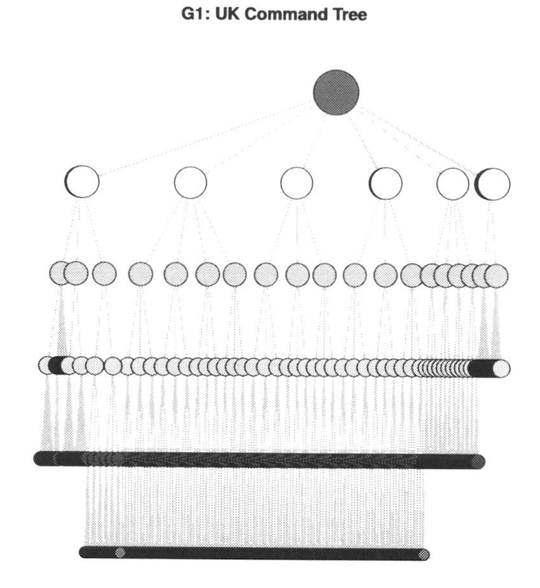

(a) G_1

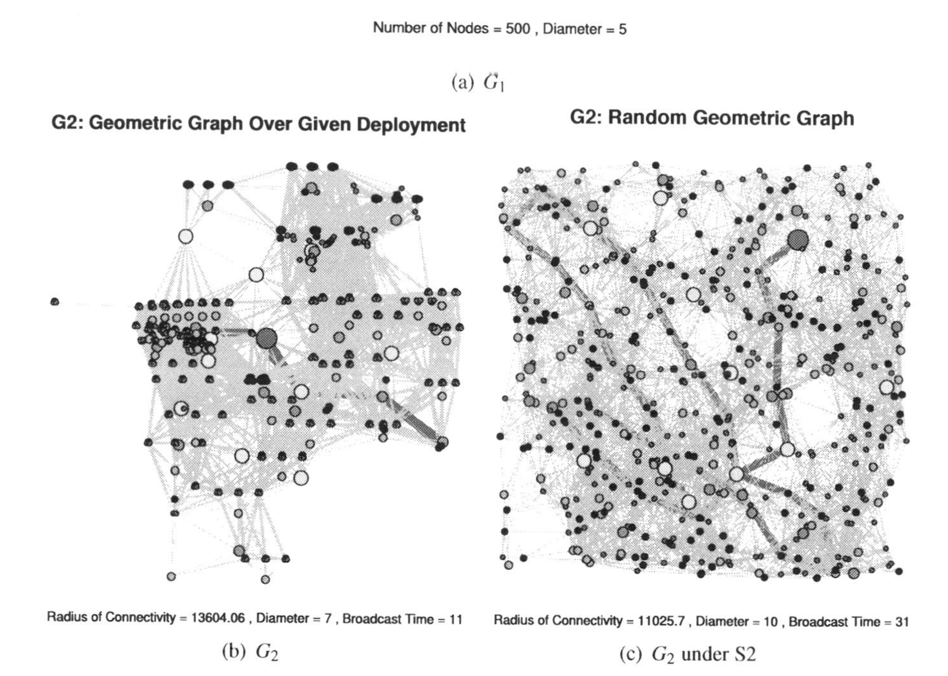

(b) G_2

(c) G_2 under S2

Figure 1.4: Realistic UK army *chain-of-command* networks. (a) a chain-of-command social network (tree); (b) historical deployment of the chain-of-command shown in (a); (c) one random deployment of the chain of command shown in (a) under scenario S2.

Table 1.2 Summary of Chain-Of-Command Data Sets

Data Set	$\lvert V_1 \rvert$	$\lvert E_1 \rvert$	$diam(G_1)$	$\lvert V_2 \rvert$	$\lvert E_2 \rvert$	$r(n)$	$cbtime$
US Army	514	513	5	514	12722	14.93 km	18
UK Army	500	499	5	500	54720	13.60 km	11

between two nodes if they lie within a prescribed radio transmission range (i.e., the critical radius of connectivity $r(n)$ outlined above) of each other. The coordinates of the nodes in G_2 are specified according to a historical deployment scenario over a prescribed area. G_2 is not a *random* geometric graph, but only a geometric graph with a fixed radius of communication. Summary statistics for both data sets are presented in Table 1.2.

The composite graph $\mathcal{G}^* = (G_1, G_2, R)$ that combines Figures 1.3(a) and 1.3(b) (resp. Figures 1.4(a) and 1.4(b)), along with the identity mapping R, is a realistic composite network structure for a military or disaster relief deployment. For the US instance, the broadcast time is 18 (hops), although the eccentricity in G_2 of the root node of G_1 is 9 (hops). Thus, the constraints imposed on the information flow by the chain of command require a message to travel through twice as many hops as was mandated by the actual deployment. A path that produces the broadcast time is highlighted in Figures 1.3(a) and 1.3(b) for the US and Figures 1.4(a) and 1.4(b) for the UK.

In order to put these broadcast times of 18 and 11 in context, for both data sets we simulated three different randomized scenarios, each of which could produce $\mathcal{G}^*$ as a singular outcome:

S1 Instead of R being the identity mapping, R is a random permutation.

S2 Instead of G_2 being a geometric graph over the actual coordinates of deployment, each node in G_1 was assigned a random 2D coordinate drawn from a square region of the same area as in S1. Therefore, G_2 is a *random* geometric graph (RGG). Figures 1.3(c) and 1.4(c) show examples of such a deployment.

S3 Instead of G_2 being a geometric graph over the actual coordinates of deployment, the coordinates were generated according to a random model that is a function of the chain-of-command tree G_1, with R as the identiy mapping. Details of the model are given below.

In the actual US deployment, note that the lowest-ranked nodes are collocated in bunches of four, hence Figure 1.3(b) appears sparser than Figure 1.3(c). Moreover, the broadcast time jumps from 18 to 50 for this particular RGG. This is because in the actual deployment, there is strong correlation between the location of a node and its rank in the command hierarchy (even though the maximum $cstretch$ is as high as 18), which does not exist in random deployments. Inspired by this observation, we constructed a *correlated* random deployment model for S3.

Details of Deployment Model for S3: Our model places each child in an equi-spaced, but randomly oriented, ring around its parent, with a random jitter applied in both the horizontal and vertical directions. This process is recursively applied down the chain-of-command tree, which we assume has height h. Let v_i be the node in G_1 at distance h_i from the root node, and having n_i children. Then the location of v_i's children are determined as follows:

1. Find the n_i roots on unity $\omega_1, \ldots, \omega_{n_i}$ and associate one with each of the n_i children.
2. Draw a uniform random variable $u \in [0, 1]$.
3. Set the default distance from parent to child to be $\rho_i = a(h - h_i)^2$, where a is a parameter determined from analysis of the actual deployment data. [In the US case $a = 1.85$.].
4. For each $j \in 1, \ldots, n_i$, set the target location $c_j = \rho_i \cdot e^{2\pi i u} \cdot \omega_j$, and draw two random coordinates x_j and y_j from normal distributions with mean $\Re(c_j)$ and $\Im(c_j)$, respectively, and standard deviation $b\bar{\rho}_{h_i}$. Here, b is a parameter determined from the data (we assume a model with a constant coefficient of variation: $b = 0.293$ for the US data set), and $\bar{\rho}_{h_i}$ is the mean distance between parent and child at height h_i.
5. Return $(x_{v_i}, y_{v_i}) + (x_j, y_j)$.

1.7.1.1 Evaluation of Basic Composite Network Metrics

Figure 1.5(a) (respectively, 1.5(b)) plots the simulated distribution of broadcast time (*cbtime*) for each of the aforementioned scenarios, for the US (UK) data set. The fact that the dashed vertical line (corresponding to the actual broadcast time) falls within the distribution of S3 (more precisely, in the 92nd percentile), suggests that our correlated deployment model is a useful one. That this correlated random deployment model captures the behavior of both the US and UK deployments without modification, strengthens our claim that the model generates realistic random deployments. Moreover, the other two scenarios, both of which were agnostic to the structure of the chain-of-command tree, performed comparatively poorly. Empirically, the probability of obtaining a broadcast time as low as that of $\mathcal{G}^*$ via either S1 or S2 appears to be negligible in both cases. This illustrates the potential efficiency dangers inherent in composite networks.

Our composite betweenness centrality metric also yields meaningful information not captured by the conventional betweenness centrality metric on a single network. In Figure 1.6(a), we reproduce the actual UK deployment, with node sizes proportional to the normalized vertex betweenness centrality in that network. It is apparent that nodes on narrow "bridges," especially those close to the center of the deployment, have greater centrality. However, this is not necessarily true for *composite* vertex betweenness centrality, as shown in Figure 1.6(b). In other words, certain nodes in G_2 might be significantly more (or less) central than the classic betweenness centrality measure would predict. This reflects the reality that information flow in the communication network is *constrained* by the structure of the chain-of-command social network.

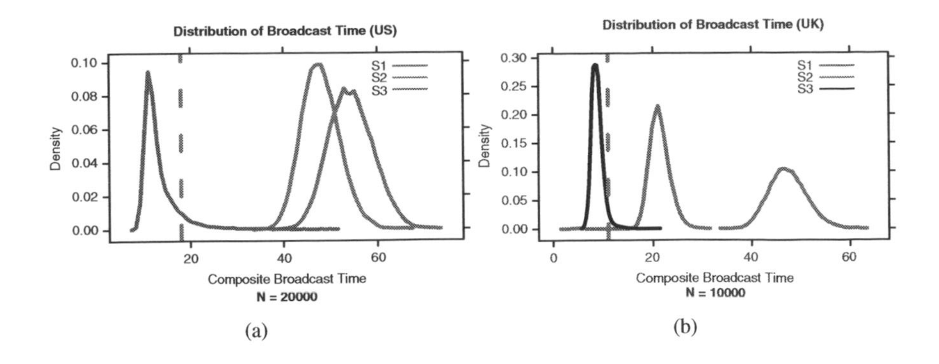

(a) (b)

Figure 1.5: ***CBtime*****: Comparison of** ***CBtime*** **for various deployments of the chains of command shown in Figure 1.3(a) and 1.4(a). Various upper and lower bounds derived in Section 1.4 are illustrated.**

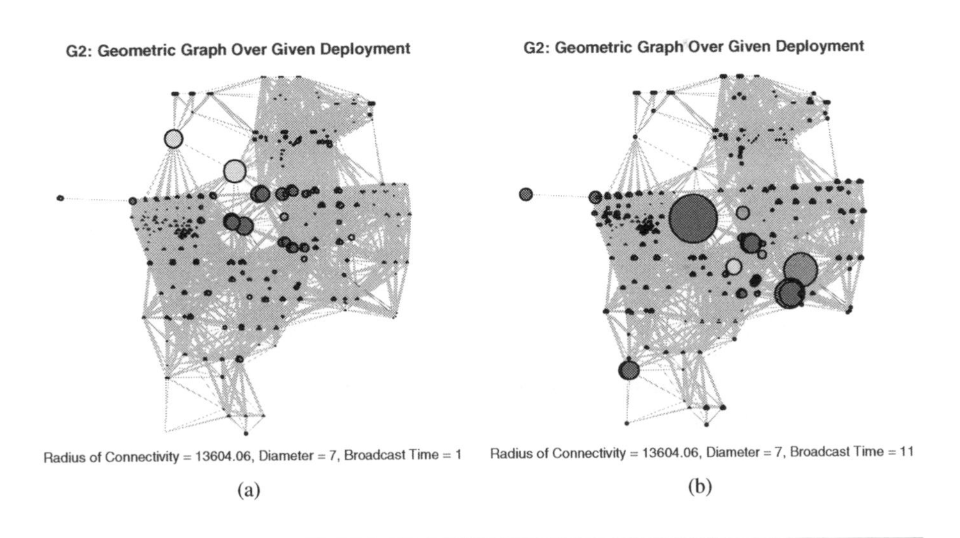

(a) (b)

Figure 1.6: ***CBetweenness*****: In (a), Figure 1.4(b) is reproduced with node size proportional to the normalized vertex betweenness centrality; in (b) the same plot is produced with node size proportional to the normalized** ***composite*** **vertex betweenness centrality.**

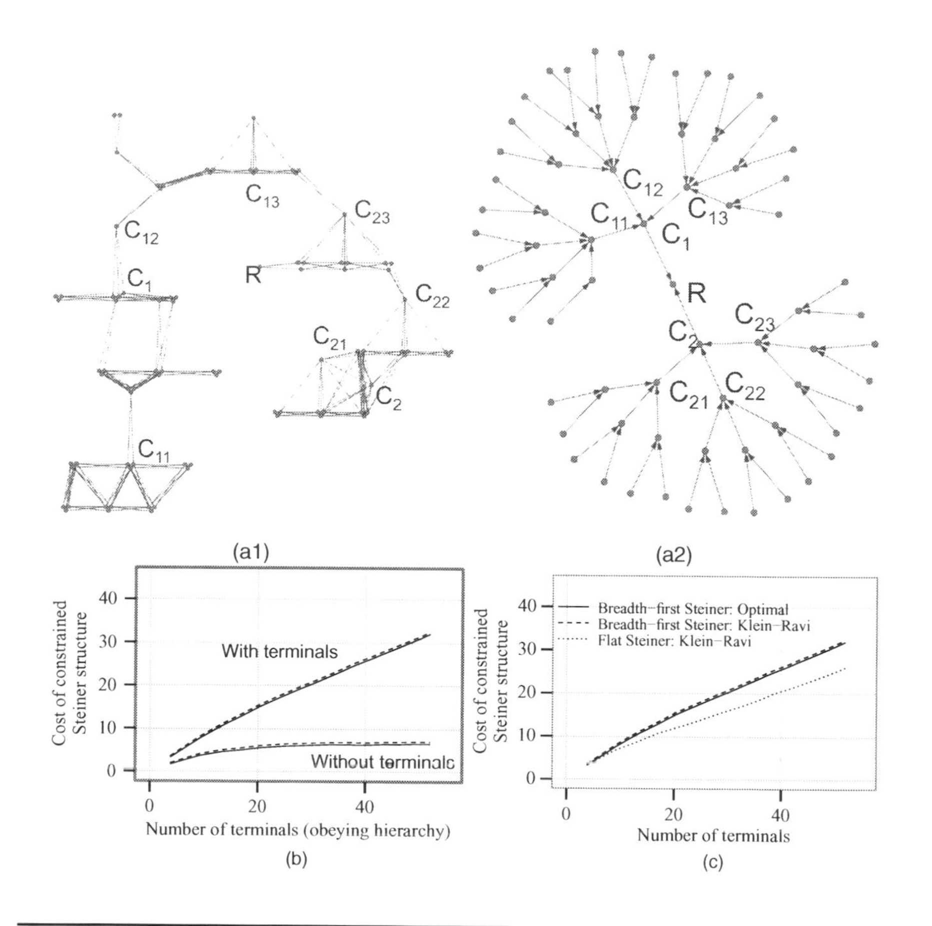

Figure 1.7: H-compliant multicast: (a1) historical deployment, and (a2) hierarchy; (b) costs: Klein-Ravi (dotted), Opt (solid); (c) with and without H-compliance.

1.7.1.2 Evaluation of Composite Network Multicast

For the evaluation of multicast algorithms in composite networks, we use a 91-node fragment of the UK chain-of-command data set depicted in Figure 1.4(a). Figure 1.7(a) illustrates the locations of various troop units and their relative positions in the military hierarchy. We simulated random sets of multicast terminals M in this network and calculated the cost of H-compliant multicast. The node costs were also chosen randomly $\in (0,1]$. Obviously, the requirement of H-compliance results in a significant increase in the number of terminals (the new ones are ancestors of M in H). We observe from Figure 1.7(b) that as $|M|$ grows, the cost of the Steiner subgraph begins to be dominated by the costs of the terminal nodes themselves. This is because, if there are many terminals, most terminals do not need help from non-terminals to get connected to each other. Also, the costs yielded by both Optimal and Klein-Ravi versions of Algorithm 2 are very close to each other, and since Optimal does not take longer than Klein-Ravi to run when executed at each level of H (and thus with a bounded number of terminals), either algorithm is a reasonable candidate to be used within Algorithm 2.

Finally, we illustrate the overhead of H-compliance on multicast in Figure 1.7(c). We executed Algorithm 2 (with both Optimal and Klein-Ravi variants) as well as the flat Klein-Ravi Steiner tree algorithm without H-compliance (for the CFM algorithm mentioned in Section 1.6) on the 91-node network. The latter obviously yields significantly lower costs since the flows do not have to be H-compliant. However, if H-compliance is a requirement, then one must be willing to pay a 25% overhead in total costs. If only non-terminal costs are measured, then the relative overhead is much more significant.

Although hierarchical constraints on multicast can result in significant *stretch* of information flows, thus resulting in significant increase in costs of Steiner structures connecting the terminals, the good news is that approximation algorithms (with worst-case guarantees) can perform close to their optimal counterparts, while executing in low polynomial time.

1.7.2 Friend-of-a-Friend (FOAF)

Now we consider a small social network data set that was extracted from the Semantic Web Billion Triples Challenge (BTC) program. The data contains unique identifiers and indicates the existence of friendship relations between pairs of users—as per the *friend-of-a-friend (FOAF) ontology*. It also contains the geographic coordinates of users.

One can imagine the IDs in the data set communicating with their *friends* over some underlying communication network. In this scenario, the communication would typically happen over the wired Internet that connects various users on the map; however, we used this node distribution data to study the composite stretch of the FOAF social network on a geographically distributed multi-hop network assuming a geometric graph model as described in Section 1.7.1—in particular, we place a node at each location in the data set and construct a graph using a transmission radius that is large enough to connect most nodes in the network (akin to the notion of critical radius in case of a random geometric graph). In Figure 1.8, we show a 237-node social network (G_1), alongside the geometric graph imposed over the same set of vertices using the connectivity radius $r(n) = 5.1$ degrees of latitude/longitude.

Since G_1 is a rich social network, all nodes can act as sources of information which can flow along random spanning trees of G_1. Hence, we pick all nodes in G_1 one by one; and assign them as roots of random spanning trees. We then measure *cbtime* for each root node and plot the results in Figure 1.8(c) (black jitter-spaced hollow circles). The dashed black line shows the average *cbtime* as a function of the height of the spanning tree. That this is approximately linear accords with Corollary 1.4.3.

On the same figure (Figure 1.8(c)) we plot the upper and lower bound formulas for $\mathbb{E}[cbtime]$ that we derived in Section 1.4. The solid lines at bottom and third from top indicate the bounds of Corollary 1.4.3, while the second line from the top

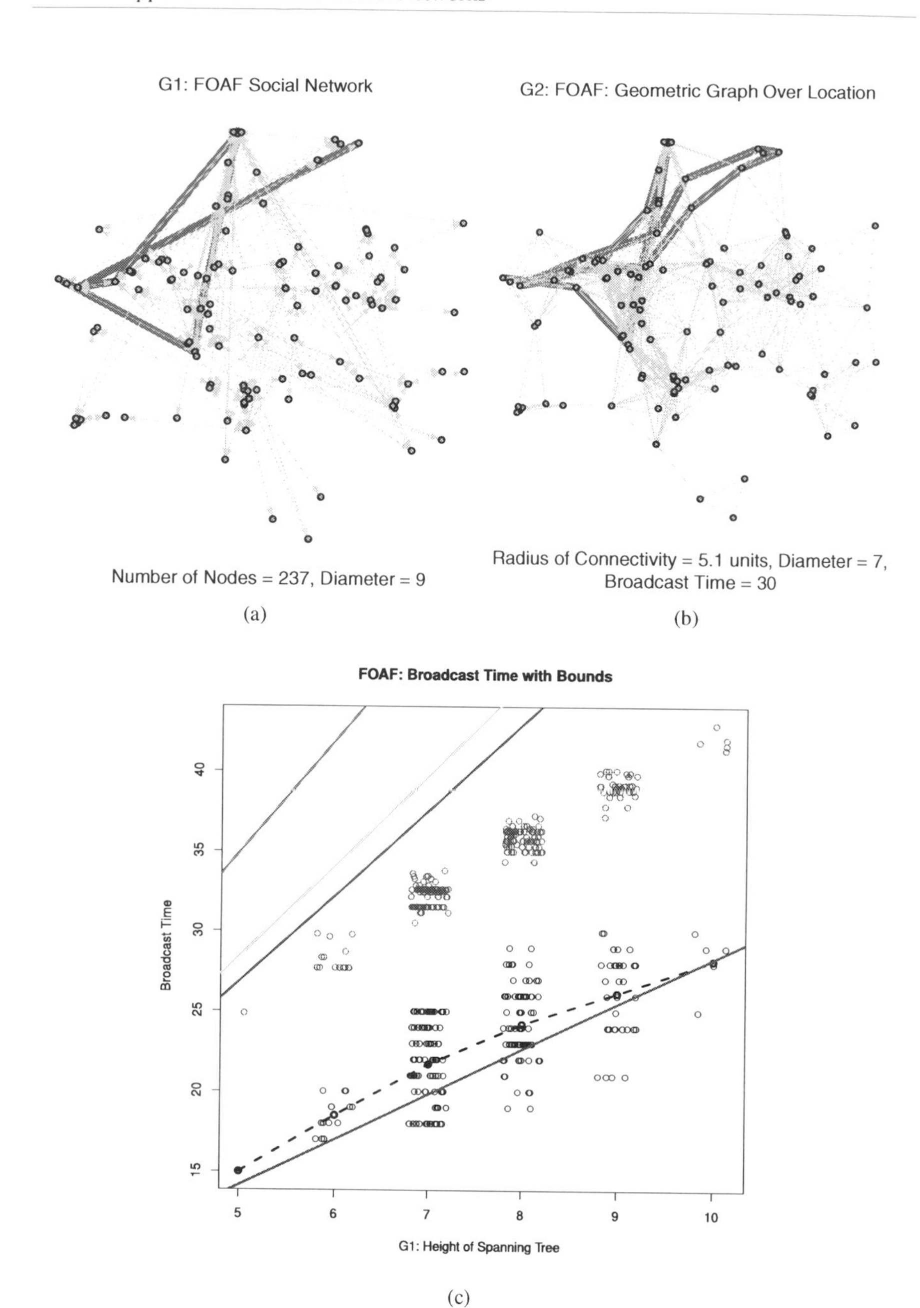

Figure 1.8: FOAF composite graph. (a) The FOAF social network in Europe; (b) a geometric graph in Europe constructed using a radius of communication of 5.1 degrees of latitude/longitude; and (c) *CBtime* in the composite graph specified in (a) and (b). Various upper and lower bounds derived in Section 1.4 are illustrated.

shows the weaker bound of Corollary 1.4.1 and the top line shows the trivial diameter bound. The jittered blue circles indicate the strongest bound for each spanning tree, derived from Observation 1.4.2. We observe that the lower bound is reasonably tight whereas the upper bounds based on Theorem 1.4.1 and average eccentricity are looser, but much better than the trivial diameter upper bound. The upper bound obtained from Observation 1.4.2 is tighter, since it takes the local structure of the spanning tree into account, rather than simply the globally maximum out-degree (46 in this case). In summary, this validates our analytical results.

1.8 Conclusion and Discussion

In this chapter, we presented an analytical modeling framework (both models and metrics) for studying the composite stretch (or elongation) suffered by information flowing on a multi-hop communication network when it is *constrained* by social network relationships. We derive scaling laws (in expected value sense) for composite stretch for random embeddings of common social network structures such as trees on a variety of multi-hop communication network models, ranging from simple linear networks to random geometric graphs.

We also analyze the constrained broadcast time metric, which measures the time taken to broadcast (or gossip) information along the edges of a social network to *all* nodes in that network while being constrained by the underlying communication network structure. We derived analytical bounds for composite broadcast time and validated them by simulations using a friendship social network data set. We also show, using simulations based on a historical military deployment data set, how the *broadcast time* of an information flow in a chain-of-command network is non-optimal, but far superior to a random deployment scenario.

We also extend the notion of classic betweenness centrality for a given graph to the composite graph domain. The new measure characterizes the fraction of time a piece of information traverses a particular node or edge in the communication network G_2 while constrained by the social network structure in G_1. Our analysis indicates that for worst-case embeddings, maximum composite betweenness centrality scales sublinearly with network size when a sparse network (e.g., social network with tree topology) is embedded on a communication grid with the same number of nodes.

We also showed that just like unicast and broadcast, multicast in composite networks suffers from significant additional *stretch* of information flows as well, thus resulting in significant increase in costs of Steiner structures that connect the terminals. However, the good news is that approximation algorithms (with worst-case guarantees) can perform close to their optimal counterparts, while executing in low-order polynomial time.

Other interesting variants of the constrained stretch metric are possible. Suppose one is only allowed to use direct communication links that exist between friends (since these are supposed to be trusted); this will result in a routing which is likely to have a higher stretch since shorter communication paths between friends (through non-friends) are disallowed. How can one characterize this highly constrained stretch? Such insights should help us *design* better communication networks (or facilitate intelligent deployment) that are suited to particular demands of the overlaid social networks of users.

Acknowledgments

This research was sponsored by the U.S. Army Research Laboratory and the U.K. Ministry of Defense and was accomplished under Agreement Number W911NF-09-2-0053 (US ARL) and Agreement Number W911NF-06-3-0002.l (US ARL and UK MOD). The views and conclusions contained in this document are those of the authors and should not be interpreted as representing the official policies, either expressed or implied, of the U.S. Army Research Laboratory, the U.S. government, the U.K. Ministry of Defence, or the U.K. government. The U.S. and U.K. governments are authorized to reproduce and distribute reprints for government purposes notwithstanding any copyright notation hereon.

We also thank Richard Allan (Raytheon BBN) for synthesizing the historical military deployment data sets (from various sources in the public domain) that were used in the evaluation section of this chapter. We also thank Richard Gibbens (University of Cambridge) and Saikat Guha (Raytheon BBN) for useful discussions on the hierarchy-compliant multicast problem.

References

[1] R. R. A. Agrawal and P. Klein. When trees collide: An approximation algorithm for the generalized Steiner problem on networks. *SIAM Journal on Computing*, 24(3), 1995.

[2] Y. Bartal. On approximating arbitrary metrices by tree metrics. In *Proc. of ACM STOC*, pages 161–168, 1998.

[3] P. Basu, C.-K. Chau, R. Gibbens, S. Guha, and R. Irwin. Multicasting under multi-domain and hierarchical constraints. In *Proc. of IEEE WiOpt*, 2013.

[4] P. Basu, W. Ke, and T. D. C. Little. Dynamic task-based anycasting in mobile ad hoc networks. *Mob. Netw. Appl.*, 8:593–612, October 2003.

[5] B. Baumer, P. Basu, and A. Bar-Noy. Modeling and analysis of composite network embeddings. In *Proc. of ACM MSWiM*, pages 341–350, 2011.

[6] S. Bokhari. On the mapping problem. *IEEE Transactions on Computers*, 30(3), 1981.

[7] M. Bradonjic, R. Elsässer, T. Friedrich, T. Sauerwald, and A. Stauffer. Efficient broadcast on random geometric graphs. In *Proc. of ACM-SIAM SODA*, pages 1412–1421, 2010.

[8] S. V. Buldyrev, R. Parshani, G. Paul, H. E. Stanley, and S. Havlin. Catastrophic cascade of failures in interdependent networks. *Nature*, 464:1025–1028, April 2010.

[9] G. Csardi and T. Nepusz. The igraph software package for complex network research. *InterJournal, Complex Systems*, 1695(5), 2006.

[10] M. R. Garey, R. L. Graham, and D. S. Johnson. The complexity of computing Steiner minimal trees. *SIAM Journal on Applied Mathematics*, 32:835–859, 1977.

[11] M. R. Garey and D. S. Johnson. The rectilinear Steiner tree problem is NP-complete. *SIAM Journal on Applied Mathematics*, 32:826–834, 1977.

[12] S. Guha and S. Khuller. Approximation algorithms for connected dominating sets. *Algorithmica*, 20:374–387, 1996.

[13] P. Gupta and P. Kumar. Critical power for asymptotic connectivity. In *Proc. of IEEE CDC*, volume 1, pages 1106–1110, 1998.

[14] G. Gutin and A. Punnen. *The Traveling Salesman Problem and Its Variations*, volume 12. Kluwer Academic Pub, 2002.

[15] C. C. Hui and S. T. Chanson. Allocating task interaction graphs to processors in heterogeneous networks. *IEEE Transactions on parallel and Distributed Systems*, 8(9), September 1997.

[16] P. Indyk and J. Matousek. Low-distortion embeddings of finite metric spaces. In *Handbook of Discrete and Computational Geometry*, pages 177–196. CRC Press, 2004.

[17] R. M. Karp. On the computational complexity of combinatorial problems. *Networks*, 5:45–68, 1975.

[18] P. Klein and R. Ravi. A nearly best-possible approximation algorithm for node-weighted Steiner trees. *Journal of Algorithms*, 19:104–115, 1995.

[19] R. Kleinberg. Geographic routing using hyperbolic space. In *Proc. of IEEE INFOCOM*, 2007.

[20] M. Kurant and P. Thiran. Layered complex networks. *Physical Review Letters*, 96(138701), 2006.

[21] E. A. Leicht and R. M. D'Souza. Percolation on interacting networks. *arXiv*, 2009. `http://arxiv.org/abs/0907.0894`.

[22] E. K. Lua, J. Crowcroft, M. Pias, R. Sharma, and S. Lim. A survey and comparison of peer-to-peer overlay network schemes. *IEEE Communications Surveys and Tutorials*, 7(2):72–93, 2005.

[23] R. Monien and H. Sudborough. Embedding one interconnection network in another. *Computing Suppl.*, 7:257–282, 1990.

[24] R Development Core Team. *R: A Language and Environment for Statistical Computing*. R Foundation for Statistical Computing, Vienna, Austria, 2010.

[25] S. Ratnasamy, P. Francis, M. Handley, R. Karp, and S. Shenker. A scalable content addressable network. In *Proc. of ACM SIGCOMM*, pages 161–172, 2001.

[26] G. Robins and A. Zelikovsky. Tighter bounds for graph Steiner tree approximation. *SIAM J. Discret. Math.*, 19(1):122–134, May 2005.

[27] S. Shi and J. S. Turner. Routing in overlay multicast networks. In *INFOCOM*, 2002.

[28] I. Stoica, R. Morris, D. Karger, M. F. Kaashoek, and H. Balakrishnan. Chord: A scalable peer-to-peer lookup service for Internet applications. In *Proc. of ACM SIGCOMM*, pages 149–160, 2001.

[29] E. Weisstein. Hypercube line picking, 2010. `http://mathworld.wolfram.com/HypercubeLinePicking.html`.

Chapter 2

Recent Advances in Information Diffusion and Influence Maximization of Complex Social Networks

Huiyuan Zhang
University of Florida
Gainesville, Florida

Subhankar Mishra
University of Florida
Gainesville, Florida

My T. Thai
University of Florida
Gainesville, Florida
and
Ton Duc Thang University
Ho Chi Minh City, Vietnam

CONTENTS

2.1 Abstract

Nowadays, social influence is ubiquitous in everyday life, online social networks have become a focal point for research in science. Formal mathematical models for the analysis of the spread of social influence have emerged as a major topic of interest in diverse areas such as sociology, economics, and computer science. Empirical studies of diffusion on social networks date back to the 1940s. Later on, theoretical propagation models were introduced in late 1970s. Then, motivated by the design of marketing strategy, and the formal definition of the problem of influence maximization, the field of studying social influence has received lots of research interest. In particular, the rapid growth of online social networks such as Facebook, Twitter, and Google+ has intensified interest in this field, and the past decade has seen a burgeoning network literature from the computer community.

In this chapter, our goal is to provide readers with a comprehensive review of this burgeoning literature. We begin with an overview of widely used theoretical diffusion models, in which there are three families of diffusion models: threshold models, cascading models, and epidemic models. Our subsequent discussion mainly focuses on the recent algorithmic study and analytical results of the influence maximization problem. We end with a discussion of some open problems and challenges.

2.2 Introduction

Nowadays, the development of the Internet has revolutionized the way we communicate with each other. Communication helps us better share knowledge, ideas, and beliefs, thus influencing human behaviors. The study of information diffusion and social influence has attracted scientists from sociology and economics since the early 1960's. In recent decades, the rapid growth of online social networks (OSNs) such as Facebook, Twitter, and Google+ provide a nice platform for information diffusion and fast information exchange among their users. In addition, the massive data obtained from millions of users and more than a billion social ties in those giant networks has greatly facilitated analytical works about user behavior, and even large-scale algorithmic studies by scientists from computer science have being engaged in this popular field.

Diffusion, according to Roger's definition [48], is the process by which an innovation is communicated through certain channels over time among the members of a social system. Three important elements: individual member, mutual interactions, and communication channels are introduced from this definition, which are set as the basis for a future analytical framework.

Later on, various diffusion models were proposed to study the contagion properties in a vast area such as widespread adoption in viral marketing [16, 46, 32, 36], information propagation on blogs [33, 34], and infectious disease transmission in epidemiology [15, 3].

One of the goals in studying social influence is the problem of *influence maximization*, which arises from the context of widespread adoption in viral marketing. This problem was first proposed by Kempe et al. [27], then rapidly became a hot topic in the social network field. The influence maximization problem is formally described as follows: given a social network represented by a(n) directed/undirected graph with nodes as users, edges correspond to social ties, edge weights capture influence probabilities, and a budget k, which is a integer; the goal is to find a seed set of k users such that by targeting these, the expected influence spread (defined as the expected number of influenced users) is maximized. Here, the expected influence spread of a seed set depends on the influence diffusion process, which is captured by diffusion models.

Therefore, in this book chapter, we start by providing an overview of diffusion models that have been extensively used in studying social influence. In general, all existing diffusion models can be categorized into three classes: threshold models [26, 27, 28, 42, 47, 6], cascading models [21, 22, 9, 8], and epidemic models [29, 35]. Figure 2.1 provides an overview of those models. For each model, we give a detailed description of the diffusion process, activation condition, as well as its properties and applications. With the framework in place, we move on to the algorithmic results of the influence maximization problem.

We are now interested in choosing an influential set to target in the context of the above models. Kempe et al. [27] prove that the influence maximization problem is NP-hard under both the Linear Threshold model and Independent Cascading model, and give a simple greedy algorithm with approximation ratio of $1 - 1/e$. However, the greedy algorithm suffers from severe scalability problems. Therefore, considerable work has been done to improve it. In the second half of this book chapter, we demonstrate recent algorithmic studies such as the CELF [33], CELF++ [24], Simpath[25] and LDAG [11] algorithms, which can obtain high scalability for the influence maximization problem.

In this chapter, we survey the recent advances in theoretical propagation models of online social networks, as well as the algorithms for the influence maximization problem. In Section 2.4, we give an overview of existing diffusion models, which can be categorized into three main classes: threshold models, cascade models, and epidemic models. For each kind of diffusion model, we also provide some interesting extensions. And with this framework, we move forward to Section 2.5, in which we survey various approaches for the influence maximization problem with high scalability. In the last section, we conclude the chapter with some applications of social influence and information diffusion.

2.3 Social Influence and Influence Maximization

Social influence, as defined by Rashotte [45], is the change in an individual's thoughts, feelings, attitudes, and behaviors that results from interaction with other people or groups. Social influence takes many forms and can been seen everywhere in OSNs. In the field of data mining and big data analysis, many applications such as viral marketing, recommendation systems, and information diffusion are involved in social influence.

Influence maximization (IM) is one of the fundamental problems in studying social influence. Because people are likely to be affected by decisions of their friends and colleagues, some researchers and marketers have investigated social influence and the word-of-mouth effect in promoting new products and making profitable marketing strategies. Suppose that with knowledge of individuals preferences and their influence on each other, we would like to promote a new product that will be adopted by a large number of users in this network. The strategy of viral marketing is to select a small number of influential members within this network at the beginning, convince

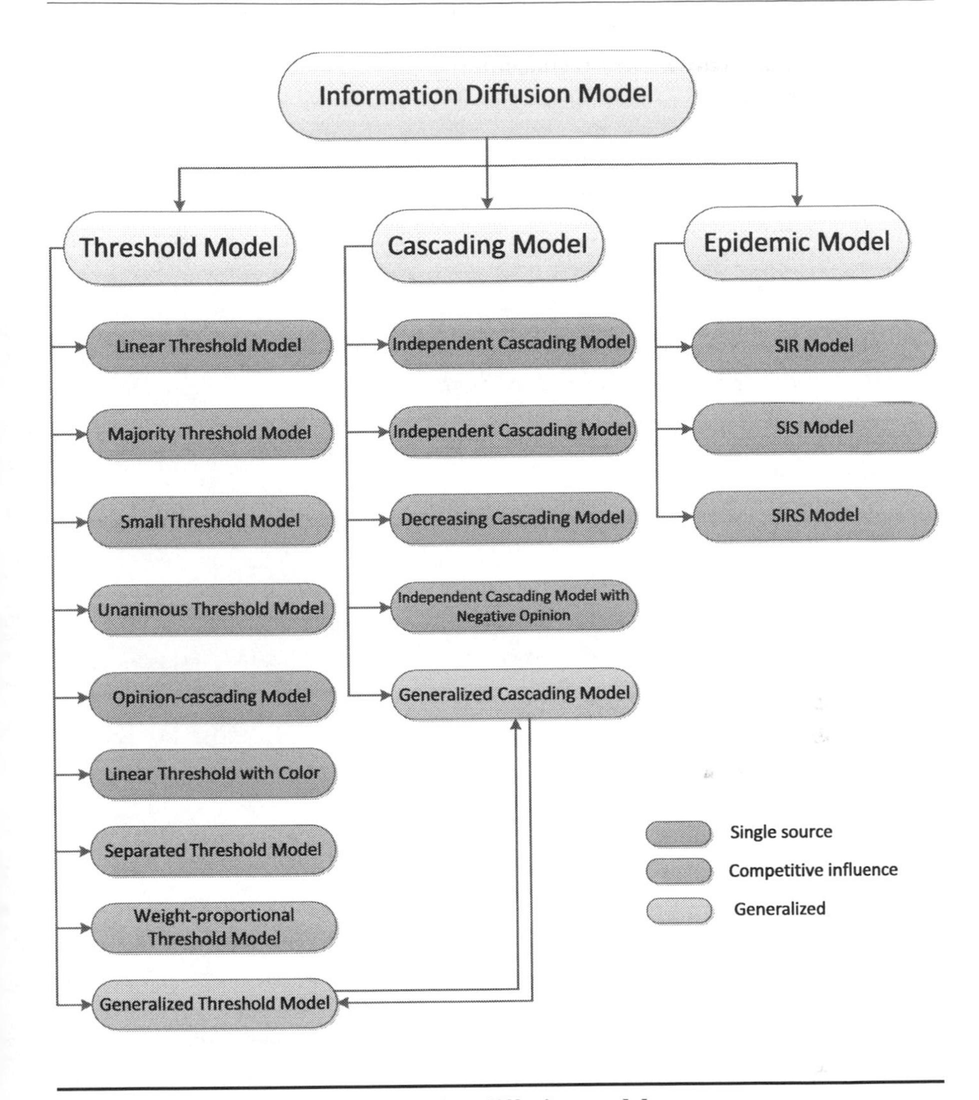

Figure 2.1: An overview of information diffusion models.

them to adopt the new product, and use the social influence effect to advertise and recommend the product to their friends, we can trigger widespread adoption. The influence maximization problem is as follows: Which key individuals should we target as the promising seeds in order to maximize the spread of influence?

In [17, 46], the influence maximization problem was studied in a probabilistic model of interaction, and selection of the most influential seeds was based on individual's overall effect on the network. In other words [27, 28, 33, 10, 53], many researchers take this seeding selection as a problem in discrete optimization. Formally, the influence maximization problem is defined as follows:

Definition 2.1 **(Influence Maximization)** Given a budget k and a social network, which is represented as a directed graph $G = (V, E)$, where users are represented as nodes and edges indicate their relationships, the goal is to select a seed set of k users such that by initially targeting them, the expected influence spread (in terms of expected number of adopted users) can be maximized.

The expected influence spread is related to its propagation process, which is captured by the diffusion models. In Section 2.4, an overview of theoretical diffusion models is provided, and for most of the models we introduce, the optimal solution for the influence maximization problem is shown to be NP-hard. A well-known greedy (1-1/e) approximation algorithm is extensively used for approximating the optimal solution of the original problem and its extensions under different models. However, the approximation algorithm requires that the influence function have two basic properties:

Definition 2.2 **(Monotonicity)** A set function f is monotone if $f(S) \leq f(T)$ such that $S \subset T \subset U$.

Definition 2.3 **(Submodularity)** A set function f is submodular if it satisfies

$$f(S \cup \{v\}) - f(S) \geq f(T \cup \{v\}) - f(T) \quad (2.1)$$

for all elements $v \subset S$ and $S \subset T$.

2.4 Information Diffusion Models

Influence diffusion is the process by which information propagates through certain intermediaries over time among the individuals of a social network. Empirical studies of diffusion in social networks began in the middle of the 20th century and Granovetter [26] was the first to introduce a formal mathematical model. Currently, there are a variety of diffusion models arising from the economics and sociology communities. The most popular models are the *Linear Threshold* model and the *Independent Cascading* model, which are widely used in studying social influence problems. Besides those two well-known models, there are many variations and extension models to reflect more complicated real-world situations. For example, in addition to the expected number of adopted users, [55] considered the expected total opinions of adopted users, which is more meaningful. Reference [9] proposed a new model, named the IC-N model, which took into account the negativity bias during the propagation process. In this section, we survey the recent literature on theoretical models of influence diffusion.

The *social network* is a kind of social structure that consists of social actors such as individual users or organizations and a complex set of relationships between each two of them. Formally, a social network is represented as a graph $G = (V, E)$, which can be either a directed or undirected graph according to its real application and network property. In graph G, each vertex $v \in V$ represents an individual user.

In a directed graph, an edge $(u,v) \in E$ represents that u has an influence on v; in an undirected graph, an edge (u,v) represents mutual influence between u and v. Particularly, an undirected graph can be viewed as a directed graph by treating each edge as a bidirectional edge with the same influence in both directions. In addition, let $N(v)$ denote v's neighbors in an undirected graph, and let $N^{in}(v)$ and N^{out} denote the sets of incoming neighbors (or in-neighbors) and outgoing neighbors (or out-neighbors), respectively.

2.4.1 Threshold Models

In this subsection, we give an overview of the concept of threshold models and show how these models characterize collective behaviors. In mathematical or statistical modeling, a *threshold model* is any model where a threshold value, or set of threshold values, is used to distinguish ranges of values where the behavior predicted by the model varies in some important way.

In threshold models, someone first breaks the silence of the network because that activity provides the individual utility. It is the distribution of individual thresholds, defined as the number of other people who must be doing the activity before a given individual joins in, that determines whether or not others would follow this activity. The threshold models were first proposed by Mark Granovetter [26] to model collective behavior, which aimed at treating binary decision problems, such as diffusion of innovations, spreading rumors and diseases, voting and so on. He used the threshold model to explain the riot, residential segregation, and the spiral of silence. In the spirit of Granovetter's threshold model, the "threshold" is "the number or proportion of others who must make one decision before a given actor does so." It is necessary to emphasize the determinants of the threshold. A threshold is different for individuals, and it may be influenced by many factors: social economic status, education, age, personality, etc. Further, Granovetter relates "threshold" with the utility that one gets from participating in collective behavior or not. By using the utility function, each individual will calculate his cost and benefit from undertaking an action. And situations may change the cost and benefit of the behavior, so the threshold is situation-specific. The distribution of the thresholds determines the outcome of the aggregate behavior (for example, public opinion). In other words, this threshold represents the number of other agents in the population or local neighborhood following that particular activity. Each agent has a threshold that, when exceeded, leads the agent to adopt an activity.

In his model, each edge (which represents a connection) (v,u) is associated with a weight $w_{v,u}$, and each node v has a threshold θ_v such that if the fraction of v's neighbors that are active exceeds v's threshold, then v will become active. Granovetter claims that minor perturbations in the standard deviation of a distribution produce massive discontinuous changes in the number of people acting, from 6% to nearly 100% of the whole group. The reason is that in threshold models, the intrinsic utility of the behavior to an individual may be more important in determining that individual's behavior than social influence. However, even a limited amount of social influence may have a strong effect on the collective outcome [26].

Threshold models are especially useful in a structural analysis of collective action, an approach that most rational theorists have avoided. Sudden changes in the level of production of a particular public good does not necessarily reflect similar changes in the overall preferences of the actors. What really matters is the distribution of thresholds and the social connections through which members could have chances to learn about the others.

2.4.1.1 Linear Threshold Model

The *Linear Threshold* (LT) model is the one that has been extensively used in studying diffusion models among the generalizations of threshold models. In this model, each node v has a threshold θ_v, and for every $u \in N(v)$, (u,v) has a nonnegative weight $w_{u,v}$ such that $\sum_{u \in N(v)} w_{u,v} \leq 1$. Given the thresholds and an initial set of active nodes, the process unfolds deterministically in discrete steps. At time t, an inactive node v becomes active if

$$\sum_{u \in N^a(v)} w_{u,v} \geq \theta_v,$$

where $N^a(v)$ denotes the set of active neighbors of v. Every activated node remains active, and the process terminates if no more activations are possible. The threshold in this model is related to a linear constraint of edge weight, which accounts for the name for the model. It is important to note that given the thresholds in advance, the diffusion process is deterministic, but we can still inject randomness by randomizing the individual threshold. For example, the thresholds selected by Kempe et al. [27, 28] are uniformly at random from the interval [0,1], which are also intended to model the lack of knowledge of their values.

Given the influence function $\sigma(\cdot)$, Kempe et al. [27] prove that:

Theorem 2.1
For an arbitrary instance of the Linear Threshold model, the objective influence function $\sigma(\cdot)$ is submodular.

Theorem 2.2
The influence maximization problem is NP-hard under the Linear Threshold model.

Granovetter and Schelling's approach is based on the use of node-specific thresholds [26, 50], but there is another class of approaches that hard-wires all thresholds at a known and fixed value. This kind of model is often used in treating binary decision problems such as voting, virus propagation networks, and so on. In this model, let $d(v)$ denote the degree of a node $v \in V$, and threshold value $\theta_v \in \mathcal{N}$, where $\theta_v \in [1, d(v)]$. This definition is adopted by the following three models.

2.4.1.2 The Majority Threshold Model

The Majority Threshold (MT) model is one of the most important and well-studied models, in which each vertex $v \in V$ becomes active if the majority of its neighbors are active, that is, the threshold $\theta_v = \frac{1}{2}d(v)$. This model has many applications in voting systems, distributive computing, and so on [43, 44]. Chen [42] shows that with the majority thresholds setting, the influence maximization problem shares the same hardness of approximation ratio as the general one. Chen [42] also provides the following inapproximability result of the majority thresholds model.

Theorem 2.3
Assume that the influence maximization problem with arbitrary thresholds cannot be approximated within the ratio of $\sigma(n)$, for some polynomial time-computable function $\sigma(n)$. Then the problem with majority thresholds cannot be approximated within the ratio of $O(\sigma(n))$.

2.4.1.3 The Small Threshold Model

The other interesting case is the Small Threshold (ST) model, in which all thresholds are small constants [47]. Intuitively, when the threshold $\theta_v = 1$, the influence maximization problem can be easily solved by selecting an arbitrary node in each connected component. However, Chen [42] shows that the hardness of the approximation result continues to hold when each vertex's threshold $\theta_v = 2$. In addition, Dreyer [18] proves that if the threshold of any vertex is θ_v for any $\theta_v \geq 3$, the problem is NP-hard as well.

Theorem 2.4
Assume that the influence maximization problem with arbitrary thresholds cannot be approximated within the ratio of $\sigma(n)$, for some polynomial time-computable function $\sigma(n)$. Then the problem cannot be approximated within the ratio of $O(\sigma(n))$ when all thresholds are at most 2.

2.4.1.4 The Unanimous Threshold Model

In the Unanimous Threshold (UT) model, the threshold for each vertex is $\theta_v = d(v)$, which is equal to its degree. With this setting, the UT model is the most influence-resistant model among all the threshold models. This model is usually used in studying complex network security and vulnerability. For example, in an ideal virus-resistant network, when the computer virus is spreading, a vertex can be affected if all of its neighbors have been infected. For this special case, the influence maximization problem is equivalent to the vertex cover problem. Thus, it admits an approximation algorithm with ratio 2 and is NP-hard as well [42].

Theorem 2.5
If all thresholds in a graph are unanimous, the influence maximization problem is NP-hard.

2.4.1.5 Other Extensions

The threshold models can be further generalized in a very natural way by replacing the activation function with an arbitrary function in relation to the set of a vertex's activated neighbors. For example, Bhagat et al. [4] propose a Linear Threshold with Color (LT-C) model that factors in user's experience with a product, in which they adapt the LT model by adding three more statuses of user activities and defining an objective function that explicitly captures product adoption, not the influence. Banerjee et al. [2] further extends the LT model to handle a more complicated case, in which each node is allowed to switch back and forth between active and inactive regarding each cascade. This model is shown to be a rapidly mixing Markov chain and the corresponding steady-state distribution is used to estimate a highly likely cascade adopted in the network. Furthermore, consider that since users now are engaged in many different social networks, information can be diffused across multiple networks simultaneously; [41, 51] adapt the LT model to deal with the IM problem under multiple networks.

2.4.2 Cascading Model

Inspired by the work on interacting particle systems [19, 37] and probability theory, dynamic cascade models are considered for the diffusion process. In the context of marketing, Goldenberg et al. [21, 22] first studied cascade models. In cascade models, the dynamics is captured in a step-by-step fashion: at time t, when a node v first becomes active, it has a single chance of influencing each previously inactive neighbor u at time $t+1$. And it successfully causes u to be activated with a probability $p_{v,u}$. In addition, if multiple neighbors of u become active at time t, their attempts to activate u are sequenced in an arbitrary order. If one of them, say w, succeeds in time t, then u becomes active in time $t+1$; however, whether w succeeds or not, it cannot make any more attempts in the following time steps. Similar to the threshold models, the process terminates when there are no more activations.

2.4.2.1 Independent Cascading Model

To better describe the cascading models, one thing we need to specify that the probability for a newly activated node v to successfully make an attempt to activate its currently inactive neighbors u. The simplest case is the Independent Cascading (IC) model, in which the probability is a constant $p_u(v)$, independent of the history of the diffusion process thus far. In addition to that, to better define the model, we also need to introduce *order independence* here. Let S denote the set of nodes that have already attempted and failed to activate u, and the probability for v to successfully active u is denoted by $p_u(v|S)$. Let $v_1, v_2, ...v_k$, and $v'_1, v'_2, ...v'_k$ be two different permutations of

S, and $T_i = \{v_1, v_2, ...v_i\}$, $T_i' = \{v_1', v_2', ...v_i'\}$. The order independence indicates that the order of attempts made by each node in S does not affect the probability for u to be active in the end, which is

$$\Pi_{i=1}^{k}(1 - p_u(v_i|S \cup T_i)) = \Pi_{i=1}^{k}(1 - p_u(v_i'|S \cup T_i')),$$

where $S \cap T = \emptyset$.

2.4.2.2 Decreasing Cascading Model

Compared with the IC model, the Decreasing Cascading (DC) model [28] is more general and practical. (We adopt all the definitions in the IC model here.) The DC model naturally incorporates a restriction that the function $p_u(v|S)$ is non-decreasing in S, which indicates that $p_u(v|S) \leq p_u(v|T)$, where $S \subset T$. This better reflects the information saturation problem in the real world: the probability of a successful activation of a node u decreases if more people have already made attempts. The DC model contains the IC model as a special case.

2.4.2.3 Independent Cascading Model with Negative Opinion

In [9], Chen proposed the Independent Cascading model with Negative Opinion (IC-N), which incorporates the negative opinions into the propagation process. The IC-N model associates a new parameter q, called the *quality factor*, which models the natural behavior of users adopting negative opinions due to defects of the product/service. In this model, each activated user can be either positive or negative, and with probability q, each newly active node becomes positive and with probability $1-q$, it becomes negative. In addition, when a node u is negatively activated, it becomes negative with probability 1 and remains negative in the following rounds. This reflects the negativity bias and dominance phenomenon in social psychology [49].

2.4.3 Generalized Threshold and Cascade Models

We have thus far introduced two families of widely studied propagation models. Before heading to the next kind of diffusion model, we want to introduce a more general and broader framework that generalizes the classic LT model and IC model in this subsection. In particular, under such setting, Kempe et al. [27] prove that the general cascade model and general threshold model are equivalent. And because of this equivalence, we can unify these two different views of diffusion in social networks.

- **Generalized threshold model.** In the general threshold model, each node v has a threshold θ_v, and associates with a function f_v that maps the set of its neighbors $N(v)$ to the range [0,1] and subject to the condition $f_v(\emptyset) = 0$. This function could be an arbitrary monotone function. The dynamics of the diffusion process follows the LT model. But a node v becomes active at time t if and only if $f_v(N^a(v)) \geq \theta_v$, where $N^a(v)$ is the subset of active neighbors of

v at time $t-1$. It is easy to see that the generalized threshold model contains the LT model as a special case, in which the threshold function is subject to $f_v = \sum_{u \in N^a(v)} w_{u,v}$, and $\sum_{u \in N(v)} w_{u,v} \leq 1$.

- **Generalized cascade model.** Compared with the specific cascade models, we generalize the cascade model by allowing the probability that u successfully activates its neighbor v to depend on the other active neighbors of v that have tried. Thus, we change the activation probability $P_{u,v}$ to an incremental function $p_v(u,S) \in [0,1]$, where u and S are two disjoint subsets of $N(v)$. In each discrete time stamp, when a newly activated node u attempts to activate a currently inactive node v, it succeeds with probability $p_v(u,S)$, where S denotes the set of nodes that have already made their attempts. The IC model can be viewed as a special case of the generalized cascade model, in which $p_v(u,S)$ is set to a constant $p_{u,v}$. Furthermore, the *order independence*, which was introduced in the IC model, is also adopted here.

Next, we show that if the threshold function *theta*$_v$ is chosen independently and uniformly at random, then those two generalized models are equivalent as shown by the following conversion.

Let f_v be a threshold function of the general threshold model, and S be the set of nodes that have already tried to activate v. Then in order to define an equivalent cascade model, we need to know the probability that an additional node u can activate v if all the nodes in S have failed. Once the node in S failed, node v's threshold θ_v should be in the range $(f_v(S), 1]$. Therefore, with the constraint that it should be uniformly distributed, the probability that a neighbor $u \notin S$ successfully activates v is

$$p_v(u,S) = \frac{f_v(S \cup \{u\}) - f_v(S)}{1 - f_v(S)}$$

where nodes in S failed to activate v. It is easy to see that the generalized cascade model can be converted to the generalized threshold model with this function.

On the other side, let v be a node in the cascade model, with its neighbor set denoted by $S = \{u_1, u_2, ...u_k\}$. All the nodes in S have tried to activate v in an order T and let us assume $T = \{u_1, u_2, ...u_k\}$, and $S_i = \{u_1, u_2, ...u_i\}$; then the probability that v hasn't been influenced is $\prod_{i=1}^{k}(1 - p_v(u_i, S_{i-1}))$. According to order independence, this value is not affected by the order of u_i, but only depends on the set S, thus we can obtain that

$$f_v(S) = 1 - \prod_{i=1}^{k}(1 - p_v(u_i, S_{i-1})).$$

In this way, the threshold model can be shown be to equivalent to the cascade model.

2.4.4 Epidemic Model

The Epidemic model has had a major impact on life and politics. Modeling the infectious diseases became a matter of general interest in the 19th century. An epidemic model describes the transmission of contagious disease through individuals. In the recent century, it has been widely used to model computer virus infections and information propagations such as news and rumors.

2.4.4.1 SIR Model

The SIR (Susceptible-Infectious-Recovered) model was first proposed by Kermack and McKendrick [29]. This model considers a fixed population which is divided into three distinct classes: Susceptible (S), Infectious (I), and Recovered (R). The individual goes through consecutive states:

$$S \to I \to R.$$

The dynamics of the model cascades in this way: given a fixed population at a particular time t, there exist three groups of people, $S(t)$ represents the number of people who are susceptible to the contagion, $I(t)$ represents the number of people who have been infected and are capable of infecting those who are susceptible, and $R(t)$ is the number of people who have been infected and recovered, which means they are immune to being infected again in the future. Using the contact rate β from S to I, and $1/\gamma$ as the average infectious period, Kermack and McKendrick [30] derived the following equations:

$$\frac{dS}{dt} = -\beta SI$$

$$\frac{dI}{dt} = \beta SI - \gamma I$$

$$\frac{dR}{dt} = \gamma I$$

The critical parameter $R_0 = \beta S_0/\gamma$ is called the basic reproduction number. We can see that $R = 1$ is the critical value; $R < 1$ implies no epidemic and $R > 1$ that an epidemic is possible.

In this model, several assumptions are made in the formulation of the equations. First of all, each individual is considered as having the same probability of contracting the disease with a rate of β, which is also the infection rate of the disease. Therefore, an infected individual can transmit the disease with βN other susceptible people per unit time, and the fraction of contacts by an infected person with a susceptible is S/N. In addition, given the rate of new infections as $\beta N(S/N)I = \beta SI$ [7], the number of newly infected people per unit time is $\beta N(S/N)$. Secondly, consider the population leaving the susceptible group as equal to the number of newly infected people, we can get the second and third equations above. Specifically, a number equals the fraction of infective people who are leaving this class per unit time to enter the removed group. These processes, which occur simultaneously, are known as the *Law of Mass Action* [12], which is a widely accepted idea that the rate of contact between two groups in a population is proportional to the size of each of the groups concerned.

2.4.4.2 SIS Model

The SIS model considers a fixed population with only two compartments—susceptible S(t) and infected I(t), thus the flow of this model may be considered as follows:

$$S \rightarrow I \rightarrow R.$$

The SIS an be easily derived from the SIR model by simply considering that the individuals recover with no immunity to the disease, that is, individuals are immediately susceptible once they have recovered.

Thus removing the equation representing the recovered population from the SIR model and adding those removed from the infected population into the susceptible population, we can get the following differential equations:

$$\frac{dS}{dt} = -\beta SI + \gamma I$$
$$\frac{dI}{dt} = \beta SI - \gamma I.$$

2.4.4.3 SIRS Model

The SIRS model is an extension of the SIR model. An individual can go through consecutive states:

$$S \rightarrow I \rightarrow R.$$

The difference between this model and the SIR model is that it allows an individual of a recovered group to leave and rejoin the susceptible group. Thus, we can get the following equations:

$$\frac{dS}{dt} = -\beta SI + fR$$
$$\frac{dI}{dt} = \beta SI - \gamma I$$
$$\frac{dR}{dt} = \gamma I - fR,$$

where f is the average loss of immunity rate of recovered individuals.

2.4.5 Competitive Influence Diffusion Models

All of the above models have primarily focused on diffusion of a single cascade, but when multiple innovations are competing within a social network, things become different yet interesting. Carnes et al. in [8] consider the problem faced by a company that would like to spread its new product in the market while a competing product is already being introduced. There are two assumptions: first, consumers use only one of the two products and influence their friends in their decision of which product to use; second, the follower has a fixed budget available that can be used to target a subset of consumers. In [8], they propose two models for describing how two technologies simultaneously diffuse over a given network.

2.4.5.1 Distance-Based Model

The first model, *a distance-based model*, is related to competitive facility location [20] on a network. In this model, the location of a node in the network is important, as well as the connectivity of a node. The central idea is that a consumer will be more likely to mimic the behavior of an early adopter if their distance in the social network is relatively small. It is pointed out in [8] that the expected number of nodes which adopt A will be denoted by

$$\rho(I_A|I_B) = \mathbf{E}[\sum_{u\in V} \frac{v_u(I_A, d_u(I,E_a))}{v_u(I_A, d_u(I,E_a)) + v_u(I_B, d_u(I,E_a))}],$$

where the expectation is over the set of active edges. I_A and I_B are the initial sets of adopters of A and B, respectively, and I is their union set. $d_u(I,E_a)$ denotes the shortest distance from u to I along the edges in E_a. Fixing I_B and trying to determine I_A to maximize the expected number of nodes that adopt technology A would be:

$$max\{\rho(I_A|I_B) : I_A \subseteq (V - I_B), |I_A| = k\}.$$

The following theorem gives an approximation bound for this equation.

Theorem 2.6
For any given I_B with $|V - I_B| \geq k$, the Hill Climbing algorithm gives a $(1-1/e-\varepsilon)$-approximation algorithm for the above result.

2.4.5.2 Wave Propagation Model

The second model, a *wave propagation model*, regards the propagation as happening in discrete steps. In step d, all nodes that are at a distance of at most $d-1$ from some node in the initial sets have adopted technology A or B, and all nodes for which the closest initial node is farther than $d-1$ do not have a technology yet. Similar to the distance-base model, it gives the solution:

$$max\{\pi(I_A|I_B) : I_A \subseteq (V - I_B), |I_A| = k\}$$

where

$$\pi(I_A|I_B) = \mathbf{E}[\sum_{v\in V} P(v|I_A, I_B, E_a)].$$

The authors provide another theorem that gives the same approximation ratio as above:

Theorem 2.7
For any given I_B with $|V - I_B| \geq k$, the Hill Climbing algorithm gives a $(1-1/e-\varepsilon)$-approximation algorithm for the above result.

Consequently, by computational experiments, the authors point out that although it is NP-hard to select the most influential subset to target, it is possible to give an efficient algorithm that is within 63% of optimal. Lastly, using the distance-based model with edge probabilities equal to 1, these problems can also be seen in the context of competitive facility location [1, 14] on a network.

2.4.5.3 Weight-Proportional Threshold Model

Consider the real-world scenarios where different kinds of innovations or products are competing with each other; competitive threshold models are suggested by Borodin et al. in [6]. Under the competitive setting, the goal is to maximize the spread of one cascade in the presence of one or more competitors.

In order to describe the process, we use the following notation for the next two models.

Definition 2.4 In discrete time stamp t, let Φ^t denote the set of active nodes, in particular, let Φ_A^t and Φ_B^t be the sets of A-active and B-active nodes in time stamp t, respectively.

Given two different seeds S_A and S_B at the beginning, in each time stamp, every inactive node v changes its status according to the incoming influence from its currently active neighbors as follows: v becomes active when $\sum_{u \in \Phi^t} w_{u,v} \geq \theta_v$ is satisfied; in addition, v becomes an A-active node with probability

$$Pr[v \in \Phi_A^t | v \in \Phi^t \setminus \Phi^{t-1}] = \frac{\sum_{u \in \Phi_A^t} w_{u,v}}{\sum_{u \in \Phi^t} w_{u,v}}.$$

It adopts cascade B, otherwise.

The problem of maximizing the spread of cascade A can be easily reduced to the original influence maximization problem by setting $S_B = \varnothing$. Thereby, this problem is also NP-hard, as proved in [27].

Intuitively, by adding one more node to the initial set S_A, the spread of cascade A could be expended. However, the influence function $\sigma(\cdot)$ is neither monotone nor submodular under the Weight-Proportional Threshold (WT) model, as shown by a count example in [6].

2.4.5.4 Separated Threshold Model

In the previous model, a node v changes its status from inactive to active whenever the influence from all of its currently active neighbors exceeds its threshold θ_v. However, nodes may not have the same threshold toward each competitor, and the influence strength between each pair of nodes could be different regarding each cascade. Formally, each node v has two thresholds θ_v^A, θ_v^B, and each edge (u,v) is associated with two weights $w_{u,v}^A$, $w_{u,v}^B$ corresponding to cascades A and B, respectively. Both weights satisfy the constraints of the LT model. In time stamp t, every inactive node v will be A-active when $\sum_{u \in N^a(v) \cap \Phi_A^{t-1}} w_{u,v}^A \geq \theta_v^A$, and will be B-active when $\sum_{u \in N^a(v) \cap \Phi_B^{t-1}} w_{u,v}^B \geq \theta_v^B$. If both thresholds are exceeded during the same stamp t, then v adopts a cascade uniformly at random.

However, unlike the previous model, the probability that cascade A will be adopted by a node cannot be increased by adding an additional B-activated node. Therefore, under the Separated Threshold (SepT) model, the influence function $\sigma(\cdot)$ is monotone, but not submodular, as proved in [6] by a counting example.

Summary

In this section, we provide an overview of diffusion models that have been extensively used in studying social influence: threshold models [26, 27, 28, 42, 47, 6], cascading models [21, 22, 9, 8], and epidemic models [29, 35]. Table 2.1 summarizes the activation condition, model properties, and applications of each model. And with this framework in place, we move on to the next section, which focuses on the algorithmic results of the influence maximization problem.

2.5 Influence Maximization and Approximation Algorithms

2.5.1 Influence Maximization

A social network is the graph of relationships and interactions within a group of individuals that plays a fundamental role as a medium for the spread of information, ideas, and influence among its members. *Influence maximization(IM)* is the problem of choosing the most potential of individuals in a network to spread information in order to trigger the widespread adoption of a product. Domingos and Richardson [17] model the problem as a Markov random field. Kempe et al. [27, 28] assume a fixed marketing budget sufficient to target k individuals and study the problem of finding the optimal k individuals in the network to target. This problem has applications in viral marketing, where a company may wish to spread the rumor of a new product via the most influential individuals in popular social networks. With online social networking sites such as Facebook, LinkedIn, Myspace, etc., attracting hundreds of millions of people, online social networks are also viewed as important platforms for

Table 2.1: Models Listing and Comparisons

Name	*Activation Condition*	*Application*	*Property*	*Reference*
LT	$\sum_{u\in N^a(v)} w_{u,v} \geq \theta_v$	Collective behavior, spreading rumors and diseases	The objective $\sigma(\cdot)$ is submodular, and IM is NP-hard	[30]
MT	$\theta_v = \frac{1}{2}d(v)$	Voting system, distributed computing	IM is NP-hard	[46, 49, 4]
ST	$\sum_{u\in N^a(v)} w_{u,v} \geq \theta_v$ where θ_v is a small constant		$\theta_v = 1$, select an arbitrary node in each connected component; $\theta_v \geq 2$, IM is NP-hard	[20, 50]
UT	$\sum_{u\in N^a(v)} w_{u,v} \geq \theta_v$ where $\theta_v = d(v)$	Network security and vulnerability	IM is NP-hard, 2-approximation algorithm	[46]
WT	$Pr[v \in \Phi_A^t \mid v \in \Phi^t \setminus \Phi^{t-1}] = \frac{\sum_{u\in \Phi_A^t} w_{u,v}}{\sum_{u\in \Phi^t} w_{u,v}}$	Deal with two competitive influence	IM is NP-hard. $\sigma(\cdot)$ is neither monotone nor submodular	[30]
SepT	$i-active: \sum_{u\in N^a(v)\cap \Phi_A^{t-1}} w_{u,v}^i \geq \theta_v^i$	Network with competitive sources	IM is NP-hard. $\sigma(\cdot)$ is monotone, but not submodular	[30, 6]
LT-C	$\theta_v = \frac{\sum_{u\in N^a(v)} w_{u,v}(r_{u,i}-r_{min})}{r_{max}-r_{min}}$	Distinguish product adoption from influenced users	NP-hard, $\sigma(\cdot)$ is monotone and submodular	[4]
OC	$\sum_{u\in N^a(v)} \geq \theta_v$	Incorporate user opinions	IM is NP, $\sigma(\cdot)$ is neither monotone nor submodular	[55]
IC	$\Pi_{i=1}^k(1-p_u(v_i\mid S\cup T_i)) = \Pi_{i=1}^k(1-p_u(v_i'\mid S\cup T_i'))$	Collective behavior, promote new products	The objective $\sigma(\cdot)$ is submodular, and IM is NP-hard	[27]
DC	$p_u(v\mid S) \leq p_u(v\mid T)$	Collective behavior, spreading information	IC is a special case of DC, and the objective $\sigma(\cdot)$ is submodular, and IM is NP-hard	[28]
IC-N	$\Pi_{i=1}^k(1-p_u(v_i\mid S\cup T_i)) = \Pi_{i=1}^k(1-p_u(v_i'\mid S\cup T_i'))$	Incorporate negative opinions	With probability q, each newly active node become positive and with probability $1-q$	[9]
SIR				

effective viral marketing practice. This further motivates the research community to conduct extensive studies on various aspects of the influence maximization problem.

2.5.2 Approximation Algorithm

We are now in a position to choose a good initial set of nodes to target in the context of the above models. Based on the basic models we introduced above, in this section, we introduce the hardness of influence maximization problems on above models, and prove that the influence maximization problem with budget k under both of LT and IC models is NP-hard.

In addition, the influcnce function $f(\cdot)$ is submodular and monotone increasing. Exploiting these properties, Kempe et al. [27] present a simple greedy algorithm that approximates the problem with the ratio of $1-1/e-\varepsilon$ for any $\varepsilon>0$. However, the running time of the worst case of the naive greedy algorithm is $O(n^2(m+n))$, which is prohibitive for large-scale networks. Thus, considerable work has been done to improve it. In this section, we demonstrate recent algorithmic studies of the CELF [33], CELF++ [24], Simpath [25], and LDAG [11] algorithms, which can obtain high scalability for influence maximization problems.

2.5.2.1 *Greedy Algorithm*

Following the definition in Section 1.3.1.1, we now provide the definitions and notations as follows. An influence graph is a weighted graph $G=(V,E,w)$ with a weight function w, where V is a set of n nodes and $E\subseteq V\times V$ is a set of m directed edges. And the weight function $w:\ V\times V\rightarrow[0,1]$ holds that $w(u,v)=0$ if and only if $(u,v)\notin E$, and $\sum_{u\in N(v)}w(u,v)=0$ where $N(v)$ means that u is the neighbor of v. In the LT model, when given a seed set $S\subseteq V$, influence cascades in graph G in discrete steps. At time t, each inactive node v becomes active if the weighted number of its activated in neighbors reaches its threshold, i.e. $\sum_{u\in N^a(v)}w_{u,v}\geq\theta_v$, where $N^a(v)$ denotes the set of active neighbors of v. The process stops at a step t when the seed set becomes empty. Each activated node remains active, and the process terminates if no more activations are possible.

The influence maximization problem under the linear threshold model is, when given the influence graph G and an integer k, finding a seed set S of size k such that its influence spread $\sigma_L(S)$ is the maximum where we call $\sigma_L(S)$ the *influence spread* of seed set S [10]. It is shown in [27] that finding the optimal solution is NP-hard, but because σ_L is monotone and submodular, a greedy algorithm has a constant approximation ratio. A generic greedy algorithm for any set function f is shown as Algorithm 3.

Algorithm 3 simply executes in k rounds, and in each round a new entry that gives the largest marginal gain in f will be selected. It is shown in [40] that for any monotone and submodular set function f with $f(\emptyset)=0$, the greedy algorithm has an approximation ratio $f(S)/f(S*)\geq 1-1/e$, where S is the output of the greedy algorithm and $S*$ is the optimal solution. However, the generic greedy algorithm

Algorithm 3: Greedy Algorithm

Input: G, k, f
Output: Seed set S
1 initialize $S \leftarrow \emptyset$;
2 **while** $|S| \leq k$ **do**
3 select $u \leftarrow argmax_{w \in V \setminus S}(f(S \cup \{w\}) - f(S))$;
4 $S \leftarrow S \cup \{u\}$;
5 **return** S ;

requires the evaluation of $f(S)$. In the context of influence maximization, the exact computation of $\sigma_L(S)$ was left as an open problem in [27] and it was later proved that the exact computation of $\sigma_L(S)$ is #P-hard in [10].

The running time of worst case of this naive greedy algorithm is $O(n^2(m+n))$, which is prohibitive for large-scale networks. Thus, considerable work has been done to improve it. We will introduce them in the following several subsections.

2.5.2.2 CELF Selection Algorithm

Relatively little work has been done on improving the quadratic nature of the greedy algorithm. The most notable work is [33], where submodularity is exploited to develop an efficient algorithm, called the Cost-Effective Lazy Forward (CELF) selection algorithm, based on a lazy-forward optimization in selecting seeds. The idea is that marginal gain of a node in the current iteration cannot be better than its marginal gain in the previous iterations. CELF maintains a table $< u, \Delta_u(S) >$ sorted on $\Delta_u(S)$ in decreasing order, where S is the current seed set and $\Delta_u(S)$ is the marginal gain of u w.r.t. S. The $\Delta_u(S)$ here corresponds to $\sigma_L(S)$ in the previous subsection. $\sigma_L(S)$ is re-evaluated only for the top node at each step and the table is resorted to only when it is necessary. If a node remains at the top, it will be picked as the next seed. In real implementation, a heap Q is employed to represent the priority of each node and maintain the sorted table information.

In [33], the authors empirically shows that CELF dramatically improves the efficiency of the greedy algorithm. Algorithm 4 shows the skeleton of the CELF algorithm. In the algorithm, $\sigma_m(S)$ denotes the expected influence spread of seed set S under the propagation model m (like IC or LT). This m could be omitted if there is no confusion in the context. As clearly explained in [23], the optimization works as follows. Maintain a heap Q with nodes corresponding to users in the network G.

The node of Q corresponding to user u stores a tuple of the form $< u.mg, u.round >$ where $u.mg = \sigma_m(S \cup \{u\}) - \sigma_m(S)$ represents the marginal gain of u w.r.t. the current seed set S while $u.round$ is the iteration number when $u.mg$ was last updated. In the first iteration, marginal gains of each node are computed and added to Q in decreasing order of marginal gains (the first *for* loop). Later, in each iteration, look at the top node u in Q and see if its marginal gain was last computed in the current iteration (using the *round* attribute). If yes, then, due to submodularity, u must be the node that provides maximum marginal gain in the current iteration,

Algorithm 4: Greedy algorithm optimized with CELF

```
   Input: G, k, σ_m
   Output: Seed set S
 1 initialize S ← ∅, Q ← ∅;
 2 for each u ∈ V do
 3     u.mg = σ_m({u}) ;
 4     u.round = 0 ;
 5     Add u to Q in decreasing order of mg.
 6 while |S| ≤ k do
 7     u ← root element in Q ;
 8     if u.round == |S| then
 9         S ← S ∪ {u} ;
10         Q ← Q − {u} ;
11     else
12         u.mg = σ_m(S ∪ {u}) − σ_m(S) ;
13         u.round = |S| ;
14         Reinsert u into Q and heapify.
15 return S ;
```

hence, it is picked as the next seed. Otherwise, recompute the marginal gain of u, update its round flag and reinsert into Q such that the order of marginal gains is maintained. This process is realized in the *while* loop in the algorithm.

It is easy to see that this optimization avoids the recomputation of marginal gains of all the nodes in any iteration, except the first one. Therefore, from the experimental results, the CELF optimization leads to a 700-times speedup in the greedy algorithm shown in [33].

2.5.2.3 CELF++ Algorithm

In [24], Goyal et al. introduce CELF++, which further optimized CELF by exploiting submodularity. Algorithm 5 describes the CELF++ algorithm. The setup is similar to CELF: $\sigma(S)$ is used to denote the spread of seed set S. A heap Q has nodes corresponding to users in the network G.

The improvement is that instead of a tuple of two attributes, they offer that the node of Q corresponding to user u stores a tuple of the form $< u.mg1, u.prev\_best, u.mg2, u.flag >$. Here $u.mg1 = \Delta_u(S)$, the marginal gain of u w.r.t. the current seed set S; $u.prev\_best$ is the node that has the maximum marginal gain among all the users examined in the current iteration, before user u; $u.mag2 = \Delta_u(S \cup \{prev\_best\})$, and $u.flag$ is the iteration number when $u.mg1$ was last updated.

The central idea is that if the node picked in the last iteration is still at the root of the heap, they don't need to recompute the marginal gains. This does save a lot of computations. It is important to note that in addition to computing $\Delta_u(S)$,

Algorithm 5: Greedy algorithm optimized with CELF++

```
   Input: G, k, σ_m
   Output: Seed set S
 1 initialize S ← ∅, Q ← ∅, last_seed ← NULL, cur_best ← NULL ;
 2 for each u ∈ V do
 3     u.mg1 ← σ({u}) ;
 4     u.prev_best ← cur_best ;
 5     u.mg2 ← Δ_u{cur_best} ;
 6     u.flag ← 0 ;
 7     Q ← Q ∪ {u} ;
 8     Update cur_best based on u.mg1 ;
 9 while |S| ≤ k do
10     u ← root element in Q ;
11     if u.flag == |S| then
12         S ← S ∪ {u} ;
13         Q ← Q − {u} ;
14         last_seed ← u ;
15         cur_best ← NULL ;
16         Continue ;
17     else if u.prev_best == last_seed and u.flag == |S| − 1 then
18         u.mg1 ← u.mg2 ;
19     else
20         u.mg1 ← Δ_u(S) ;
21         u.prev_best ← cur_best ;
22         u.mg2 ← Δ_u(S ∪ {cur_best}) ;
23     u.flag = |S| ;
24     Update cur_best ;
25     Heapify Q ;
26 return S ;
```

it is not necessary to compute $\Delta_u(S \cup \{prev\_best\})$ from scratch. In other words, the algorithm can be implemented in an efficient manner such that both $\Delta_u(S)$ and $\Delta_u(S \cup \{prev\_best\})$ are evaluated simultaneously in a single iteration of Monte Carlo simulation. In that sense, the extra overhead is relatively insignificant compared to the huge runtime gains they can achieve, as shown in the experimental results [24], leading to an improvement of CELF by 17–61%.

Algorithm 5 uses the variable S to denote the current seed set, *last_seed* to track the id of last seed user picked by the algorithm, and *cur_best* to track the user having the maximum marginal gain w.r.t. S over all users examined in the current iteration. The algorithm starts by building the heap Q initially. Then, it continues to select seeds until the budget k is reached. The optimization of CELF++ comes from where they update $u.mg1$ without recomputing the marginal gain. Clearly, this can be done since $u.mg2$ has already been computed efficiently w.r.t. the last seed node picked.

If none of the above cases applies, they recompute the marginal gain of u. From the experiments carried out in [24], one can note that although CELF++ maintains a larger data structure to store the look-ahead marginal gains of each node, the increase of the memory consumption is insignificant while the optimization on performance w.r.t. time is increased from CELF by 17–61%.

2.5.2.4 SPM and SP1M

The *shortest-path model (SPM)* and *SP1 model (SP1M)* were developed by Kimura et al. in [31]. These two models are special cases of the IC (independent cascade) model. In SPM, each node v has the chance to become active only at step $t = d(A, v)$. In other words, each node is activated only through the shortest paths from an initial active set. Namely, SPM is a special type of the ICM where only the most efficient information spread can occur. And SP1M, which slightly generalize SPM, instead considers the top 2 shortest paths from u to v.

The idea is that the majority of the influence flows through shortest paths. For these models, the influence $\sigma(A)$ of each target set A can be exactly and efficiently computed, and the provable performance guarantee for the natural greedy algorithm can be obtained. In [31], the approximation ratio is guaranteed as $\sigma(B_k) \geq (1 - 1/e)\sigma(A_k^*)$.

The experimental results show that SP1M outperforms SPM. However, a critical issue with this approach is that it ignores the influence probabilities among users. Only considering the shortest paths is not enough.

2.5.2.5 Maximum Influence Paths

From the above contribution in SPM and SP1M, Chen et al. [10] extended this idea by considering maximum influence paths instead of shortest paths. A maximum influence path between a pair of nodes (u, v) is the path with the maximum propagation probability from u to v. The main idea of this heuristic scheme is to use local arborescence structures of each node to approximate the influence propagation.

The maximum influence paths between every pair of nodes in the network can be computed by the Dijkstra shortest-path algorithm. Then we ignore MIPs with probability smaller than a influence threshold θ; this can help us effectively restrict influence to a local region. Next, we unite the MIPs beginning or ending at each node into arborescence structures, which represent the local influence regions of each node. When considering the influence propagation through these local arborescences, the diffusion model refers to the Maximum Influence Arborescence (MIA) model [10].

It is shown in [10] that the influence spread in the MIA model is submodular (i.e. having a diminishing marginal return property), and thus the simple greedy algorithm that selects one node in each round with the maximum marginal influence spread can guarantee an influence spread within $(1 - 1/e)$ of the optimal solution in the MIA model, while any higher ratio approximation is NP-hard.

The complete greedy algorithm for the basic MIA model is presented in Algorithm 6. Before the process was introduced, the authors in [10] defined sev-

Algorithm 6: Greedy algorithm optimized with MIA

Input: G, k, θ
Output: Seed set S
1 initialize $S \leftarrow \emptyset$, $IncInf(v) \leftarrow 0$for each node $v \in V$;
2 **for** *each node* $v \in V$ **do**
3   compute $MIIA(v,\theta)$ and $MIOA(v,\theta)$;
4   set $ap(u,S,MIIA(v,\theta)) = 0, \forall u \in MIIA(v,\theta)$;
5   compute $\alpha(v,u), \forall u \in MIIA(v,\theta)$;
6   **for** *each node* $u \in MIIA(v,\theta)$ **do**
7     $IncInf(u) += \alpha(v,u) \cdot (1 - ap(u,S,MIIA(v,\theta)))$;
8 **while** $|S| \leq k$ **do**
9   pick $u = argmax_{v \in V \setminus S}\{IncInf(v)\}$;
10   /* update incremental influence spreads */ ;
11   **for** $v \in MIOA(u,\theta) \setminus S$ **do**
12   /* subtract previous incremental influence */ ;
13   **for** $w \in MIIA(v,\theta) \setminus S$ **do**
14       $IncInf(w) -= \alpha(v,w) \cdot (1 - ap(w,S,MIIA(v,\theta)))$;
15   $S = S \cup \{u\}$;
16   **for** $v \in MIOA(u,\theta)$ **do**
17   compute $ap(w,S,MIIA(v,\theta)), \forall w \in MIIA(v,\theta)$;
18   compute $\alpha(v,w), \forall w \in MIIA(v,\theta)$;
19   **for** $w \in MIIA(v,\theta) \setminus S$ **do**
20       $IncInf(w) += \alpha(v,w) \cdot (1 - ap(w,S,MIIA(v,\theta)))$;
21 **return** S ;

eral methods. The maximum influence in-arborescence of a node $v \in V$ is defined as $MIIA(v,\theta) = \cup_{u \in V, pp(MIP_G(u,v)) \geq \theta} MIP_G(u,v)$, and the maximum influence out-arborescence $MIOA(v,\theta) = \cup_{u \in V, pp(MIP_G(v,u)) \geq \theta} MIP_G(v,u)$. Further, let the activation probability of any node u in $MIIA(v,\theta)$, denoted as $ap(u,S,MIIA(v,\theta))$, be the probability that u is activated when the seed set is S and influence is propagated in $MIIA(v,\theta)$. Due to the limit of pages, we do not discuss these methods, but one can easily find the definitions and details in [10].

The whole MIA algorithm works as follows. First, it evaluates the incremental influence spread $IncInf(u)$ for any node u when the current seed set is empty. The evaluation is described using the linear coefficients $\alpha(v,u)$. Second, the algorithm updates the incremental influences whenever a new seed is selected. Suppose u is selected as the new seed in an iteration. The influence of u in the MIA model only reaches nodes in $MIOA(u,\theta)$. Thus the incremental influence spread $IncInf(w)$ for some w needs to be updated if and only if w is in $MIIA(v,\theta)$ for some $v \in MIOA(u,\theta)$. This means that the update process is relatively local to u. The update is done by first subtracting $\alpha(v,w) \cdot (1 - ap(w,S,MIIA(v,\theta)))$ before adding u into the seed set, and then adding u into the seed set outside the loop. Recompute $ap(w, S, MIIA(v, \theta))$ and $\alpha(v, w)$ under the new seed set, and add $\alpha(v, w) \cdot (1 - ap(w, S, MIIA(v, \theta)))$ into $IncInf(w)$.

The authors later proposed an extension model *prefix excluding MIA (PMIA)*. Intuitively, in the PMIA model, the seeds have an order. For any given seed s, its maximum influence paths to other nodes should avoid all seeds in the prefix before s. The major technical difference is the definition of the maximum influence in(out)-arborescence for the PMIA model, especially if one would like to design an efficient greedy algorithm in the framework of Algorithm 6. From experiments with all four real networks with different scales, the authors argue that their algorithms are scalable and the running time is efficient. However, these heuristics would not perform well on high-influence graphs, as pointed out by [23], that is, when the influence probabilities through links are large.

Wang et al. [53] proposed an alternative approach. The focus of their study was the IC model. They argue that most of the diffusion happens only in small communities, even though the overall networks are huge. Taking this as an assumption, they first split the network into communities, and then use a greedy dynamic programming algorithm to select seed nodes. To compute the marginal gain of a prospective seed node, they restrict the influence spread to the community to which the node belongs.

2.5.2.6 SIMPATH

SIMPATH, proposed by Goyal et al. in [25], is an efficient and effective algorithm for the influence maximization problem under the linear threshold model. According to the experiments in [25], SIMPATH consistently outperforms the state of the art w.r.t. running time, memory consumption, and the quality of the seed set chosen, measured in terms of expected influence spread achieved.

SIMPATH builds on the CELF optimization that iteratively selects seeds in a lazy forward manner. However, instead of using expensive MC simulations to estimate the spread, it is shown in [25] that under the LT model, the spread can be computed by enumerating the simple paths starting from the seed nodes. It is known that the problem of enumerating simple paths is #P-hard [52]. However, the majority of the influence flows within a small neighborhood, since probabilities of paths diminish rapidly as they get longer. Thus, the spread can be computed accurately by enumerating paths within a small neighborhood. In addition to the Simpath-Spread algorithm used by SIMPATH, two other optimizations reduce the number of spread estimation calls in SIMPATH. The first one, vertex cover optimization, addresses a key weakness of the simple greedy algorithm: The spread of a node can be computed directly using the spread of its out-neighbors. Thus, in the first iteration, a vertex cover of the graph is constructed and the spread is obtained only for these nodes using the spread estimation procedure. The spread of the rest of the nodes is derived from this. This significantly reduces the running time of the first iteration. Second, they observe that as the size of the seed set grows in subsequent iterations, the spread estimation process slows down considerably. They provide the optimization called the look-ahead optimization-which addresses this issue and keeps the running time

Algorithm 7: SIMPATH

Input: $G = (V, E, b), k, \delta, l$
Output: Seed set S
1 Find the vertex cover C of input graph G. ;
2 **for** *each* $u \in C$ **do**
3     $U \leftarrow (V - C) \cap N^{in}(u)$;
4   Compute $\sigma(u)$ and $\sigma^{V-v}(u), \forall v \in U$ in a single call to the $SIMPATH - SPREAD(u, \delta, U)$;
5   Add u to CELF queue. ;
6 **for** *each* $v \in V - C$ **do**
7   Compute $\sigma(v)$;
8   Add v to CELF queue ;
9 $S \leftarrow \emptyset$, $spd \leftarrow 0$;
10 **while** $|S| \leq k$ **do**
11     $U \leftarrow$top-l nodes in CELF queue Compute $\sigma^{V-x}(S), \forall x \in U$, in a single call to the $SIMPATH - SPREAD(u, \delta, U)$;
12   **for** *each* $x \in U$ **do**
13   **if** *x is previously examined in the current iteration* **then**
14             $S \leftarrow S + x$;
15   Update spd ;
16   Remove x from CELF queue, break out of the loop;
17   Call $BACKTRACK(x, \delta, V - S, \emptyset)$ to compute $\sigma^{V-S}(x)$. ;
18   Compute $\sigma(S + x)$. ;
19   Compute marginal gain of u as $\sigma(S + x) - spd$. ;
20   Re-insert u in CELF queue such that its order is maintained. ,
21 **return** S ;

of subsequent iterations small. These three inventions are quite helpful for speeding up the SIMPATH algorithm; one can find details about these in [25], and we will not discuss them but rather present the complete algorithm in Algorithm 7.

The whole algorithm is presented in Algorithm 7. First, the algorithm find a vertex cover C, then for every node $u \in C$, its spread is computed on required subgraphs needed for the optimization. This is done in a single call to $SIMPATH - SPREAD$. Next, for the nodes that are not in the vertex cover, the spread is computed. The CELF queue is built accordingly, and sorted in decreasing order of marginal gains. Next, by using look-ahead optimization, the algorithm selects the seed set in a lazy forward fashion. The spread of the seed set S is maintained using the variable spd. Later they take a batch of top-l nodes, call it U, from the CELF queue. In a single call to $SIMPATH - SPREAD$, the spread of S is computed on required subgraphs needed for the optimization. For a node $x \in U$, if it is processed before in the same iteration, then it is added in the seed set as it implies that x has the maximum marginal gain w.r.t. S. Recall that the CELF queue is maintained in decreasing order of marginal gains and thus, no other node can have a larger marginal gain [23]. Ifx is not seen

Algorithm 8: VirAds—viral advertising in OSNs

Input: $G = (V, E), 0 \leq \rho \leq 1, d \in \mathbb{N}^+$
Output: A small d-seeding

1 $n_v^e \leftarrow d(v), n_v^a \leftarrow \rho \cdot d(v), r_v \leftarrow d+1, v \in V$;
2 $r_v^i = 0, i = 0..d, P \leftarrow \emptyset$;
3 **while** *there exist inactive vertices* **do**
4   **while** $u \neq argmax_{v \notin P}\{n_v^e + n_v^a\}$ **do**
5         $u \leftarrow argmax_{v \notin P}\{n_v^e + n_v^a\}$;
6   Recompute n_v^e as the number of new active edges after adding u. ;
7     $P \leftarrow P \cup \{u\}$;
8   Initialize a queue: $Q \leftarrow \{(u, r_v)\}$;
9     $r_u \leftarrow 0$;
10   **for** *each* $x \in N(u)$ **do**
11         $n_x^{(a)} \leftarrow max\{n_x^{(a)}\}$;
12   **while** $Q \neq \emptyset$ **do**
13         $(t, \tilde{r}_t) \leftarrow Q.pop()$;
14   **for** *each* $w \in N(t)$ **do**
15   **for** *each* $i = r_t \rightarrow min\{\tilde{r}_t - 1, r_w - 2\}$ **do**
16                 $r_w^{(i)} = r_w^{(i)} + 1$;
17   **if** $(r_w^{(i)} \geq \rho \cdot d_w) \wedge (r_w \geq d) \wedge (i+1 < d)$ **then**
18   **for** *each* $x \in N(w)$ **do**
19                         $n_x^{(a)} \leftarrow max\{n_x^{(a)} - 1, 0\}$;
20                     $r_w = i + 1$;
21   **if** $w \notin Q$ **then**
22                         $Q.push((w, r_w))$;
23 **return** P ;

before, its marginal gain needs to be recomputed, then the CELF queue is updated accordingly.

2.5.2.7 VirAds

In recent studies, researchers have discovered that the propagation in a social network often fades quickly within only a few hops from the sources, counteracting the assumption of self-perpetuating of influence considered in some literature. Dinh et al. [13] investigated the cost-effective, massive, and fast propagation (CFM) problem and proposed an algorithm, VirAds, to minimize the seeding cost and to tackle the problem on large-scale networks.

This scalable algorithm is shown as Algorithm 8, where r_v is the round in which v is activated, $n_v^{(e)}$ represents the number of new active edges after adding v into the seeding and $n_v^{(a)}$ refers to the number of extra active neighbors v needs in order to activate v. In addition, $r_v^{(i)}$ is the number of activated neighbors of v up to round i

where $i = 1...d$. Generally, the VirAds algorithm favors the vertex which can activate the most number of edges. This could distinguish between good and bad seeds. In the early stages, the algorithm behaves similar to the degree-based heuristics that favors vertices with high degree. However, after a certain number of vertices have been selected, VirAds will make the selection based on the information within the d-hop neighbor around the considered vertices, which is different from the degree-based heuristic that considers only one-hop neighborhoodship.

Given those measures, VirAds selects, in each step, the vertex u with the highest $effectiveness$, which is defined as $n_u^{(e)} + n_u^{(a)}$. After that, the algorithm needs to update the measures for all the remaining vertices.

It is introduced in [13] that the cost-effective, massive, and fast propagation problem (CFM) can be easily shown to be NP-hard by a reduction from the set cover problem. It is also proved that an approximation algorithm with a factor less than $O(logn)$ is unlikely. However, if we assume the network is power-law, their algorithm is an approximation algorithm for this problem with a constant factor.

2.6 Conclusion

Social networks are graphs of individuals and their relationships [5], such as friendships, collaborations, or advice-seeking relationships. With the increasing popularity of social network services, more and more people communicate with each other through such networks. This survey mainly conveys a framework for studying information diffusion problems and their approximations as well as optimizations. It provides readers with a number of interesting models and effective algorithms on social networks. However, these techniques and models only form the foundation for further research; there are many open questions that need to be answered.

We reviewed novel and interesting questions based on the initial work from Domingos and Richardson [17, 46], inspires Kempe et al. [28, 27], Mossel and Roch [38], and many others to develop a solid theoretical foundation of literature resources on the influence maximization problem. The main challenge now is to find solutions that are applicable in real viral marketing environments. Working toward various models and algorithms, with comprehensive experiments, researchers are trying to find a way that could really give a satisfying result without requiring too much data load or making unrealistic independence assumptions. In order to achieve this goal and to determine the real applicability of the existing approaches, more effective designs and empirical studies are needed, and the test of approximation techniques are also required.

The more recent work of Leskovec et al. [54] gives us insight in modeling the diffusion through implicit networks, in which the underlying network structure is unknown, and all the predicting of activation and influence spread is focusing on a global view. Furthermore, in [39], Myers et al. propose a new model that takes into account influence from outside of the network. Inspired by those works, for future works, it would be interesting to relax the assumption of uniform influence inside the network to seek better strategy to maximize influence. Furthermore, in contrast to the influence maximization problem, for misinformation or computer viruses that spread through in the networks, how to efficiently prevent the audience from getting infected is also very attractive to us. Formulating and solving those problems with more practical models and efficient algorithms is a fascinating challenge with great potential.

References

[1] H. K. Ahn, S. W. Cheng, O. Cheong, M. Golin, and R. Oostrum. Competitive facility location: the Voronoi Game. *Theoretical Computer Science*, 310(1–3):457–467, 2004.

[2] A. Banerjee, N. Pathak, and J. Srivastava. A generalized linear threshold model for multiple cascades. In *ICDM*, page 965970, 2010.

[3] N. Berger, C. Borgs, J. T. Chayes, and A. Saberi. On the spread of viruses on the Internet. In *Proceedings of the 16th ACM-SIAM Symposium on Discrete Algorithm (SODA)*, 2005.

[4] S. Bhagat, A. Goyal, and L. V. Lakshmanan. Maximizing product adoption in social networks. In *Proceedings of the Fifth ACM International Conference on Web Search and Data Mining, WSDM*, pages 603–612, 2012.

[5] S. Bharathi, D. Kempe, and M. Salek. Competitive influence maximization in social networks. In *WINE*, pages 306–311, 2007.

[6] A. Borodin, Y. Filmus, and J. Oren. Threshold models for competitive influence in social networks. In *Proceedings of the 6th International Conference on Internet and Network Economics*, WINE '10, 2010.

[7] F. Brauer and C. Castillo-Chvez. Mathematical models in population biology and epidemiology. *Springer*, 2001.

[8] T. Carnes, R. Nagarajan, S. M. Wild, and A. V. Zuylen. Maximizing influence in a competitive social network: a follower's perspective. In *Proceedings of the Ninth International Conference on Electronic Commerce*, pages 351–360. ACM, 2007.

[9] W. Chen, A. Collins, R. Cummings, T. Ke, Z. Liu, D. Rincon, X. Sun, Y. Wang, W. Wei, and Y. Yuan. Influence maximization in social networks when negative opinions may emerge and propagate, In *Proceedings of the 11th SIAM International Conference on Data Mining*, 2011.

[10] W. Chen, C. Wang, and Y. Wang. Scalable influence maximization for prevalent viral marketing in large-scale social networks. In *16th ACM SIGKDD International Conference on Knowledge Discovery and Data Mining, KDD 10*, pages 1029–1038, New York, NY, USA, 2010.

[11] W. Chen, Y. Yuan, and L. Zhang. Scalable influence maximization in social networks under the linear threshold model. In *2010 IEEE International Conference on Data Mining, ICDM 10*, pages 88–97, Washington, DC, USA, 2010.

[12] D. J. Daley and J. Gani. *Epidemic Modeling: An Introduction.* New York: Cambridge University Press, 2005.

[13] T. N. Dinh, D. T. Nguyen, and M. T. Thai. Cheap, easy, and massively effective viral marketing in social networks: Truth or fiction? In *ACM Conference on Hypertext and Social Media (Hypertext)*, 2012.

[14] G. Dobson and U. S. Karmarkar. Competitive location on a network. *European Journal of Operational Research*, 35(4):565574, 1987.

[15] P. S. Dodds and D. J. Watts. Universal behavior in a generalized model of contagion. *Physical Review Letters*, 92, 2004.

[16] P. Domingos. Mining social networks for viral marketing. In *IEEE Intelligent Systems*, 2005.

[17] P. Domingos and M. Richardson. Mining the network value of customers. In *Seventh ACM SIGKDD International Conference on Knowledge Discovery and Data Mining, KDD 01*, pages 57–66, New York, NY, USA, 2001.

[18] P. A. Dreyer. Applications and variations of domination in graphs. Ph.D. Thesis, Rutgers University, 2000.

[19] R. Durrett. *Lecture Notes on Particle Systems and Percolation.* Wadsworth Publishing, 1988.

[20] H. Eiselt and G. Laporte. Competitive spatial models. *European Journal of Operational Research*, 39:231–242, 1989.

[21] J. Goldenberg, B. Libai, and E. Muller. Talk of the network: A complex systems look at the underlying process of word-of-mouth. *Marketing Letters*, 3(211–223), 2001.

[22] J. Goldenberg, B. Libai, and E. Muller. Using complex systems analysis to advance marketing theory development. *Academy of Marketing Science Review*, 2001.

[23] A. Goyal. *Social Influence and Its Applications.* PhD thesis, University of British Columbia, 2005–2013.

[24] A. Goyal, W. Lu, and L. V. S. Lakshmanan. Celf++: optimizing the greedy algorithm for influence maximization in social networks. In *20th International Conference Companion on World Wide Web, WWW 11*, New York, NY, USA, 2011.

[25] A. Goyal, W. Lu, and L. V. S. Lakshmanan. Simpath: An efficient algorithm for influence maximization under the linear threshold model. In *2011 IEEE 11th International Conference on Data Mining, ICDM 11*, pages 211–220, Washington, DC, USA, 2011.

[26] M. S. Granovetter. Threshold models of collective behavior. *The American Journal of Sociology*, 83(6):1420–1443, 1978.

[27] D. Kempe, J. Kleinberg, and E. Tardos. Maximizing the spread of influence through a social network. In *Ninth ACM SIGKDD International Conference on Knowledge Discovery and Data Mining, KDD 03*, pages 137–146, New York, NY, USA, 2003.

[28] D. Kempe, J. Kleinberg, and E. Tardos. Influential nodes in a diffusion model for social networks. In *32nd International Conference on Automata, Languages and Programming, ICALP05*, pages 137–146, Berlin, Heidelberg, 2005. Springer-Verlag.

[29] M. Kermack. Contributions to the mathematical theory of epidemics. *Royal Society of Edinburgh. Section A. Mathematics*, volume 115, 1972.

[30] W. Kermack and A. McKendrick. A contribution to the mathematical theory of epidemics. *Royal Society of London*, 115:700721, 1927.

[31] M. Kimura and K. Saito. Approximate solutions for the influence maximization problem in a social network. *Knowledge-Based Intelligent Information and Engineering Systems*, 4252 (Lecture Notes in Computer Science):937–944, 2006.

[32] J. Leskovec, L. Adamic, and B. Huberman. The dynamics of viral marketing. In *Proceedings of the Seventh ACM Conference on Electronic Commerce (EC)*, 2006.

[33] J. Leskovec, L. A. Adamic, and B. A. Huberman. The dynamics of viral marketing. In *ACM Trans*, volume 1, May 2007.

[34] J. Leskovec, M. McGlohon, C. Faloutsos, N. Glance, and M. Hurst. Cascading behavior in large blog graphs. In *SIAM International Conference on Data Mining (SDM)*, 2006.

[35] J. Leskovec, M. McGlohon, C. Faloutsos, N. Glance, and M. Hurst. Cascading behavior in large blog graphs. In *SDM*, 2007.

[36] J. Leskovec, A. Singh, and J. Kleinberg. Patterns of influence in a recommendation network. In *Pacific-Asia Conference on Knowledge Discovery and Data Mining (PAKDD)*, 2006.

[37] T. M. Liggett. *Interacting Particle Systems*. Springer, 1985.

[38] E. Mossel and S. Roch. On the submodularity of influence in social networks. In *Proceedings of the 39th ACM Symposium on Theory of Computing (STOC)*, 2007.

[39] S. Myers, C. Zhu, and J. Leskovec. Information diffusion and external influence in networks. In *ACM SIGKDD International Conference on Knowledge Discovery and Data Mining (KDD)*, 2012.

[40] G. L. Nemhauser, L. A. Wolsey, and M. L. Fisher. An analysis of approximations for maximizing submodular set functions. *Mathematical Programming*, 14(1):265–294, 1978.

[41] D. T. Nguyen, S. Das, and M. T. Thai. Influence maximization in multiple online social networks. In *IEEE Globecome 2013*, 2013. (Accepted).

[42] C. Ning. On the approximability of influence in social networks. In *Proceedings of the Nineteenth Annual ACM-SIAM Symposium on Discrete Algorithms*, 2008.

[43] D. Peleg. Local majority voting, small coalitions and controlling monopolies in graphs: A review. In *Proceedings of the 3rd Colloquium on Structural Information and Communication Complexity*, pages 170–179, 1996.

[44] D. Peleg. Size bounds for dynamic monopolies. *Discrete Applied Mathematics*, 86:263–273, 1998.

[45] L. Rashotte. Social influence. In *Blackwell Encyclopedia of Sociology*, IX:44264429, 2006.

[46] M. Richardson and P. Domingos. Mining knowledge-sharing sites for viral marketing. In *Proceedings of the Eighth ACM SIGKDD International Conference on Knowledge Discovery and Data Mining (KDD)*, 2002.

[47] F. S. Roberts. Graph-theoretical problems arising from defending against bioterrorism and controlling the spread of fires. In *DIMACS/DIMATIA/Renyi Combinatorial Challenges Conference*, 2006.

[48] E. M. Rogers. *Diffusion of Innovations*. New York: Free Press, 1962.

[49] P. Rozin and E. B. Royzman. Negativity bias, negativity dominance, and contagion. *Personality and Social Psychology Review*, 5(4):296320, 2001.

[50] T. Schelling. *Micromotives and Macrobehavior*. Norton, 1978.

[51] Y. Shen, H. Zhang, and M. T. Thai. Interest-matching information propagation in multiple online social networks. In *CIKM'12*.

[52] L. G. Valiant. The complexity of enumeration and reliability problems. *SIAM Journal on Computing*, 8(3):410–421, 1979.

[53] Y. Wang, G. Cong, G. Song, and K. Xie. Community-based greedy algorithm for mining top-k influence nodes in mobile social networks. In *16th ACM SIGKDD International Conference on Knowledge Discovery and Data Mining, KDD 10*, pages 1039–1048, New York, NY, USA, 2010.

[54] J. Yang and J. Leskovec. Modeling information diffusion in networks. In *IEEE International Conference on Data Mining (ICDM)*, 2010.

[55] H. Zhang, T. N. Dinh, and M. T. Thai. Maximizing the spread of positive influence in online social networks. In *ICDCS*, pages 317–326, 2013.

Chapter 3

Characterizing Link Connectivity in Opportunistic Networks

Chul-Ho Lee
North Carolina State University
Raleigh, North Carolina

Do Young Eun
North Carolina State University
Raleigh, North Carolina

CONTENTS

3.1 Introduction

Opportunistic networks, a.k.a., delay-or disruption-tolerant networks, have emerged as one of the most promising evolutions of mobile ad hoc networks, and are designed to operate without the need of infrastructure support and in the presence of frequent disruptions in link connectivity. In these networks, mobile nodes exploit node mobility and opportunistic contacts (based on their geographical proximity) for communications, while coping with such intermittent connectivity. The so-called store-carry-and-forward paradigm is, thus, used as a means of forwarding and routing in opportunistic networks: mobile nodes can carry messages, and copy and/or relay them to other nodes when they come into contact, thereby rendering messages eventually delivered to their destinations, even if end-to-end paths never exist [46]. In other words, direct device-to-device communications based upon opportunistic contacts and the cooperation of participating nodes (as relay nodes) enable network-wide communications.

With recent drastic growth in the number of users carrying smart mobile devices (e.g., smartphones and tablets), it becomes more interesting and important to envision opportunistic networks forming with such devices carried by humans. This is particularly essential because the increase in infrastructure capacity does not scale according to the widespread popularity of smart devices and users' ever-increasing demand for more bandwidth. As positive signs and prospects for such a direction, several recent studies show that opportunistic networking can lead to the increase of network capacity/coverage [14, 27, 43], the efficient usage of scarce radio spectrum [51], and mobile data offloading[1] [24] that is complementary to cellular traffic offloading via femtocell and WiFi. In addition with multiple interfaces of (smart) mobile devices including cellular, WiFi, and Bluetooth interfaces, there have been a number of research studies on the design of an integrated architecture with cellular and opportunistic ad hoc networks and relevant applications (e.g., [41, 44]).

In this chapter, we provide a comprehensive overview of the fundamental characteristics of link-level connectivity in opportunistic networks, which is crucial to understanding and evaluation of network performance (e.g., information spread and message delivery). We start by reviewing several important findings on node mobility and its induced link-level metrics, such as inter-contact time, as well as its resulting network performance. While node mobility is leveraged for opportunistic networking, mobility is still the fundamental source of uncertainty and randomness in predicting the performance of opportunistic networking. There has been, thus

This work was supported in part by National Science Foundation under grants CNS-0831825 and CCF-0830680.

[1]This is to reduce the burden on 3G/4G networks from high traffic demands.

far, considerable observations on the relationship between random mobility patterns and stochastic properties of mobility-induced inter-contact time, and network performance. We here discuss them exhaustively, including popular performance modeling built upon the Poisson contact assumption—the inter-contact time between any pair of nodes is exponentially distributed, and currently prevalent arguments centered around power-law or "dichotomic" inter-contact time distribution (i.e., first a power-law, followed by an exponential distribution).

We next explain the presence of another important factor for opportunistic networking—"user availability" for communication—that has been largely overlooked in the literature.[2] Our recent work [36] shows that user availability should be considered, together with mobility, to correctly understand link-level connectivity and its resulting network performance. We here present the state-of-the-art findings on the characteristics of link-level connectivity and their impacts on network performance when the process of each user's availability for communication over time (simply, user availability process) is taken into account, in addition to the mobility-driven contact/inter-contact process. In particular, we show the occurrence of three different regimes depending on the relative time scale between user availability and mobility-induced contact/inter-contact processes, in each of which (1) the contact/inter-contact process (or mobility) primarily governs network performance; (2) such an impact of mobility on performance becomes vanishingly small or its extent is not that large as commonly expected; (3) the user availability process becomes dominant. Interestingly, in the second regime, the distribution of the actual "off-duration" of the communication link, called inter-transfer time, between two mobile nodes can become arbitrarily close to an exponential distribution, even when the inter-contact time distribution is a non-exponential distribution (e.g., the "dichotomic" inter-contact time distribution). However, in the third regime, even when the original inter-contact time is exponentially distributed, the resulting inter-transfer time can significantly deviate from the exponential.

The remainder of this chapter is organized as follows. We first provide mathematical definitions for the mobility-driven link-level metrics (contact-based metrics), and give an overview of the opportunistic networking literature with focus on contact-based metrics and network performance in Section 3.2. We then present our recent findings on the characteristics of link-level connectivity under the joint effect of user availability for communication and user mobility (or its induced contact/inter-contact process) in Section 3.3. We also provide relevant simulation results in Section 3.4. We finally conclude this chapter in Section 3.5.

3.2 Mobility-Induced Link-Level Metrics and Network Performance

To set the stage for subsequent exposition, we first collect definitions for mobility-induced link-level metrics such as inter-contact time and contact time. We then re-

[2] In this chapter, we use the terms *user*, *node*, and *mobile node* interchangeably.

view several important findings in the literature for such link-level metrics, especially the characteristics of inter-contact time and their impacts on network performance.

3.2.1 Mathematical Definitions

Consider a set of mobile nodes $\{A, B_1, B_2, \ldots, B_n\}$, each of which moves independently according to some mobility model in a common domain Ω. Let $A(t), B_i(t) \in \Omega$ be the position of nodes A and B_i at time t, respectively. Let $\mathcal{C}_A(t) \subset \Omega$ be the *contact set* (or neighborhood) of node A at time t, i.e., node B_i can communicate with node A at time t if and only if $B_i(t) \in \mathcal{C}_A(t)$. The simplest and most popular model for communication (or channel interference) is the so-called Boolean model, in which the contact set becomes [10, 11, 38]

$$\mathcal{C}_A^{\text{Boolean}}(t) = \{x \in \Omega : \|x - A(t)\| \leq d\}, \tag{3.1}$$

where d is some communication/sensing range. A more realistic channel interference model with many other nodes would be the SINR (signal-to-interference noise ratio) model [22, 16], for which we have [11]

$$\mathcal{C}_A^{\text{SINR}}(t) = \left\{x \in \Omega : \frac{P\|x - A(t)\|^{-\alpha}}{N_0 + \sum_i P\|x - B_i(t)\|^{-\alpha}} \geq \beta\right\} \tag{3.2}$$

for some suitable threshold β (e.g., the minimum SINR required for successful decoding at the receiver) and path loss exponent $\alpha \in [2, 4.5]$, i.e., node B_i can communicate with node A at time t ($B_i(t) \in \mathcal{C}_A^{\text{SINR}}(t)$) if and only if the channel condition between two nodes (given by SINR) is good enough. Here, N_0 is the noise power level, and the (emitting) power level P chosen by each node is the same as the others for simplicity.

Then, the *inter-contact time* of nodes A and B_i has been defined by

$$T_I \triangleq \inf\{t > 0 : B_i(t) \in \mathcal{C}_A(t)\}, \tag{3.3}$$

given that $B_i(0^-) \in \mathcal{C}_A(0^-)$ and $B_i(0) \notin \mathcal{C}_A(0)$. In words, the inter-contact time between two nodes is the duration of time these two nodes stay "out of contact" before getting in contact with each other. Similarly, if we replace $\mathcal{C}_A(t)$ in (3.3) by $\overline{\mathcal{C}_A}(t) = \Omega \setminus \mathcal{C}_A(t)$, we obtain the *contact time* T_C (or contact duration) of A and B_i. The contact time is the length of time during which the two nodes stay in contact, i.e., there exists a temporary link (channel) between the two nodes until they get out of contact again.

3.2.2 The Status Quo for Mobility-Induced Link-Level Dynamics

The inter-contact time between mobile nodes has been considered as a crucial mobility-driven link-level metric and the main determinant of the performance of opportunistic networks, as it denotes that how long it takes one node to encounter the other node to have any chance for data communications (or to relay/forward messages). For example, larger inter-contact time leads to longer end-to-end delay. Thus far, there have been many studies to uncover the characteristics of inter-contact time and their impacts on network performance. Nonetheless, we have not yet noticed a conclusive agreement on such characteristics, but rather seen the debate between the sufficiency of network modeling based on the exponentially distributed inter-contact time (or Poisson contact assumption) and the necessity of capturing the non-exponential inter-contact time distribution, e.g., the so-called dichotomic or mixture behavior of exponential and power-law distributions, discovered from measurements.

On the one hand, the Poisson contact assumption has been widely adopted (implicitly or explicitly) and is still popular in most analytical studies ranging from the capacity-delay trade-off [21] to the cost-delay trade-off [23, 42] as well as the design and analysis of various forwarding algorithms/policies [20, 54, 45, 4, 3, 2] thanks to its tractable analysis. Under the Poisson contact assumption, opportunistic contacts between *any pair* of nodes occur according to a Poisson process with the same rate λ, or the inter-contact time between any two mobile nodes is independently drawn from an exponential distribution with mean $1/\lambda$, i.e., $\mathbb{P}\{T_I > t\} = e^{-\lambda t}$. This assumption entails a "memoryless" property, i.e., each node is likely to meet any other node at any time instant *regardless of* the past contact experience (e.g., when the last contact with other nodes was), which greatly simplifies mathematical analysis and also enables us to consult Markovian techniques [20, 23, 45] or fluid approximations based on a set of ordinary differential equations [23, 54, 3, 2]. Also, the final expression of a performance metric under consideration becomes simply a function of the contact rate λ and the number of nodes in the network (network size), thereby providing more accessible fundamental insights.

As an example, consider the average message delays for an epidemic routing scheme [50] and a multicopy two-hop relay scheme [20]—widely used reference routing schemes (forwarding algorithms) for opportunistic networks [20, 54, 11, 7, 24]. In the epidemic routing scheme, every node can copy a message and forward its copy ("infect") to any other node that does not have the copy already upon encounter. In contrast, in the multicopy two-hop relay scheme, only the source node can replicate a message and forward its copy to any relay node that does not have the copy when they come into contact. In both schemes, once any of the message copies (including the original message) reaches the destination, the message delivery is done. Assuming that there are $n+1$ nodes in the network, the average message delays under the epidemic routing and multicopy two-hop relay protocol schemes (say, $\mathbb{E}\{D_{\text{ER}}\}$ and $\mathbb{E}\{D_{\text{MTR}}\}$, respectively) are given by [20, 54]

$$\mathbb{E}\{D_{\text{ER}}\} = \frac{1}{\lambda n}\sum_{i=1}^{n}\frac{1}{i} \approx \frac{1}{\lambda n}\log(n), \tag{3.4}$$

$$\mathbb{E}\{D_{\text{MTR}}\} = \frac{1}{\lambda n}\sum_{i=1}^{n}\frac{i^2(n-1)!}{(n-i)!n^i} \approx \frac{1}{\lambda}\sqrt{\frac{\pi}{2n}}. \tag{3.5}$$

Then, we can readily see that the performance gain of allowing message replications beyond two hops (consuming more resources), or the relative performance of the epidemic routing and multicopy two-hop relay schemes, is given by the following simple form [20, 54]:

$$\frac{\mathbb{E}\{D_{\mathrm{ER}}\}}{\mathbb{E}\{D_{\mathrm{MTR}}\}} \approx \frac{\log(n)}{\sqrt{n}}\sqrt{\frac{2}{\pi}},$$

which is independent of the contact rate λ.

Such a Poisson contact assumption has also been extended to incorporate the *heterogeneous* node mobility, or more precisely, the heterogeneity of contact behaviors among different nodes as follows. The pairwise inter-contact time between mobile nodes i and j, denoted by T_{ij}, is still exponentially distributed but with *different* rates λ_{ij} over different pairs (i, j), i.e., $\mathbb{P}\{T_{ij} > t\} = e^{-\lambda_{ij}t}$. In particular, this extended model can capture social community structures. For example, after dividing mobile nodes into multiple social groups and assigning higher values to the contact rates for nodes in the same group (than those between different groups), we can emulate human social behavior—people are more likely to meet their friends or others from the same community than some strangers. It is worth noting that from a *given* node-pair's point of view, this heterogeneous model is still in the realm of the "Poisson contacts" or exponential inter-contact time. When it comes to the analysis of network performance, however, the model presents non-trivial challenges, since the network performance depends on the underlying forwarding/routing algorithm as to how to exploit such heterogeneity. Thus, the heterogeneous model is actively employed for recent studies on opportunistic networks [19, 47, 34, 35, 55, 31, 53].

Only a few works, however, have partially provided empirical and theoretical support on justification for the Poisson assumption. The exponentially distributed inter-contact time is typically supported by numerical simulations (e.g., [20, 54]) based on (synthetic) mobility models such as the random waypoint model.[3] Cai and Eun [10] prove that a *finite domain* (finite boundary), on which most mobility models are defined, is one of the key factors that give rise to an exponential *tail* (exponential decay) of the inter-contact time. They then show that the *entire* inter-contact time distribution can be even close to an exponential distribution (or the Poisson contact process arises) in a *small bounded domain* when coupled with proper time-scale of networking operation. Informally speaking, consider a given (random) mobility model defined in a small bounded domain so that mobile nodes *quickly* hit the boundary or reach everywhere in the domain, which leads to the (near-)complete mixing of the trajectories of the nodes as in the *i.i.d.* jump mobility model.[4] Then, when the time scale of interest is much longer than the time scale to traverse the domain, the whole shape of the inter-contact time distribution (not just tail) becomes close to an

[3] In the random waypoint model [9, 12], each node first selects a destination (random waypoint) uniformly in a bounded domain and a speed uniformly from $[v_{min}, v_{max}]$ $(0 < v_{min} \leq v_{max})$, and then travels to its destination at its chosen speed. After reaching the destination, the node may pause for a random amount of time and then select a new destination and speed combination, and repeat the same procedure independently.

[4] In the *i.i.d.* jump mobility model, each node moves in a square of size $N \times N$ consisting of N^2 unit cells. At each time slot, the node independently jumps to any one of N^2 cells with equal probability. Note that if we assume that two nodes "meet" whenever they are in the same cell, then the inter-contact time distribution is a geometric distribution with mean N^2, which is the discrete analogue of the exponential distribution.

exponential. Note that the appearance of an exponential *tail* of the inter-contact time itself does not justify the Poisson contact assumption.

Banerjee et al. [7] also empirically obtain a similar finding from contact traces collected in UMass DieselNet. They show that a significant portion (90%) of inter-contact samples gathered in a *subset* of the whole network domain (a confined area) approximately follow an exponential distribution, but with different rates over different subsets. In addition, La [33] demonstrates the convergence of inter-contact time distribution to an exponential distribution under a *hybrid random walk* model, although the hybrid random walk model is built upon the *i.i.d.* jump mobility model inside, thus lacking continuity of mobile trajectories. Further, it is shown in [15, 48] that there exist a non-negligible portion of node pairs in real contact traces (such as those collected in the MIT Reality Mining and RollerNet experiments) whose inter-contact time distributions can be well fitted by exponential distributions.

On the other hand, many measurement studies [25, 13, 28, 37] have indicated a common observation that the *entire* distribution of the inter-contact time is *no longer* a pure exponential. The authors in [25, 13] empirically observe, using a diverse set of real mobility traces ranging from WiFi connectivity traces (access point association records for laptops and PDAs) to Bluetooth contact traces, that the inter-contact time distribution follows a power-law (or heavy-tailed) distribution over a large range of values, i.e., $\mathbb{P}\{T_I > t\} \sim t^{-a}$ for some $a > 0$. In particular, there it is shown that the power-law exponent a is less than 1 in most data sets, implying the *infinite mean* of the inter-contact time. Then, they draw a conclusion that in such a case ($a < 1$), the mean end-to-end delay of any forwarding algorithm including flooding (epidemic routing) also becomes *infinite* [13]. This argument is, however, later substituted by the following observation made by Karagiannis et al. in [28]. They examine a similar set of real mobility traces and show that the inter-contact time distribution is largely characterized by first a power-law (power-law "head") up to a certain characteristic time, order of half a day, beyond which it decays exponentially fast (exponential "tail"). This is called a "dichotomy" in the distribution of the inter-contact time [28]. This observation indicates that the argument based on the power-law *tail* of the inter-contact time [13] is rather over-pessimistic, since the time scale of interest for opportunistic networking may be of the same order as the characteristic time.

The mixture (or dichotomic) behavior of power-law and exponential distributions then becomes a critical feature that newly proposed mobility models, e.g., the self-similar least-action walk model [37] and the small world in motion model [40], need to reproduce in order to be more *realistic*, while [28, 10, 11] explain such a phenomenon using a class of random walk models. For example, the authors in [28] show that already simple mobility models such as one-dimensional (discrete-time) random walk on a ring can exhibit the same dichotomy in the distribution of the inter-contact time. In addition, Cai and Eun [10] demonstrate that if the boundary for a given mobility model is *large enough* so that mobile nodes *rarely* hit the boundary under the time scale of interest or they usually take a long time to travel from one place to another, then the inter-contact time distribution has a power-law "head" followed by an exponential "tail." (Here the exponential tail of the inter-contact time appears due to the finite boundary as explained above.) This is somewhat opposite of

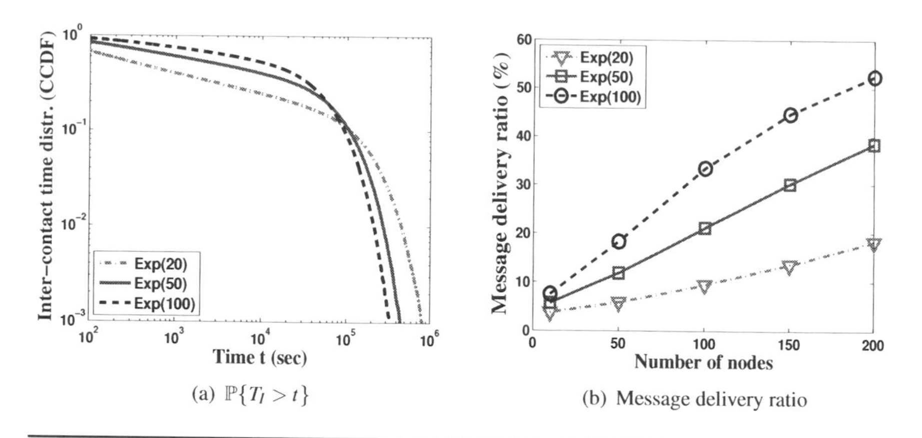

(a) $\mathbb{P}\{T_I > t\}$

(b) Message delivery ratio

Figure 3.1: For RW models on a $2000m \times 2000m$ domain with exponentially distributed step-length with means $20,\ 50,\ 100m$: (a) $\mathbb{P}\{T_I > t\}$ in a log-log scale; (b) message delivery ratio of epidemic routing scheme when varying the number of nodes in the network.

the aforementioned situation where "Poisson contacts" naturally arise within a small bounded domain. It is also presented in [11, 38, 32] that the network performance under node mobility leading to non-exponential inter-contact time distribution differs substantially from that under the Poisson contact assumption. To see the importance of non-exponential inter-contact time distribution on network performance and assess its impact, we explain below an observation made by Cai and Eun in [11]. Here we reproduce similar simulation results as those in [11].

Suppose that there are a number of mobile nodes in a network domain of size $2000m \times 2000m$, each of which moves according to a (two-dimensional isotropic) random walk (RW) mobility model[5] with constant speed 1 m/s whose step-length distribution is exponential with mean 20, 50, and $100m$, respectively. Note that longer mean step-length implies that the mobile node has a stronger tendency of moving in the same direction (stronger correlations in its trajectory). The Boolean model in (3.1) is used for "contact events" with communication range $d = 50m$, and no pause-time effect is considered. Figure 3.1(a) shows the ccdf of inter-contact time under the RW mobility models. As expected from [28, 10], the inter-contact time distribution always first follows a power-law and then decays exponentially fast later on, but with different "knee points" for different step-length distributions. Then, the epidemic routing scheme is considered in evaluating the impact of such non-exponential inter-contact time distributions on network performance. For the RW models, it turns out that the stationary distributions of node positions (being uniform) and the average

[5]In the RW mobility model [12, 11], each node first chooses a random step-length L (chosen from some suitable distribution) and a random angle ϕ uniformly drawn from $[0, 2\pi)$. Then, the node travels for the step-length L with the angle ϕ at some given (possibly random) speed $v > 0$. Once the node finishes this step, the whole procedure is repeated, independent of all others. Note that by choosing different step-length distributions, the RW model can generate many different mobility patterns.

inter-contact times are all the same, meaning that their "contact rate" must be also the same [11].[6] In other words, existing results on performance analysis, in which all the derivations depend only on the contact rate between a pair of nodes, necessarily predict the same performance [20, 54]. See (3.4) for example. However, as seen in Figure 3.1(b), the message delivery ratio (measured at the same time instant) under the epidemic routing scheme differs by up to 3 times for the different mobility patterns, whose average inter-contact time (thus the contact rate) and stationary distributions are all the same.

In summary, the fundamental basis of the Poisson assumption for the vast number of analytical studies in the literature so far has been made with *partial* empirical and theoretical support, but not yet firmly established, while there has also been the debate on the "true" characteristics of inter-contact time beyond the Poisson regime and its network implication.

3.3 Impact of User Availability on Link-Level Dynamics: Model and Analysis

Despite active research efforts for opportunistic networking, especially the *mobility-driven* link-level metrics and network performance as we reviewed above, our recent work [36] reveals that important *but often overlooked* constraints pose a new challenge to the link-level metrics and rather make them problematic. In particular, our work demonstrates that mobility alone does not suffice to characterize the link-level connectivity and there exists another equally important dimension—user availability for communication that has been largely overlooked in the literature, but is necessary, together with mobility, to correctly understand the link-level dynamics and its resulting network performance. We will explain our findings in [36] in detail in the rest of this chapter.

3.3.1 *Motivation*

The prevalent assumption of opportunistic ad hoc communications (or opportunistic networking) in the literature is that any two of mobile nodes can exploit contact opportunities based on their geographical proximity. As long as two nodes are close enough to communicate with each other or able to communicate with high SINR, we often say that they are in contact, though it depends on the underlying forwarding algorithm whether to utilize the contact opportunity for data transfer. As shown in Section 3.2, this has been precisely the way the contact duration and the inter-contact time are defined and empirically measured from real traces [13, 28, 15] as well as from synthetic mobility models [10, 11, 40, 37]. Also, the performance of

[6] More specifically, it is shown in [11] as an *invariance* property that the *mean* contact/inter-contact time remains the same under a class of random walk models, each of which has the same uniform stationary distribution of node positions, but with varying temporal dynamics (the degree of motion correlation).

most forwarding/routing algorithms [50, 20, 6, 45, 26, 27] are all evaluated under this hypothesis on the contact opportunities.

However, this implies that each mobile node has to carry a "special" communication device tailored (and also dedicated) to opportunistic ad hoc communications, which is not the case for smartphones. Clearly, not all the humans carrying smart mobile devices are such dedicated participants. More importantly, the basic premise that the link for communication is considered to be *available* solely based on nodes' positions via their mobility, is simply *not true*.

First, those devices are known to be power-hungry. One of the biggest concerns for smartphone users is battery lifetime, as their increasing energy demands are far outpacing improvements in battery technology [1]. In this regard, reducing WiFi power consumption, one primary source of battery drainage, has been an active research area for many years,[7] via enhancing the standard WiFi power save mode or a new WiFi duty cycling scheme putting the WiFi radio in the sleep mode for a longer time span (see, e.g., [5, 39] and references therein), although their focus is more on single-hop communication (or direct Internet access) through WiFi. In smartphones, a low battery can easily deny a link that would otherwise be available. Second, when the WiFi interface is used for Internet access, it is unavailable for other peering users with smartphones via WiFi. Third, such devices are subject to the (time-varying) behavior of each user in cooperation with other users for opportunistic communications. Finally, there are other issues such as lack of incentives to participate, privacy [24], neighbor/device discovery [52, 8], and possible failures/delay to set up connections (or links) [43].

3.3.2 User Availability Comes into Picture

As seen in Section 3.2, the adoption of link-level metrics based on nodes' proximity via the Boolean (or SINR) model has been prevalent, generating abundant research studies in the literature so far. However, as noted above, not every "contact" between mobile nodes leads to actual formation of a link. For two mobile nodes to have a chance to communicate, not only they should be "close" to each other in proximity, but also their wireless interfaces both be on. To properly capture the notion of on/off wireless interfaces, let $I_A(t), I_B(t)$ be 0–1 valued processes representing the availability of nodes (or users) A and B at time t, respectively. Note that we hereafter drop the subscript i for node B_i (simply, node B), since we are focusing on the link formation between two mobile nodes A and B. Specifically, $I_A(t) = 1$ if A is available for opportunistic communication (e.g., its wireless interface is "on" or performing idle listening); and $I_A(t) = 0$ if otherwise. Also, let $A_{\text{on}}, A_{\text{off}}$ be random variables to denote "on" durations (1-period lengths) and "off" durations (0-period lengths) for the availability of node A, respectively. Similarly for $I_B(t)$, B_{on}, and B_{off}. See Figure 3.2 for illustration.

[7] WiFi is often the preferred way for opportunistic ad hoc communication due to its higher data rate and longer transmission range, and also simpler peering/service detection procedure, when compared with Bluetooth (albeit its lower power consumption) [24, 49].

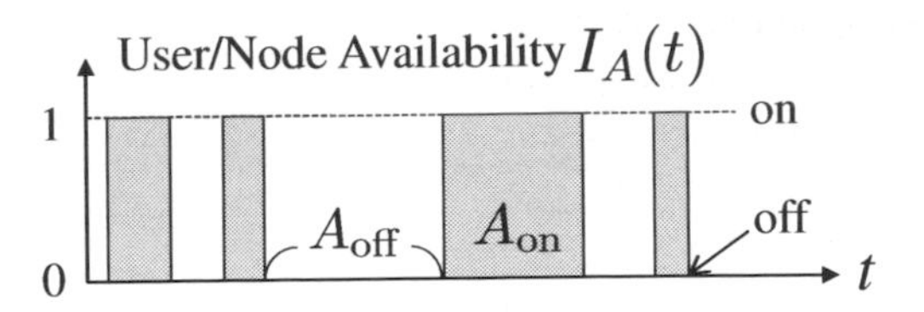

Figure 3.2: 0–1 valued process for user availability.

In this setup, the notion of contact set under the Boolean model in (3.1) (or under the SINR model in (3.2)) should be revised accordingly. For a given *A*–*B* pair, we say that they have a link, or event $L(A-B)$ occurs, at time t if and only if all the following three conditions are satisfied:

$$\text{(i) } B(t) \in \mathcal{C}_A(t), \text{ (ii) } I_A(t) = 1, \text{ and (iii) } I_B(t) = 1. \tag{3.6}$$

Thus, the link formation process between *A* and *B* is yet another 0–1 valued process, given by the indicator function of the event $L(A-B)$ at time t. To be precise,

$$\mathbf{1}_{\{L(A-B)\}}(t) = \mathbf{1}_{\{B(t)\in\mathcal{C}_A(t)\}} \cdot I_A(t) \cdot I_B(t), \tag{3.7}$$

and the notion of the inter-contact time between *A* and *B* can be generalized into the following:

$$L_{\text{off}} \triangleq \inf\{t > 0 : \mathbf{1}_{\{L(A-B)\}}(t) = 1\}, \tag{3.8}$$

given that $\mathbf{1}_{\{L(A-B)\}}(0^-) = 1$ and $\mathbf{1}_{\{L(A-B)\}}(0) = 0$, which we call *inter-transfer time* for the *A*–*B* pair. In words, it is the "off" duration of the process $\mathbf{1}_{\{L(A-B)\}}(t)$. Again, if we similarly replace $\mathbf{1}_{\{L(A-B)\}}(t) = 1$ in (3.8) by $\mathbf{1}_{\{L(A-B)\}}(t) = 0$ (with properly reversed conditional event), then we obtain the *transfer time* L_{on} (or the link "on" duration) for the *A*–*B* pair. Figure 3.3 depicts the link formation process between *A* and *B* with its resulting transfer/inter-transfer time.

In summary, the actual link formation dynamics should be described by jointly considering users' mobility and their availability processes together. While the conventional usage of contact-based metrics is *only* valid for the scenario with all "dedicated" users for communication, our link definitions correctly capture both users' mobility and availability, and are general enough to cover a wide range of scenarios that take into account all the constraints (including limited batter power) affecting the link formation dynamics.

3.3.3 Analysis of Link-Level Dynamics

3.3.3.1 Transfer-Time Distribution and Mean Inter-transfer Time

We first provide exact formulas for the transfer-time distribution $\mathbb{P}\{L_{\text{on}} > t\}$ and the mean inter-transfer time $\mathbb{E}\{L_{\text{off}}\}$, in terms of distributions of contact/inter-contact time and on/off duration in the user availability. To proceed, we consider the superposition of three on/off processes that appear in the definition of link indicator as in

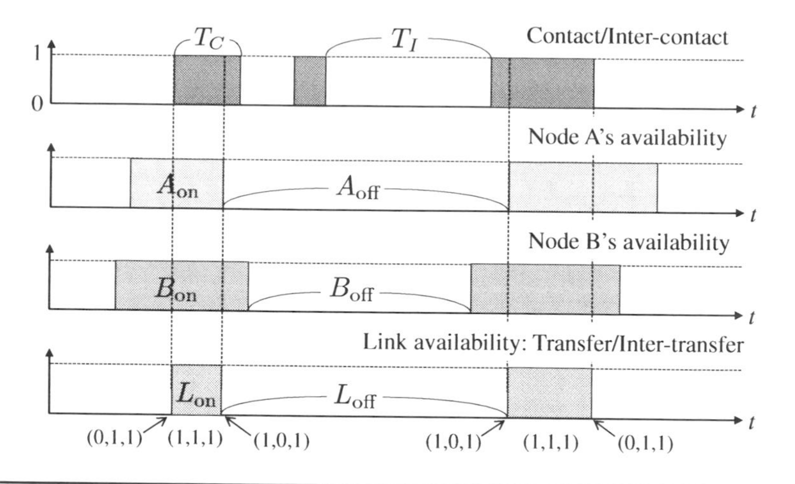

Figure 3.3: Link formation process for A–B pair.

(3.7). We assume that these on/off processes are independent of each other and the process $Z(t) \triangleq (\mathbf{1}_{\{B(t)\in\mathcal{C}_A(t)\}}, I_A(t), I_B(t)) \in \{0,1\}^3$ is stationary, ergodic, and right-continuous such that its state transitions occur at distinct time instants (and so is each of three 0–1 valued processes). Here, a state transition means the change of *any one* element (but not multiple elements) of the 3-dimensional vector, as $Z(t)$ is the superposed process. For example, transition from $(0,0,0)$ to $(1,1,1)$ occurs only with probability zero. Note that $\mathbf{1}_{\{L(A-B)\}}(t) \equiv \mathbf{1}_{\{Z(t)=(1,1,1)\}}$. Then, from stationarity, we have

$$P_C \triangleq \mathbb{P}\{\mathbf{1}_{\{B(t)\in\mathcal{C}_A(t)\}} = 1\} = \frac{\mathbb{E}\{T_C\}}{\mathbb{E}\{T_C\}+\mathbb{E}\{T_I\}},$$

and similarly for $P_A \triangleq \mathbb{P}\{I_A(t)=1\}$ and $P_B \triangleq \mathbb{P}\{I_B(t)=1\}$. Also, from independence, we have

$$\mathbb{P}\{Z(t) = (1,1,1)\} = P_C P_A P_B = \frac{\mathbb{E}\{L_{\text{on}}\}}{\mathbb{E}\{L_{\text{on}}\}+\mathbb{E}\{L_{\text{off}}\}}, \tag{3.9}$$

where the last equality is from $\mathbb{P}\{Z(t)=(1,1,1)\} = \mathbb{P}\{\mathbf{1}_{\{L(A-B)\}}(t)=1\}$ and stationarity. To avoid triviality, we assume that $P_C, P_A, P_B \in (0,1)$. If any one of these 0–1 valued processes is always one over time t, then we can simply ignore the process in evaluating the link formation dynamics. Also, if there is a process giving zero for all t, then the link for the A–B pair is never formed over t.

In addition, we define by T_C^e the residual (or remaining) contact time from a random time instant when A and B are in contact, whose distribution is given by

$$\mathbb{P}\{T_C^e > t\} = \frac{1}{\mathbb{E}\{T_C\}} \int_t^\infty \mathbb{P}\{T_C > s\} ds.$$

This is often called the *equilibrium* distribution of T_C. Similarly, we can define by T_I^e, A_{on}^e, A_{off}^e, B_{on}^e, and B_{off}^e as the residual time for T_I, A_{on}, A_{off}, B_{on}, and B_{off}, respectively. Then, we have the following [36].

Theorem 3.1
The transfer-time distribution is

$$\mathbb{P}\{L_{\text{on}}>t\} = \tau \cdot \left[\frac{\mathbb{P}\{T_C>t\}}{\mathbb{E}\{T_C\}} \mathbb{P}\{A^e_{\text{on}}>t\}\mathbb{P}\{B^e_{\text{on}}>t\} + \frac{\mathbb{P}\{A_{\text{on}}>t\}}{\mathbb{E}\{A_{\text{on}}\}} \mathbb{P}\{T^e_C>t\}\mathbb{P}\{B^e_{\text{on}}>t\} \right.$$
$$\left. + \frac{\mathbb{P}\{B_{\text{on}}>t\}}{\mathbb{E}\{B_{\text{on}}\}} \mathbb{P}\{T^e_C>t\}\mathbb{P}\{A^e_{\text{on}}>t\} \right] \tag{3.10}$$

with $\tau \triangleq \mathbb{E}\{L_{\text{on}}\} = \left[1/\mathbb{E}\{T_C\}+1/\mathbb{E}\{A_{\text{on}}\}+1/\mathbb{E}\{B_{\text{on}}\}\right]^{-1}$. *Also, the mean inter-transfer time is given by* $\mathbb{E}\{L_{\text{off}}\} = \tau \cdot [1/(P_C P_A P_B) - 1]$.

It is worth noting that Theorem 3.1 does not require any distributional or independence assumption on the joint distribution of the *successive* contact and inter-contact times (resp. on and off durations) for the contact process (resp. user availability process). In contrast to the transfer-time distribution, we cannot obtain the inter-transfer time distribution in any useful form, since the inter-transfer time can span over infinitely many transition points (e.g., $(1,1,1) \to (1,1,0) \to \cdots \to (1,0,1) \to (1,1,1)$) of the process $\mathbf{1}_{\{Z(t)=(1,1,1)\}}$.

Invariance principle: Theorem 3.1 says that the mean transfer/inter-transfer time is solely determined by the mean contact/inter-contact time and the mean on/off duration for each node's availability. Also, as mentioned in Section 3.2.2, it is shown in [11] that the mean contact/inter-contact time is invariant under a class of random walk models whose stationary distributions are uniform but with varying temporal dynamics (the degree of motion correlation). Thus, by Theorem 3.1, such an *invariance property* is still preserved for the transfer/inter-transfer time under the same class of mobility models and various distributions of on/off duration for user availability, as long as their mean values remain the same.

Nonetheless, as seen from the folklore about the impact of different mobility patterns on network performance (see Figure 3.1 for example), the inter-transfer time L_{off} (corresponding to the conventional notion of inter-contact time) would still be a critical factor governing the network performance, while its mean value (first-order) is *insufficient* to predict the performance. Thus, in the following section, we will investigate the distributional shape of the inter-transfer time to better understand its higher-order behaviors.

3.3.3.2 Inter-transfer Time Distribution

While it is impossible to obtain the exact form of the inter-transfer time distribution, we here present how the *shape* of the distribution changes according to different relative time scales between contact/inter-contact dynamics and user availability dynamics, and explain its impact on network performance.

To this end, we consider that the contact/inter-contact process for the A–B pair is an alternating renewal process, i.e., the (successive) inter-contact times of the pair are *i.i.d.* and their contact times are also *i.i.d.* It is known that the contact time between two nodes is much smaller than their inter-contact time and thus the contact duration is often ignored in the performance analysis of many forwarding/routing algorithms [13, 20, 54, 45, 4, 55]. This implies that the stationary probability that nodes A and B are in contact, or P_C, is very small, and also $\mathbb{E}\{T_C\} \ll \mathbb{E}\{T_I\}$. We also note that the inter-contact time distribution is typically a mixture of power-law and exponential functions [10, 28, 11], and utilize the following [18].

Lemma 3.1

For any completely monotone function $g : \mathbb{R}_+ \to \mathbb{R}_+$, i.e., $(-1)^n g^{(n)}(t) \geq 0$ for all $t > 0$ and $n = 0, 1, \ldots$, where $g^{(n)}$ is the nth derivative of g, there exists a non-negative function $h(s)$ such that $g(t) = \int_0^\infty e^{-st} h(s)ds$.

In words, a completely monotone function is a (generalized) mixture of exponential functions. Note that both power-law and exponential functions are completely monotone, and so is their mixture. It follows that the ccdf of the inter-contact time T_I showing "dichotomic" behavior (first power-law followed by exponential) in [28, 11] is also completely monotone. In addition, it is presented in [17] that a *finite* mixture of exponential distributions (completely monotone) can approximate a long-tail (or power-law) distribution over a wide range of interest, when the tail of the approximated distribution eventually decays exponentially fast. We thus consider T_I with completely monotone density (or ccdf) having finite moments.

For simplicity and uncluttered exposition, from now on, we consider the user-pair availability for communication rather than considering the availability of users A and B separately. Specifically, let $I_{AB}(t) = I_A(t)I_B(t)$, indicating whether or not *both* A and B are on or available for communication (regardless of their geographical proximity). Similarly as before, let AB_{on} and AB_{off} be the random variables denoting on duration (1-period length) and off duration (0-period length) for the process $I_{AB}(t)$, respectively. We assume that this process is also an alternating renewal process. Clearly, the stationary probability that both nodes A and B are available for communication, say P_{AB}, is

$$P_{AB} = \frac{\mathbb{E}\{AB_{\text{on}}\}}{\mathbb{E}\{AB_{\text{on}}\} + \mathbb{E}\{AB_{\text{off}}\}} = P_A P_B.$$

As there are many factors that prevent the wireless interface of each user from being always on for communication, we focus on the case that P_{AB} is small.

We next identify the distributional *shape* of the inter-transfer time L_{off} (or the actual off-duration of the link) depending on how quickly the node-pair on/off dynamics changes as compared with the contact/inter-contact dynamics, while P_{AB} remains fixed.

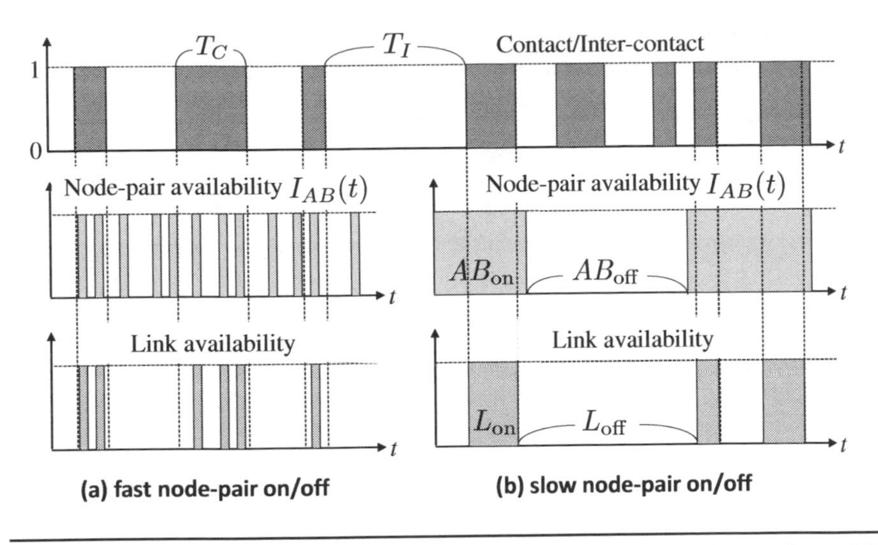

Figure 3.4: Examples of user-pair on/off dynamics over (a) faster time scale, and (b) slower time scale, than that of contact/inter-contact dynamics.

Fast node-pair on/off dynamics: The A–B pair becomes on/off for communication on a (fast) time scale, faster than the contact/inter-contact time dynamics, as shown in Figure 3.4(a). This regime includes the case that $AB_{\text{on}}+AB_{\text{off}}<T_C$ and $AB_{\text{on}}+AB_{\text{off}}<T_I$ with high probability. In this regime, most samples of T_I will appear almost "as is" in L_{off} samples, while there are some newly created small-valued L_{off} samples coming from the small-valued AB_{off}. Thus, the distribution of L_{off} is largely determined by that of T_I, i.e., the usual notion of "impact of mobility" prevails. In other words, almost all contact opportunities can translate into 'transfer' opportunities. One extreme example in this regime would be the usual measurement scenario to collect contact traces with "always-on" communication devices whose sensing interval (or the granularity of the experiment) is relatively small such that almost every contact event is properly captured and recorded.

Slow node-pair on/off dynamics: The 0–1 valued $I_{AB}(t)$ process changes very slowly as compared with the contact/inter-contact dynamics, as depicted in Figure 3.4(b). This regime also includes the case that $T_C+T_I<AB_{\text{on}}$ and $T_C+T_I<AB_{\text{off}}$ with high probability. Then, one can see that the large-valued samples of L_{off} are mostly from AB_{off}, while relatively smaller-valued samples of L_{off} are from T_I. That is,

$$L_{\text{off}} \stackrel{d}{\approx} \begin{cases} T_I & \text{if } T_I < AB_{\text{on}}, \\ AB_{\text{off}} & \text{otherwise.} \end{cases} \tag{3.11}$$

As an example, suppose the inter-contact time T_I is exponentially distributed with small average value (i.e., the typical Poisson contact model in the literature) and AB_{off} follows some wildly fluctuating distribution (e.g., heavy-tail or dichotomic distribution). Then, even when the mobility-induced contact process is a pure Poisson, we expect that the distribution of the inter-transfer time L_{off}, can significantly deviate from an exponential, as governed by that of AB_{off}.

Intermediate node-pair on/off dynamics: Both the node-pair availability process and the contact/inter-contact process operate on a similar time scale, and so there is no dominant process between them in deciding the link formation dynamics. This regime is highly non-trivial, but could be relevant in reality. Under this regime, we show, in considerable generality, that the distribution of the inter-transfer time L_{off} becomes very close to an *exponential*, which will also be supported by extensive simulations later on.

We consider that $I_{AB}(t)$ is likely to change at least once over a period of the inter-contact time T_I, while AB_{off} is larger than the contact duration T_C with high probability. Recall that $\mathbb{E}\{T_C\} \ll \mathbb{E}\{T_I\}$, and so this condition can hold for a large range of $AB_{\text{on}}, AB_{\text{off}}$. An example is when $T_C < AB_{\text{on}} < T_I$ and $T_C < AB_{\text{off}} < T_I$ with high probability. Then, the (stationary) probability that the transfer opportunity is granted, or the actual link is established, upon a physical contact, denoted as P_{link}, becomes

$$P_{\text{link}} = P_{AB} + (1 - P_{AB})\mathbb{P}\{T_C > AB^e_{\text{off}}\}, \tag{3.12}$$

where AB^e_{off} denotes the residual time until the A–B pair becomes "on" for communication, measured from a random time instant at which the pair is off. To see this, first note that when a contact opportunity arises, the probability that both A and B are ready for communication is simply the stationary probability that both are on, or P_{AB}. Also, even if at least one of A and B is *not available* for communication *at* a physical contact (the beginning of contact duration T_C), both can become available over the course of contact duration T_C, with probability $\mathbb{P}\{T_C > AB^e_{\text{off}}\}$ as shown in (3.12).

Let T_i $(i = 1, 2, \ldots)$ be *i.i.d.* copies of T_I. Then, if we ignore the contact duration T_C toward the inter-transfer time L_{off} (recall that T_C is typically much smaller than the inter-contact time T_I and often ignored in the literature), then the inter-transfer time L_{off} can be well approximated by $L_{\text{off}} \overset{d}{=} \sum_{i=1}^{N} T_i$, where N is an independent geometric random variable with parameter P_{link}. This is because the process $I_{AB}(t)$, under the (relative) time scale of interest, does not create dependency over successive contact epochs, implying that the opportunity of the link formation between A and B upon their physical contact with probability P_{link} is mostly independent of their previous/next opportunities. Then, we have the following [36].

Theorem 3.2

Under the aforementioned setting with intermediate time scale for node-pair on/off dynamics,

$$\sup_{t>0} \left| \mathbb{P}\{L_{\text{off}} > t\} - e^{-\frac{t}{\mathbb{E}\{L_{\text{off}}\}}} \right| \le 1 - \frac{2}{P_{\text{link}}(c^2_{T_I} - 1) + 2},$$

where $c^2_{T_I} \triangleq \text{Var}\{T_I\}/\mathbb{E}\{T_I\}^2$ is the squared coefficient of variation (CoV) of the inter-contact time T_I. □

Theorem 3.2 says the maximum gap between the inter-transfer time distribution and its exponential counterpart can be measured by how close $P_{\text{link}}(c^2_{T_I}-1)$ is to zero. Note that $c^2_{T_I} \geq 1$, as the distribution of the inter-contact time T_I is completely monotone [36]. If T_I itself is exponentially distributed with $c^2_{T_I}=1$, the upper bound becomes zero. In other words, the exponential behavior of T_I carries over to the inter-transfer time L_{off}. More importantly, even when T_I follows a non-exponential distribution (say, a mixture of power-law and exponential distributions [28, 10]) with $c^2_{T_I}>1$, $P_{AB}=P_A P_B$ is typically small valued and under the intermediate time scale of interest for pair on/off dynamics, $\mathbb{P}\{T_C > AB_{\text{off}}\}$ is also small, and so is $\mathbb{P}\{T_C > AB^e_{\text{off}}\}$.[8] All of these make P_{link} in (3.12) smaller, suggesting that the difference between the actual distribution of L_{off} and its corresponding exponential distribution becomes negligible over entire t. Therefore, in this regime, we expect that the so-called impact of mobility on network performance disappears or its extent is not significant.

In summary, for a given contact/inter-contact dynamics and a small, but fixed P_{AB}, our findings imply that the impact of mobility on network performance first prevails when the node-pair on/off dynamics $I_{AB}(t)$ is operating on a relatively faster time scale. Then, such impact becomes vanishingly small as the time scale for $I_{AB}(t)$ starts to grow, but still comparable to that of the contact/inter-contact process. However, if the time scale for $I_{AB}(t)$ gets arbitrarily larger (slow pair on/off dynamics), the on/off dynamics for communication availability itself will eventually take over in deciding the network performance.

3.4 Impact of User Availability on Link-Level Dynamics: Simulation Results

We provide simulation results using ONE simulator [29] with proper modification and pre-/post-processing to support our findings. We consider an alternating renewal process as the availability process of each user (say, A) for which $\mathbb{P}\{A_{\text{on}} > t\} = e^{-t/\mathbb{E}\{A_{\text{on}}\}}$. We test two different distributions for A_{off};

(i) Exponential distribution: $\mathbb{P}\{A_{\text{off}} > t\} = e^{-t/\mathbb{E}\{A_{\text{off}}\}}$

(ii) Hyper-exponential distribution: $\mathbb{P}\{A_{\text{off}} > t\} = q_1 e^{-\mu_1 t} + q_2 e^{-\mu_2 t}$,

where

$$q_1 = 0.5 \cdot \left[1 - \sqrt{(c^2_{A_{\text{off}}}-1)/(c^2_{A_{\text{off}}}+1)}\right] = 1 - q_2,$$

$\mu_1 = 2q_1/\mathbb{E}\{A_{\text{off}}\}$, and $\mu_2 = 2q_2/\mathbb{E}\{A_{\text{off}}\}$. We use the same $\mathbb{E}\{A_{\text{off}}\}$ for both cases and set the squared CoV of the hyper-exponential distribution to $c^2_{A_{\text{off}}}=16$, to see any impact of more variability of A_{off} around its mean than the exponential counterpart. In

[8] When the distribution of AB_{off} is completely monotone (reasonable as for the case of inter-contact time T_I), AB^e_{off} is stochastically larger than AB_{off}, implying that $\mathbb{P}\{T_C > AB^e_{\text{off}}\} < \mathbb{P}\{T_C > AB_{\text{off}}\}$.

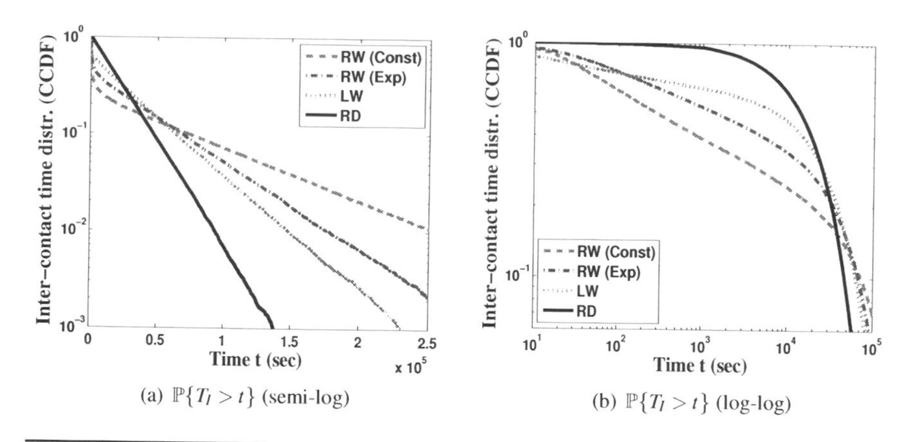

(a) $\mathbb{P}\{T_I > t\}$ (semi-log)

(b) $\mathbb{P}\{T_I > t\}$ (log-log)

Figure 3.5: The ccdf of the inter-contact time of a given pair in (a) a semi-log scale, and (b) a log-log scale to clearly show the power-law "head" behavior.

all cases, we fix $P_A = \frac{\mathbb{E}\{A_{\text{on}}\}}{\mathbb{E}\{A_{\text{on}}\}+\mathbb{E}\{A_{\text{off}}\}} = 0.1$ to emulate the low chance of communication availability per node (caused by many factors), while varying $\mathbb{E}\{A_{\text{on}}\}$ (and also $\mathbb{E}\{A_{\text{off}}\} = (1/P_A - 1)\mathbb{E}\{A_{\text{on}}\} = 9\mathbb{E}\{A_{\text{on}}\}$) to generate different relative time scales between user availability and mobility-induced contact/inter-contact processes.

We consider a class of random walk (RW) models and random direction (RD) models [12, 20, 54, 10, 11, 38]. As mentioned in Section 3.2.2, in the former, a mobile node follows a randomly chosen direction for some random amount of time—a randomly chosen step-length (or distance) divided by a random node speed—and then chooses another direction and repeats the same process, with a reflective boundary condition. We consider constant and exponential step-length distributions with mean $20m$, each of which approximates Brownian motion. We also consider power-law distribution for the step length (say, L), i.e., $\mathbb{P}\{L > l\} = l^{-a}$ defined over $l \geq 1$ with $a \in (0, 2)$, which is known as the *Lévy walk* (LW) model [30]. We here use $a = 1.2$ for the exponent, implying that, theoretically, the step-length has finite mean and infinite variance. In the RD model, a mobile node moves in a randomly chosen direction until it hits the boundary, and then chooses another direction and repeats.[9] In all cases, the initial node locations are drawn from uniform distribution over the domain whose size is $1000m \times 1000m$, and the node speed is set to $1\ m/s$. We use the Boolean model for contact with transmission range $d = 20m$, and do not consider the pause of mobile nodes in order to focus on the random mobility pattern itself.

Before going into the details, we measure the distribution of inter-contact time between two mobile nodes, each of which moves according to one of the above models. As seen from Figure 3.5, the ccdf of the inter-contact time first follows a power-law then decays exponentially fast later on—the so-called dichotomic distribution [28, 11], for RW models with constant and exponential step-length distribu-

[9] Here we use the definition of random direction model given in [12, 11]. The name "random direction" model is sometimes used to refer to a kind of random walk model, e.g., [20, 54].

tions as well as LW model, while the inter-contact time distribution under the RD model is exponential. These are well expected from the existing results [10, 11, 38]. (See Figure 3.1(a) for example.) For all of these models, empirical measurements give $\mathbb{E}\{T_I\} \approx 2 \times 10^4$ and $\mathbb{E}\{T_C\} \approx 25$, as also expected from the aforementioned invariance property in [11] where the mean contact/inter-contact time does not depend on the choice of step length under the RW models.

For network performance evaluation, we next measure the average message delay of the epidemic routing scheme. The total number of nodes in the domain is 100, and each of them independently moves according to one of the aforementioned mobility models. To ensure a fair comparison under different relative time scales between user (or user-pair) availability and mobility-induced contact/inter-contact processes, we evaluate how much the average message delay deviates from the case where every inter-transfer time L_{off} sample is now independently drawn from an exponential distribution with the same mean. This way, we can identify under which regime L_{off} becomes close to an exponential, even when the original T_I is highly non-exponential (dichotomic). Then, when L_{off} is exponentially distributed for each pair, we can reuse the existing theoretical result for the average delay of the epidemic routing scheme given in (3.4). Note that this is obtained when the transfer opportunity is always granted upon encounter. Here we use the formula in (3.4) but with $\lambda' = 1/\mathbb{E}\{L_{\text{off}}\}$ instead of the contact rate λ there, where $\mathbb{E}\{L_{\text{off}}\}$ is empirically measured for each of our simulation settings.[10]

Figure 3.6(a) shows the performance ratio of the average delay obtained under each mobility pattern to the analytical prediction by (3.4), while varying $\mathbb{E}\{A_{\text{on}}\}$ from 1 to 32805, for the case of exponential off-duration A_{off}. The inset in Figure 3.6(a) shows the performance ratio when all nodes are always "on" for communication, for which we use $\lambda' = 1/\mathbb{E}\{T_I\}$. We clearly see the impact of mobility on network performance, as these ratios are quite different over different choices of mobility models. For example, the average delay under the RW model with constant step length is almost 5 times worse than the analytical prediction based on the Poisson contact assumption. Note that this is consistent with the observation in Figure 3.1. When $\mathbb{E}\{A_{\text{on}}\}$ (and also $\mathbb{E}\{A_{\text{off}}\}$) is relatively small (say, $\mathbb{E}\{A_{\text{on}}\} = 1$, the first point of x-axis), the impact of mobility still remains effective. However, as $\mathbb{E}\{A_{\text{on}}\}$ starts to increase, the measured average delays under all mobility models become almost identical to the analytical prediction (the ratio is close to one). This corresponds to the intermediate node-pair on/off dynamics, as expected from Theorem 3.2. Clearly, in this regime, the impact of mobility disappears and interestingly the usual Poisson contact assumption (now for the transfer opportunities) prevails. On the other hand, when $\mathbb{E}\{A_{\text{on}}\}$ is very large (say, $\mathbb{E}\{A_{\text{on}}\} = 32805$, the ending point of x-axis), then the measured delay performance is much different from the analytical prediction, implying that the user availability takes over in deciding network performance as also suggested in Section 3.3.3.2 (the slow on/off dynamics).

[10]To avoid any possible bias, we exclude the consecutive small samples of AB_{off} during the course of same contact duration T_C, in estimating $\mathbb{E}\{L_{\text{off}}\}$. Such situation arises when two nodes are in contact under the fast node-pair on/off dynamics, in which only the first L_{on} matters while all the many subsequent short-lived samples do not contribute toward any message transfer.

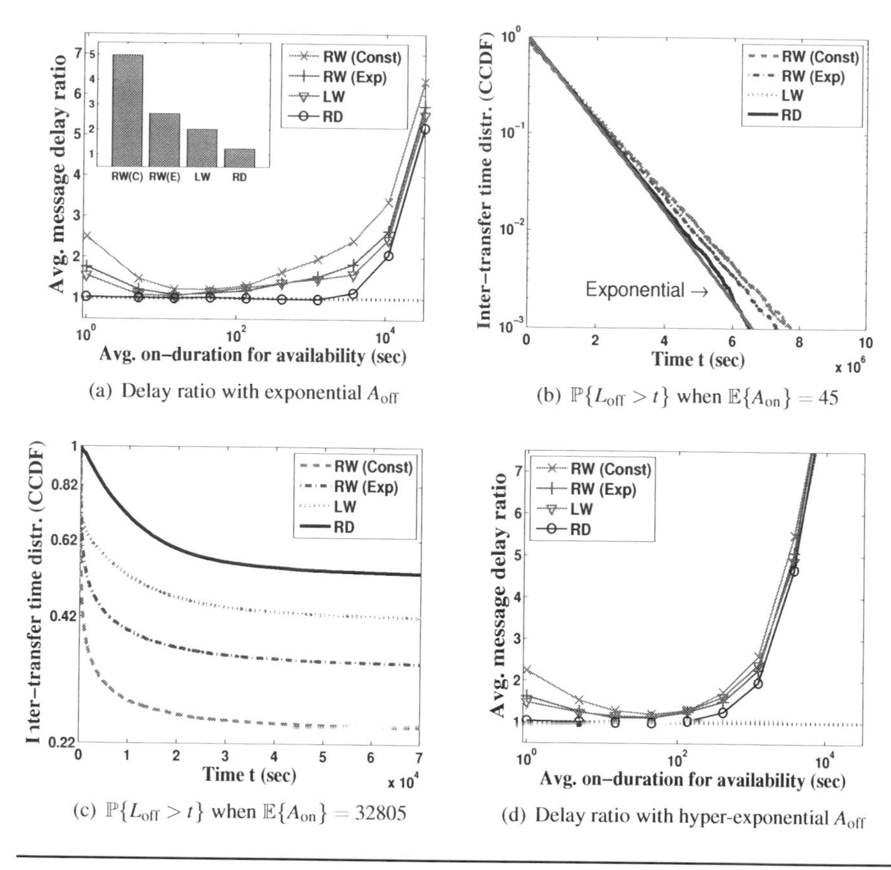

(a) Delay ratio with exponential A_{off}

(b) $\mathbb{P}\{L_{\text{off}} > t\}$ when $\mathbb{E}\{A_{\text{on}}\} = 45$

(c) $\mathbb{P}\{L_{\text{off}} > t\}$ when $\mathbb{E}\{A_{\text{on}}\} = 32805$

(d) Delay ratio with hyper-exponential A_{off}

Figure 3.6: For random mobility models: (a) average message delay ratio for the case of exponential on/off-durations A_{on} and A_{off}, while varying $\mathbb{E}\{A_{\text{on}}\}$ (and also $\mathbb{E}\{A_{\text{off}}\}$); $\mathbb{P}\{L_{\text{off}} > t\}$ measured when (b) $\mathbb{E}\{A_{\text{on}}\} = 45$ and (c) $\mathbb{E}\{A_{\text{on}}\} = 32805$ (corresponding to the fourth and last data points in the figure (a), respectively); (d) average delay ratio under the case of exponential on-duration A_{on} and hyper-exponential off-duration A_{off}.

Figure 3.6(b) shows $\mathbb{P}\{L_{\text{off}} > t\}$ when $\mathbb{E}\{A_{\text{on}}\} = 45$ on a semi-log scale. The inter-transfer time distributions become very close to an exponential distribution with the same mean, under all mobility choices, as expected from Theorem 3.2. When $\mathbb{E}\{A_{\text{on}}\} = 32805$, however, the inter-transfer time distributions highly deviate from the exponential and also depend on mobility models, as seen in Figure 3.6(c). We also test the hyper-exponentially distributed off-duration A_{off} and report the delay ratio in Figure 3.6(d). Again, we observe a similar trend as in Figure 3.6(a), while the actual ratio values are slightly different from the case of exponentially distributed off-duration for each node.

Further, we obtain similar simulation results under a map-based mobility model with a geographical constraint (Helsinki downtown area) and real GPS mobility

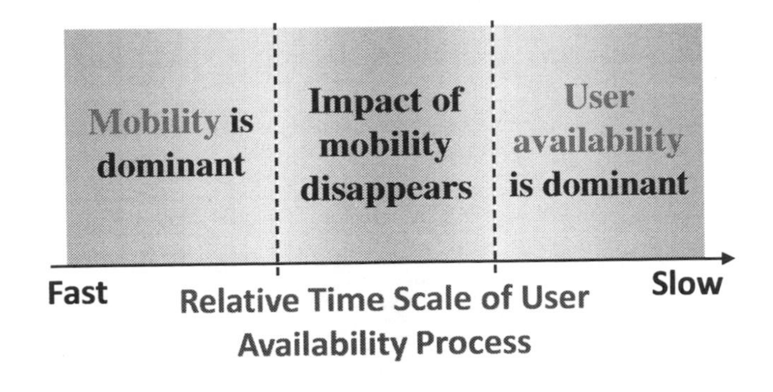

Figure 3.7: Three different regimes depending on the relative time scale of the user availability process.

traces, confirming our findings. Interested readers are referred to [36] for more details.

3.5 Discussion and Conclusion

In this chapter, we have first given a review of the opportunistic networking literature with focus on the contact-based metrics such as inter-contact time, defined based solely on node mobility, and network performance. Many important observations have been made both empirically and theoretically, which greatly extends our understanding of the random, complicated nature of opportunistic networking due to node mobility. However, we have also indicated that mobility alone, even with fine-grained GPS coordinates for every user, is not sufficient to fully characterize the link-level dynamics and there exists another equally important and orthogonal dimension, namely, *user availability* for communication, which has been largely overlooked in the literature. Specifically, we have identified the presence of three different regimes depending on the relative time scale between user availability and mobility-induced contact/inter-contact processes, as summarized in Figure 3.7. In particular, the user availability dynamics alone can drive the link-level dynamics and its resulting network performance into many other ways possible, which are completely different from the prediction when only the mobility (or contact/inter-contact dynamics) is taken into account.

To gauge how much the user availability process can affect the link-level dynamics, recall the link formation process $\mathbf{1}_{\{L(A-B)\}}(t) = \mathbf{1}_{\{B(t)\in\mathcal{C}_A(t)\}} \cdot I_A(t) \cdot I_B(t)$ in (3.7). On the one hand, if all mobile nodes are always "on" for communication, as commonly considered in the current literature, then $\mathbf{1}_{\{L(A-B)\}}(t) = \mathbf{1}_{\{B(t)\in\mathcal{C}_A(t)\}}$ and so the mobility-induced contact/inter-contact process fully characterizes the link-level dynamics. On the other hand, if all mobile nodes are always in "contact" with each other, then $\mathbf{1}_{\{L(A-B)\}}(t) = I_A(t)I_B(t)$, implying that the user availability process is now the sole factor. Indeed, the link formation process is simply the product of $\mathbf{1}_{\{B(t)\in\mathcal{C}_A(t)\}}$ and $I_A(t)I_B(t)$, with equal and indistinguishable contribution from

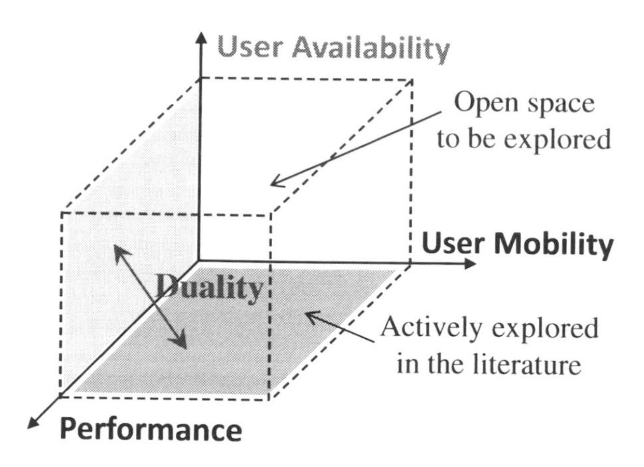

Figure 3.8: The role of user availability.

each of these 0–1 processes. That is, the mobility-induced contact/inter-contact dynamics and the user-pair availability can be thought of as "duals" of each other. See Figure 3.8 for illustration. Thus, the currently prevalent arguments centered around power-law or dichotomic inter-contact time distribution, mainly obtained in the former case, can also be *equally applicable* for the latter case. Clearly, the impact of user availability can be significant to the same extent as that of user mobility.

While mobility-related research studies abound in the literature, most of them assumed the wireless interfaces of mobile nodes to be always on and dedicated for communication. However, as mentioned before, when it comes to reality with mobile devices carried by humans, a number of constraints, including limited battery power, prevent the wireless interface of each user from being always on, and make the common assumption questionable.[11] More importantly, our findings caution that such an assumption can lead to largely misleading results, but at the same time, suggest that there exist many uncharted territories for further exploration, as depicted in Figure 3.8. We envision that our findings stimulate active research on more in-depth studies of user availability and its impact on network performance. For example, it would be interesting to develop a more accurate and realistic user-availability process and identify a relevant time scale for the availability process, which necessitates careful measurement studies. It can also be possible to analyze the *joint* impact of user mobility and availability on the scaling properties of network capacity and delay.

[11]The low chance of "true" transfer opportunities, for the performance of opportunistic networking, can be compensated by the large node populations carrying mobile devices—the power of the crowd [55].

References

[1] PCWorld, Why your smartphone battery sucks, May 2011. `http://www.pcworld.com/article/228189/why_your_smartphone_battery_sucks.html`.

[2] E. Altman, T. Basar, and F. De Pellegrini. Optimal monotone forwarding policies in delay tolerant mobile ad hoc networks. *Performance Evaluation*, 67(4):299–317, Apr. 2010.

[3] E. Altman and F. De Pellegrini. Forward correction and fountain codes in delay tolerant networks. In *Proc. of IEEE INFOCOM*, pages 1899–1907, 2009.

[4] E. Altman, G. Neglia, F. De Pellegrini, and D. Miorandi. Decentralized stochastic control of delay tolerant networks. In *Proc. of IEEE INFOCOM*, pages 1134–1142, 2009.

[5] G. Ananthanarayanan and I. Stoica. Blue-Fi: enhancing WiFi performance using Bluetooth signals. In *Proc. of ACM MobiSys*, pages 249–262, 2009.

[6] A. Balasubramanian, B. N. Levine, and A. Venkataramani. DTN routing as a resource allocation problem. In *Proc. of ACM SIGCOMM*, pages 373–384, 2007.

[7] N. Banerjee, M. D. Corner, D. Towsley, and B. N. Levine. Relays, base stations, and meshes: enhancing mobile networks with infrastructure. In *Proc. of ACM MobiCom*, pages 81–91, 2008.

[8] J. A. Bitsch Link, C. Wollgarten, S. Schupp, and K. Wehrle. Perfect difference sets for neighbor discovery: energy efficient and fair. In *Proc. of ExtremeCom*, pages 1–6, 2011.

[9] J. Broch, D. A. Maltz, D. B. Johnson, Y.-C. Hu, and J. Jetcheva. A performance comparison of multi-hop wireless ad hoc networking routing protocols. In *Proc. of ACM MobiCom*, pages 85–97, 1998.

[10] H. Cai and D. Y. Eun. Crossing over the bounded domain: from exponential to power-law inter-meeting time in MANET. In *Proc. of ACM MobiCom*, pages 159–170, 2007.

[11] H. Cai and D. Y. Eun. Toward stochastic anatomy of inter-meeting time distribution under general mobility models. In *Proc. of ACM MobiHoc*, pages 273–282, 2008.

[12] T. Camp, J. Boleng, and V. Davies. A survey of mobility models for ad hoc network research. *Wireless Communications and Mobile Computing*, 2(5):483–502, Aug. 2002.

[13] A. Chaintreau, P. Hui, J. Crowcroft, C. Diot, R. Gass, and J. Scott. Impact of human mobility on the design of opportunistic forwarding algorithms. In *Proc. of IEEE INFOCOM*, pages 1–13, 2006.

[14] A. Chaintreau, J.-Y. Le Boudec, and N. Ristanovic. The age of gossip: spatial mean field regime. In *Proc. of ACM SIGMETRICS/Performance*, 109–120, 2009.

[15] V. Conan, J. Leguay, and T. Friedman. Characterizing pairwise inter-contact patterns in delay tolerant networks. In *Proc. of Autonomics*, pages 1–9, 2007.

[16] O. Dousse, F. Baccelli, and P. Thiran. Impact of interferences on connectivity in ad hoc networks. *IEEE/ACM Transactions on Networking*, 13(2):425–436, Apr. 2005.

[17] A. Feldmann and W. Whitt. Fitting mixtures of exponentials to long-tail distributions to analyze network performance models. In *Proc. of IEEE INFOCOM*, pages 1096–1104, 1997.

[18] W. Feller. *An Introduction to Probability Theory and Its Applications.* John Wiley & Son, 1968.

[19] W. Gao, G. Li, B. Zhao, and G. Cao. Multicasting in delay tolerant networks: a social network perspective. In *Proc. of ACM MobiHoc*, pages 299–308, 2009.

[20] R. Groenevelt, G. Koole, and P. Nain. Message delay in mobile ad hoc networks. In *Proc. of IFIP Performance*, pages 210–228, 2005.

[21] M. Grossglauser and D. N. C. Tse. Mobility increases the capacity of ad hoc wireless networks. *IEEE/ACM Transactions on Networking*, 10(4):477–486, Aug. 2002.

[22] P. Gupta and P. R. Kumar. The capacity of wireless networks. *IEEE Transactions on Information Theory*, 46(2):388–404, Mar. 2000.

[23] Z. J. Haas and T. Small. A new networking model for biological applications of ad hoc sensor networks. *IEEE/ACM Transactions on Networking*, 14(1):27–40, Feb. 2006.

[24] B. Han, P. Hui, V. S. A. Kumar, M. V. Marche, G. Pei, and A. Srinivasan. Cellular traffic offloading through opportunistic communications: a case study. In *Proc. of CHANTS*, pages 31–38, 2010.

[25] P. Hui, A. Chaintreau, J. Scott, R. Gass, J. Crowcroft, and C. Diot. Pocket switched networks and human mobility in conference environments. In *Proc. of WDTN*, pages 244–251, 2005.

[26] P. Hui, J. Crowcroft, and E. Yoneki. BUBBLE Rap: social-based forwarding in delay tolerant networks. In *Proc. of ACM MobiHoc*, pages 241–250, 2008.

[27] S. Ioannidis, A. Chaintreau, and L. Massoulie. Optimal and scalable distribution of content updates over a mobile social network. In *Proc. of IEEE INFOCOM*, pages 1422–1430, 2009.

[28] T. Karagiannis, J.-Y. Le Boudec, and M. Vojnovic. Power law and exponential decay of inter contact times between mobile devices. In *Proc. of ACM MobiCom*, pages 183–194, 2007.

[29] A. Keränen, J. Ott, and T. Kařkkäinen. The ONE simulator for DTN protocol evaluation. In *Proc. of SIMUTools*, pages 1–10, 2009.

[30] S. Kim, C.-H. Lee, and D. Y. Eun. Super-diffusive behavior of mobile nodes and its impact on routing protocol performance. *IEEE Transactions on Mobile Computing*, 9(2):288–304, Feb. 2010.

[31] Y. Kim, K. Lee, N. B. Shroff, and I. Rhee. Providing probabilistic guarantees on the time of information spread in opportunistic networks. In *Proc. of IEEE INFOCOM*, pages 2067–2075, 2013.

[32] D. J. Klein, J. Hespanha, and U. Madhow. A reaction-diffusion model for epidemic routing in sparsely connected MANETs. In *Proc. of IEEE INFOCOM*, pages 884–892, 2010.

[33] R. J. La. Distributional convergence of intermeeting times under the generalized hybrid random walk mobility model. *IEEE Transactions on Mobile Computing*, 9(9):1201–1211, Sep. 2010.

[34] C.-H. Lee and D. Y. Eun. Exploiting heterogeneity in mobile opportunistic networks: an analytic approach. In *Proc. of IEEE SECON*, pages 502–510, 2010.

[35] C.-H. Lee and D. Y. Eun. On the forwarding performance under heterogeneous contact dynamics in mobile opportunistic networks. *IEEE Transactions on Mobile Computing*, 12(6):1107–1119, Jun. 2013.

[36] C.-H. Lee, J. Kwak, and D. Y. Eun. Characterizing link connectivity for opportunistic mobile networking: does mobility suffice? In *Proc. of IEEE INFOCOM*, pages 2076–2084, 2013.

[37] K. Lee, S. Hong, S. J. Kim, I. Rhee, and S. Chong. SLAW: a new mobility model for human walks. In *Proc. of IEEE INFOCOM*, pages 855–863, 2009.

[38] U. Lee, S. Y. Oh, K.-W. Lee, and M. Gerla. Scaling properties of delay tolerant networks with correlated motion patterns. In *Proc. of CHANTS*, pages 19–26, 2009.

[39] J. Manweiler and R. R. Choudhury. Avoiding the rush hours: WiFi energy management via traffic isolation. In *Proc. of ACM MobiSys*, pages 253–266, 2011.

[40] A. Mei and J. Stefa. SWIM: a simple model to generate small mobile worlds. In *Proc. of IEEE INFOCOM*, pages 2106–2113, 2009.

[41] M. Motani, V. Srinivasan, and P. S. Nuggehalli. PeopleNet: engineering a wireless virtual social network. In *Proc. of ACM MobiCom*, pages 243–257, 2005.

[42] G. Neglia and X. Zhang. Optimal delay-power trade-off in sparse delay tolerant networks: a preliminary study. In *Proc. of CHANTS*, pages 237–244, 2006.

[43] A.-K. Pietiläinen and C. Diot. Experimenting with opportunistic networking. In *Proc. of MobiArch*, pages 1–6, 2009.

[44] A.-K. Pietiläinen, E. Oliver, J. LeBrun, G. Varghese, and C. Diot. MobiClique: Middleware for mobile social networking. In *Proc. of WOSN*, pages 49–54, 2009.

[45] T. Spyropoulos, K. Psounis, and C. S. Raghavendra. Efficient routing in intermittently connected mobile networks: the multiple-copy case. *IEEE/ACM Transactions on Networking*, 16(1):77–90, Feb. 2008.

[46] T. Spyropoulos, R. N. B. Rais, T. Turletti, K. Obraczka, and A. Vasilakos. Routing for disruption tolerant networks: taxonomy and design. *Wireless Networks*, 16(8):2349–2370, Nov. 2010.

[47] T. Spyropoulos, T. Turletti, and K. Obraczka. Routing in delay-tolerant networks comprising heterogeneous node populations. *IEEE Transactions on Mobile Computing*, 8(8):1132–1147, Aug. 2009.

[48] P.-U. Tournoux, J. Leguay, F. Benbadis, V. Conan, M. D. de Amorim, and J. Whitbeck. The accordion phenomenon: analysis, characterization, and impact on DTN routing. In *Proc. of IEEE INFOCOM*, pages 1116–1124, 2009.

[49] S. Trifunovic, B. Distl, D. Schatzmann, and F. Legendre. WiFi-Opp: ad hoc-less opportunistic networking. In *Proc. of CHANTS*, pages 37–42, 2011.

[50] A. Vahdat and D. Becker. *Epidemic Routing for Partially-Connected Ad Hoc Networks*. Technical report, Duke University, Apr. 2000.

[51] V. Vukadinović and G. Karlsson. Spectral efficiency of mobility-assisted podcasting in cellular networks. In *Proc. of MobiOpp*, pages 51–57, 2010.

[52] W. Wang, V. Srinivasan, and M. Motani. Adaptive contact probing mechanisms for delay tolerant applications. In *Proc. of ACM MobiCom*, pages 230–241, 2007.

[53] M. Xiao, J. Wu, C. Liu, and L. Huang. TOUR: time-sensitive opportunistic utility-based routing in delay tolerant networks. In *Proc. of IEEE INFOCOM*, pages 2085–2091, 2013.

[54] X. Zhang, G. Neglia, J. Kurose, and D. Towsley. Performance modeling of epidemic routing. *Computer Networks*, 51(10):2867–2891, Jul. 2007.

[55] G. Zyba, G. M. Voelker, S. Ioannidis, and C. Diot. Dissemination in opportunistic mobile ad hoc networks: the power of the crowd. In *Proc. of IEEE INFOCOM*, pages 1179–1187, 2011.

Chapter 4

Discovering and Predicting Temporal Patterns of WiFi-Interactive Social Populations

Xiang Li

Adaptive Networks and Control Laboratory
Department of Electronic Engineering
Fudan University
*Shanghai, China**

Yi-Qing Zhang

Adaptive Networks and Control Laboratory
Department of Electronic Engineering
Fudan University
*Shanghai, China**

Athanasios V. Vasilakos

Department of Electrical and Computer Engineering
National Technical University of Athens
Athens, Greece

CONTENTS

4.1 Introduction

Since the publication of Moreno and Jennings' sociometry book in 1934 [1], network (graph) theory has become one of the most powerful tools to characterize social interactions and population dynamics [2, 3, 4]. The discoveries of small-world [5] and scale-free networks [6] in the late of 1990s focused world-wide attention on complex networks and network science. We have witnessed fruitful and exciting advances to understand the hidden patterns behind complex connectivity features and characteristics of diverse large-scale networking systems. Such natural and/or man-made examples range from the Internet, the World Wide Web, biological brain networks, protein-to-protein interaction networks, power grids, and wireless communication networks, to categories of social, economic, and financial networks at different levels of human society. Extensive efforts and elegant attempts have been devoted to answering a fundamental question: how does the fascinating complex topological features affect or determine the collective behaviors and performance of the corresponding complex networked system [7, 8, 9, 10, 11, 12, 13, 14, 15, 16, 17, 18, 19, 20, 21]. This widely accepted key question still remains open in the fast-developing field of network science, especially in the newly focused situation that temporal information as an explicit element defines the edges of such a so-called temporal network [22] when they are active.

Nowadays, many incredible products of flourished information and communication technologies (ICTs) are unobtrusively embedded into the physical world of human daily activities: we communicate with our friends/colleagues with emails and/or mobile phones, we shop on the web and check out with credit cards, and travel/commute in public transportation networks with payment by transit cards, and we surf online 24 hours a day covered by WiFi or 3G signals everywhere. Such creative digital devices/instruments not only reshape our daily life, but also record tremendous digital traces produced by human activities. These digital records offer unparalleled

*The authors were grateful to the Informatization Office of Fudan University for the WiFi Data collection, and the Archives of Fudan University for providing the blueprint of all six teaching buildings. This work was partly supported by the National Key Basic Research and Development Program (No.2010CB731403), the National Natural Science Foundation (No.61273223), the Research Fund for the Doctoral Program of Higher Education (No. 20120071110029), and the Key Project of National Social Science Fund (No. 12&ZD18) of China.

opportunities to explosively digitize physical human interactions and provide fresh temporal clues to shed light on human behavioral patterns.

In the literature, the successful stories mainly include Bluetooth, active Radio Frequency Identification (RFID), wireless sensors, and WiFi technology [23, 24, 25, 26, 27, 28, 29, 30, 31, 32, 33, 34, 35, 36]. For instance, Eagle and his colleagues took advantage of Bluetooth embedded in mobile phones, and collected proxy data of person-to-person interactions in the program of Reality Mining [23, 24]. Barrat et al. built a flexible framework based on RFID technology, and recorded the volunteers' face-to-face interactions in different rendezvous such as conference, museum, and primary school [25, 26, 27]. Salathé et al. utilized wireless sensors to trace person-to-person encounters among the members of a high school, and further evaluated the respiratory disease transmission risk in school campuses [28]. In addition many researchers explored the inter-contact intervals of mobile users' interactions by Bluetooth and WiFi technology, and designed more efficient algorithms to improve the performance of data dissemination among a large-scale population of mobile users [29, 30, 31, 32]. Moreover, such fruitful research pushes the desire to depict human social networking populations with interdependent collective dynamics as well as behavioral patterns.

WiFi, as one of the most ubiquitous wireless access techniques, has been widely deployed in human daily circumstances. Actually, "Free WiFi" signs can be found in nearly every corner of urban areas, and the notion of a "WiFi-City" becomes reality. For example, since 2010, a significant increase in the global number of WiFi certified product launches and hotspots (both public and private) have been reported, while more than 9,000 product certifications have been issued by the WiFi Alliance, and the global number of hotspots is over 280 million [37]. Therefore, the flood of commercial WiFi systems comes as a powerful proxy tool to collect digital access traces of a huge population equipped with WiFi devices. As a snapshot of modern society, a university covered by WiFi signals, where the WiFi system records the digital access logs of the authorized WiFi users when they access the campus wireless service. Such WiFi access records, as the indirect proxy data of a large-scale population's social interactions without artificial interference, are the targets to explore in this chapter.

As a response to the above motivations, this chapter comes as a synthetical review of our extensive efforts devoted to temporal networks and human population dynamics in the past years [33, 34, 35, 36, 38, 39], and we target a cyber-social population with the example of a university campus with the WiFi coverage. We conduct the study based on a dataset from a Chinese university, whose findings have also been verified with other open-source data sets in different venues, and this chapter only reviews this series of works in detail with the focus of WiFi interactions. The first part of this chapter is organized as follows. As a preliminary section, Section 4.2 introduces the background of the WiFi data-collection setting in an involved Chinese university, i.e., Fudan University, and defines the related concepts as well as the dataset preprocessing. Section 4.3 presents a simplified temporal network version for such a cyber-social population with the proxy of campus WiFi access, where the reachability of such individual-level temporal networks affects the efficiency of information spreading. Section 4.4 defines the so-called event interaction to capture the concurrent interactive patterns among a group of individuals, and constructs the transmission graph to embed more temporal information, where both the vertex and edge dynamics of transmission graphs present rich temporal patterns. In Section 4.5, we introduce the outlier performance of the involved low-degree vertices' role in such temporal networks, which has been underestimated in the literature of static networks. We define a temporal quantity as the participation activity potential to feature the role of individuals in a population, which achieves the prediction accuracy of ranking the individual centrality as high as 100% with the verification of the WiFi data set. Finally, Section 4.6 concludes the whole chapter.

4.2 Background

Fudan University covers four campuses, and Handan Campus is the oldest and also the largest campus, which accommodates around 75% of students of the whole university. Handan Campus is the first campus of Fudan University covered by WiFi signal, and in 2009 there were 517 wireless access points (WAPs) distributed in all of the main buildings, which continuously provide dual-band (802.11g/n (2.4-GHz) and 802.11a (5-GHz)) wireless access services to all authorized campus members, i.e., students, teachers, staffs, and visiting scholars.

In the campus WiFi system, all WAPs share the same service set identifier (SSID), which guarantees that wireless devices/clients (e.g., laptops, pads, video game consoles, smartphones, and digital audio players) can smoothly roam from one WAP to another. Each WAP owns a unique IP address. When a client is roaming, the WiFi system automatically records the access logs without notifying the WiFi user. The assess logs of all authorized users are recorded in a database administered by the Informatization Office of Fudan University. With the permission and assistance of the Informatization Office, we collected the campus WiFi users' access logs during the 2009–2010 fall semester (10/18/09–01/09/10) as the so-called data set called FudanWiFi09, each piece of which contains four parts: the Media Access Control (MAC) address of the wireless device, the device's connecting and disconnecting time stamps, and the MAC address of the WAP that the device accesses. Totally, the data set of FudanWiFi09 contains the records of 22,050 WiFi devices' 423,422 behavioral trajectories in the period of nearly three months.

Since we target the proxy data of those wireless devices to feature the users' interactive behaviors, we only focus on the WiFi data recorded in the buildings which are open to the public without safeguards for the WiFi users' devices, i.e., when users leave such a place, they take the device with them. Therefore, the data set of 18,715 individuals' 262,109 behavioral trajectories from six teaching buildings with WiFi coverage have been employed in this work. Figure 4.1 presents the spatial distribution of all 129 WAPs deployed in the six buildings from the vertical view of a part of Handan Campus of Fudan University.

For simplicity, we assume that every WiFi device with the MAC address recorded in the WiFi system represents one WiFi user, neglecting the few cases that one person

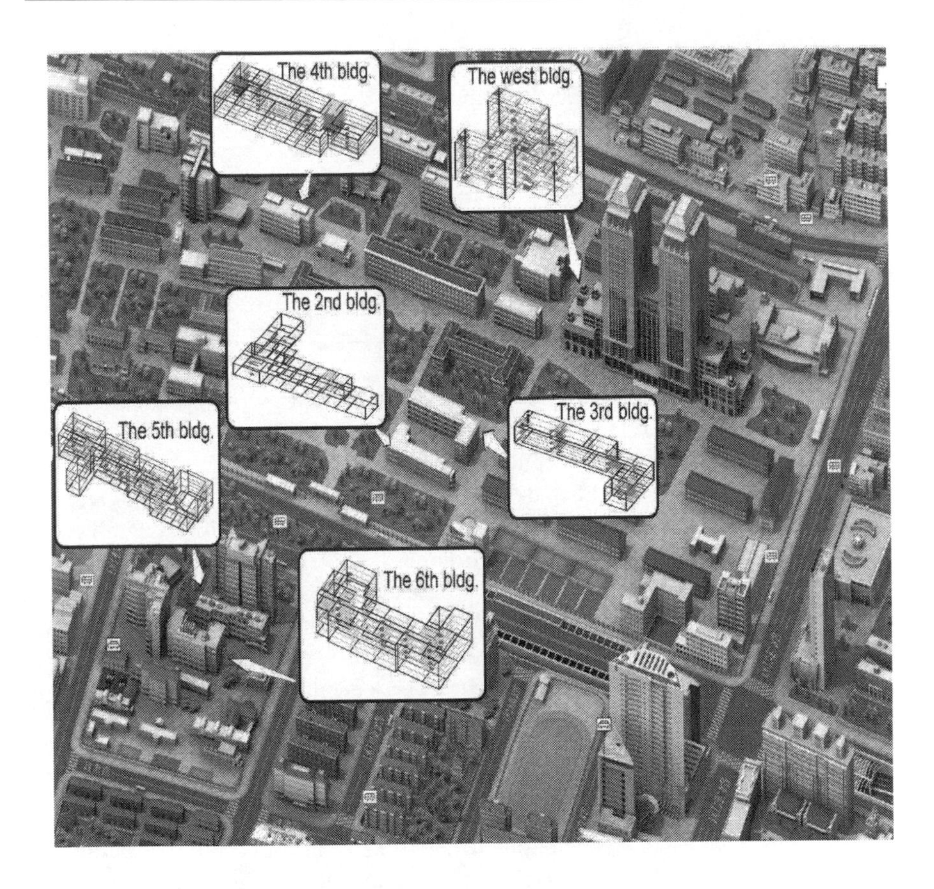

Figure 4.1: The vertical view of Handan Campus of Fudan University, and the internal structures of six teaching buildings with detailed deployments of wireless access points (dots).

may use several WiFi devices at the same time, and several WiFi users may share the same WiFi device. Before transforming the access logs of WiFi data as the proxy of a large-scale population of WiFi users' interactions, we state our main assumption that WiFi devices "seeing" the same WAP imply co-located interactions among these device' owners, which is motivated by the assumption of "geographic coincidences" proved in the work of Crandall and et al. in 2010 [40]. They defined a spatiotemporal co-occurrence between two Flickr users as an instance where they both took photos at approximately the same place and at approximately the same time, which infers social ties in the human population.

Therefore, to verify the assumption of geographic coincidences in the campus WiFi system, the Euclidean distance between two WiFi users is critical. Due to the difficulty of locating the WiFi devices' indoor positions with a single WAP, we cannot exactly calculate the Euclidean distance between two WiFi users in the campus.

Table 4.1 The Estimated Average Euclidean Distances and Standard Deviations between any Two WiFi Users

Building	No. of WAPs	$\langle d \rangle$(m)	δ(m)
the 2nd bldg.	16	6.26	1.96
the 3rd bldg.	21	7.14	0.96
the 4th bldg.	15	7.86	1.71
the 5th bldg.	28	6.98	1.37
the 6th bldg.	33	7.11	1.59
the west bldg.	14	11.75	0.98

However, we can estimate the average Euclidean distances by Voronoi decomposition with the detailed distribution of WAPs shown in Figure 4.1 and the number of students in every classroom from the Fudan University Curriculum Schedule of the 1st Semester in 2009–2010. Each building has several floors, and the building materials between floors can dramatically attenuate wireless signals, which guarantees that the coverage regions of two WAPs on different floors do not overlap. On each floor, wireless devices automatically connect to the closest WAP that offers the strongest signals. Only when the WAP is overloaded, will the device switch to another adjacent WAP. In the campus WiFi system of Fudan University, each WAP can serve 50 users at most, and all the involved WAPs were not overloaded in the 3 months of this study. Therefore, we decompose each floor by the distribution of deployed WAPs into the corresponding Voronoi tessellations. Given a Voronoi cell, we apply the Monte Carlo method that randomly locates the persons in the classrooms of the cell to calculate the Euclidean distances of any two WiFi users, and estimate the average Euclidean distances and the standard deviations of any two WiFi users in every building, as summarized in Table 4.1. We conclude that the average Euclidean distances, along with their standard deviations, do not change much in different buildings, indicating that the assumption of geographic coincidences is valid. Therefore, in such a close indoor circumstance, people have a high probability to directly communicate with each other or have some relationship via indirect interactions. In the following sections, we utilize the collected data of campus WiFi assess logs to explore categories of interactive patterns in a cyber-social population, which is illustrated with the population of WiFi users in Handan Campus of Fudan University.

4.3 Pairwise Interactive Patterns of Temporal Contacts and Reachability

We start from a simplified version of WiFi assess logs as shown in Figure 4.2 (a), where five users (a,b,c,d,e) accessing to two different WAPs (AP 1 and AP 2) during the interval of $[t_0,t_3]$ are illustrated. Obviously, as time evolves, user a interacts with user b before user b interacts with users c and d. Since we assume that the WiFi devices "seeing" the same WAP infers co-located interactions among these device owners, such geographic coincidences indicate temporal contacts among the involved WiFi users co-located in the same WAP. Therefore, a temporal contact net-

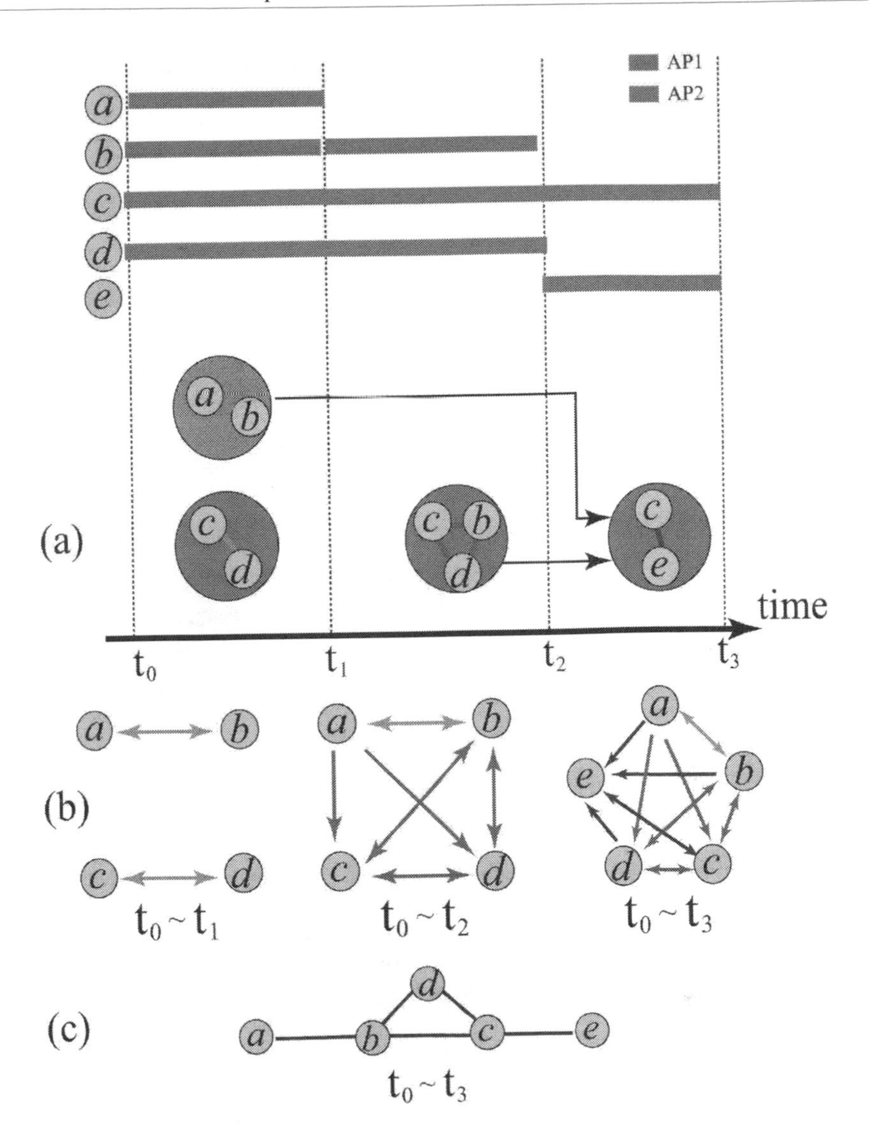

Figure 4.2: (a) WiFi access logs are translated into the corresponding interaction events, where the black arrowed lines show the temporal contacts between users in different interaction events. (b) The construction of three temporal contact networks during the intervals $[t_0, t_1]$, $[t_0, t_2]$, $[t_0, t_3]$, respectively, where the edges are colored according to the latest time that they are contacted. (c) The construction of the aggregated contact network during the interval $[t_0, t_3]$.

work comes as a direct extension of contact networks by embedding the temporal information.

To construct a temporal contact network, we define $e_i = (V_i, t_{i1}, t_{i2})$ as an interaction event, where V_i denotes the set of users connecting to the same WAP simultaneously, and any user $v \in V_i$ interacts with other users in V_i during the interval from t_{i1} to t_{i2}. We define a temporal contact (TC): user i makes a temporal contact with user j if there exists a sequence of interaction events with non-decreasing time between them. Therefore, a temporal contact network is constructed at the level of individual users, i.e., define a temporal contact network $\mathcal{G} = \{\mathcal{V}, \mathcal{E}\}$ with $\mathcal{V}$ the set of vertices (WiFi users) and $\mathcal{E}$ the collections of all temporal contacts. Note that $\mathcal{E} = \mathcal{E}_{\rightarrow} \bigcup \mathcal{E}_{\rightleftharpoons}$ is the union set of directed edges $\mathcal{E}_{\rightarrow}$ and bidirectional edges $\mathcal{E}_{\rightleftharpoons}$, which stand for the time-respecting paths and pairwise interactions, respectively. Given an observed interval, as shown in Figure 4.2 (b), there exist two time-respecting paths: one path is from user a to user c, and the other is from user a to user d. In addition, there exist two pairwise interactions between two pairs of users b, c, and users b, d. Therefore, we generate three different temporal contact networks in different intervals $[t_0, t_1], [t_0, t_2], [t_0, t_3]$, respectively. As a comparison, we aggregate all interaction events in the interval of $[t_0, t_3]$, and generate an aggregated contact network as shown in Figure 4.2 (c), which clearly shows the difference between two different contact networks even in the same interval $[t_0, t_3]$.

Such different connectivity patterns between temporal contact networks and aggregated contact networks are also reflected in the basic property of network connectivity: reachability. Traditionally, in the case of aggregated (static) networks, since the edges are continuously active, the size of the component, N_i, that a given vertex i belongs to, indicates an upper bound of the number of vertices influenced by vertex i. Therefore, N_i characterizes the reachability of vertex i in the static contact network aggregated within a given time interval. In the case of temporal (contact) networks, however, the edges are no longer continuously active, whose active durations and time stamps dominate the network transitivity. Since a temporal contact network defined here is rather simplified, and only keeps the contact order of time stamps, the reachability of vertex i in a temporal (contact) network is the number of vertices that can be temporally influenced by i during the time length Δt [41]. More specifically, the reachability of vertex i of a temporal contact network equals the number of elements in the set $\{\phi_{i,j}(t_2) \mid \phi_{i,j}(t_2) > t_1,\ j \in V\}$, where $\phi_{i,j}(t)$ is the time of the inception of the latest temporal contact from i to j before t. The average (maximum) network reachability is the mean (maximum) value of all vertices' reachability.

Figure 4.3 compares the network reachability (normalized by the network size) between temporal contact networks and aggregated contact networks with the same time length. Both the average and maximum values show that the aggregated contact networks' reachability is much larger than those of temporal contact networks, especially when the time length $\Delta t \rightarrow 0$. In addition, the reachability of aggregated contact networks quickly attains its saturation with increasing the time length Δt, while the reachability of temporal contact networks saturates much more slowly. Since the saturated value of the average reachability of temporal contact networks is much smaller than that of aggregated contact networks, the temporal dimension

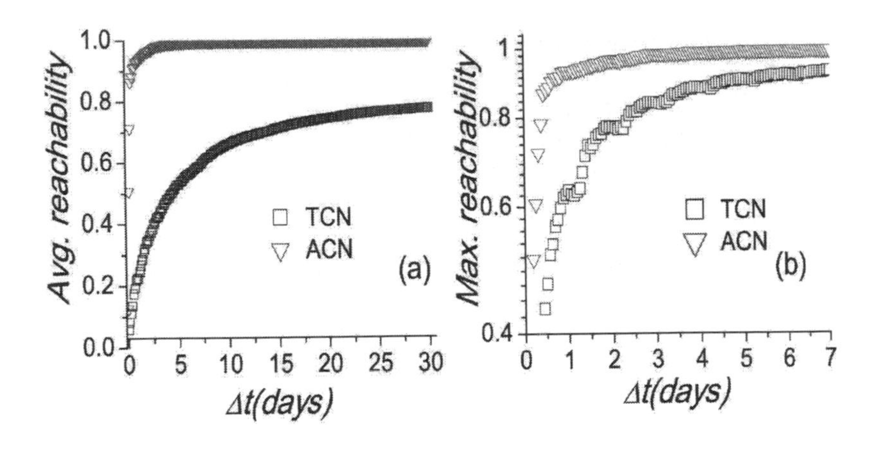

Figure 4.3: (a) The average network reachability of the aggregated contact networks (ACNs) and temporal contact networks (TCNs) with the time length Δt=30 days. (b) The maximum network reachability of the ACNs and TCNs with the time length Δt=7 days.

provides a more precise upper bound to the reachability of a network, which may avoid the overestimation of the information efficiency over such networks.

In a temporal contact network, the out-degree $d_{out,i}$ of vertex i quantifies the number of receptors temporally affected by i, while the in-degree $d_{in,i}$ specifies the number of its potential inciters. In an information-spreading process on a temporal (contact) network, for example, except the source/destination vertices as well as those leaf vertices, all other vertices in the network play both roles of receiving and forwarding information in the process. Therefore, we put more attention on the correlated distribution of $d_{out,i}$ and $d_{in,i}$ of all vertices in the temporal contact networks with different time lengths Δt. Figure 4.4 presents the joint probability distributions $C^{\Delta t}(d_{out}, d_{in})$ of out-degrees and in-degrees with different time lengths $\Delta t = 1,2,3,5,7,8$ days, respectively. Each joint probability distribution is averaged by the generated networks from the data set with a given Δt. In the case of the smallest daily time length, i.e., $\Delta t = 1$ day, Figure 4.4(a) shows that almost all data points reside in the lower triangular matrix, and many data points are even sticking on the axes. That is to say, in a short observation period, e.g., 1 day, only a very limited number of users keep frequently online during the whole day, while most individuals act according to their curriculum schedules, which seldom cover the whole day. With the increase of time length $\Delta t = 2,3,5$ days, there are more and more vertices emerging in the upper region of two axes (Figures. 4.4(b) (d)), which shows that the forward and backward temporal influence of each user increases as time proceeds. When the time length is more than one week (Figures. 4.4(e) and (f)), there are two evidently nontrivial clusters of temporal hubs: one contains the vertices along the diagonal, and the other contains the vertices anchoring in the upper right corner. In the former cluster of hubs, the users balance to receive and forward the influence from other users, while

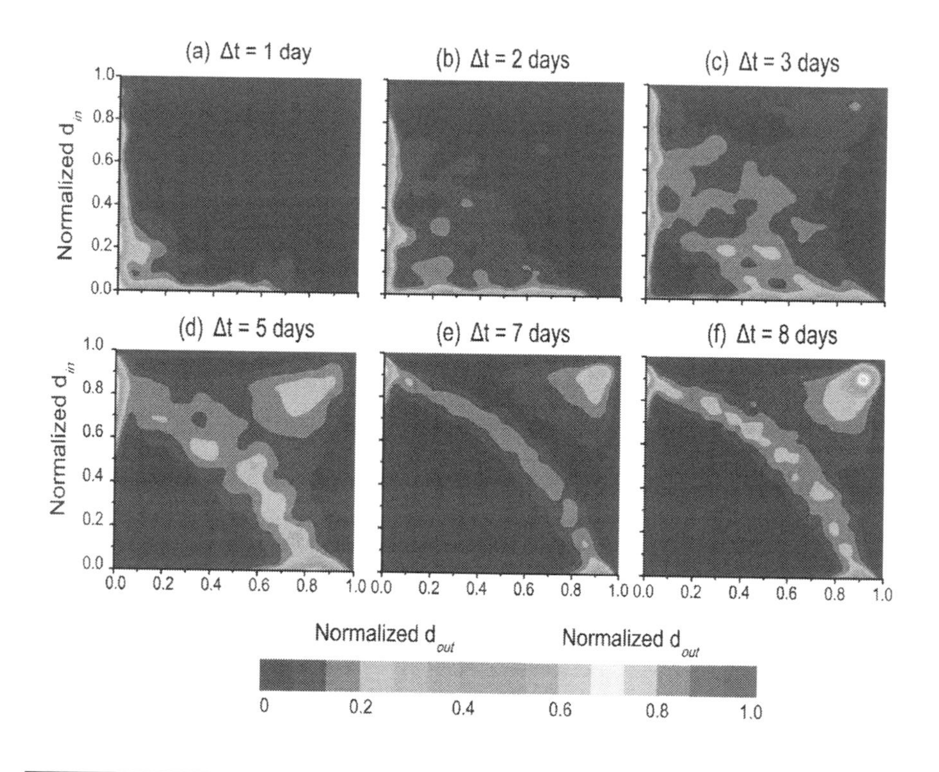

Figure 4.4: The joint probability distribution of the temporal out-degrees d_{out} and in-degrees d_{in} with the given time length Δt as (a) 1 day, (b) 2 days, (c) 3 days, (d) 5 days, (e) 7 days, (f) 8 days. When the time length Δt is larger than 1 week (7 days), a set of data points always present at the right-up corner.

those users presented in the upper right corner are more important due to their role as relay hubs, which are critical to facilitate spreading processes on temporal networks. Obviously, the weekly cycle of the curriculum schedules of all campus WiFi users accelerates the emergence of such temporal patterns of relay hubs.

4.4 Concurrent Interactive Patterns of Event Interactions and Temporal Transmission Graphs

Temporal contact networks as a category of reachability graph are generated based on the "time-respecting" paths and pairwise interactions of temporal contacts, which focus on the temporal interactive sequence between a pair of individuals while neglecting the concurrent interactive dynamics of more involved individuals. For example, recall Figure 4.2 (b), the three users b, c, d concurrently access AP 2 in the overlapped interval $[t_1, t_2]$, while this concurrent interaction among three users has to be simplified to two pairwise interactions (*bd* and *bc*) with the definition of temporal contacts. To characterize such concurrent interactions, we extend the temporal contacts at the level of individual users to a group of users with the newly defined event interactions and transmission graphs to embed more temporal information.

Figure 4.5 (a) gives an instance of five WiFi users' access logs as the corresponding individual behavioral traces, where the bold lines pertain to their online durations, i.e., the five users stay in the coverage of the same WAP during $[t_0, t_9]$, and only these five users access this WAP during the interval. As we assume that the WiFi devices "seeing" the same WAP infers the geographic co-located interactions among these devices' owners (i.e., co-locating at the same time in the same small region), the five users participate in several access events when different users access and disconnect in the WAP at different time stamps. Therefore, we define an event interaction as the concurrent interaction of multiple users accessing from the same WAP within a given interval, e.g., EI $E_{AB}^{t_0}$ characterizes the concurrent interactions of users A and B at the beginning time t_0. Figure 4.5 (b) illustrates the process of translating the five users' individual behavioral trajectories into event interactions (EIs). Since a user is only involved in one unique event interaction at any time, all illustrated events in Figure 4.5 (b) involve two users for simplicity, and an event interaction with more users involved is still valid, which is reflected in the later defined quantity of event size.

Define an event interaction (EI) as a vertex, and two vertices (event interactions) are connected in a transmission graph with the following three defined rules:

1. At least one user exists in both the source and sink EIs.
2. In the time sequence, the beginning time of the source EI is the closest one prior to the beginning time of the sink EI.
3. Given a sink EI, when there exist several source EIs, the set of shared users in a pair of source EI and the sink EI does not intersect with other pairs of source and sink EIs.

The first two rules follow the principle of temporal nearest adjacency, and the third rule follows the principle of minimum number of transmission edges (paths) generated. Therefore, we construct a transmission graph as illustrated in Figure 4.5 (c) with the above three rules, yielding five vertices (event interactions of Figure 4.5 (b)) and seven directed edges, where the time stamps, such as t_0, t_6 over the directed edge between vertex *AB* and vertex *AC*, stand for the beginning time stamps of the two (source and sink) event interactions. Also, we aggregate all event interactions following the three rules of transmission edges with the same data set to generate an aggregated transmission graph, which is illustrated in Figure 4.5 (d).

To quantify the temporal information of event interactions (EIs) and the transmission graph (TG), we further define the following quantities:

- The size, s, of a given EI is defined as the number of involved individuals.
- Given an EI active at the epoch of $[t^{begin}, t^{end}]$, its active duration is defined as $\Delta t^{EI} = t^{end} - t^{begin}$.

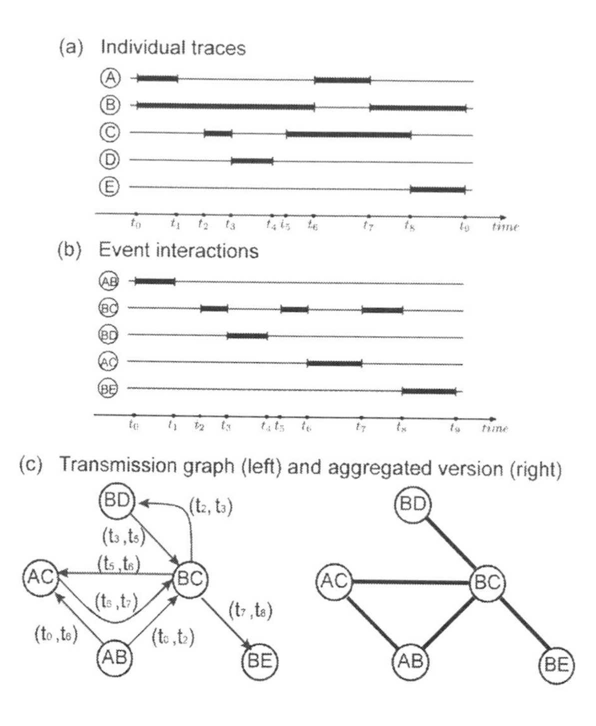

Figure 4.5: (a)WiFi access logs as the proxy of the users' behavioral trajectories. The bold lines pertain to their online durations. (b) Each bold line possesses an exclusive time interval, where the corresponding individuals are assembled into a contact clique. (c) In a transmission graph, the vertices are event interactions (EIs), and the edges between two EIs are the transmission paths defined by the three rules. The aggregated version is derived from the transmission graph, where no multiple edges and time labels are allowed.

- Given a pair of source and sink EIs, the time-stamped label in the directed edge is (t_{source}, t_{sink}). The transmission duration δ is defined as the interval between the beginning time of the source and sink EIs, i.e., $\delta = t_{sink} - t_{source}$.

Since the vertices of a transmission graph are the event interactions, we first present the vertex dynamics of transmission graphs to unveil the statistical characteristics of concurrent interactions among WiFi users. As shown in Figure 4.6 (a), the probability distribution of all event interactions' active durations collected during the three months falls into a truncated power law, whose exponent of the power-law part approximates to 1. This feature indicates that long-lasting concurrent interactions can hardly survive due to those newly emergent events with short active durations, while the non-Poisson distribution of concurrent interactive durations tells that WiFi users do not randomly break their interactions with other users. On the other hand, the

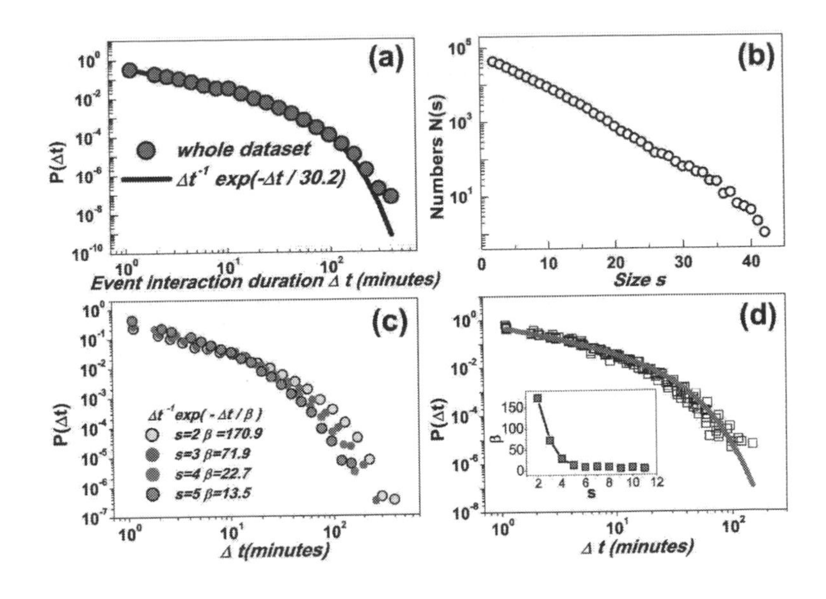

Figure 4.6: (a) The probability distribution of event interactions' active durations Δt^{EI}_{all}. (b) The size distribution of event interactions. (c) The probability distributions of event interactions' active durations Δt^{EI}_{s} with the given size s = 2,3,4,5, respectively. (d) The probability distributions of event interactions' active durations Δt^{EI}_{s} with the given size s=5,6,...11, respectively. The solid bold line is the distribution with s=5. The inset shows the dependence between size s, and the exponent of exponential cutoff β.

number of event interactions with a given size is exponentially distributed as shown in Figure 4.6 (b), i.e., the users randomly co-locate with each other at the same time.

Note that such an exponential distribution also implies that large event interactions are rare, and we may conjecture that there exists some interdependence between the probability distribution of event interactions' active durations Δt^{EI} and their sizes s, which, however, does not exist as shown in Figures. 4.6 (c)-(d). Although the probability distributions of active durations Δt^{EI}_{s} with size $s = 2,3,4,5$ keep the truncated power laws, whose power-law parts still obey the exponents close to 1, we observe that their exponential cutoffs gradually decay with the increased event size from 2 to 5 (Figure 4.6(c)). Furthermore, with the increased event size from 6 to 11 as shown in Figure 4.6(d), the probability distributions of the active durations with size $s \geq 5$ present the same shape $\Delta t^{-\gamma}\exp(-\frac{\Delta t}{\beta})(\gamma \approx 1)$. At the same time, the inset of Figure 4.6(d) shows that the exponent of exponential cutoff β decreases with the growth of size s from 2 to 5, which keeps invariant with increasing s from 5 to 11. Therefore, we conclude that the active durations of event interactions have the *size-free* feature, which, counter-intuitively, is independent from the event sizes.

On the edge dynamics of transmission graphs, we focus on the transmission durations to uncover the temporal relations between concurrent interactions. As shown in Figure 4.7 (a), the probability distribution of transmission durations obeys a bifold

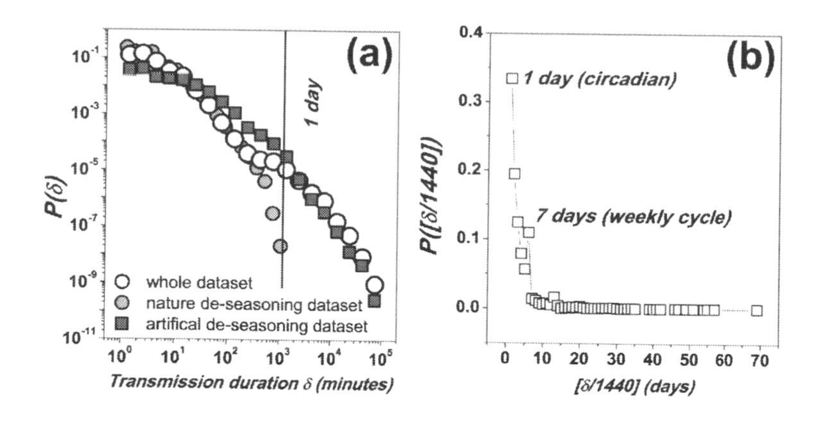

Figure 4.7: (a) The probability distribution of transmission durations, δ, from the whole dataset (white circle), and "filtered" by the natural de-seasoning method (gray circle) and the artificial de-seasoning method (square). (b) The probability distribution of integral days spent by transmission duration $[\frac{\delta}{1440}]$ (days).

power-law with two turning points: the first turning point is at around 120 minutes, i.e., the length of two teaching courses; the other turning point is at around 24 hours, which indicates the existence of daily circadian rhythms. To test the existence of periodic rhythms, we calculate the integral days spent by each δ ($[\frac{\delta}{1440}]$), as shown in Figure 4.7(b), where there are two peaks at the 1st day and the 7th day. That is to say, the daily burst behavior of "adjacent" event interactions evolves with the weekly rhythm. We further employ two (natural and artificial) de-seasoning methods to remove the circadian and weekly rhythms. In the natural de-seasoning method, we conserve the source and sink event interactions taking place in the same day. In the artificial de-seasoning method, we randomly select two event interactions and exchange their beginning active time (their active durations are ignored), and check all individuals to ensure this round of time-shuffle operations does not generate event interactions with the same involved individuals with have the same beginning time. As shown in Figure 4.7(a), the probability distributions of the transmission durations 'filtered' by two de-seasoning methods both fall into the truncated power-laws, and we witness that such burst behaviors of populations take place *independently* of the daily and weekly rhythms.

4.5 Temporal Degrees and Hubs: Ranking and Prediction

In Section 4.3, those users located in the upper right corner of Figure 4.4 (f), i.e., users with both large in-degrees and out-degrees, function as the relay hubs of temporal contact networks, which emerge with the weekly rhythms of WiFi users' curriculum schedules. In this section, we further explore the temporal hubs with more temporal information embedded into event interactions and transmission graphs as defined in Section 4.4. When neglecting temporal information, traditionally, a static contact network with the same data set of WiFi campus access logs is generated, where the vertices are WiFi users, and an edge connects two vertices if the two WiFi users have at least one round of interaction (i.e., geographic coincidence) during the whole 3 months. As a comparison, we aggregate all transmission edges of Figure 4.5 (c) (left) to generate an aggregated transmission graph, as illustrated in Figure 4.5 (c) (right).

Both temporal and static aggregated networks present some differences. The static contact network defines the vertices at the individual level of WiFi users, which is topological homogeneous with the degrees' coefficient of variation (CV) as 1.59. While the aggregated transmission graph defines the vertices at the level of event interactions, which concurrently involve a group of WiFi users, it is more heterogeneous with the degrees' coefficient of variation as $7.13 \gg 1$. Figure 4.8 (a) visualizes a small sample of the aggregated transmission graph. Notice that the degree of event interactions (vertices) in the aggregated transmission graph is the sum of the out-degree and in-degree in the corresponding transmission graph, and we define the degree of an event interaction as the degree of the event interaction's involved users. Generally, a (WiFi) user may participate in several event interactions with different degrees, and we define the temporal degree of a user as the maximal degree of all the involved event interactions. Therefore, we rank the temporal degrees of WiFi users as shown in Figure 4.8(b), where the top 10 users with the largest temporal degrees are identified with solid circles. Interestingly, these temporal-hub users are not as attractive in the static contact network, which, on the contrary, are almost the low-degree vertices as identified with solid circles in Figure 4.8(c). That is to say, many low-degree vertices, which generally were neglected in the literature of static contact networks, may participate to play the dominant role as "hubs" in the version of temporal networks. Such difference also highlights the temporal significance of concurrent interactions to characterize a typical cyber-social population with the society example of a WiFi campus, and the outlier nontrivial performance has also been verified in other public data sets [35].

Note that we cannot directly identify the WiFi users who function as temporal hubs from the traditional version of static contact networks, as we have visualized that the low-degree individuals in the static human contact networks gather together to form temporal hubs in the aggregated transmission graphs. Therefore, to identify (and even predict) such important users who have high temporal degrees in the WiFi campus, a natural means is constructing such an aggregated transmission graph before ranking all involved users with their temporal degrees, which, however, is a task with very high computational load. For example, given a temporal data set, which yields M event interactions with time stamps, we construct a transmission graph having H vertices and P edges, whose computational complexity in the worst case is $O(M^2)$, while the process of aggregating such a transmission graph requires the computational load of $O(H * P)$. Therefore, with the WiFi access data addressed in this chapter, we totally generate $M = 260,925$ event interactions and a transmission graph

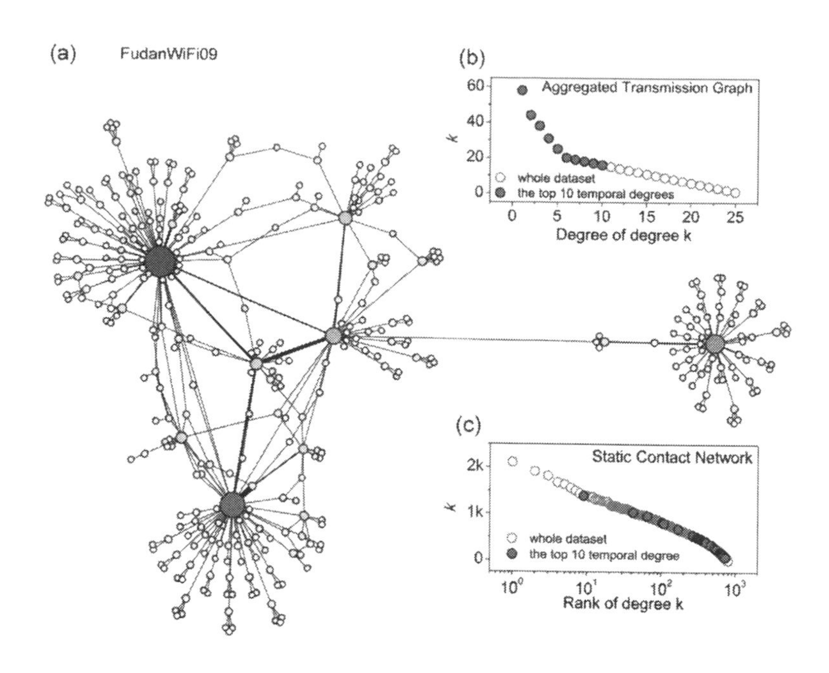

Figure 4.8: (a) A small sample of the aggregated transmission graph consisting of several hubs that WiFi users are involved in. (b) The degrees versus their ranks of all WiFi users in the aggregated transmission graph. The solid circles represent the WiFi users who have the top 10 temporal degrees. (c) The degrees versus their ranks of all WiFi users in the static contact network with the same WiFi data set. The solid circles represent the same WiFi users in (b).

having $H = 248,663$ vertices and $P = 260,513$ edges, whose computational load is as high as $O(1.32 * 10^{11})$ to generate the aggregated transmission graph, excluding the search cost to rank temporal degrees. Obviously, a new and efficient method to rank and predict the temporal hubs without constructing such a transmission graph is desirable and challenging, and we give a satisfactory solution as follows.

As shown in Figure 4.9 (a), the vertices of an aggregated transmission graph are event interactions (EIs), which contain several WiFi users followed by the definitions of transmission graph in Section 4.4. The active duration of an event interaction, e_i, is defined as $\Delta t^{EI}(e_i)$. Given a finite length of time period ΔT and an event interaction, e_i, the total active (WiFi users') number of the event interaction is $n(e_i)$, and we define the total active duration of the event interaction as

$$\Delta t^{EI}_{sum}(e_i) = \sum_{m=1}^{n(e_i)} \Delta t^{EI}_{m}(e_i).$$

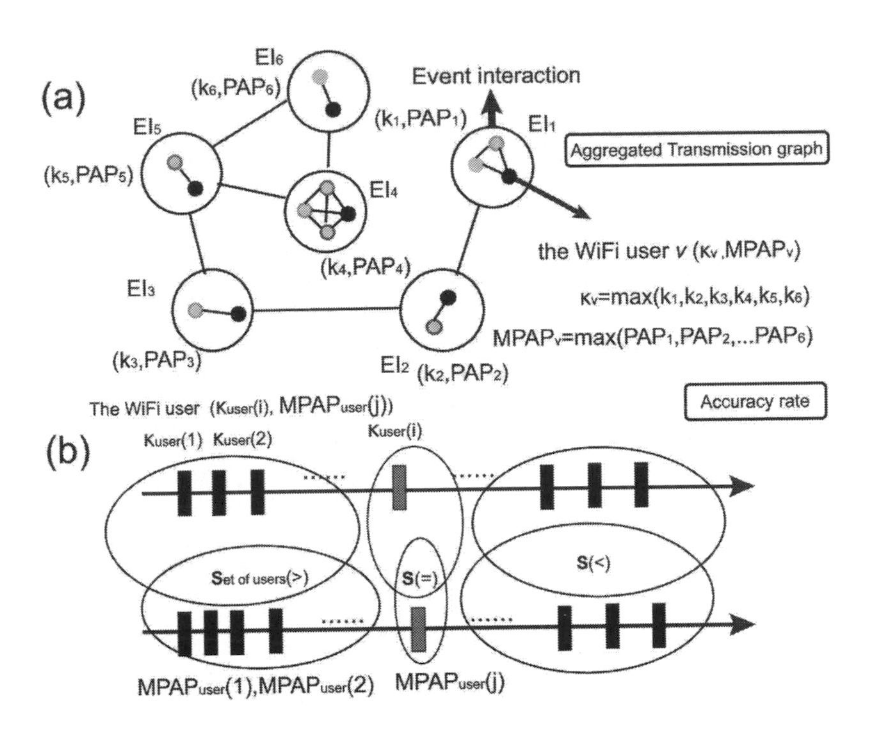

Figure 4.9: (a) In an aggregated transmission graph, each vertex is an event interaction (EI), and each EI contains several WiFi users. The temporal degree, κ, of a WiFi user is the maximum degree of the involved EIs, and the corresponding maximum participation activity potential (MPAP) is the maximum value of all participation activity potentials (PAPs) of the involved EIs. (b) The temporal degree, κ, and MPAP are both in descending order. Given a WiFi user, its accuracy rate is the ratio of the other WiFi users who are in the same positions predicted by both κ and MPAP.

Given that the WiFi user j participates in event interaction e_i, we define the *participation activity potential (PAP)* of user j involving event interaction e_i as:

$$PAP(j,e_i) = (\Delta t^{EI}_{sum}(e_i))^{\alpha}(n(e_i))^{1-\alpha}, \alpha \in [0,1]. \tag{4.1}$$

As we know, a WiFi user may involve several different event interactions, therefore, given the set of event interactions involving WiFi user j as $\Gamma(j) = \{e_i | j \in e_i\}$, we define the *maximum participation activity potential (MPAP)* of user j as:

$$MPAP(j) = \max_{e_i \in \Gamma(j)} (PAP(j,e_i)). \tag{4.2}$$

Different from temporal degrees to characterize WiFi users' temporal role, which require a heavy computation cost as we have discussed, the new proposed quantity of maximum participation activity potential (MPAP) does not need any preliminary information from the aggregated transmission networks. For instance, in Figure 4.9(a), WiFi user v is involved in six event interactions with the corresponding degree k_i and the participation activity potential PAP_i, where $i = 1,2,...6$. The temporal degree of user v is the maximal degree of the involved six event interactions, $\kappa(j) = \max_{i=1,...,6} k_i$, which is dependent on constructing such an aggregated transmission graph. To calculate the maximum participation activity potential of user j, the maximum value of all PAPs of the involved event interactions, $MPAP(j) = \max_{i=1,...,6} PAP_i$, is directly calculated from the information of each event interaction itself.

Therefore, we propose the maximum participation activity potential (MPAP) of a user as the candidate quantity of "temporal degrees," with which we rank all WiFi users and predict the temporal hubs in the campus population. As shown in Figure 4.9(b), we list all temporal degrees and all MPAPs of the WiFi users in the descending order, where "$\kappa_{user}(1) > \kappa_{user}(2) > ...$" and "$MPAP_{user}(1) > MPAP_{user}(2) > ...$". Given a WiFi user with temporal degree $\kappa_{user}(i)$, all the other WiFi users in the population are divided into the following three sets based on temporal degree $\kappa_{user}(i)$:

- set $S_\kappa(>)$: the users whose temporal degrees $\kappa > \kappa_{user}(i)$.
- set $S_\kappa(=)$: the users whose temporal degrees $\kappa = \kappa_{user}(i)$.
- set $S_\kappa(<)$: the users whose temporal degrees $\kappa < \kappa_{user}(i)$.

Similarly, we divide all the same users into another group of three sets based on $MPAP_{user}(j)$:

- set $S_{MPAP}(>)$: the users whose $MPAP > MPAP_{user}(j)$.
- set $S_{MPAP}(=)$: the users whose $MPAP = MPAP_{user}(j)$.
- set $S_{MPAP}(<)$: the users whose $MPAP < MPAP_{user}(j)$.

Finally, we find the part of WiFi users who are present in the following sets as shown in Figure 4.9(b):

- the set of $S(>)$: the users present in $S_\kappa(>) \bigcap S_{MPAP}(>)$.
- the set of $S(=)$: the users present in $S_\kappa(=) \bigcap S_{MPAP}(=)$.
- the set of $S(<)$: the users present in $S_\kappa(<) \bigcap S_{MPAP}(<)$.

Therefore, the users ranked by their temporal degrees and the maximum participation activity potential in the same positions describe the accuracy rate predicted by the maximum participation activity potential defined as follows:

$$AR(v) = \frac{|S(>)| + |S(=)| + |S(<)|}{n-1}.$$

In more detail, given a rank of the temporal degree $r(\kappa)$, we define the average predicative accuracy rate and its standard deviation as

$$\langle AR(r_\kappa)\rangle = \frac{\sum_{r_\kappa(v)=r} AR(v)}{|\{v|r_\kappa(v)=r\}|} \tag{4.3}$$

$$\delta(AR(r_\kappa)) = \sqrt{\frac{\sum_{r_\kappa(v)=r}(AR(v)-\langle AR(r_\kappa)\rangle)^2}{|\{v|r_\kappa(v)=r\}|}}. \tag{4.4}$$

Note that in the definition of participation activity potential *PAP*, Equation (4.1), an appropriate value of parameter $\alpha \in [0,1]$ influences the accuracy rate. To find the optimal α and achieve the highest prediction accuracy, we further define the following function $F(\alpha)$:

$$F(\alpha) = \frac{\sum \kappa * \langle AR(\kappa)\rangle}{\sum \kappa}.$$

As shown in Figure 4.10 (a), the maximum accuracy ratio is $F(\alpha) = 98.6\%$ with the corresponding $\alpha = 0.04$ as the optimum, and a random selection of $\alpha \in (0,1)$ achieves an accuracy rate higher than 96%. Substituting the optimum of $\alpha = 0.04$ to the participation activity potential *PAP* defined in Equation (4.1), the average accuracy rate (Equation (4.3)) predicts the *top 15* temporal hubs by the maximum participation activity potential *MPAP* as high as 100% accuracy rate (Figure 4.10(b)). Furthermore, to predict the different cases, for example, of the top 1, 5, 10, and 15

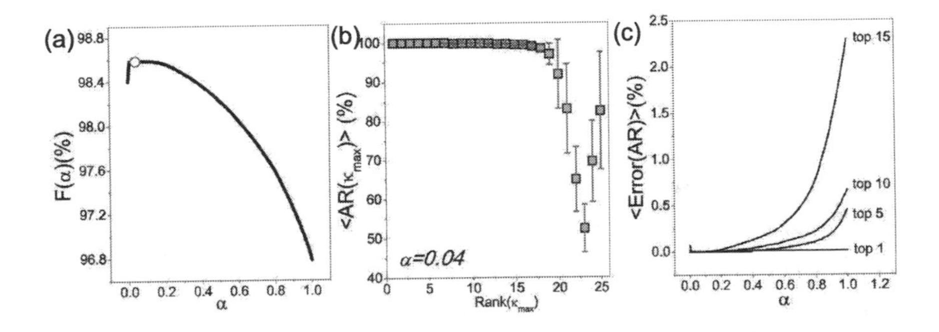

Figure 4.10: (a) The weighted accuracy rate F(α) shows that the accuracy rate of predicating temporal hubs achieves more than 96% when randomly selecting a value of α from [0,1]. The dot represents that the maximum weighted accuracy rate can achieve 98.6% with a given optimal nonlinear parameter α=0.04. (b) The average accuracy rate $\langle$AR(κ_{max})$\rangle$ of predicating temporal hubs versus the rank of their temporal degrees κ_{max} with the optimal α=0.04. The bar represents the standard deviation δ(AR(κ_{max})) of the accuracy rate. (c) The average error between the accuracy rate given a random nonlinear parameter, α, and the optimal α=0.04 to predict top 1, 5, 10, 15 WiFi users.

temporal hubs, the performance difference between a randomly selected $\alpha \in (0,1)$ and the optimal $\alpha = 0.04$ is as low as 2.5% at most, which is visualized in Figure 4.10(c). Note that such an optimal $\alpha = 0.04$ is not universal; such a value is different case by case [34]. Therefore, a randomly selected $\alpha \in (0,1)$ may achieve a sub-optimal performance of the prediction in practice.

4.6 Conclusion

To summarize this chapter, with the proxy data of WiFi access logs collected in a campus of Fudan University, we have presented the series of temporal interactive patterns of the large-scale population of WiFi users as a representative example of cyber-social populations in the networking era of today. The efforts from the viewpoint of temporal networks offer sufficiently powerful supports to discover human interactions, as we illustrated in this chapter, from both pairwise and event interactions defined at the levels of individual users and a group of concurrent users, respectively. Not limited to the rich temporal patterns of the WiFi interactive population, the outlier performance of the temporal hubs along with the roles of leaf connectivity status also shed more light on the significance of temporal networks, which, in return, help to rank and predict the important active users in the temporal framework.

Nevertheless, the understanding of network science from the viewpoints of temporal networks as well as human population dynamics still leaves many topics to explore, which is far beyond the previous addressed issues such as time-varying, time-evolving, and time-switching networks. Especially, new concepts from not only network topological connectivity but also dynamical processes and the collective performance need to bridge the gap between the traditional (static/aggregated) version and the now focused temporal dimension, for example, but not limited to, temporal degree, temporal path in previous sections, and temporal clustering coefficient [38], temporal betweenness[42], in-and out-components[43], temporal small world [44], temporal motif [45, 46, 36], etc.

Return to the key question in the beginning of this chapter. The modeling of such fascinating and yet temporal complex features of a network very recently have reported several models as a combination of human dynamics and evolving networks [47, 48, 49, 50], and the dynamical processes [51, 52, 53] as well as the collective behaviors [54, 55] and even the structural controllability [39] over such temporal networks have recently attracted more broad attention. We also note the book entitled *Temporal Networks* [56], which covers a more extensive scope of temporal networks for the interested readers' further reference, while, on the other hand, this chapter synthesizes this new emergent branch with a detailed WiFi-interactive social population.

References

[1] J. L. Moreno and H. H. Jennings, *Who Shall Survive?: A New Approach to the Problem of Human Interrelations.* Nervous and Mental Disease Publishing Co., 1934.

[2] L. C. Freeman, D. White, and A. K. Romney, *Research Methods in Social Network Analysis.* Transaction Publishers, New Brunswick, New Jersey, 1992.

[3] J. Scott, *Social Network Analysis: A Handbook.* Sage Publications, 2000.

[4] S. Wasserman and K. Faust, *Social Network Analysis: Methods and Applications.* Cambridge University Press, 1994.

[5] D. J. Watts and S. H. Strogatz, "Collective dynamics of small-worldnetworks," *Nature*, 393(6684): 440–442, 1998.

[6] A.-L. Barabási and R. Albert, "Emergence of scaling in random networks," *Science*, 286(5439): 509–512, 1999.

[7] R. Albert and A.-L. Barabási, "Statistical mechanics of complex networks," *Review Modern Physics*, 74(1): 47–97, 2002.

[8] A. Barrat, M. Barthlemy, and A. Vespignani, *Dynamical Processes on Complex Networks.* Cambridge University Press, 2008.

[9] G. Chen, X. F. Wang, and X. Li, *Introduction to Complex Networks: Models, Structures and Dynamics.* Higher Education Press, 2012.

[10] L. d. F. Costa, O. N. Oliveira Jr., G. Travieso, F. A. Rodrigues, L. Antiqueira, and et al., "Analyzing and modeling real-world phenomena with complex networks: A survey of applications," *Advances in Physics*, 60(3): 329–412, 2011.

[11] L. d. F. Costa, F. A. Rodrigues, G. Travieso, and P. R. Villas Boas, "Characterization of complex networks: A survey of measurements," *Advances in Physics*, 56(1): 167–242, 2007.

[12] S. N. Dorogovtsev and J. F. F. Mendes, "Evolution of networks," *Advances in Physics*, 51(4): 1079–1187, 2002.

[13] D. Easley and J. Kleinberg, *Networks, Crowds and Markets: Reasoning about a Highly Connected World.* Cambridge University Press, 2010.

[14] T. Gross and H. Sayama (Ed.), *Adaptive Networks: Theory, Models and Applications.* Springer, 2009.

[15] M. O. Jackson, *Social and Economic Networks.* Princeton University Press, 2010.

[16] M. E. J. Newman, "The structure and function of complex networks," *SIAM Review*, 45(2): 167–256, 2003.

[17] ——, *Networks: An Introduction.* Oxford University Press, 2009.

[18] R. Pastor-Satorras and A. Vespignani, *Evolution and Structure of the Internet: A Statistical Physics Approach.* Cambridge University Press, 2007.

[19] X. F. Wang and G. Chen, "Complex networks: Small-world, scale-free and beyond," *IEEE Circuits and Systems Magazine*, 3(1): 6–20, 2003.

[20] X. F. Wang, X. Li, and G. Chen, *Complex Networks: Theories and Applications (in Chinese).* Tsinghua University Press, 2006.

[21] ——, *Network Science: An Introduction (in Chinese).* Higher Education Press, 2012.

[22] P. Holme and J. Saramäki, "Temporal networks," *Physics Reports*, 519(3): 97–125, 2012.

[23] N. Eagle and A. Pentland, "Reality mining: Sensing complex social systems," *Personal and Ubiquitous Computing*, 10(4): 255–268, 2006.

[24] N. Eagle, A. Pentland, and D. Lazer, "Inferring friendship network structure by using mobile phone data," *Proceedings of the National Academy of Sciences of the United States of America*, 106(36): 15 274–15 278, 2009.

[25] C. Cattuto, W. Van den Broeck, A. Barrat, V. Colizza, J.-F. Pinton, and V. A., "Dynamics of person to person interactions from distributed RFID sensor networks," *PloS ONE*, 5(7): e11596, 2010.

[26] L. Isella, J. Stehlé, A. Barrat, C. Cattuto, J.-F. Pinton, and W. Van den Broeck, "What's in a crowd? Analysis of face-to-face behavioral networks," *Journal of Theoretical Biology*, 271(1): 166–180, 2011.

[27] J. Stehlé, N. Voirin, A. Barrat, C. Cattuto, L. Isella, and et al., "High-resolution measurements of face-to-face contact patterns in a primary school," *PLoS ONE*, 6(8): e23176, 2011.

[28] M. Salathé, M. Kazandjieva, J. W. Lee, P. Levis, M. W. Feldman, and J. H. Jones, "A high-resolution human contact network for infectious disease transmission," *Proceedings of the National Academy of Sciences of the United States of America*, 107(51): 22 020–22 025, 2010.

[29] H. Cai and D. Y. Eun, "Crossing over the bounded domain: From exponential to power-law intermeeting time in mobile ad hoc networks," *IEEE/ACM Transactions on Networking*, 17(5): 1578–1591, 2009.

[30] A. Chaintreau, P. Hui, J. Crowcroft, C. Diot, R. Gass, and J. Scott, "Impact of human mobility on opportunistic forwarding algorithms," *IEEE Transactions on Mobile Computing*, 6(6): 606–620, 2007.

[31] P. Hui and J. Crowcroft, "Human mobility models and opportunistic communications system design," *Philosophical Transactions of the Royal Society A: Mathematical, Physical and Engineering Sciences*, 366(1872): 2005–2016, 2008.

[32] T. Karagiannis, J.-Y. Le Boudec, and M. Vojnovic, "Power law and exponential decay of intercontact times between mobile devices," *IEEE Transactions on Mobile Computing*, 9(10): 1377–1390, 2010.

[33] Y. Zhang, L. Wang, Y.-Q. Zhang, and et al., "Towards a temporal network analysis of interactive WiFi users," *Europhysics Letters*, 98(6): 68002, 2012.

[34] Y.-Q. Zhang and X. Li, "Characterizing large-scale population's indoor spatio-temporal interactive behaviors," in *Proceedings of the ACM SIGKDD International Workshop on Urban Computing*. ACM, 2012, pp. 25–32.

[35] ——, "Temporal dynamics and impact of event interactions in cyber-social populations," *Chaos*, 23(1): 013131, 2013.

[36] Y.-Q. Zhang, J. Xu, X. Li, and A. V. Vasilakos, "Exploring mesoscopic patterns of human interaction activities with temporal motifs," *submitted*, 2013.

[37] I. T. M. B. Alliance, "WBA industry report 2011: Global developments in public WiFi," Available at: www.wballiance.com/resourcecentre/global-developmentswifireport.html, 2011.

[38] J. Cui, Y.-Q. Zhang, and X. Li, "On the clustering coefficients of temporal networks with extension to epidemic dynamics," in *IEEE International Symposium on Circuits and Systems*, 2013, pp. 2299–2302.

[39] Y. Pan, X. Li, and J. Zhan, "On the priority maximum matching of structural controllability of temporal networks," in *China Control Conference*, 2013, pp. 1164–1169.

[40] D. J. Crandall, L. Backstrom, D. Cosley, S. Suri, D. Huttenlocher, and J. Kleinberg, "Inferring social ties from geographic coincidences," *Proceedings of the National Academy of Sciences of the United States of America*, 107(52): 22 436–22 441, 2010.

[41] P. Holme, "Network reachability of real-world contact sequences," *Physics Review E*, vol. 71, no. 4, p. 046119, 2005.

[42] R. Pfitzner, I. Scholtes, A. Garas, C. J. Tessone, and F. Schweitzer, "Betweenness preference: Quantifying correlations in the topological dynamics of temporal networks," *Physics Review Letter*, 110(19): 198701, 2013.

[43] M. Konschake, H. H. K. Lentz, F. J. Conraths, P. Hövel, and T. Selhorst, "On the robustness of in- and out-components in a temporal network," *PLoS ONE*, 8(2): e55223, 2013.

[44] J. Tang, S. Scellato, M. Musolesi, C. Mascolo, and V. Latora, "Small-world behavior in time-varying graphs," *Physics Review E*, 81(5): 055101(R), 2010.

[45] L. Kovanen, M. Karsai, K. Kaski, J. Kertész, and J. Saramäki, "Temporal motifs in time-dependent networks," *Journal of Statistical Mechanics: Theory and Experiment*, arXiv: 1107.5646v2.

[46] L. Kovanen, K. Kaski, J. Kertész, and J. Saramäki, "Temporal motifs reveal homophily, gender-specific patterns and group talk in mobile communication networks," *arXiv preprint arXiv:1302.2563*, 2013.

[47] A. Barrat, B. Fernandez, K. K. Lin, and L.-S. Young, "Modeling temporal networks using random itineraries," *Physics Review Letter*, 110(15): 158702, 2013.

[48] H. H. Jo, R. K. Pan, and K. Kaski, "Emergence of bursts and communities in evolving weighted networks," *PloS ONE*, 6(8): e22687, 2011.

[49] B. Min, K. I. Goh, and I. M. Kim, "Waiting time dynamics of priority-queue networks," *Physics Review E*, 79(5): 056110, 2009.

[50] N. Perra, B. Gonçalves, R. Pastor-Satorras, and A. Vespignani, "Activity driven modeling of time varying networks," *Nature Scientific Reports*, 2: 469, 2012.

[51] S. Lee, L. E. C. Rocha, F. Liljeros, and P. Holme, "Exploiting temporal network structures of human interaction to effectively immunize populations," *PLoS ONE*, 7(5): 2012.

[52] L. E. C. Rocha and V. D. Blondel, "Bursts of vertex activation and epidemics in evolving networks," *PLoS Computational Biology*, 9(3): e1002974, 2013.

[53] A. Machens, F. Gesualdo, C. Rizzo, A. E. Tozzi, A. Barrat, and C. Cattuto, "An infectious disease model on empirical networks of human contact: Bridging the gap between dynamic network data and contact matrices," *BMC Medicine*, 13: 185, 2013.

[54] N. Perra, A. Baronchelli, D. Mocanu, B. Gonçalves, R. Pastor-Satorras, and A. Vespignani, "Random walks and search in time-varying networks," *Physics Review Letter*, 109(23): 238701, 2012.

[55] M. Starnini, A. Baronchelli, and R. Pastor-Satorras, "Modeling human dynamics of face-to-face interaction networks," *Physics Review Letter*, 110(16): 168701, 2013.

[56] P. Holme and J. Saramäki (Ed.), *Temporal Networks*. Springer, 2013.

Chapter 5

Behavioral and Structural Analysis of Mobile Cloud Opportunistic Networks

Anh-Dung Nguyen
Department of Mathematics Computer Science and Automatic Control
ISAE, University of Toulouse
Toulouse, France
LAAS-CNRS, Toulouse, France

Patrick Senac
Department of Mathematics Computer Science and Automatic Control
ISAE, University of Toulouse
Toulouse, France
LAAS-CNRS, Toulouse, France

Michel Diaz
LAAS-CNRS, Toulouse, France

CONTENTS

5.1 Introduction

Opportunistic networks are dynamic networks based on nodes' intermittent contacts. This type of network was shown to be the sole communication means in many extreme situations, especially when an infrastructure-based communication is not possible, for instance, deep space communications, disaster recovery networks, and battlefield networks. Opportunistic networks pave the way to a pervasive and universal communication environment in which opportunistic communication can play an important role by its capacity to free users from infrastructure.

"Cloud computing" has recently appeared as a buzzword in many medias in which the term refers both to the technology advancement and also to the business model. The idea is not new, but has its roots in already developed technologies such as distributed computing, autonomic computing, hardware virtualization, and web services. It's the maturation and convergence of all these technologies that makes cloud computing viable today. By virtualizing the aggregated computing resources in order to offer users on-demand utilities (e.g., computing, storage, software as service) in a pay-as-you-go fashion, much like the power distribution grid system, cloud computing appears as a main actor in the information industry today. This can be seen through the explosion of cloud computing services deployed via the Internet in recent years.

With the advances of electronic technologies, mobile wireless devices have gradually become more and more powerful in terms of processing, storage, and communication capacity. The ever-increasing density of mobile wireless devices and sensors that populate the edge of the Internet raises the question of the efficient cooperative use of these vast resources by leveraging the spontaneous communication capacity offered for free by the "mobile cloud" of edge devices. These "mobile clouds," which leverage opportunistic contacts between users, can potentially deliver free communication, storage, and processing services shared between users according to peer-to-peer resource sharing policies.

Although the application perspective sounds interesting, the underlying technology challenges are not negligible due to the difficulties raised by the dynamic nature of such networks. The first obstacle comes from the mobile nature of such networks and raises the question of how the mobility impacts the distributed processing performances of the mobile clouds. Indeed, the unstable network topology makes continuous end-to-end communication unguaranteed and hence the service delivery may be disrupted. Indeed, in the context of spontaneous and infrastructureless networks, with a kind of disruption/delay-tolerant network, nodes must rely on intermittent contacts leading to the use of the store-carry-and-forward communication paradigm for inter-node communication. Therefore, although the role of mobility on communication performances such as end-to-end delay and bandwidth has been already studied [9], the impact of mobility schemes on the global processing power delivered by a mobile network cloud has not been studied yet.

In this chapter, we discuss this issue in depth and show how the mobility and the dynamic network structure impact on the processing capacity of mobile cloud opportunistic networks. In Section 5.2, we propose STEPS—a simple parametric model which covers a large spectrum of human mobility patterns [24]. The model implements two seminal features of human mobility, i.e., preferential attachment and attractors. We show that this model can capture various characteristics of human mobility usually observed in real traces, e.g., inter-contact/contact time distribution. In Section 5.3, based on the STEPS model, we study opportunistic networks in both the space and time domains to extract their salient structural characteristics. Specifically, we show that highly dynamic nodes can play the role of bridge between disconnected components and help to significantly reduce the characteristic path length of the network, hence contributing to the emergence of the small-world structure in opportunistic networks [22]. In STEPS, starting from a regular dynamic network in which nodes tend to stay in their preferential zones, we rewire it by progressively increasing the percentage of nomadic nodes in the network. We show that, as soon as the ratio of nomadic nodes passes 10%, the network exhibits small-world properties with a high dynamic clustering coefficient and a low shortest dynamic path length. In Section 5.4, by applying particle swarm optimization techniques and STEPS, we show that mobility and the small-world structure can dramatically increase the processing capacity of mobile cloud opportunistic networks [23]. Moreover, we show that these properties can enhance the resilience of mobile clouds against network churn. Finally, Section 16.8 will conclude the chapter and discusses open questions.

5.2 Understanding and Modeling Opportunistic Networks

Mobility is the seminal process which underlines dynamic networks composed of human portable devices. In this chapter, we are interested in the modeling of such dynamic networks. We introduce Spatio-TEmporal Parametric Stepping (STEPS)—a simple parametric mobility model which can cover a large spectrum of human mobility patterns. STEPS abstracts spatio-temporal preferences in human mobility by using a power law to rule the nodes' movement coupled with two fundamental mobility principles. Nodes in STEPS have preferential attachment to favorite locations where they spend most of their time. Via simulations, we show that STEPS is able, not only to express peer-to-peer properties such as inter-contact/contact time distribution, but also to express the structural properties of the underlying interaction graph such as the small-world phenomenon. Moreover, STEPS is easy to implement, flexible to configure, and also theoretically tractable.

5.2.1 Introduction

Human mobility is known to have a significant impact on performance of opportunistic networks. Unfortunately, there is no model that is able to capture all the characteristics of human mobility due to its high complexity. We introduce Spatio-TEmporal Parametric Stepping (STEPS)—a powerful formal model for human mobility or mobility inside social/interaction networks. The introduction of this new model is justified by the lack of modeling and expressive power, in the currently used models, for the spatio-temporal correlation usually observed in human mobility.

We show that preferential location attachment and location attractors are invariant properties, at the origin of the spatio-temporal correlation of mobility. Indeed, as observed in several real mobility traces, while few people have a highly nomadic mobility behavior, the majority have a more sedentary one.

We assess the expressive and modeling power of STEPS by showing that this model succeeds in expressing easily several fundamental human mobility properties observed in real traces of dynamic networks:

- The distribution of human traveled distance follows a truncated power law.
- The distribution of pause time between travels follows a truncated power law.
- The distribution of inter-contact/contact time follows a truncated power law.
- The underlying dynamic graph can emerge as a small-world structure.

5.2.2 Related Works

Human mobility has attracted a lot of attention from not only computer scientists but also epidemiologists, physicists, etc., because its deep understanding may lead to many other important issues in different fields. The lack of large-scale real mobility traces means that research is initially based on simple abstract models, e.g., Random Waypoint, Random Walk (see [3] for a survey). These models, whose parameters are usually drawn from an uniform distribution, although good for simulation, can-not reflect the reality and even are considered counterproductive in some cases [32]. In these models, there is no notion of spatio-temporal preferences.

Recently, available real data allowed researchers to more deeply understand the nature of human mobility. The power law distribution of the traveled distance was initially reported in [1], in which the authors study the spatial distribution of human movement based on bank note traces. The power law distribution of the inter-contact time was initially studied by Chaintreau et al. in [4]. In [11], Karagiannis et al. confirm this and also suggest that the inter-contact time follows a power law up to a characteristic time (about 12 hours) and is then cut off by an exponential decay.

In [8], the authors show that people have a significant probability to return to a few highly frequented locations. Another study [26] shows that humans show significant propensity to return to the locations they visited frequently before, like home or workplace. These two intrinsic human mobility characteristics, i.e., location preference and attractor, are implemented in our model.

Some more sophisticated mobility models have been proposed. In [7], the authors have proposed a universal model that is able to capture many characteristics of human daily mobility by combining different sub-models. With a lot of parameters to configure, the complexity of this type of model makes them hard to use.

In [14], the authors propose SLAW—a Random Direction like model—except that the traveled distance and the pause time distributions are ruled by a power law. An algorithm for trajectory planning was added to mimic the human behavior for always choosing the optimal path. Although being able to capture statistical characteristics like inter-contact, contact time distribution, the notion of spatio-temporal preferential attachment is not covered by this model.

Another modeling stream aims to integrate social behaviors. In [17], a community-based model was proposed in which the movement depends on the relationship between nodes. The network area is divided in zones and the social attractivity of a zone is based on the number of friends in the same zone. The comparisons of this model with real traces show a difference for the contact time distribution.

Time Varying Community [10] is another interesting model in which the authors try to model the location preference and periodic re-appearance of human mobility by creating community zones. Nodes have different probabilities to jump in different communities to capture the location preference. Time structure was built on the basis of night/day and days in a week to capture the periodic re-appearance. However, fundamental features of human trajectories such as traveled distance distribution and inter-contact/contact time distributions, were not highlighted.

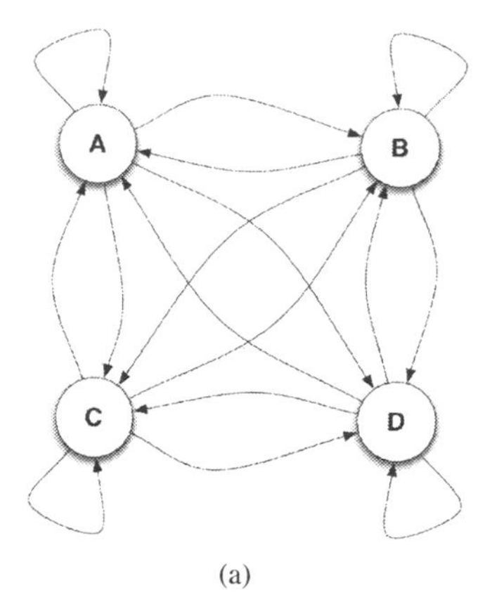

(a)

2	2	2	2	2
1	1	1	2	2
1	*A*	1	2	2
1	1	1	2	2
2	2	2	2	2

(b)

Figure 5.1: (a) Human mobility modeling under a Markovian view: States represent different localities, e.g., House, Office, Shop, and Other places, and transitions represent the mobility pattern. (b) 5×5 **torus representing the distances from location A to the other locations.**

Recent research shows that some mobility models (including [10]), despite of their capacity of capturing spatio-temporal characteristics, deviate significantly in routing performances when compared to the ones obtained with real traces [29]. This aspect, which has not always been considered in existing models, is indeed really important because it shows the capacity of a model to reproduce a fundamental feature of a dynamic network.

5.2.3 Characterizing and Modeling Human Mobility

STEPS is inspired by observable characteristics of human mobility behavior, specifically the spatio-temporal correlation. Indeed, people share their daily time between some specific locations at some specific times, e.g., home/office, night/day. These spatio-temporal patterns repeat at different scales and have been recently observed on real traces [10].

On a short-term basis, i.e., a day, a week, we can assume that one has a finite space of locations. According to these observations, we define two mobility principles:

- Preferential attachment: the probability for a node to move in a location is inversely proportional to the distance from his preferential location.
- Attractor: when a node is outside of his preferential location, he has a higher probability to move closer to this location than moving farther.

From this point of view, human mobility can be modeled as a finite state space Markov chain in which the transition probability distribution expresses a movement pattern. Figure 5.1(a) illustrates a Markov chain with 4 states which correspond to 4 locations of interest.

In STEPS, a location is modeled as a zone which corresponds to a Markov chain state. Inside a zone, nodes can move freely according to a random mobility model such as Random Waypoint. The displacement between zones is drawn from a power law distribution whose exponent value expresses the more or less nomadic behavior. By simply tuning the power law exponent, STEPS can cover a large spectrum of mobility patterns from purely random ones to highly localized ones. In addition, complex heterogeneous mobility behavior can be described by combining nodes with different mobility patterns as defined by their preferential zones and the related attraction power. Group mobility is also supported in our implementation. Hereafter, we give the details of this model.

5.2.3.1 STEPS

Assume that the network area is a square torus[1] divided in N square zones. The distance between zones is defined according to a metric. We use Chebyshev distance in this case. Figure 5.1(b) illustrates an example of a 5×5 torus with the distances from zone A. One can imagine a zone as a geographic location (e.g., home, school, supermarket) or a logical location (i.e., a topic of interest such as football, music, philosophy, etc.). Therefore, we can use the model to study both human geographic mobility and human social behaviors. In this thesis, we deal only with the first case.

In such structured space, each mobile node is associated with a preferential zone z_{pref}. For the sake of simplicity, we assume that each node is attached to one zone. However, this model can be extended by associating a node with several preferential zones. The node movement between zones is driven by a power law satisfying the two mobility principles described above. The probability mass function of this power law is given by

$$P[D = d] = \frac{\beta}{(1+d)^{\alpha}}, \tag{5.1}$$

where D is a discrete random variable representing the distance from z_{pref}, α is the power law exponent that represents the attraction power of z_{pref}, and β is a normalizing constant.

From Equation 5.1 we can see that, the farther a zone is from the preferential zone, the less likely a node is to move in. This is the principle of preferential attachment. On the other hand, when a node is outside its preferred zone, it has a higher probability to move closer to this one than moving farther. This is the principle of attraction. In consequence, with a power law, the model is able to capture two basic characteristics of human mobility.

This small set of modeling parameters allows the model to cover the full spectrum of mobility behaviors. Indeed, according to the value of the α exponent a node can have a more or less nomadic behavior. For instance,

- when $\alpha < 0$, nodes have a higher probability to choose a long distance than a short one and so the preferential zone plays a repulsion role instead of a attraction one;

[1] The choice of a torus allows us to have a homogeneous area.

- when $\alpha > 0$, nodes are more localized, and hence tend to stay in their preferential zones;
- when $\alpha = 0$, nodes move randomly toward any zone with a uniform probability.

We summarize the details of STEPS in Algorithm 9. A MATLAB implementation of STEPS can be downloaded at [21].

Algorithm 9: STEPS algorithm for a node

```
1  Initial zone ← z_pref;
2  repeat
3     Choose a random distance d from the probability distribution (5.1);
4     Select uniformly at random a zone z_i among all zones that are d distance
      units away from z_pref;
5     Choose uniformly at random a point in z_i;
6     Go linearly to this point with a speed chosen uniformly at random from
      [v_min, v_max], 0 < v_min ≤ v_max < +∞;
7     Choose uniformly at random a staying time t from
      [t_min, t_max], 0 ≤ t_min ≤ t_max < +∞;
8     while t has not elapsed do
9        Perform Random Waypoint movement in z_i
10    end while
11 until End of simulation
```

5.2.3.2 *The Underlying Markov Chain*

In this section, we give the details and the properties of the Markovian model underlying STEPS. Wherever a node's preferential zones are, the torus structure gives different nodes the same spatial structure, i.e., they have the same number of zones with a given distance from their preferential zones. More specifically, for a distance d, we have $8d$ zones with equal distances from z_{pref}. Consequently, the probability for a node to choose one among these zones is

$$P\left[z_i, d_{z_i z_{\text{pref}}} = d\right] = \frac{1}{8d} P\left[D = d\right] = \frac{\beta}{8d(1+d)^{\alpha}}. \tag{5.2}$$

Because the probability for a node to select a destination zone z_i depends only on the distance between z_i and z_{pref}; wherever the node is residing, it has the same probability to select z_i. If we define the transition probabilities of the Markov chain as a stochastic matrix, this one will have identical rows. The matrix is then idempotent, i.e., multiplication by itself gives the same matrix. This matrix is given by

$$\begin{bmatrix} P(z_0) & P(z_1) & \dots & P(z_{N-1}) \\ \vdots & \vdots & \ddots & \vdots \\ P(z_0) & P(z_1) & \dots & P(z_{N-1}) \end{bmatrix}. \tag{5.3}$$

It is straightforward to deduce the stationary state of the Markov chain which is

$$\Pi = \begin{pmatrix} P(z_0) & P(z_1) & \dots & P(z_{N-1}) \end{pmatrix}. \tag{5.4}$$

From this result, it is interesting to characterize the inter-contact time[2] of STEPS because this network property was shown to have important impacts on the forwarding algorithm in opportunistic networks. To simplify the problem, let us assume that a contact occurs if and only if two nodes are in the same zone and a node's movement is limited to jumping between zones. Let two nodes A and B move according to the STEPS model, starting initially from the same zone. The probability that the two nodes are in the same zone at a given time is

$$\begin{aligned} p_{\text{contact}} &= P_A(z_0)P_B(z_0) + \ldots + P_A(z_{n-1})P_B(z_{n-1}) \\ &= \sum_{i=0}^{N-1} P(z_i)^2 = \sum_{d=0}^{d_{\max}} \frac{1}{8d}\left[\frac{\beta}{(d+1)^{\alpha}}\right]^2 \end{aligned}, \tag{5.5}$$

where $d_{\max} = \lfloor \frac{\sqrt{N}}{2} \rfloor$ is the maximum distance between zones in the torus.

Let ICT be the discrete random variable which represents the time elapsed before A and B are in contact again. One can consider that ICT is the number of trials until the first success of an event with probability of success $p_{contact}$. Hence ICT follows a geometric distribution with the parameter $p_{contact}$. Therefore, the probability mass function of ICT is given by

$$P[ICT = t] = (1 - p_{\text{contact}})^{t-1} p_{\text{contact}}. \tag{5.6}$$

It is well known that the continuous analog of a geometric distribution is an exponential distribution. Therefore, the inter-contact time distribution for i.i.d. nodes can be approximated by an exponential distribution. This is true when the attractor power α is equal to 0, i.e., when nodes move uniformly, because there is no spatio-temporal correlation between nodes. But when $\alpha \neq 0$, there is a higher correlation in their movement and in consequence, the exponential distribution is not a good approximation. Indeed, in [2], the authors report this feature for the Correlated Random Walk model where the correlation of nodes induces the emergence of a power law in the inter-contact time distribution. In the following, we provide simulation results related to this feature for STEPS.

[2]The time interval between two consecutive contacts of the same node pair.

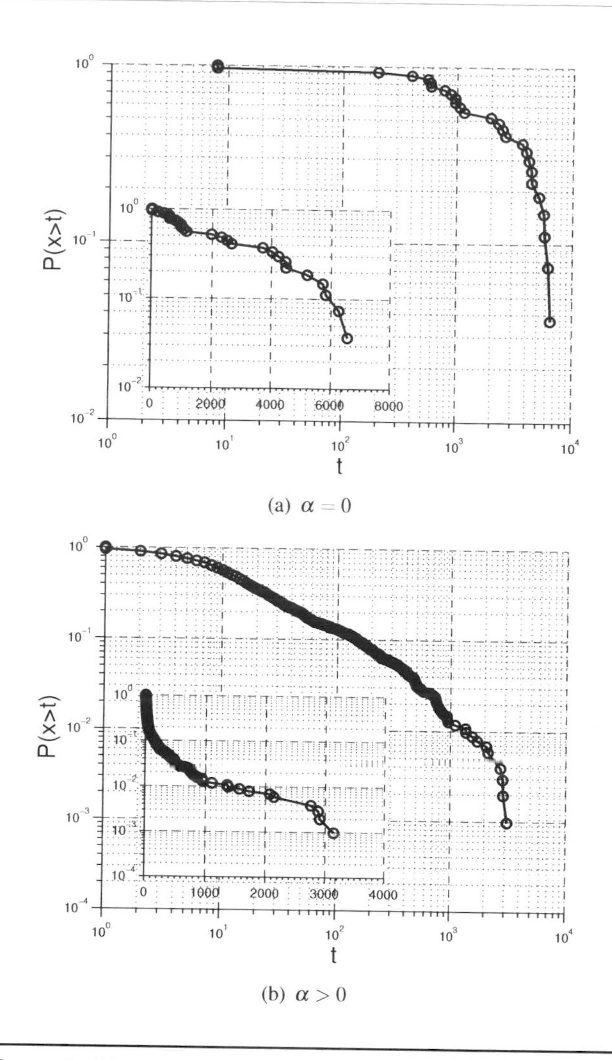

Figure 5.2: Theoretical inter-contact time distribution of STEPS.

Figure 5.2 gives the log-log and linear-log plots of the complementary cumulative distribution function or CCDF of the inter-contact time that results from a simulation of the Markov chain described above when $\alpha = 0$ and $\alpha > 0$. In the first case, the inter-contact time distribution fits an exponential distribution (i.e., is represented by a linear function in the linear-log plot) while in the second case it fits a power law distribution (i.e., is represented by a linear function in the log-log plot) with an exponential decay tail. These results confirm the relationship between the spatio-temporal correlation of nodes and the emergence of a power law in inter-contact time distribution.

Table 5.1 Infocom 2006 Trace

Characteristic	Value
Number of nodes	98
Duration	4 days
Connectivity	Bluetooth
Average inter-contact time	1.9 hours
Average contact duration	6 minutes

5.2.4 Fundamental Properties of Opportunistic Networks in STEPS

It is worth mentioning that a mobility model should express the fundamental properties observed in real opportunistic networks. In this section, we show that STEPS can, indeed, capture the salient characteristics of human mobility.

5.2.4.1 Inter-Contact Time Distribution

The inter-contact time is defined as the delay between two encounters of a pair of nodes. Real traces analysis suggests that the distribution of inter-contact time can be approximated by a power law up to a characteristic time (i.e., about 12 hours) followed by an exponential decay [11]. In the following we will use the set of traces presented by Chaintreau et al. in [4] as a base of comparison with STEPS mobility simulations. Figure 5.3(a) shows the aggregate CCDF of the inter-contact time[3] for different traces. In order to demonstrate the capacity of STEPS to reproduce this feature, we configured STEPS to exhibit the results observed in the Infocom 2006 conference trace, which is the largest trace in the dataset. Table 5.1 summarizes the characteristics of this trace.

To simulate the conference environment, we create a 10×10 torus of size $120 \times 120m^2$ that mimics rooms in the conference. The radio range is set to $10m$ which corresponds to Bluetooth technology. The node speed is chosen as human walking speed which ranges from $[3,5]$ km/h. Figure 5.3(b) shows the CCDF of the resulting inter-contact time in log-log and lin-log plots. We observe that the resulting inter-contact time distribution given by the STEPS simulations fits the one given by the real trace.

5.2.4.2 Contact Time Distribution

Because of the potential diversity of node behaviors, it is more complicated to reproduce the contact duration given by real traces. Indeed, the average time spent for each contact depends on the person. For instance, some people spend a lot of time talking while the others just shake hands. We measured the average contact duration and the average node degree, i.e., the global number of neighbor nodes, of the Info-

[3]The CCDF of inter-contact time samples over all distinct pairs of nodes.

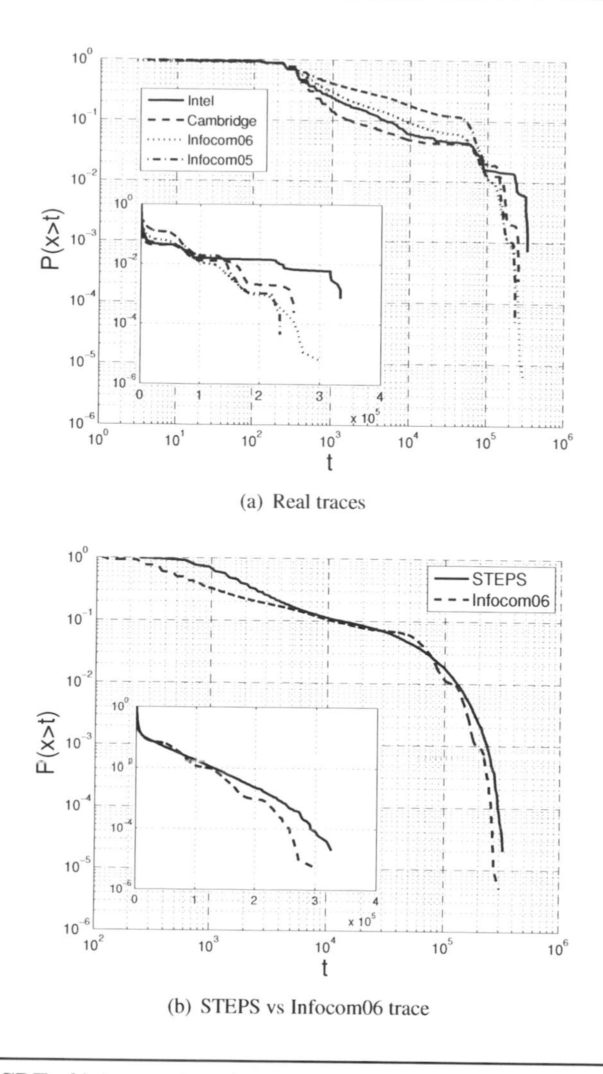

(a) Real traces

(b) STEPS vs Infocom06 trace

Figure 5.3: CCDF of inter-contact time.

com06 nodes and ranked them according to their average contact duration. The result is reported in Figure 5.4(a). According to this classification, it appears that the more (resp. less) popular the person is, the less (resp. more) time he or she spends for each contact. Because the contact duration of STEPS depends principally on the pause time of the movement inside the zone (i.e., the pause time of RWP movement), to mimic this behavior with STEPS, we divided nodes in four groups. Each group corresponds to a category of mobility behavior: highly mobile, mobile, slightly mobile, and rarely mobile. The pause time for each group is summarized in Table 5.2.

With this configuration, we succeeded the behavior observed in mimicking Infocom06 trace where a large percentage of nodes have short contacts and a few nodes

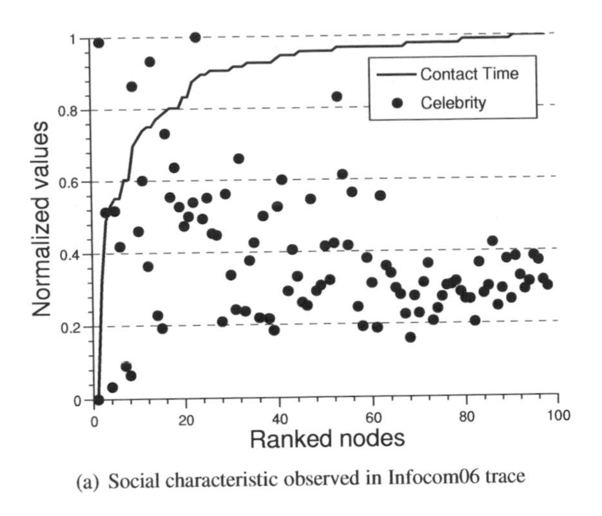

(a) Social characteristic observed in Infocom06 trace

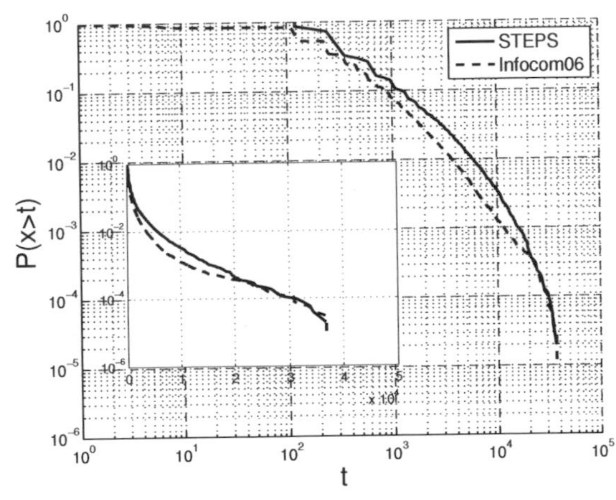

(b) CCDF of contact time of STEPS vs Infocom06 trace

Figure 5.4: Contact time behavior of STEPS vs. real trace.

have long to very long contacts. The CCDFs of the contact time of STEPS and Infocom06 trace, as shown in Figure 5.4(b), show that STEPS can also capture with a good accuracy this mobility behavior.

Table 5.2 Node Categories for the Contact-Time Measure

Mobility degree	RWP pause time range (s)	Number of nodes
Very high	[0, 60]	65
High	[60, 900]	15
Low	[900, 3600]	10
Very low	[3600, 43200]	8

5.3 Small-World Structure of Opportunistic Networks

The small-world phenomenon first introduced in the context of static graphs consists of graphs with high clustering coefficient and low shortest path length. This is an intrinsic property of many real complex static networks. Recent research has shown that this structure is also observable in dynamic networks but how it emerges remains an open problem. In this chapter, we first formalize the metrics to qualify the small-world structure and then investigate in-depth various real traces to highlight the phenomenon. With the help of STEPS, we point out that highly mobile nodes are at the origin of the emergence of the small-world structure in dynamic networks. Finally, we study information diffusion in such small-world networks. Analytical and simulation results with the epidemic model show that the small-world structure significantly increases the information spreading speed in dynamic networks.

5.3.1 Introduction

In his famous experiment, Milgram showed that the human acquaintance network has a diameter on the order of six, leading to the small-world qualification. Watts and Strogatz later introduced a model of small-world phenomenon for static graphs [30]. They proposed a random rewiring procedure for interpolating between a regular ring lattice and a random graph. Between these two extrema, the graph exhibits an exponential decay of the average shortest path length in contrast with a slow decay of the average clustering coefficient. Interestingly, numerous real static networks exhibit such property. From a communication perspective, Watts and Strogatz also pointed out that a small-world network behaves as a random network in terms of information diffusion capacity.

However, the great majority of studies on small-world network properties and behaviors focused on static graphs and ignored the dynamics of real mobile networks. For example, in a static graph, an epidemic cannot break out if the initial infected node is in a disconnected component of the network; conversely in a mobile network, node movements can ensure the temporal connectivity of the underlying dynamic graph. Moreover, an epidemic can take off or die out depending not only on the network structure and the initial carrier, but also on the time when the disease begins to spread. These aspects cannot be captured by a static small-world model.

In order to formalize the dynamic small-world phenomenon, we extend to dynamic networks the notions of dynamic clustering coefficient and shortest dynamic path. By studying the evolution of these metrics from our rewiring process, we show the emergence of a class of dynamic small-world networks with high dynamic clustering coefficients and low shortest dynamic path length. We then demonstrate the capacity of STEPS to capture these structural characteristics of mobile opportunistic networks by showing that, when progressively increasing the dynamicity of a network from an initial stable state, this rewiring process entails the emergence of a small-world structure.

5.3.2 Related Works

In [30], Watts and Strogatz introduce a model of the small-world phenomenon in static graphs. From a regular ring lattice, they randomly rewire edges in this graph with a probability varying from 0 (i.e., leading to a regular network) to 1 (i.e., leading a random graph). During this process they observed an abrupt decrease of the average shortest path length, leading to a short path of the same order of magnitude as observed in random graphs, while the clustering coefficient is still of the same order of magnitude as the one of a regular graph. These features suggested the emergence of the small-world phenomenon. The authors also demonstrate that this graph structure, observed in many real static networks, allows information to diffuse as fast as in a random graph.

Kleinberg [13] extended the model to 2-D lattices and introduced a new rewiring process. This time, the edges are not uniformly rewired but follow a power law $\frac{1}{d^{\alpha}}$ where d is the distance on the lattice from the starting node of the edge and α is the parameter of the model. Newman [18] proposed another definition of the clustering coefficient which has a simple interpretation and is easier to process. The authors argue that the definition in [30] favors vertices with low degree and introduces a bias toward networks with a significant number of such vertices.

Although research pays much attention to the small-world phenomenon in static networks, either through modeling or analytical analysis, the small-world phenomenon in dynamic networks like opportunistic networks is still not well understood. This is partly due to the lack of models and metrics for dynamic graphs. Recently, J. Tang et al. [27] defined several metrics for time-varying graphs, including temporal path length, temporal clustering coefficient, and temporal efficiency. They showed that these metrics are useful to capture temporal characteristics of dynamic networks that cannot be captured by traditional static graph metrics. The definition of dynamic path, introduced in this paper, is close to their definition (we had the dual metric of the number of hops). We also introduce a new definition of dynamic clustering coefficient which captures more accurately the dynamics of opportunistic networks.

In [28], the authors highlight the existence of the small-world behavior in real dynamic network traces. Using the definition of temporal correlation introduced in [5], they show that real dynamic networks have a high temporal correlation and low temporal shortest path, suggesting a dynamic small-world structure. In this modeling work, we consistently extend to dynamic networks the initial small-world metrics defined in [30] (i.e., shortest path length and clustering coefficient).

5.3.3 Small-World Phenomenon in Opportunistic Networks

In this section, we formalize the notion of the small-world phenomenon in opportunistic networks by introducing two metrics used for qualifying such phenomenon: shortest dynamic path length and dynamic clustering coefficient. Then, by extensively analyzing real opportunistic network traces, we show fundamental characteristics which are at the origin of the dynamic small-world phenomenon.

5.3.3.1 Dynamic Small-World Metrics

5.3.3.1.1 Shortest Dynamic Path Length

Basically, the shortest path problem in static graphs consists of finding a path such that the sum of the weights of its constituent links is minimized. From a graph theory point of view, a dynamic network can be described by a temporal graph, i.e., a temporal sequence of graphs that describe the discrete evolution of the network according to nodes and links creation and destruction events. A path in a dynamic network can be seen as an ordered set of temporal links that allow a message to be transferred using the store-move-forward paradigm between two nodes. Formally, let l_{ij}^{t} be a link between node i and node j at instant t. A dynamic path from node u to node v from time t_0 to time t is described by a time-ordered set $p_{uv}(t_0,t) = \left\{l_{ui}^{t_0}, l_{ij}^{t_1}, \ldots, l_{wv}^{t}\right\}$ where $t_{k+1} > t_k$. We consider two metrics of dynamic paths:

- Delay: the sum of the inter-contact times between consecutive links, which constitutes the path.
- Number of hops: the number of temporal links, which constitutes the path.

This leads to the following definition of shortest dynamic path length.

Definition 5.1 The shortest dynamic path is the path giving the minimum amount of delay.[4] If there are several paths giving the same delay, then we select the one giving the minimum number of hops. Formally, the shortest dynamic path length between i and j from time t_0 is

$$\mathcal{L}_{ij}^{t_0} = \inf\{t - t_0 | \exists p_{ij}(t_0,t)\} \ . \tag{5.7}$$

The shortest dynamic path length of a network of N nodes from time t_0 is the average of the shortest dynamic path lengths of all pairs of nodes in the network

$$\mathcal{L}^{t_0} = \frac{\sum_{ij} \mathcal{L}_{ij}^{t_0}}{N(N-1)} \ . \tag{5.8}$$

To find the shortest dynamic path length of all pairs of nodes, we propose an algorithm leveraging the following interesting property of an adjacency matrix in static graphs.

The adjacency matrix A is defined as the matrix in which the element $(A)_{i,j} \in \{0,1\}$ at row i and column j denotes the existence of a link between node i and node j. If we process the power n of such matrix, then its element $(A^n)_{i,j}$ gives the number of paths of length n between i and j. Indeed, for example, when $n = 2$, $(A^2)_{i,j} = \sum_k (A)_{i,k} \times (A)_{k,j}$ sums all the possibilities to go from i to j through an intermediate

[4] We can also have another definition for minimizing the number of hops. In this work, we focus only on delay constrained path.

node k. We extend this property to dynamic networks. Let $A_t, t = 0, 1, \ldots, n$ be the adjacency matrix of a dynamic network at time t. The matrix C_t obtained as follows

$$C_t = \mathbf{A}_t \vee \mathbf{A}_t^2 \vee \ldots \vee \mathbf{A}_t^n \ , \tag{5.9}$$

where $\mathbf{A}_t^i$ denotes the binary version of the matrix A_t^i (i.e. the element $(\mathbf{A}_t)_{i,j}$ equals 1 if $(A_t)_{i,j} > 0$ and 0 otherwise) has its elements $(C_t)_{i,j}$ which indicate if there is a direct or indirect link (up to n hops) between i and j at time t. Indeed, $(C_t)_{i,j}$ is the logical sum of all possibilities to have a direct or indirect link (up to n hops) between i and j at time t. In consequence, the product

$$D_t = C_0 C_1 \ldots C_t \tag{5.10}$$

results in a matrix in which the element $(D_t)_{i,j}$ specifies, when not null, that there is dynamic path of delay t between i and j. Therefore, the shortest dynamic path length from node i to node j is given by the smallest value of t such that $(D_t)_{i,j}$ equals to 1.

It is straight-forward to demonstrate that if a node k belongs to a shortest path between node i and node j, then the i to k sub-path gives the shortest path between i and k. Therefore, the shortest path between two nodes i and j can be easily backwardly reconstructed. Note that at time t two nodes can be connected to each other via a multiple-hop path. To find the complete spatio-temporal path with all the intermediate nodes, we simply apply a breadth-first search each time we find a spatial multiple-hop link. In practice, as it's unlikely to have a large number of nodes connected to each other at a given moment, we can optimize the algorithm by limiting the number of iterations n in Equation 5.9 to an upper bound of the network diameter. Finally, this algorithm is more efficient (time complexity $O(n^3)$) than the depth-first search approach as proposed in [27] (time complexity $O(n^4)$).

We provide a MATLAB implementation of this algorithm in [19].

5.3.3.1.2 Dynamic Clustering Coefficient

As defined in [30], the clustering coefficient measures the cliquishness of a typical friend circle. The clustering coefficient of a node is calculated as the fraction of actual existing links between its neighbors and the number of possible links between them. The clustering coefficient of a network is calculated by averaging the clustering coefficients of all the nodes in the network. In [18], Newman defines the clustering coefficient in terms of transitivity. The connection between nodes u, v, w is said to be transitive if u connected to v and v connected to w implies that u is connected to w. The clustering coefficient of a network is then calculated as the fraction of the number of closed paths of length two over the number of paths of length two, where a path of length two is said to be closed if it is a transitive path. This definition is simple to interpret and easy to calculate. Considering that the initial definition gives more weight to nodes with low degree and introduces a bias toward a graph composed of several of these nodes, in this work, we favor Newman's definition and extend it to dynamic networks.

In [27], Tang et al. first introduced a generalization of Watts and Strogatz's definition of a temporal graph. The temporal clustering coefficient of a node during a

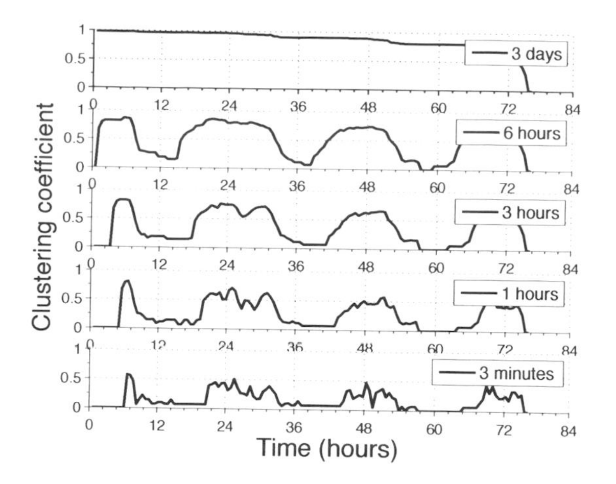

Figure 5.5: Time window size effect on the measure of temporal clustering coefficient (as defined in [11]).

time interval t is the fraction of existing opportunistic contacts (multiple contacts count once) between the neighbors of the node over the number of possible contacts between them during t. The clustering coefficient of the network is the average of the clustering coefficients of all the nodes. This is simply the application of Watts and Strogatz's definition on a time snapshot of the network. While this definition was shown to better capture temporal characteristics of a time-varying graph, it depends strongly on the length of the chosen time interval. If this interval tends to infinity, the temporal clustering coefficient tends to 1 as all the nodes meet each other with a high probability. On the other hand, if the interval tends to 0, then conversely the temporal clustering coefficient tends to 0. Figure 5.5 illustrates the influence of the time window size on measures of the clustering coefficient on a real mobility trace.

In order to avoid this temporal bias, we propose a new definition of the dynamic clustering coefficient which captures the dynamics of the degree of transitivity and is independent of the measuring time interval.

Definition 5.2 A dynamic path from node i to node j is transitive if there exists a node k and time t_1,t_2,t_3 so that i is connected to k at t_1, k is connected to j at t_2, i is connected to j at t_3, and $t_1 \leq t_2 \leq t_3$. The dynamic clustering coefficient of node i from time t_0 is measured by the inverse of the time $t - t_0$, where t is the first instant from time t_0 when the transitive path from i to j is formed, that is, $C_i^{t_0} = \frac{1}{t-t_0}$. The dynamic clustering coefficient of a network of N nodes is then calculated by

Table 5.3 Dataset of Real Opportunistic Network Traces

Trace	Intel	Cambridge	Infocom05	Infocom06
Number of nodes	9	12	41	98
Duration (days)	6	6	4	4
Granularity (seconds)	120	120	120	120
Connectivity	Bluetooth	Bluetooth	Bluetooth	Bluetooth
Environment	Office	Office	Conference	Conference

averaging over the dynamic clustering coefficient of all the nodes from time t_0

$$\mathcal{C} = \frac{1}{N}\sum_i \mathcal{C}_i \ . \tag{5.11}$$

We also provide a MATLAB code for computing the dynamic clustering coefficient in [20].

5.3.3.2 *Opportunistic Network Traces Analysis*

In this section, we extensively analyze real opportunistic network traces to understand how small-world behavior emerges in dynamic networks. We first apply the above definitions to highlight the existence of the dynamic small-world phenomenon on these traces. For that, we use the data sets from the Haggle project [4, 15] which consist of the recording of opportunistic Bluetooth contacts between users in conference or office environments. The settings of these data sets are summarized in Table 5.3.

For each trace, we measure the shortest dynamic path length and the dynamic clustering coefficient every 2000 seconds to see the evolution of these metrics over time. The obtained values are then normalized and plotted in Figure 5.6. We can observe a periodic pattern of both metrics with a typical period of 24 hours and a phase change every 12 hours. This can be easily explained by the fact that human daily activity is periodic with night/day phases. Indeed, people are more nomadic during the day while they are mostly sedentary at night. Besides, it is interesting to note that the dynamic clustering coefficient and the shortest dynamic path length evolve in opposite phases. Despite a slight diversity in different traces due to differences in node number and density, trace durations, etc., during the dynamic phase (i.e., day, for instance, the period around 24 hours in Infocom2005 trace), these networks always exhibit high dynamic clustering coefficients and low shortest dynamic path lengths, suggesting the existence of a dynamic small-world phenomenon.

To explain the emergence of this phenomenon, let us focus on and analyze the structure of these networks during the dynamic phases. In the Watts and Strogatz model, the small-world phenomenon emerges when shortcut edges are randomly added to a regular graph. These shortcuts allow the average shortest path length to be reduced significantly while conserving the network nodes's cliquishness. We argue that in dynamic networks, mobile nodes are implicitly at the origin of these shortcuts. Indeed, it is known that people spend their daily life among different social commu-

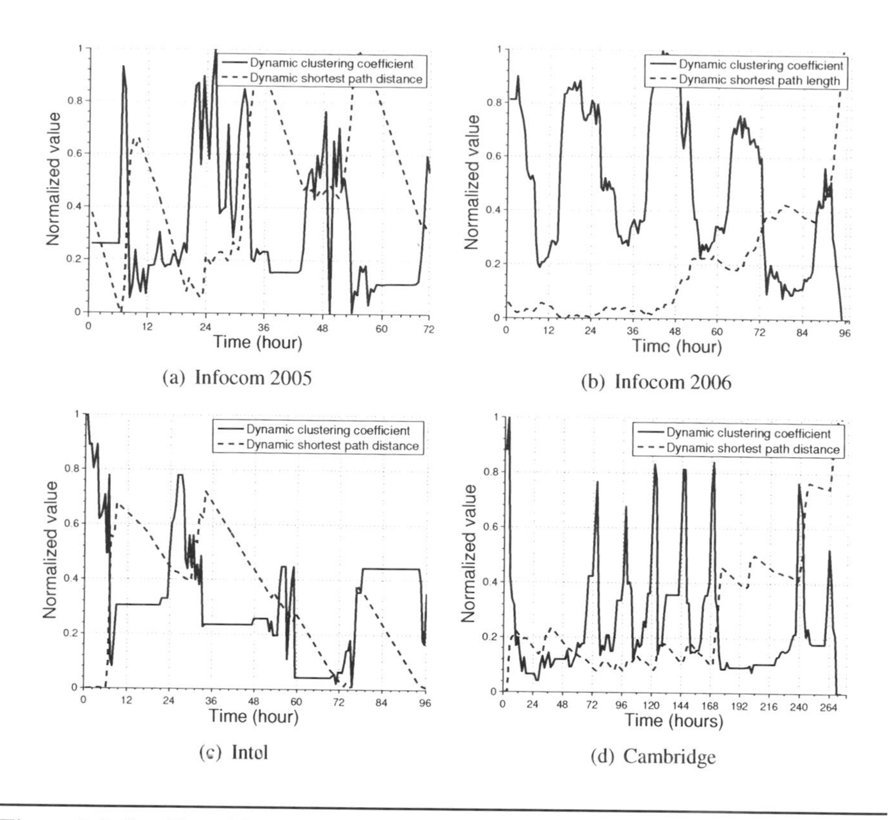

(a) Infocom 2005

(b) Infocom 2006

(c) Intel

(d) Cambridge

Figure 5.6: Small world phenomenon observed in real mobility traces.

nities at specific locations at different times (e.g., colleagues at office in the morning, family at home in the evening). A community can be disconnected from the others in space and/or in time. In addition, some people are more "mobile" than others they have contacts in many communities and move often between these communities or areas. These "nomadic" nodes contribute to reduce significantly the shortest dynamic path length from nodes in a disconnected component to the rest of network and hence contribute to the emergence of the dynamic small-world phenomenon.

To identify the spatio-temporal shortcuts in dynamic networks, we introduce a metric that measures the influence a node has on the characteristic dynamic path length of the network. The nodes with highest influence are the ones whose removal from the network increases the average shortest dynamic path length of the network. To identify these nodes, one may adapt to dynamic networks the notion of betweenness centrality already introduced for static networks. In the context of dynamic networks, we call it *dynamic betweenness centrality*. Consider a node i; first of all, we measure the average of the shortest dynamic path lengths between all pairs of nodes s,t except paths from and to i. Then, we remove i from the network and perform the same measure. The dynamic betweenness centrality of i is defined as the ratio

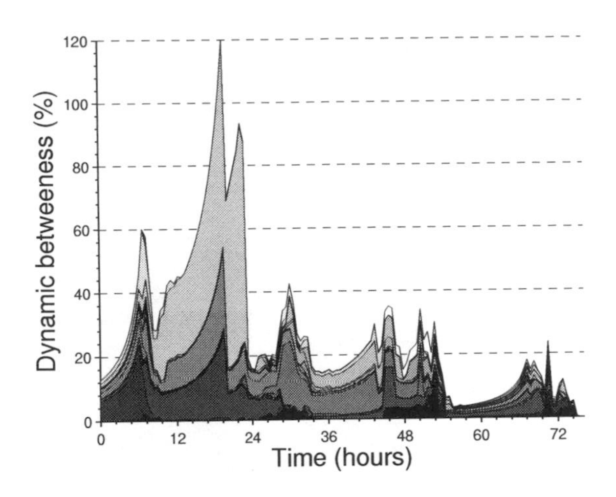

Figure 5.7: Evolution of dynamic betweenness centrality of nodes in Infocom05 trace.

between these two measures. Formally, that is

$$x_i = \frac{\sum_{st} \mathcal{L}'_{st}}{\sum_{st} \mathcal{L}_{st}}, \tag{5.12}$$

where $\mathcal{L}'_{st}$ and $\mathcal{L}_{st}$ are, respectively, the shortest dynamic path lengths from s to t after and before removing i.

We measure the dynamic betweenness of all nodes at different times on the traces. Figure 5.7 shows the temporal evolution of the dynamic betweenness for the Infocom05 trace. On the figure, the different areas represent the measured metric (in %) of different nodes. We can observe that the dynamic betweenness of a node's changes over time and that there are some nodes with very high influence whose removal would dramatically increase the dynamic characteristic path length of the network. Indeed, at the time of the highest pick, redrawing the most influential node would result in a 70% increase of the average shortest dynamic path length of the network. We can also see the abrupt decrease of the average shortest dynamic path length at time 24 hours on Figure 5.6(a). The time average of the dynamic betweenness centrality of each network node allows one to rank the influence of each node during a time window. For instance, in Figure 5.7 the highest value for the daytime period corresponds to a node with ID 34.

Now let us look at the highest pick to see what happened in the network at that moment. We measure the impact of node 34 on the shortest dynamic paths from and to each other node (respectively denoted out/in shortest dynamic path on Figure 5.8(a)). Figure 5.8(a) shows that node 34 has very high influence on node 31. We then apply our all-pairs-shortest-dynamic-path algorithm to find all the shortest dynamic paths from node 31 to others before and after removing node 34. Figure 5.8(b) shows that node 34 is crucial for node 31 to efficiently communicate with other nodes at

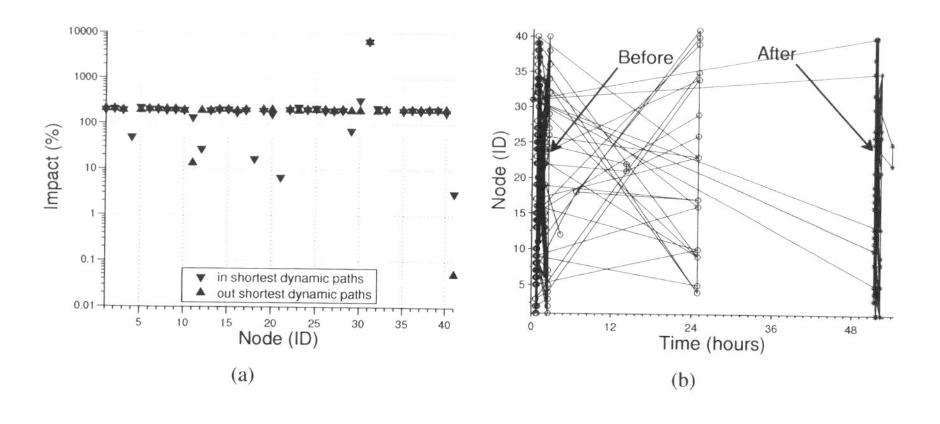

Figure 5.8: (a) Impact of node 34 on the shortest dynamic paths from and to the other nodes at $t = 22$h (Infocom05 trace); (b) shortest dynamic paths from node 31 at $t = 22$h in Infocom05 trace, before and after removal of node 34.

that instant. Therefore, node 34 plays the role of a spatio-temporal shortcut between node 31 and the other nodes. This result can be observed for any node with high dynamic betweenness. Indeed, if the so disconnected network part is a giant component, then the impact of the removal of a dynamic shortcut can dramatically increase the characteristic dynamic path length of the network.

5.3.4 Modeling Dynamic Small-World Structure with STEPS

The previous section showed that the dynamic small-world phenomenon is intrinsic to a great diversity of opportunistic network traces. Therefore, the definition of a model that abstracts this phenomenon in order to study the impact of its emergence on dynamic network communications performances is a significant issue. In this section, we show the modeling and expressive capacity of STEPS for capturing the dynamic small-world structure in opportunistic networks. As evoked previously, the dynamic small-world phenomenon results from human mobility behaviors:

- People move between communities.
- Some people are nomadic while the others are more sedentary. These nomadic nodes move from zone to zone, contributing to reduce the network diameter.

Consider an opportunistic network modeled by STEPS in which each node belongs to a preferential attachment zone. We introduce two types of nodes: sedentary nodes, which move only inside their preferential zone, and nomadic nodes, which can move between zones. Nodes moving in two different zones cannot be connected and therefore nodes in different zones can communicate only through the movement

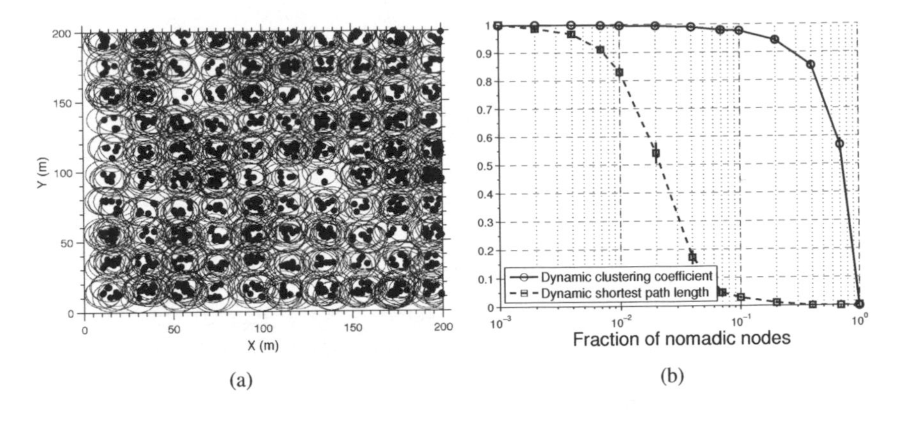

Figure 5.9: (a) Small-world network configuration; (b) small-world phenomenon in dynamic networks.

of nomadic nodes. Initially, nomadic nodes are uniformly distributed over all zones. We introduce another parameter of STEPS p, which is the fraction of nomadic nodes.

Obviously, when p equals 0, there is no possible communication between nodes in different zones, and hence the network is totally partitioned in disconnected zones. On the contrary, when p equals 1, all the nodes are nomadic and hence there are dynamic paths between clusters and in consequence the network is highly connected. In consequence, the *rewiring process* consists of varying p from 0 to 1. We are interested in the properties of networks which are formed between these two extrema.

We simulate in MATLAB an opportunistic network of 1000 nodes. The simulated network area of size 200×200 m^2 divided into 20×20 zones as shown in Figure 5.9(a) gives a density of 25000 nodes/km^2, which corresponds to the population density of a large city like Paris. We uniformly distribute nodes over zones and set the radio range to 10 m, which corresponds to Bluetooth technology. We set a very high value of the power law exponent for sedentary nodes and 0 for nomadic nodes. While varying the fraction of nomadic nodes from 0 to 1, we measure the dynamic clustering coefficient and shortest dynamic path length of resulting networks.

Figure 14.7 plots the resulting average over 10 simulations with the corresponding 95% confidence intervals. We observe that the shortest dynamic path length drops rapidly as soon as we introduce a small percentage of nomadic nodes (i.e., less than 10%) into the network while the clustering coefficient remains very high. This result is consistent with the original static small-world model described in [30] and shows how the small-world phenomenon can emerge in opportunistic networks.

5.3.5 Information Diffusion in Dynamic Small-World Networks

Following the understanding and modeling of the small-world phenomenon in dynamic networks, this section studies the effect of the phenomenon on information diffusion performance. We present here the analytical and simulation results obtained with the Susceptible Infected (SI) epidemic model applied to the small-world model. We use a simplified version of the dynamic small-world model in which nodes' displacement inside sites is abstracted and nodes' movements are limited to jumping from one site to another. Simulation results for the complete version of the model are also provided.

Consider an N-sites square lattice with initially 1 node in each site. Node i is associated to site number i. Assume that when 2 nodes are in the same site, they are connected. At each time t, mobile nodes jump to another site while sedentary nodes stay where they are. Initially, the network has 1 infected node and all other nodes are susceptible. We formulate the dynamics of the network's infection through a differential equation as follows.

Consider a node i. Let $x_i(t)$ be the random variable that denotes the probability of node i being infected at time t. Let's express the probability that i becomes infected between time t and $t+dt$. As a static node i can only be infected by a mobile node m, this infection occurs with probability δ_{mi} (i.e., probability that node m jumps to site i). As a consequence, the probability that the infection (or information) is transmitted during the interval dt is βdt where β is the transmission rate, which is a standard parameter of the SI epidemic model. Summing over all the mobile nodes and then multiplying by the probability $1-x_i$ that i is not infected at time t, we obtain the differential equation

$$\frac{dx_i}{dt} = \beta(1-x_i)\sum_m \delta_{mi} x_m \ . \tag{5.13}$$

On the other hand, for a mobile node to be infected, the node must be in contact with a mobile or static node that is already infected. If a mobile node i jumps to a site j where a static node resides, the infection occurs with probability $x_j + \sum_{m \neq i} \delta_{mj} x_m$ because i can receive the information from either j or another mobile node. We obtain

$$\frac{dx_i}{dt} = \beta(1-x_i)(x_j + \sum_{m \neq i} \delta_{mj} x_m) \ . \tag{5.14}$$

On the contrary, if the destination site j is associated with a mobile node, then the probability of infection is $\sum_{m \neq i} \delta_{mj} x_m$ and hence

$$\frac{dx_i}{dt} = \beta(1-x_i)\sum_{m \neq i} \delta_{mj} x_m \ . \tag{5.15}$$

Combining all these cases, we obtain the following matrix differential equation describing the dynamics of the system

$$\begin{aligned} \frac{d\mathbf{x}}{dt} =& \beta(1-\mathbf{x}) \circ \langle (1-m) \circ [\mathbf{D}(m \circ \mathbf{x})] + \\ &+ m \circ \{\mathbf{D}[(1-m) \circ \mathbf{x} + \mathbf{D}(m \circ \mathbf{x})]\}\rangle \ , \end{aligned} \tag{5.16}$$

where $\mathbf{x} = [x_1 \ldots x_N]'$ is the random vector containing node infection probabilities

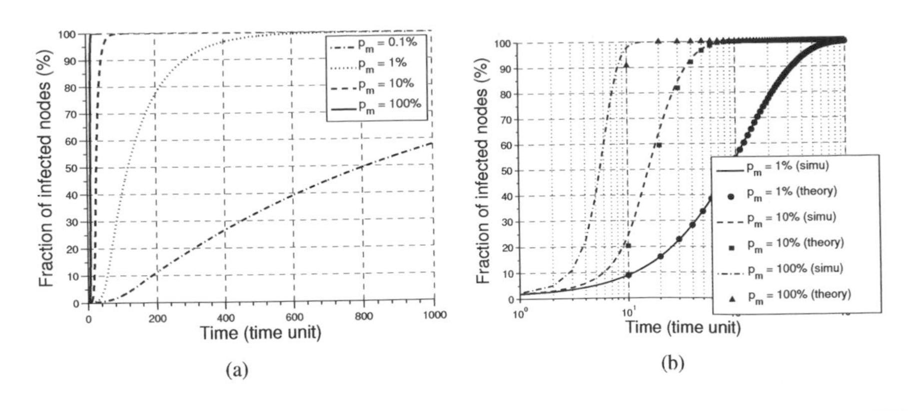

Figure 5.10: (a) Evolution of fraction of infected nodes over time (theory). (b) Comparison of analytical and simulation results.

and m is the binary vector in which $m_i = 1$ for a mobile node and $m_i = 0$ for a static node. The matrix

$$\mathbf{D} = \begin{bmatrix} \delta_{11} & \dots & \delta_{1N} \\ \vdots & \ddots & \vdots \\ \delta_{N1} & \dots & \delta_{NN} \end{bmatrix} = \lambda \begin{bmatrix} 1 & \dots & \frac{1}{(1+d_{1N})^{\alpha}} \\ \vdots & \ddots & \vdots \\ \frac{1}{(1+d_{N1})^{\alpha}} & \dots & 1 \end{bmatrix}$$

is the stochastic matrix describing the stationary state of the system.

To have an approximated solution of Equation 5.16, we integrate it by numerical method. Figure 5.10(a) shows the fraction of infected nodes as a function of time for a 10000-node network. The 4 curves correspond to 4 values of p_m, which are, respectively $0.1\%, 1\%, 10\%$, and 100%. The STEPS parameter α for mobile nodes is set to 0 and the transmission rate β is set to 1. The results show that, for a low fraction of mobile nodes, epidemic spreads very slowly. As soon as the small-world structure emerges, i.e., with 10% mobile nodes, the epidemic breaks out rapidly.

We also ran simulations with a network of 100 nodes, with α equal to 0, β equal to 1, and different values for the fraction of mobile nodes. The average over 100 simulations is then compared with the analytical results. Figure 5.10(b) shows the strong accuracy and compliance of our analytical model with the simulation results.

Finally, we run a simulation with the complete version of the small-world model. The obtained result is the average over 10 simulations of a 100-node network with the same configurations described above. Figure 5.11 shows a similar result compared to the simplified case. The effect of displacement time between zones only spreads the curves so that the epidemic takes much longer to take off.

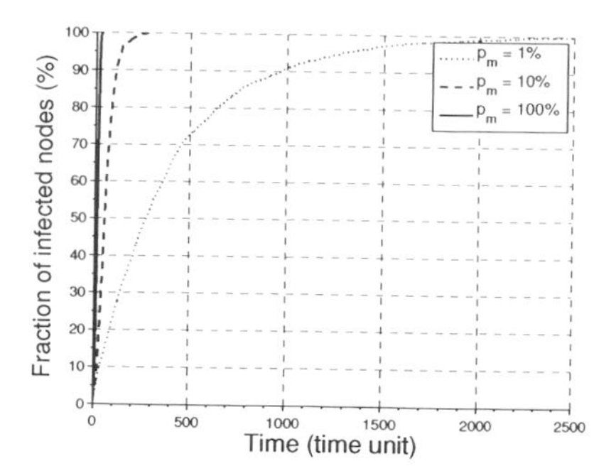

Figure 5.11: Evolution of fraction of infected nodes over time with the complete version of the model (simulation result).

5.4 Mobile Cloud Opportunistic Networks

In this section, we address an important and still unanswered issue in mobile cloud computing—how mobility impacts the distributed processing power of network and computing clouds formed from mobile opportunistic networks. Indeed, mobile opportunistic networks potentially offer an aggregate cloud of resources delivering collectively processing, storage and networking resources. We demonstrate that mobility can significantly increase the performance of distributed computation in such networks. In particular, we show that this improvement can be achieved more efficiently with mobility patterns that entail a dynamic small-world network structure on mobile clouds. Moreover, we show that the dynamic small-world structure can significantly improve the resilience of mobile cloud computing services.

5.4.1 Related Works

Mobile cloud computing is still a young field and there is still discussion on its definition. In its infancy, mobile cloud computing has been considered as a derived branch of cloud computing with two schools of thought (see [31] for a survey). The first refers to performing computing activities (i.e., data storage and processing) in an infrastructured cloud and let mobile devices be simple terminals to access service. This centralized approach has the advantage that mobile devices don't need to have a powerful computing capacity, but the drawback is that users depend strongly on the infrastructure network and on its performance.

The second school of thought defines mobile cloud computing as performing computing activities on a mobile platform. Therefore, a mobile cloud network is an infrastructureless extension of the traditional infrastructure-based cloud network. Mobile devices are clients of service but are also part of the cloud, providing hardware and software resources. The benefit of this distributed approach is the omnipresence and the speed of service accessibility, the support of mobility and locality, the freedom of deployment, and the use of new services as well as reduced hardware maintenance costs. Although the approach is promising, its main challenge resides in the network dynamics which poses difficulties in communication and hence service access. In this chapter, we focus on this definition of mobile cloud computing.

To the best of our knowledge, very few contributions have been proposed for mobile cloud computing. Hyrax [16] is a mobile-cloud infrastructure that enables smartphone applications that are distributed both in terms of data and computation. Hyrax allows applications to conveniently use data and execute computing jobs on smartphone networks and heterogeneous networks of phones and servers. Its implementation is based on Hadoop and tested on the Android platform. But since Android doesn't support ad hoc networks yet, the phones have to communicate through a WIFI central router.

Satyanarayanan et al. [25] present the cloudlet concept. In this approach, a mobile client is seen as a thin client with respect to a service which is customized over a virtual machine in the wireless LAN. Hence the cloudlet is a proxy representation of a real service enhanced for the mobile device. The main motivation is how bandwidth limits and latency over wireless networks impact user services.

5.4.2 Impact of Mobility on Mobile Cloud Computing

In this section, we evaluate the impact of mobility on mobile cloud computing. Let us consider that a mobile cloud network created by several human portable devices offers a distributed processing service such as optimizing a function via a Particle Swarm Optimization (PSO) algorithm. According to the PSO, each node in the network has a local solution to the optimization problem. Through intermittent contacts, mobile nodes learn others' solutions to improve their local optimum and hence accelerate the convergence toward the global optimum. For the sake of simplicity, we make the following assumptions:

1. Each node in the network knows the goal function and its solution in advance.
2. An external system (e.g., WIFI hot-spots) is responsible for results retrieval from mobile nodes.
3. The service is considered delivered when the global optimum reaches a goodness predefined by the user.

In practice, the goal function as well as its solution is usually unknown in advance and therefore we have to rely on a diffusion technique to disseminate the information of the goal function into the network. The obtained global optimum and stopping condition in that case will depend on the current solutions found by nodes (e.g., a node stops the computation when its local solution no longer changes for a while). In this theoretical work, we focus only on the impact of mobility on computing delay and therefore the previous assumptions seem reasonable.

5.4.2.1 Mobility Model

In order to reproduce at the simulation level realistic human mobility patterns, we use the STEPS mobility model introduced in Section 5.2. As we have shown in Section 5.2, this flexible parametric model can express a large spectrum of mobility patterns from highly nomadic ones to localized ones. Therefore, STEPS makes it possible to evaluate the impact of different mobility contexts on mobile cloud computing. In this evaluation, the network area is modeled as a torus divided into several zones. STEPS implements the notion of preferential attachment usually observed in human mobility in which each node is attached to one or several preferential zones. Inside zones, mobile nodes move according to the Random Waypoint model. The movement of nodes between zones follows a Markov chain of which the transition probability is given by a power law distribution. This distribution is driven by a parameter of the STEPS model which allows the nodes' nomadism to be enforced or reduced (i.e., the probability that a node moving outside its preferential zone has to return to that zone).

5.4.2.2 Particle Swarm Optimization

The Particle Swarm Optimization (PSO) algorithm [12] is an optimization method based on swarm intelligence—a sub-field of artificial intelligence which studies the collective intelligent behavior emerging from the interactions between individuals of a swarm of autonomous agents. Swarm intelligence considers intelligence as the combination of the knowledge acquired by individuals through experiences in the past and the knowledge acquired from the others through social interactions. In the PSO algorithm, a set of candidate solutions called particles move around in the search space according to a simple mathematical formula that involves the particle's position and velocity. Each particle's movement is influenced by its local best-known position and also by the global best-known position found by other particles. The swarm is expected to move collectively toward the optimal solution. Moreover, this method is able to solve a multimodal optimization problem.

In its simplest form, let $\vec{x}_i$ be the multidimensional vector of the particle i position; the position of the particle is updated according to the formula

$$\vec{x}_i(t+1) = \vec{x}_i(t) + \vec{v}_i(t), \tag{5.17}$$

where $\vec{x}_i(t)$ is the position of particle i at time t.

The velocity of the particle is updated according to the formula

$$\vec{v}_i(t) = \vec{v}_i(t-1) + \phi_1 \left[\vec{p}_i - \vec{x}_i(t-1)\right] + \phi_2 \left[\vec{p}_g - \vec{x}_i(t-1)\right], \tag{5.18}$$

where

- ϕ_1, ϕ_2 are uniform random variables taking values in $[0,1]$. These variables represent the relativity between the effect of individual experience and of social influence.
- $\vec{p}_l$ denotes the best-known position of particle i ("l" for local).

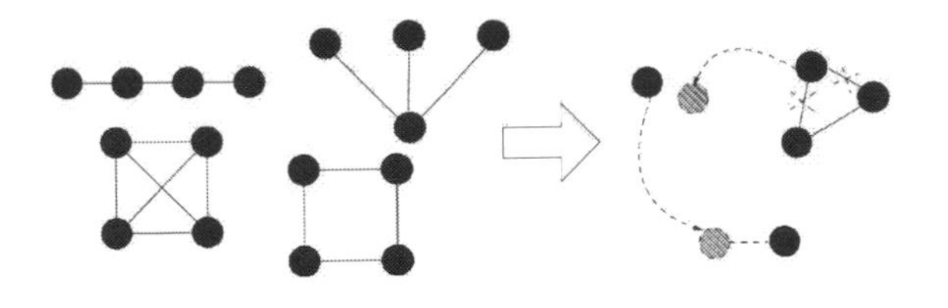

Figure 5.12: Typical static neighborhood topologies vs. dynamic neighborhood topology generated by the mobility model.

- $\vec{p}_g$ denotes the best-known position of i's neighbors ("g" for global).

This formula entails wider and wider oscillations of particles in the search space. One solution to this issue is based on velocity damping, that is, if $v_{id} > V_{max}$ then $v_{id} = V_{max}$, else if $v_{id} < -V_{max}$, then $v_{id} = -V_{max}$ where v_{id} is the dimension d of $\vec{v}_i$. In consequence, the particles move only in a restricted search space.

In PSO, agents can be connected to each other according to a great number of neighborhood topologies. Figure 5.12 illustrates the most used schemes. Each neighborhood topology, traditionally considered as static as opposed to our analysis, results in different behaviors and performances for the PSO algorithm.

In this work, since nodes move, the neighborhood topology is no longer static but dynamic. Indeed, when the mobility degree is low, links between nodes are stable and the network is nearly static. On the contrary, when the mobility is high, links change rapidly over time and so does the neighborhood topology. Therefore, the goal of this study is to evaluate the effect of mobility on the convergence delay of the algorithm.

5.4.2.3 Simulation Results and Discussion

We implemented the mobility model and the PSO algorithm on MATLAB. At the beginning of each simulation, 100 nodes are uniformly distributed over the network area, which is divided in 10×10 zones representing preferential attachment according to the STEPS model. The movement of nodes between zones is driven by the locality degree parameter α of the STEPS model. We vary α between 0 and 8 to obtain a large spectrum of mobility patterns. When $\alpha = 0$, nodes are highly nomadic, moving from one zone to another in a random manner that makes the network highly dynamic. On the contrary, when $\alpha = 8$, nodes are highly localized (i.e., sedentary) and therefore there is less information exchange between distant zones.

The PSO algorithm is implemented in every mobile node so that each node contains 1 particle. The position of the particle is randomly initialized, taking values in the range $[-x_{max}, x_{max}]$. The particle's position is updated at each contact with another node according to formula 5.18.

As the goal function, we used the Sphere function from the De-Jong test suite [6]. This suite consists of goal functions with different difficulties to measure the

Table 5.4 Simulation Settings

Characteristic	Value
Number of nodes/particles	100
Number of zones	10×10
Network size	100×100 m^2
Radio range	10 m
Node speed	$3 - 5$ km/h
De Jong function	Sphere
Number of dimensions	2
Stopping condition	error $<= 10^{-6}$

performances of optimizers. The Sphere function is the first and easiest function of the suite. It is symmetric, unimodal, and is often used to measure the general efficiency of optimizers. The sphere function is defined as

$$f(\vec{x}) = \sum_{i=1}^{D} {x_i}^2,$$

where D is the number of components of $\vec{x}$. The Sphere function has a global optimum $f(\vec{x}) = 0$ at $\vec{x} = (0,0,0,\ldots,0)$.

We used root-mean-square error to measure the goodness of the solution. The algorithm stops when the error is smaller than a predefined threshold. The simulation settings are summarized in Table 5.4

Figure 5.13 shows the optimization convergence delay according to nodes' locality degree. These results are averaged over 10 simulation runs. In the figure, we

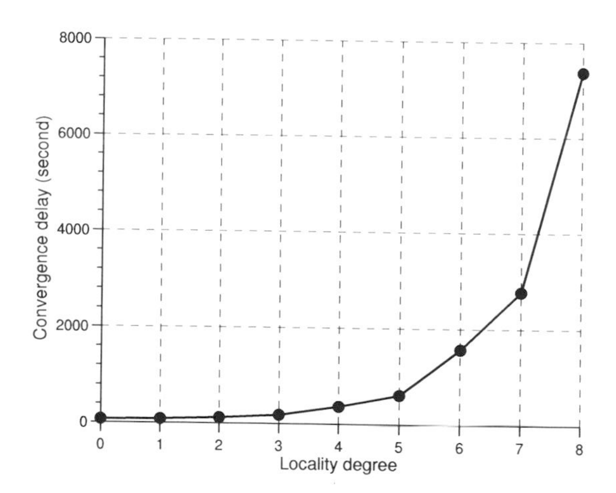

Figure 5.13: Impact of mobility on the convergence delay of the PSO algorithm.

can see that the more mobile nodes are, the smaller the convergence delay is. This result shows that node mobility can dramatically increase the processing capacity of mobile cloud networks.

5.4.3 Impact of Network Structure on Mobile Cloud Computing

In this section, we evaluate the processing capacity of mobile cloud computing under various dynamic network structures. With the same approach as introduced in Section 5.4.2, we measure the convergence delay of a PSO algorithm implemented on a mobile cloud network and show that this delay can be significantly minimized if the network has a dynamic small-world structure.

The small-world phenomenon introduced by Watts and Strogatz [30] refers to static graphs with a high clustering coefficient and low shortest path length. Through a process which consists of randomly rewiring edges of a graph, by varying the rewiring ratio, the authors showed that for an interval, a rewiring rations the resulting static graph, exhibits a small-world structure that combines the short path observed in random graphs with the high clustering coefficient intrinsic to regular lattices. In Section 5.3, we have shown that this small-world behavior can be observed in dynamic graphs too. We have shown that in dynamic networks, the analog of the rewiring process in static graph is done by varying the ratio and intensity of nomadic nodes. Moreover, we have shown that the STEPS model is capable of exhibiting this small-world phenomenon in dynamic networks.

Indeed, starting from the same network configuration as in Section 16.3.1, we divide mobile nodes in 2 categories. The first consists of highly localized nodes which stay in their preferential zones almost all the time. The second category consists in highly nomadic modes which move constantly from zone to zone. At the beginning of simulation, the nodes are distributed over the network area so that nodes in different preferential zones cannot communicate with each other. We vary the fraction of mobile node p_m from 0 to 1. When p_m equals 0, the network consists of disconnected islands with only intra-zone communications that entails a regular structure similar to the one in static graphs. On the contrary, when p_m equals 1, the inter-zone movement of highly mobile nodes cause the network topology to change constantly, which entails a random network structure.

We processed the PSO algorithm over all these network structures and then measured the resulting convergence delay. Figure 5.14 shows the results averaged over 10 simulations. These simulation results show that the convergence delay of PSO decreases rapidly down to an asymptotic part started when the network exhibits a small world structure. This original result is significant because the small-world structure, which as shown in this chapter improves distributed processing, was shown to emerge naturally in the great majority of real dynamic networks, as underlined in Section 5.3.

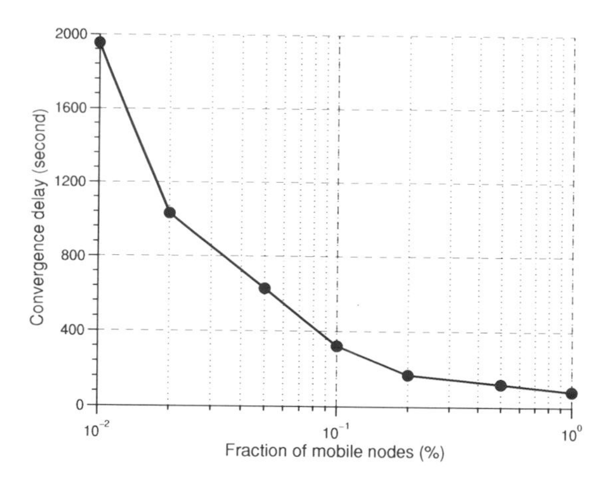

Figure 5.14: Mobile cloud computing in small-world networks.

5.4.4 Resilience of Mobile Cloud Computing Service

Nowadays, mobile devices still have limited energy capacity, and communication and processing are two important sources of energy waste. Therefore, node churn is intrinsic to dynamic network clouds. Nodes running out of battery cannot contribute to distributed processing anymore and in consequence, mobile cloud networks may suffer unpredictable nodes failures. Besides, mobile cloud networks may be the target of attacks, for instance DDOS, which can potentially make unavailable parts of the network. In this section, we evaluate, under various mobility contexts, the resilience of distributed services deployed on such networks.

First, we assume that the evolution of the number of inactive nodes (i.e., attacked or out of battery) follows a Poisson process. Therefore, the number of inactive nodes during a time interval τ is distributed according to a Poisson distribution

$$P[(N(t+\tau)-N(t))=k]=\frac{\exp -\lambda\tau(\lambda\tau)^k}{k!},$$

where $k=0,1,2,\ldots$ and λ is the arrival rate of inactive nodes.

With the same simulation settings as introduced in Section 5.4.3, we perform simulations with various values of λ $(1/15, 1/12.5, 1/10, 1/5)$ and under various mobility contexts (i.e., by varying the fraction of mobile nodes). In these simulations, we stop the PSO algorithm when 95% of the nodes reach the optimum. If this threshold is not reached before all the nodes become inactive, as there is no recovery possible in this case, the service will never be delivered and hence we assign the simulation duration time to the convergence delay.

Figure 5.15 shows the results averaged over 20 simulations. These simulations show that with dynamic small-world networks (Figure 5.15(b) and 5.15(c)), the dis-

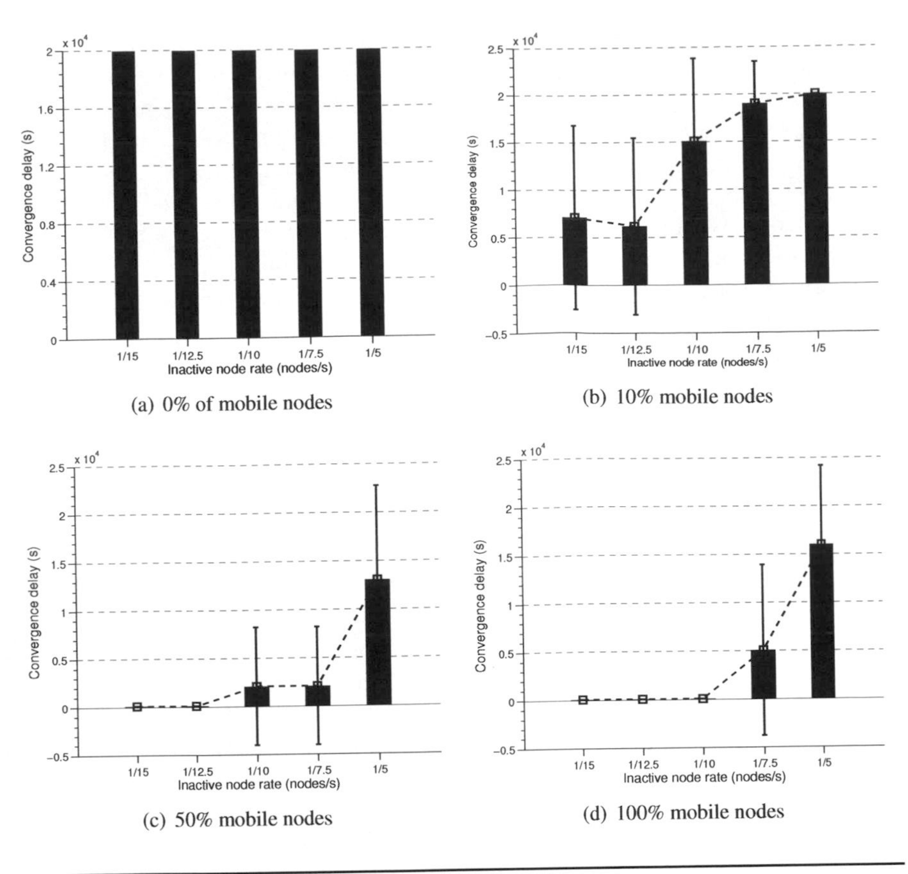

Figure 5.15: Resilience of mobile cloud network distributed services under various mobility contexts.

tributed service is much more resistant to departed nodes compared to a highly localized network (Figure 5.15(a)) and offers approximately the same resilience level as random networks (Figure 5.15(d)). These results suggest that a small-world structure not only contributes to enhancing distributed performances but also offers good resilience properties.

5.5 Conclusion

Throughout this chapter, we have presented an in-depth analysis of the nature of dynamic networks through the modeling, structural analysis, and mobile cloud computing performance analysis of such networks. We first proposed STEPS—a novel human mobility model for dynamic networks. In this model, two intrinsic properties of human mobility are implemented: preferential attachment, i.e., people have very high probability to visit few specific locations, and attraction, i.e., they have high probability to return to these locations when moving outside. We define the network as a torus divided into square zones. Inside each zone, nodes move according to a random mobility model. The movement between zones is ruled by a power law such that the probability for a node to select a new zone is inversely proportional to the distance between that zone and the preferential zone of the node. By tuning the power law exponent, we vary the nomadic degree of nodes. We show that this simple parametric model can capture characteristics of human mobility usually observed in real traces. For instance, the contact/inter-contact time distribution follows a power law.

Based on this model, we then study the structure of dynamic networks. We show that dynamic networks exhibit a small-world structure in which nodes are highly clustered while the network diameter is very small. We extend the notion of clustering coefficient and shortest path length in static networks to dynamic networks and use these metrics to investigate real traces. The results show that human contact networks exhibit a high dynamic clustering coefficient and low shortest dynamic path length during day phases, while behaving in the opposite manner in night phases, suggesting the small-world phenomenon in dynamic phases. We demonstrated that highly dynamic nodes are responsible for this phenomenon. Indeed, these nodes play the role of bridges between disconnected components of dynamic networks, hence contributing to reduce significantly the diameter of the network. We model this behavior with STEPS. Starting from a highly localized network where nodes tend to stay in their preferential locations, we tuned the ratio of nomadic nodes and measured the two metrics. As soon as we have at least 10% of such nodes, the shortest dynamic path length drops as low as in a random network, while the dynamic clustering coefficient still remains as high as in a fully localized network. We then showed that this structure allows information to spread as fast as in a fully dynamic network. These results mean that, to boost information dissemination speed in a dynamic network, it is enough to inject only 10% dynamic nodes into the network.

Finally, by leveraging these behavioral and structural properties of opportunistic networks, we not only showed that nodes' mobility enhances the processing capacity of opportunistic network cloud computing, but we also show how mobility impacts the performance and the resilience of these mobile clouds. In particular, we have shown that significant performance improvement can be obtained when opportunistic networks exhibit a small-world structure and moreover, this particular structure can improve the resilience of the network against inactive nodes. This means that by introducing even a small percentage of highly mobile (about 10%) nodes in a high localized network, we can significantly improve the processing capacity and resilience of mobile cloud computing. These results open the way to adaptive strategies that would aim to adapt dynamic network topology and behavior according to their processing load and constraints. Moreover, these strategies have to consider storage and energy consumption, which are critical in the context of handheld systems.

References

[1] D. Brockmann, L. Hufnagel, and T. Geisel. The scaling laws of human travel. *Nature*, 439(7075):462–465, May 2006.

[2] H. Cai and D.Y. Eun. Toward stochastic anatomy of inter-meeting time distribution under general mobility models. In *Proceedings of the 9th ACM International Symposium on Mobile Ad Hoc Networking and Computing*, MobiHoc '08, pages 273–282, New York, New York, USA, 2008. ACM.

[3] T. Camp, J. Boleng, and V. Davies. A survey of mobility models for ad hoc network research. *Wireless Communications and Mobile Computing*, 2(5):483–502, 2002.

[4] A. Chaintreau, P. Hui, J. Crowcroft, C. Diot, R. Gass, and J. Scott. Impact of human mobility on opportunistic forwarding algorithms. *IEEE Transactions on Mobile Computing*, 6(6):606–620, June 2007.

[5] A. Clauset and N. Eagle. Persistence and periodicity in a dynamic proximity network. In *DIMACS/DyDAn Workshop on Computational Methods for Dynamic Interaction Networks*, 2007.

[6] J. De and A. Kenneth. Analysis of the behavior of a class of genetic adaptive systems. PhD thesis, 1975.

[7] F. Ekman, A. Keranen, J. Karvo, and J. Ott. Working day movement model. In *Proceedings of the 1st ACM SIGMOBILE Workshop on Mobility Models*, pages 33–40, New York, New York, USA, May 2008. ACM.

[8] M.C. González, C.A. Hidalgo, and A.L. Barabási. Understanding individual human mobility patterns. *Nature*, 453(7196):779–782, June 2008.

[9] M. Grossglauser and D.N.C. Tse. Mobility increases the capacity of ad hoc wireless networks. *Wireless Networks*, 10(4):477–486, 2002.

[10] W.J. Hsu, T. Spyropoulos, K. Psounis, and A. Helmy. Modeling spatial and temporal dependencies of user mobility in wireless mobile networks. *IEEE/ACM Transactions on, Networking*, 17(5):1564–1577, 2009.

[11] T. Karagiannis, J.Y. Le Boudec, and M. Vojnovic. Power law and exponential decay of inter-contact times between mobile devices. *IEEE Transactions on Mobile Computing*, pages 183–194, 2010.

[12] J. Kennedy and R.C. Eberhart. *Swarm Intelligence*, volume 78. Springer, January 2006.

[13] J.M. Kleinberg. Navigation in a small world. *Nature*, 406(6798):845, August 2000.

[14] K. Lee, S. Hong, S.J. Kim, I. Rhee, and S. Chong. Slaw: A mobility model for human walks. In *Proceedings of the 28th Annual Joint Conference of INFOCOM*, 2009.

[15] J. Leguay, A. Lindgren, J. Scott, T. Friedman, and J. Crowcroft. Opportunistic content distribution in an urban setting. In *Proceedings of the 2006 SIGCOMM Workshop on Challenged Networks—CHANTS '06*, pages 205–212, New York, New York, USA, September 2006. ACM Press.

[16] E.E. Marinelli. *Hyrax: Cloud Computing on Mobile Devices Using MapReduce*. Technical Report September, DTIC Document, 2009.

[17] M. Musolesi and C. Mascolo. A community based mobility model for ad hoc network research. In *Proceedings of the Second International Workshop on Multi-hop Ad Hoc Networks: From Theory to Reality—REALMAN '06*, page 31, New York, New York, USA, May 2006. ACM Press.

[18] M.E.J. Newman. The structure and function of complex networks. *SIAM Review*, 45(2):167, November 2003.

[19] A.D. Nguyen. All-pairs shortest dynamic path length algorithm. In *MATLAB Central File Exchange*. http://www.mathworks.com/matlabcentral/fileexchange/39739

[20] A.D. Nguyen. Dynamic clustering coefficient algorithm. In *MATLAB Central File Exchange*. http://www.mathworks.com/matlabcentral.fileexchange/39740

[21] A.D. Nguyen. STEPS mobility model. In *MATLAB Central File Exchange*. http://www/mathworks.com/matlabcentral/fileexchange/38854-steps-mobility-model

[22] A.D. Nguyen, P. Senac, and M. Diaz. Understanding and modeling the small-world phenomenon in dynamic networks. In *ACM MSWiM 2012*, page 377, New York, New York, USA, October 2012. ACM Press.

[23] A.D. Nguyen, P. Senac, and V. Ramiro. How mobility increases mobile cloud computing processing capacity. In *First International Symposium on Network Cloud Computing and Applications*, pages 50–55. IEEE, November 2011.

[24] A.D. Nguyen, P. Sénac, V. Ramiro, and M. Diaz. STEPS: An approach for human mobility modeling. *IFIP NETWORKING 2011*, pages 254–265, 2011.

[25] M. Satyanarayanan, V. Bahl, R. Caceres, and N. Davies. The case for VM-based cloudlets in mobile computing. *IEEE Pervasive Computing*, 1, 2011.

[26] C. Song, T. Koren, P. Wang, and A.L. Barabási. Modelling the scaling properties of human mobility. *Nature Physics*, 6(10):818–823, September 2010.

[27] J. Tang, M. Musolesi, C. Mascolo, and V. Latora. Temporal distance metrics for social network analysis. In *Proceedings of the 2nd ACM Workshop on Online Social Networks—WOSN '09*, page 31, New York, New York, USA, August 2009. ACM Press.

[28] J. Tang, S. Scellato, M. Musolesi, C. Mascolo, and V. Latora. Small-world behavior in time-varying graphs. *Physical Review E*, 81(5):5, May 2010.

[29] G.S. Thakur, U. Kumar, A. Helmy, and H. Wei-jen. On the efficacy of mobility modeling for DTN evaluation: Analysis of encounter statistics and spatio-temporal preferences. In *Wireless Communications and Mobile Computing Conference (IWCMC), 2011 7th International*, pages 510–515, 2011.

[30] D.J. Watts and S.H. Strogatz. Collective dynamics of small-world networks. *Nature*, 393(6684):440–442, June 1998.

[31] F. Xiaopeng, C. Jiannong, and M. Haixia. A Survey of Mobile Cloud Computing. ZTE Corporation, 2011.

[32] J. Yoon, M. Liu, and B. Noble. Random waypoint considered harmful. In *IEEE Societies INFOCOM 2003. Twenty-Second Annual Joint Conference of the IEEE Computer and Communications*, volume 2, 2003.

Chapter 6

An Overview of Routing Protocols in Mobile Social Contact Networks

Cong Liu
Sun Yat-Sen University
Guangzhou, Guangdong, P.R. China

Chengyin Liu
Sun Yat-Sen University
Guangzhou, Guangdong, P.R. China

Wei Wang
Sun Yat-Sen University
Guangzhou, Guangdong, P.R. China

CONTENTS

6.1 Preliminaries and Network Models

6.1.1 Ad Hoc Wireless Networks

A wide range of novel applications have been enabled with the advance in inexpensive wireless network devices. Portable devices, such as cell phones and laptops, are equipped with wireless modules, which enable the access of information via Internet access points anywhere, anytime. A difficulty in the deployment of wireless networks is the need to have a large number of base stations that cover the whole network area. The energy consumption for data transmission of the battery-powered devices is also another challenge. These problems are addressed by ad hoc wireless networking through allowing mobile nodes to communicate directly with each other without the help of any communication infrastructure. Intermediate nodes between two nodes that are not in direct transmission range of each other can serve as voluntary routers to route messages.

6.1.2 Opportunistic Networks

A common assumption of existing routing protocols in ad hoc wireless network is that, for any pair of source and destination, there is always a contemporary connected path between them. However, this assumption is not always true with the advent of short-range communication protocols, such as ZigBee and Bluetooth. Also, the wide physical range of the network and the high degree of mobility of the nodes make the above assumption further unrealistic. When the network is sparse due to the above reasons, it can be viewed as a set of disconnected clusters of nodes, and these clusters are only time-varying due to the constant mobility of the nodes. Many real networks fall into this category, and these networks belong to the general category of opportunistic networks or delay-tolerant networks (DTNs) [7].

Therefore, the current ad hoc routing protocols fails whenever the network is partitioned or a connected path between the source and the destination does not appear at any instance. In this case, the conventional ad hoc network routing protocols, such as DSR[14] and AODV [22], would fail. Specifically, reactive routing algorithms will fail to discover a contemporary connected path, and proactive routing protocols will fail to converge due to high network dynamics and will result in a deluge of topology update packets.

Fortunately, the non-existence of simultaneously connected paths does not mean that the messages cannot be delivered. As the topology of the network changes over time, links come up and down, and form serials of connectivity graphs. If in the sequence of connectivity graphs, a path exists over time, messages can be sent over the links when they are up, and get buffered at the next hop and wait for the next link to come up. Messages can be forwarded in this manner until they reach their destinations.

Although certain applications, such as instant messages, require timely delivery of the messages, which requires a connected path for real-time communication, a large number of other applications, such as file transfer and email, can tolerate a certain amount of delay, and therefore, can be benefit from the eventual delivery of the messages in network.

6.2 Oblivious Routing Algorithms

6.2.1 Epidemic Routing

6.2.1.1 Basic Idea

The goal of Epidemic routing [56] is to a routing strategy for the delivery of data even when the network partition of a path between the source and the destination never exists during the routing process of the message. Epidemic routing is an oblivious routing protocol because it makes the minimum assumption about the connectivity of the nodes in the network.

Specifically, in Epidemic routing, a source does not know where the destination is located, nor does it know which route is the best for the message to follow in order to reach the destination. Nodes randomly come into contact with each other through mobility.

Epidemic routing relies on the mobile nodes, which carry the message and move to another portion of the network. In this way, messages can spread to the nodes on the other islands of the network. The theoretical foundation of Epidemic routing is the theory of epidemic algorithms, which assures that all nodes in a network will receive a copy of a given message in a bounded amount of time, given that the nodes in the network exchange data randomly.

Experimental results show that Epidemic is able to deliver almost all messages in network scenarios where existing ad hoc routing protocols fail due to network partitions or the absence of spontaneous connected paths. To maximize message delivery rate and to minimize the aggregate forwarding overhead, Epidemic can place an upper bound on the hop-count of the message and the per-node buffer space.

6.2.1.2 Design Issues

The goal of Epidemic routing is twofold. The first objective is to route messages with high delivery rate in the partially connect ad hoc network in a probabilistic way. The second objective is to minimize the amount of resource consumed in the delivery of the messages in terms of the average number of copies forwarded per message. The following design issues are raised for Epidemic routing.

Message carriers have no knowledge of the location of the trajectory of the other nodes in the network. When a carrier comes into contact with another node randomly, a key issue is to determine whether the carrier should forward the message to the node and the latter can become a potential carrier.

The system needs to balance the conflicting objectives by proper resource allocation. For instance, multiple copies of a message need to be buffered at multiple nodes to increase the likelihood of the eventual delivery of the message. On the other hand, each message should not consume the buffer space of all nodes just to ensure its own delivery.

6.2.1.3 Implementation Details

When two nodes come into contact, the node with the smaller ID initializes an anti-entropy process, in which both nodes exchange their summary vector of the message they contain. To avoid redundant anti-entropy processes, each node maintains a list of nodes that it encountered recently, and the anti-entropy process is not re-initialed with the nodes that have been contacted within a configurable period of time.

In the anti-entropy process, two nodes determine which messages are common for both nodes by examining the summary vector of the other. Advance algorithms like the Bloom filter can be applied to reduce the size of the summary vectors. Each node then sends requests for copies of the messages that they did not buffer. By placing a maximum buffer size associated with each node, Epidemic routing can determine the maximum number of messages that a host can carry on behalf of their nodes.

6.2.2 Spray and Wait

6.2.2.1 Basic Idea

Spray and Wait is proposed to reduce the message forwarding overhead of flooding-based routing protocols and try to maintain good performance with respect to delivery rate and delay. Like Epidemic, Spray and Wait does not require the use of any network information, such as the past encounter history. Analytical studies are conducted to compute the number of copies per message that Spray and Wait requires to achieve a given delivery delay.

6.2.2.2 Design Goals

The design goals of Spray and Wait in opportunistic networks are listed as follows:

- Spray and Wait should forward significantly fewer message copies than Epidemic and other flooding-based routing protocols in any situation.
- Spray and Wait should generate low contention even under high traffic load.
- Spray and Wait should achieve a higher delivery delay than single-copy and multi-copy schemes, and it should generate close-to-optimal performance.
- Spray and Wait should be highly scalable under any network size and node density.
- Spray and Wait should be simple and it requires as little knowledge of the network as possible.

6.2.2.3 Implementation Details

Spray and Wait routing consists of two phases: the spray phase and the wait phase. In the spray phase, L copies of each message are initially spread (forwarded), by the source node and the nodes that subsequently receive copies of the message. After the L copies are sprayed, each of the L nodes carrying a copy performs a direct transmission to the destination.

Spray and Wait combines the speed and simplicity of Epidemic routing and the thriftiness of direct transmission. After enough L copies have been sprayed into the network, which will guarantee a quick delivery of the message with a high probability, Spray and Wait stops to save energy consumption by excessive forwarding of message copies.

The binary spraying strategy is adopted by Spray and Wait, in which a node that carried L copies of the message, upon encountering another node, will hand over $\lfloor L/2 \rfloor$ copies to the node and keep $L - \lfloor L/2 \rfloor$ copies. It is easy to see that the forwarding tree which has the maximum number of nodes at every level has the maximum number of active nodes at every step.

6.3 Encounter-Based Uni-cast Algorithms

6.3.1 Encounter-Based Routing

6.3.1.1 Basic Idea

Nelson et al. [19] indicate a observed mobility property of certain networks: *the future rate of node encounters can be roughly predicted by past data.* This property means that nodes that experience more encounters are more likely to meet destination nodes in the future. By taking advantage of this property, Nelson et al. propose a quota-based DTN routing protocol, which is named EBR, that achieves high delivery ratios comparable to flooding-based protocols, while maintaining low network overhead. EBR is quota-based, which means that it limits the number of replicas of any message in the system. Thus, it minimizes network resource usage.

6.3.1.2 Encounter-Based Routing

Every node running EBR is responsible for maintaining their past rate of encounter average, which is used to predict future encounter rates. Nodes maintain two pieces of local information:

- **Encounter value (EV)**: Encounter value represents the node's past rate of encounters as an exponentially weighted moving average.
- **Current window counter (CWC)**: Current window counter is used to obtain information about the number of encounters in the current time interval.

EV is periodically updated to account for the most recent CWC in which rate of encounter information was obtained. Updates to EV are computed as follows:

$$EV \leftarrow \alpha \cdot CWC + (1-\alpha) \cdot EV,$$

where α is the weight given to recent complete CWC. Nelson et al. experiments reveal that an α of 0.85 and update interval of around 30 seconds allow for reasonable results in a variety of networks [19].

The number of replicas of a message transferred during a contact opportunity is proportional to the ratio of the EVs of the nodes. When two nodes A and B contact, for every message M_i, node A sends

$$m_i \cdot \frac{EV_B}{EV_A + EV_B}$$

replicas of M_i, where m_i is the total number of M_i replicas stored at node A.

When L, the maximum number of replicas of a message, is discrete, it is clearly that EBR is quota-based. However, L values are not limited to a discrete maximum number of replicas. Nelson et al. also consider the situation that the discrete structure is relaxed into a probabilistic structure, while maintaining meaningful (yet probabilistic) bounds. Under the assumption that L values follow the Gaussian distribution, Nelson et al. prove that EBR is also quota-based. The limitation is that if a node A wishes to spilt the message M into two replicas, M_A and M_B, node A must follow the following rule:

If $M \sim N(\mu, \sigma^2)$, then it can only be split into $M_A \sim N(\mu_A, \sigma_A^2)$ and $M_B \sim N(\mu_B, \sigma_B^2)$ such that $\mu = \mu_A + \mu_B$ and $\sigma^2 = \sigma_A^2 + \sigma_B^2$.

6.3.1.3 Securing EBR

Nelson et al. also consider the possibility if a denial-of-service (DoS) attack. The decision regarding how many replicas of a message a node should transmit to a contact depends completely upon the ratio of both parties' encounter values. Therefore, malicious nodes may fake an ultra-high encounter value and convince other nodes

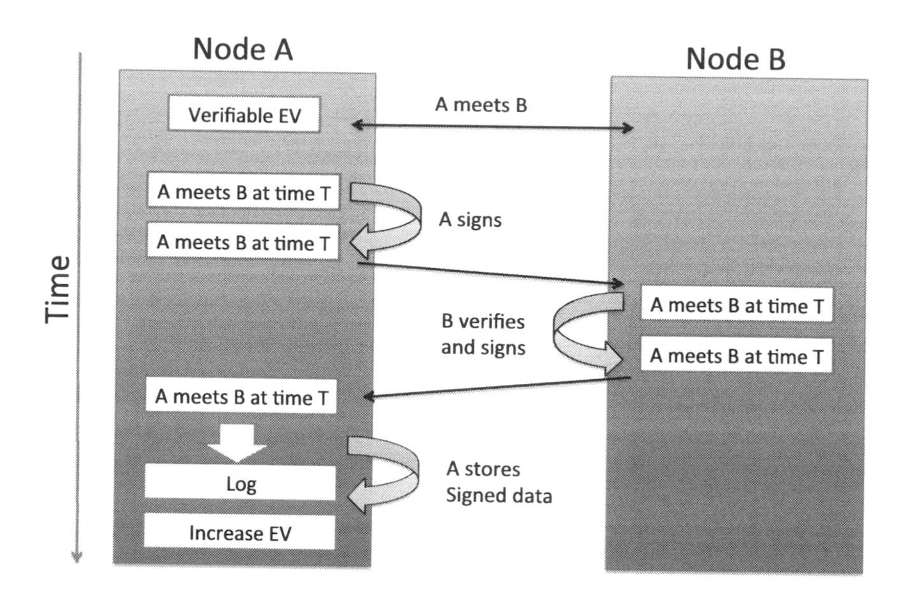

Figure 6.1: Timestamp protocol.

following EBR to transmit almost all replicas to them. Then they simply delete these messages, which results in network degradation.

It is observed that an encounter value can never be altered unless an external event occurs. Therefore, nodes must assure other nodes that the encounter value was only altered during external events and is not individually forged. Nelson et al. design a sign protocol based on the timestamp to support this mechanism. The protocol works as follows:

When two nodes A and B meet, they digitally sign timestamped messages stating that "node A met B at time T" and exchange the messages. Nodes keep these messages as evidences of transactions. When node A meets node C, and node C wishes to send data to node A, node A can then offer all of these messages to node C, and allow node C to recompute node A's encounter value. If the recomputed value is equal to the value provided by node A, then node C can confidently transmit replicas to node A. A graphical illustration of this is given in Figure 6.1.

6.3.2 Optimal Opportunistic Forwarding

6.3.2.1 Preliminary and Motivation

In most opportunistic forwarding protocols, each node is associated with a forwarding metric for each destination, which signifies the quality of the node as a forwarder.

Existing forwarding metrics are usually: (1) direct (1-hop) metrics between the nodes and the destination, such as encounter frequency [16] and the time elapsed since the last encounter [26, 1, 8, 10], or (2) the expected forwarding metric along the expected forwarding path, such as expected cost [4] and expected delay [28]. When node i meets node j, node i forwards a message to node j depending on whether or not the direct forwarding quality of i is better than j.

We found two drawbacks in such strategies. The first drawback is that a forwarding decision based on comparing the direct or multi-hop forwarding qualities of nodes i and j cannot guarantee good forwarding for the following reasons. (1) The forwarding quality of j being better than i does not necessarily mean that j is a good forwarder. (2) Even though the quality of j is high, i might encounter better nodes in the near future. (3) Similarly, even though the quality of j is lower than i, j might still be the best forwarder that i could encounter in the future.

The second drawback is that the forwarding quality of a node is regarded as a constant. However, the forwarding quality of a node may change significantly at different stages. For example, in hop-count-limited forwarding, two important states of the copy are: *remaining hop-count* and *residual time-to-live*. Remaining hop-count is an important factor: a node can be a bad 1-hop forwarder for having a large mean inter-meeting time with the destination, but it can still be an excellent 2-hop forwarder if it has a node that it frequently contacts, which is also a node that frequently contacts the destination. On the other hand, residual time-to-live is important because it affects a node's direct delivery probability, as well as its chance of contacting high-quality intermediate nodes.

6.3.2.2 Basic Idea

To rectify these drawbacks, we use a comprehensive forwarding metric, which reflects: (1) not the relative forwarding quality between two nodes (node i and the next node j that would hold custody of a new copy of the message), but the relative forwarding qualities among all possible next nodes j, and (2) not the quality (1-hop or multi-hop delivery probability or delay) of a particular message copy, but the joint delivery probability or delay, of all copies when multiple copies of the message can be forwarded along multiple paths.

We define a delivery probability P_{i,d,K,T_r} for each copy in i and each destination d. This metric is comprehensive because it represents the joint probability of all descendant copies, and it is also dynamic since it is a function of the remaining hop-count K and residual time-to-live T_r. With P_{i,d,K,T_r}, our optimal forwarding rule is presented as follows: we logically regard a forwarding from a node i to another node j as replacing a message copy with two new copies in the two nodes, respectively. Whether i should forward the copy to j depends on whether replacing the copy in i with two logically new copies increases the joint delivery probability: the copy is forwarded only if the joint probability of $P_{i,d,K-1,T_r-1}$ and $P_{j,d,K-1,T_r-1}$ (in case of forwarding) is greater than the probability P_{i,d,K,T_r-1} (in case of no forwarding).

6.3.3 Energy-Efficient Opportunistic Forwarding

6.3.3.1 Basic Idea

Efficient algorithms and policies for opportunistic forwarding are crucial for maximizing the message delivery probability while reducing the delivery cost. Li et al. [15] investigate the problem of energy-efficient opportunistic forwarding for DTNs.

In the network, it is assumed that every message transmission costs constant energy consumption γ, which includes both the reception energy at the receiving node and the sending energy at the transmitting node. That is to say, the energy consumption for delivering a message to the destination is proportional to the expected number of transmission times the message's lifetime. Let Q denote the total energy constraint for delivering a message, and $X(T)$ denote the number of the message copies in all nodes, the energy constraint can be expressed as follows [15]:

$$\gamma(E(X(T))-1) \leq Q.$$

Let $F(t)$ denote the probability that the message has been delivered to the destination at time t, and $\Psi = Q\,\gamma + 1$, the problem of the optimal opportunistic forwarding policy with the energy constraint for message delivery, can be expressed as the following optimization problem:

$$\begin{array}{ll} \text{Maximize} & F(T) \\ \text{Subject to} & E(X(T)) \leq \Psi. \end{array}$$

6.3.3.2 Optimization Formulation

Li et al. investigate two typical forwarding algorithms:

- **Two-hop forwarding:** In two-hop forwarding, the source node can forward messages to any other nodes, but other nodes can only forward messages to the destination. When a source node encounters any other nodes without the message, except the destination, at time t, the source node forwards the message with probability $p(t) \in [0,1]$.

- **Epidemic forwarding:** In epidemic forwarding, when one node with the message encounters another node without the message, except the destination, at time t, it forwards the message to the node with probability $p(t) \in [0,1]$.

First, Li et al. model the message dissemination by introducing a continuous time Markov framework. Based on this framework, Li et al. formulate the optimization problem of opportunistic forwarding, with the constraint of energy consumed by the message delivery for both two-hop and epidemic forwarding.

For probabilistic two-hop forwarding, the optimization problem can be specified as:

$$\begin{array}{ll} \text{Maximize} & \int_0^T E(X(s))ds \\ \text{Subject to} & \begin{cases} \int_0^T p(\tau)d\tau \leq \frac{1}{\lambda} ln\frac{N-X(0)}{N-\Psi} \\ 0 \leq p(t) \leq 1 \end{cases}, \end{array} \tag{6.1}$$

the for probabilistic epidemic forwarding, the optimization problem can be specified as:

$$\begin{array}{ll} \text{Maximize} & \int_0^T E(X(s))ds \\ \text{Subject to} & \begin{cases} \int_0^T p(\tau)d\tau \le \frac{1}{\lambda N} ln \frac{\Psi(N-X(0))}{X(0)(N-\Psi)} \\ 0 \le p(t) \le 1 \end{cases}, \end{array} \quad (6.2)$$

where N is the number of nodes in the network, and λ is the exponential parameter of the inter-contact time between any two nodes.

6.3.4 Policy Design and Optimal Policy

Based on the solution of the optimization problem, Li et al. design different kinds of forwarding policies such as static and dynamic policies.

When static policy is used, $p(t)$ is assumed to be constant, i.e., $p(t) = \alpha$. Given the solution of the optimization problem, the calculation of α is trivial.

Dynamic policy includes continuous dynamic policy and threshold dynamic policy. The continuous dynamic policy is more complex than the static policy, because $p(t)$ varies with time. To investigate the property of the dynamic optimal policy, Li et al. study two typical continuous functions:

- **The power function**: The power policy is set as $p(t) = \alpha t^{\beta}$.
- **The exponential function**: The exponential policy is set as $p(t) = \alpha e^{\beta t}$.

Given Equation 6.1 and Equation 6.2, the value of α and β can be solved.

When threshold dynamic policy is used, it is set that $p(t) = 1$ if $t < t_0$, where t_0 is the threshold. When $t > t_0$, $p(t)$ is set as 0.

Among these policies, Li et al. also prove that the threshold dynamic policy is optimal for both two-hop and epidemic forwarding.

6.3.5 NUS Student Contact Trace Model

6.3.5.1 Basic Idea

Many efforts have been made to understand human mobility or contact patterns in scenarios such as a campus environment. As shown by the National University of Singapore (NUS) student contact trace model [27], when the class schedules and class rosters for each class are obtained from a university-wide intranet learning portal, an authentic trace reflecting contact patterns between students over large time scales can be constructed without long-term contact data collection.

In this trace, 22,341 students and 4,485 classes are involved, and students attend several classes weekly. The contacts between students are inferred based on assumptions that two students are in contact with each other if and only if they are in the same venue at the same time, while two students who are in different classrooms are out of range of each other, even if one classroom is just next door to the other. To simplify the model, the trace has been processed in two ways: First, the durations of classes are normalized to one or several business hours. That is to say, business hour is the unit of time for the contact duration. Second, any idle time slots between sessions are removed so that only class hours are concerned. The advantages of the trace synthesized in this model are that it provides contact patterns of a large population which reflect characteristics of those observed in the real world, and the relatively long durations of contacts between students provide stable environments for students to exchange content of large size.

6.3.5.2 Delay-Tolerant Networking

Srinivasan et al. investigate the characteristics of the model from the DTN perspective, where the inter-contact time and time distance between pairs of students are studied. Srinivasan et al. show that in a campus environment, arbitrary pairs of students can communicate with each other in less than two business days on average. They also show that with random hubs present, this number reduces to 7.6 hours. It is assumed that at a given hour, if a student is not attending any session, with probability p_{hub} the student is in one of the random hubs. Srinivasan et al. also point out that the distribution of the inter-contact time of this trace does not follow the power law distribution, as the inter-contact time for in-class contacts is bounded by the fact that students are expected to see each other at least once every week.

6.3.6 Location-Based Routing (PER)

6.3.6.1 Basic Idea

Most previous routing schemes make forwarding decisions by predicting whether two nodes would contact each other, without considering the delay of the contact. The inspiration of [32] is that the nodes occur in a small number of landmarks periodically in some environment. The observation is that the node is roaming among some locations ([32], the author uses landmark instead of location to represent the location of the node). The routing scheme, called PER, proposed in [32] uses the probability that two nodes will contact with each other within some time period as forwarding metric to route messages efficiently.

Example 1 *Nodes show preference for some locations and would move less often to other locations. A campus DTN example is shown in Figure 6.2. A student may visit some of the locations such as dormitory, classroom, and lab frequently in his daily life. The arrow denotes the mobility behavior of the student.*

6.3.6.2 TH-SMP Model

In this article, the authors make the following assumptions:

- There is enough bandwidth for nodes to transfer all messages at each contact.
- Each location is associated with a unique ID and nodes are aware of their current location at any time.

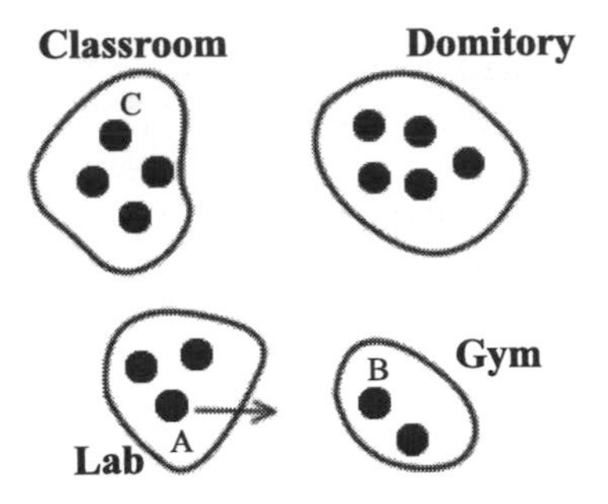

Figure 6.2: An example of a location-based mobility model.

Reference [32] models the mobility of a node m with a time-homogeneous semi-Markov (TH-SMP), (X_n^m, T_n^m) with discrete time, where the x_n^m is the location of node m at time unit n, and T_n^m represents the time instant of the transition $x_n^m \to x_{n+1}^m$. Random variable $T_{n+1}^m - T_n^m$ describes the location sojourn time.

The associated time-homogeneous semi-Markov kernel Q is defined by:

$$Q_{ij}^m(t) = P(X_{n+1}^m = j, T_{n+1}^m - T_n^m \leq t | X_n^m = i). \tag{6.3}$$

Suppose $P^m = |p_{ij}^m|$ is the transition probability matrix of the (X_n^m) embedded Markov chain for node m. Then the transition probability from state i to j is $p_{ij}^m = \lim_{t\to\infty} Q_{ij}^m(t)$, where $i, j \in L$, and L is the set of landmarks (in the following, we will drop the m superscript to simplify the notation).

Also, we derive the probability $S_i^m(t)$ that node m will leave the neighborhood of landmark i on or before time unit t:

$$S_i^m(t) = P(T_{n+1}^m - T_n^m \leq t | X_n^m = i) = \sum_{j=1}^{l} Q_{ij}^m(t). \tag{6.4}$$

Note that $S_i^m(t)$ also indicates the distribution of the dwell time at landmark i for node m, regardless of the next landmark.

Let Z^m (Z_t^m) be another TH-SMP that describes the state (landmark) occupied by node m at time t. The transition probabilities for process Z are defined by $\phi_{ij}^m = P(Z_t^m = j | Z_0^m = i)$. If we know that a node is currently in state i, after t time units, it will be in state j with probability $\phi_{ij}(t)$. ϕ provides the prediction of the nodes location at a landmark at an arbitrary time $t > 0$ knowing its current location.

There is a relationship between $Q_{ij}(k)$, p_{ij} and $S_{ij}(k)$:

$$Q_{ij}(k) = p_{ij} S_{ij}(k). \tag{6.5}$$

If the node transitions at least once between times 0 and t, we consider on the time k of the transition from i, and on the state r to which the process moves immediately after state i. $\phi_{ij}(t)$ is calculated by:

$$\phi_{ij}(t) = (1 - S_i(t))\delta_{ij} + \sum_{r=1}^{l} \sum_{k=1}^{t-1} \dot{Q}_{ir}(k)\phi_{rj}(t-k), \quad (6.6)$$

where $\dot{Q}_{ir}(k) = Q_{ir}(k) - Q_{ir}(k-1)$ is the time derivative of Q, and $\phi_{ij}(0) = \delta_{ij}$, δ is the Kronecker symbol. Note that ϕ can be calculated iteratively.

6.3.6.3 Contact Probabilities

With sojourn time probability distributions S_{ij}^m and the transition probability matrix P^m, we can predict the future landmark location of node m based on its current location using probability distributions $\phi_{ij}^m(k)$.

Assuming that the most recent known state of node a is s_a (at time k_a), and for node b is s_b (at time k_b, with $k_a, k_b < k, k > 0$), the probability of contact between a and b at a landmark i at time k is

$$C_{ab}^i(k) = \phi_{s_a i}^a(k - k_a) \cdot \phi_{s_b i}^b(k - k_b). \quad (6.7)$$

Then, the probability that a and b are in contact at a time k at any landmark is

$$C_{ab}(k) = \sum_{i \in L} C_{ab}^i(k). \quad (6.8)$$

The probability of the first contact between a and b at time k is dened as

$$R_{ab}(k) = C_{ab}(k) \prod_{t=0}^{k-1} (1 - C_{ab}(t)). \quad (6.9)$$

6.3.6.4 Delivery Probability Metrics

Let n_c be the chosen neighbor for evaluating the delivery probability metrics and d is the destination. Then according to the contact probability between two nodes ($C_{ab}(k)$) at a period of time, the author proposed three delivery probability metrics to guide message forwarding:

- The maximum probability of contact in time $[1, D]$, which is defined as

$$f_1 = \max_k C_{n_c d}(k), \ 1 \le k \le D. \quad (6.10)$$

- The maximum average probability of contact in time $[1, D]$, which is

$$f_2 = \sum_{k=1}^{D} C_{n_c d}(k). \quad (6.11)$$

- The maximum probability of the rst contact before the deadline is defined as

$$f_3 = \sum_{k=1}^{D} R_{n_c d}(k). \tag{6.12}$$

At each contact between s (the message hosting node) and a node n_c, which is also called a neighbor of s, the forwarding decision process is presented as follows: using the metric f_i ($i \in \{1,2,3\}$) to calculate the delivery probability between s and d, the result is denoted by $f_i(s,d)$, and then calculate that between n_c and d ($f_i(n_c,d)$), the message is forwarded to the neighbor n_c if $f_i(s,d) < f_i(n_c,d)$; and not forwarded, otherwise.

6.3.7 Delegation Forwarding

Delegation forwarding may use a wide range of forwarding metrics (qualities). To illustrate, we assume that the mean inter-meeting time $I_{k,d}$ of node k with destination d are used as the forwarding quality of a node k. If $I_{j,d} < I_{i,d}$, then node j has a higher forwarding quality than node i. Each message maintains a threshold τ, which is initially the forwarding quality of the destination. Whenever the current custody node of the message encounters another node, the current node queries the encountered node for its quality regarding the destinations of the messages. For each message in the custody node, if the acquired quality is better than the current threshold, the threshold of the message is updated. For each message whose threshold is updated, if the encounter does not have a copy of the message, a copy of the message will be sent to the encountered node. The Delegation algorithm is listed in Algorithm 10.

Algorithm 10: SOFA

```
1  n1 ← the current node
2  n2 ← a node that n1 encounters
3  for (each message m in n1's buffer) {
4     if (m is newly created) {
5        m.τ = quality(m.src, m.dest)
6     }
7     if (sim(n2, m.dest) > m.τ) {
8        m.τ ← quality(n2, m.dest)
9        send a copy of m to n2
10    }
11 }
```

6.3.8 Encounter-Based Routing (RAPID)

A DTN routing scheme called RAPID is proposed in [2]. It focuses on routing in those networks where resources, such as bandwidth and storage, are limited. While

Table 6.1 List of Commonly Used Variables [2]

D(i)	Packet *is* expected delay $= T(i) + A(i)$
T(i)	Time since creation of i
a(i)	Random variable that determines the remaining time to deliver i
A(i)	Expected remaining time $= E[a(i)]$

PER, which we discussed in the previous section, assumes unlimited bandwidth. The idea of RAPID is to model DTN routing as a utility-driven resource allocation problem. Table 6.1 lists the notations that we will use in the following.

One of the key issues in RAPID is how to estimate the expected delivery delay. RAPID assumes that the inter-meeting time between nodes follows an exponential distribution,

$$P(a(i) < t) = 1 - e^{-k\lambda t}. \tag{6.13}$$

It can estimate the expected delivery delay $D(i)$ using the properties of exponential distribution. When there are k replicas, the expected delivery delay is the mean meeting time divided by k, i.e.,

$$A(i) = \frac{1}{k\lambda}. \tag{6.14}$$

The key question for RAPID is: If the bandwidth is limited, how do we replicate a packet in the node's buffer so as to optimize a specified routing metric. RAPID derives a per-packet utility function from the routing metric? At a transmitting opportunity, it replicates a packet that locally results in the highest increase in utility [2].

RAPID consists of three core components, which we will introduce in the next subsection:

- *Selection algorithm*
- *Inference algorithm*
- *Control channel*

6.3.8.1 Selection Algorithm

When two nodes come into contact with each other, the RAPID protocol will be executed. The packets exchanged at each contact are limited due to the limited bandwidth. If a node exhausts all available storage, packets with the lowest utility are deleted first as they contribute least to the performance of the whole system. RAPID defines utility functions from different routing metrics. At each forwarding opportunity, it replicates a packet that results in the highest increase in utility and forwards the packet.

Take the minimized average delay of packets as an example, which is a commonly used routing metric in many protocols. In this case the corresponding utility

U_i of a packet i is the negative of the expected delay to deliver i. δU_i denotes the increase in U_i by replicating i and s_i denote the size of i. Then, RAPID replicates the packet with the highest value of $\frac{\delta U_i}{s_i}$ among packets in the buffer of i's host node; in [2], the $\frac{\delta U_i}{s_i}$ is also called marginal utility. RAPID always replicates the packet that results in the highest marginal utility. We can see that the marginal utility takes both the reduction in delivery delay and the size of the packet into consideration.

6.3.8.2 Inference Algorithm

This is the key part of RAPID. RAPID support various routing metrics. With these metrics, RAPID can determine which packet should be replicated and sent first.

- **Metric 1: Minimizing average delay**

  To minimize the average delay of packets in the network, we define the utility of a packet as

$$U_i = -D(i). \tag{6.15}$$

  With this metric, the protocol attempts to greedily replicate the packet whose replication reduces the delay among all packets in its buffer.

- **Metric 2: Minimizing missed deadlines**

  To minimize the number of packets that miss their deadlines, the utility is defined as the probability that the packet will be delivered within its deadline:

$$U_i = \begin{cases} P(a(i) < L(i) - T(i)), \ L(i) > T(i) \\ 0, \quad otherwise, \end{cases} \tag{6.16}$$

  where $L(i)$ is the packet lifetime. With this metric, the protocol replicates the packet that yields the highest improvement among packets in its buffer.

- **Metric 3: Minimizing maximum delay**

  To minimize the maximum delay of the packets in the network, we define the utility U_i as

$$U_i = \begin{cases} -D(i), \ D(i) \geq D(j) \ \forall j \in S \\ 0, \quad otherwise, \end{cases} \tag{6.17}$$

  where S denotes the set of all packets in X's buffer. Thus, U_i is the negative expected delay if i is a packet with the maximum expected delay among all packets held by X. Hence, replication is useful only for the packet whose delay is the maximum.

The packets in the host node's buffer are stored in a decreasing order of $\frac{\delta U_i}{s_i}$ at any time. At each contact, the packet at the head of the buffer will be replicated and sent first to optimize the corresponding routing metric with limited bandwidth. That is how RAPID manages packets as a resource allocation problem.

6.3.8.3 Control Channel

RAPID uses an in-band control channel to exchange acknowledgments for delivered packets as well as metadata about every packet learned from past exchanges.

6.4 Social-Based Uni-cast Algorithms

Routing in social networks is a hot topic recent years. We will discuss some social-based routing algorithms such as [13, 36, 24] in this section. Unlike previous work, the social-based routing schemes focus on routing messages efficiently using extensive social information. Extensive simulation results show that this kind of routing scheme achieves good performance.

6.4.1 Bubble Rap

A social-based forwarding scheme [13] is proposed, which improves forwarding efficiency significantly compared to oblivious forwarding schemes. Since the message routing works like a bubble flowing up in water, this routing scheme is called *Bubble* forwarding. It focuses on the following two specific aspects of society:

- *Community*: In social networks, a community is a group of people who have some social relations, for example, share interests, activities, backgrounds, or real-life connections.
- *Centrality*: Within a community, for those people who are more popular (i.e., more active), and interact with more people than others, we say that they have high centrality.

Popularity ranking [13] is used to measure the popularity of a node in the network. The forwarding process acts as real-life experiences. First, the source forwards the message via surrounding people (nodes) more popular than itself, and then the message is bubbled up to well-known popular people in the wider community, such as a postman. When the postman meets a member of the destination community, the message will be passed to that community. The first community member who receives the message will try to identify more popular members within this community, and bubble the message up again within the local hierarchy, until the message reaches a more popular member, and until the message reaches the destination [13].

6.4.1.1 Community Detection

The following two centralized communities are exploited in detection algorithms to measure the node centrality [13].

- *K-clique*: First proposed by Palla et al. [21].
- *Weighted network analysis* (WNA): Proposed by Newman [20].

Details of *K-clique* and WNA are presented in [20, 21], respectively.

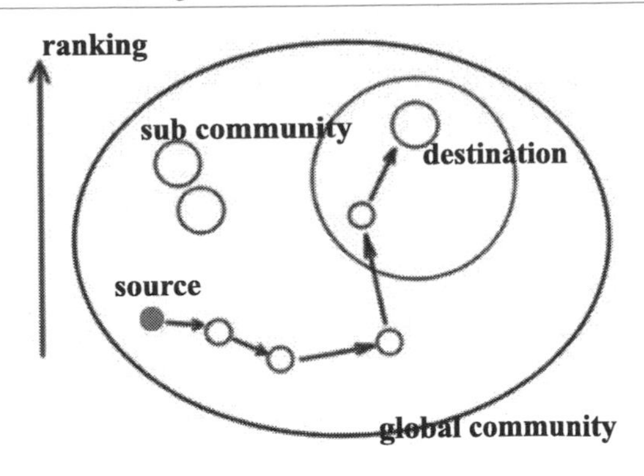

Figure 6.3: Illustration of the Bubble algorithm [13]. The oval is the global community; the small cycles are sub-communities. The message is routed from the source community to that of the destination.

6.4.1.2 Bubble Forwarding

In Bubble forwarding, two assumptions are made as follows:

- Each node belongs to at least one community.
- Each node is associated with a global ranking (in other words, global centrality), and also with a local ranking within its local community. A node may belong to multiple communities and thus may have more than one local ranking.

Bubble forwarding uses the Label scheme [12]. Each node is assumed to have a label that indicates the community to which it belongs. A message is forwarded to the encountered node if one of the following two conditions is satisfied: (1) the encountered node belongs to the same community (i.e. has the same label as destination) as the destination; (2) they are not in the same community, but the encountered node is more popular than the message hosting node (i.e. the encountered node has a higher ranking than the hosting node).

The forwarding process is carried out in the following two steps.

- Suppose the source node has a message that heads for another node (destination). The source first bubbles the message up the hierarchical ranking tree using the global ranking, until it reaches a node which is in the same community as the destination node (i.e. a node with the same label as the destination). The process is shown as Figure 6.3.
- Then the local ranking system is used instead of the global ranking, and the message continues to bubble up through the local ranking tree until it reaches the destination or it times out.

6.4.2 Social Feature-Based Routing

6.4.2.1 Basic Idea

Unlike location-based routing and encounter-based routing, [30] exploits more information, i.e. social features, to route messages. In social contact networks, where nodes (individuals) interact at each contact based on their common interests, social features play an important role. The social features represent either physical features, such as gender, or logical ones, such as a membership in a social group.

A multi-path routing approach [30] is proposed (denoted by SFMR) in DTNs based on the social features of nodes. The approach includes two processes:

- Social feature extraction
- Multi-path routing

In social feature extraction, the authors use entropy to extract the m most informative social features to create a feature space (F-space): $(F_1, F_2, \cdots, F_i)$, where F_i corresponds to a feature. The routing method then becomes a hypercube-based feature matching process where the routing process is a step-by-step feature difference resolving process.

Two routing schemes are offered in [30] based on the social features:

- Node-disjoint-based routing
- Delegation-based routing

6.4.2.2 Preliminaries

Feature Space: Each individual can be represented by a social feature profile, a representation of her/his social features within a feature space, also called the F-space. Figure 6.4 represents a $4 \times 3 \times 2$ F-space. There are three different social features in the F-space, represented by four, three, or two distinct values, respectively. There are 24 groups in the example. In Figure 6.4, dimension 1 (the leftmost position) corresponds to a city with four distinct values: New York (0), London (1), Paris (2), and Shanghai (3); dimension 2 (the second leftmost position) shows position with three distinct values: professor (0), researcher (1), and student (2); dimension 3 represents gender with two distinct values: male (0) and female (1). Two groups have a connection if they differ in exactly one feature.

Hypercubes: The users can be represented by a node in a hypercube as shown in Figure 6.4 according to the social feature profile. Thus, the closeness between two individuals (virtual similarity) can be measured by the feature distance in the hypercube.

The binary hypercube is a special cube where each feature has a binary value: 0 and 1. Nodes can be specified to group, which can be represented by a binary vector, in the binary hypercube.

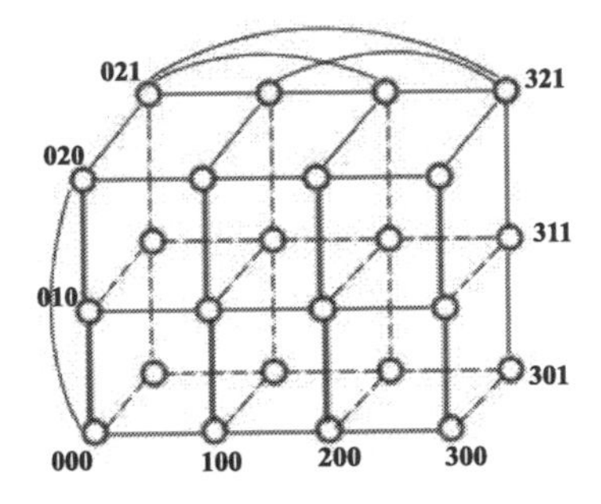

Figure 6.4: A 3-dimensional hypercube.

Example 2 *Node A can be represented by* $(a_1, a_2, \cdots, a_m)$*; if the value of feature* a_i *(*$i \leq m$*) is the same as the corresponding feature of the destination, then* $a_i = 1$*. The vector of the destination's group is* $(1, 1, \cdots, 1)$*.*

In a binary cube, the feature distance between two individuals, A and B, is denoted as H_{AB}, which is the Hamming distance between A and B. Since an individual is specified to a group which corresponds to a node of a binary cube and the social closeness between node pairs can be measured by the feature distance, the routing problem is to forward from groups to groups until the message reaches the destination group.

6.4.2.3 *Feature Extraction*

Reference [30] uses Shannon entropy to extract key features.

$$E(F_i) = -\sum_{i=1}^{n} p(x_i) \log_2 p(x_i), \; (j = 1, 2, \cdots, m'), \tag{6.18}$$

where $E(F_i)$ denotes the entropy of the feature F_i, and p denotes the probability mass function of F_i, and $\{x_1, x_2, \cdots, x_n\}$ are the possible values of feature F_i.

6.4.2.4 *Routing Schemes*

Node-disjoint-based routing: The group of the node is represented by a binary vector. We suppose that the source group is $(0,0,0,0)$ (there are four features for each node in the network). The destination vector is $(1,1,1,1)$. Therefore, the routing process attempts to resolve the differences between (0,0,0,0) and (1,1,1,1) via intermediate nodes. A possible path, represented by nodes' social feature vectors, would be (0,0,0,0)→(1,0,0,0)→(1,0,0,1)→(1,1,0,1)→(1,1,1,1).

Theorem 6.1
We assume that source S has a packet for destination D with feature distance k in an m-dimension binary cube. There are exactly m node-disjoint paths from S to D [30].

Since there are exactly m node-disjoint paths in the m dimensions cube, the source node achieves a high delivery rate by forwarding messages to the m paths, and that is the social feature-based multi-path routing.

Delegation-based routing: Delegation-based routing forwards the copies of a packet only to an individual with a smaller feature distance to the destination.

There are two values to determine packet forwarding:

- Quality value: The quality value Q_{AD} of an individual A with destination D is inversely proportional to the feature distance between A and D. That is, $Q_{AD} = 1/H_{AD}$.
- Level value: The highest quality value that A ever met in the history. When A meets a node B with a higher level value than that of A, a copy of the message is forwarded to B.

6.4.3 SimBet Routing

6.4.3.1 Basic Idea

A key challenge of message delivery in MANETs is to find a route that can provide good delivery performance and low end-to-end delay. Daly et al. [26] present a multi-disciplinary solution, the SimBet Routing, based on nodes' 'betweenness' centrality metrics and locally determined social similarity to the destination node.

6.4.3.2 Betweenness Centrality and Similarity

SimBet exploits nodes' betweenness centrality metrics and social similarity to the destination node to help make forwarding decisions. Typical betweenness centrality is defined as [21]:

$$C_B(p_i) = \sum_{j=1}^{N}\sum_{k=1}^{j-1} \frac{g_{jk}(p_i)}{g_{jk}},$$

where g_{jk} is the total number of geodesic paths linking p_j and p_k, and $g_{jk}(p_i)$ is the number of those geodesic paths that include p_i. It measures the extent to which a node lies on the paths linking other nodes. However, this metric is difficult to evaluate in networks with a large node population. Instead, SimBet uses ego network analysis, which can be performed locally, to calculate the sociocentric betweenness centrality [17]. The social similarity to the destination node is calculated based on the number of their common neighbors.

In SimBet routing, each node maintains a $n \times n$ adjacency matrix A:

$$A = \{a_{ij}\},$$

where n is the number of contacts a given node has encountered, and $a_{ij} = 1$ if there is a contact between i and j, $a_{ij} = 0$ otherwise.

The betweenness of a given node is the sum of the reciprocals of the entries of $A^2[1 - A]_{i,j}$. Node similarity is also calculated using this matrix. The number of common neighbors between the current node i and destination node j is a simple count of the non-zero equivalent row entries in the matrix [21].

6.4.3.3 SimBet Routing

SimBet defines the SimBet utility to help select appropriate relay nodes. The SimBet utility is a value between 0 and 1 and it is based on two components: similarity utility and betweenness utility. When node n encounters node m, the similarity utility $SimUtil_n$ and the betweenness utility $BetUtil_n$ of node n for delivering a message to destination node d compared to node m is given by:

$$SimUtil_n(d) = \frac{Sim_n(d)}{Sim_n(d) + Sim_m(d)}$$

$$BetUtil_n = \frac{Bet_n}{Bet_n + Bet_m}.$$

The $SimUtil_n(d)$ is given by combing the normalized relative weights of the attributes given by:

$$SimBetUtil_n(d) = \alpha SimUtil_n(d) + \beta BetUtil_n,$$

where α and β are tunable parameters and $\alpha + \beta = 1$ [21]. These parameters allow for the adjustment of the relative importance of the two utility values.

The SimBet utility is used to help make forwarding decisions. When two nodes m and n contact, they first exchange their betweenness values and the destination nodes they are currently carrying message, for. For each destination node, node m calculates the corresponding SimBet utilities of itself and node n. If node n has a higher SimBet utility for a given destination node, it sends all messages for this destination node to node n, and then removes these messages from its memory. Also node n sends corresponding messages to node m if m has higher SimBet utilities for some destinations.

6.4.4 Homing Spread

6.4.4.1 Basic Idea

Recently, mobile social networks (MSNs) are attracting more attention. Wu et al. [31] consider an MSN in which nodes visit some locations, called *community homes* or simply *homes*, frequently, while the other locations are visited less frequently. It is assumed that each home supports a *virtual* throw box, a mechanism that can store a message at a local storage device, or at another node currently at the same home. Wu et al. propose a multi-copy routing algorithm, *homing spread* (HS), to address message delivery in this kind of MSN.

6.4.4.2 Homing Spread

HS includes three phases: *homing*, *spreading*, and *fetching*.

1) **Homing phase**: In the homing phase, the source sends copies quickly to homes. Upon reaching the first home, the message holder dumps all copies to the home. When roaming occurs (i.e., a message holder meets another node at another location), copies are equally split between the two nodes and both become message holders.

2) **Spreading phase**: In the spreading phase, homes with multiple copies spread them to other homes and mobile nodes. The home gives one copy to each node located at the same home, subject to the availability of the copies. However, the last copy is kept at the home through a virtual throw box. Each new message holder with one copy starts its homing phase.

3) **Fetching phase**: In the fetching phase, the destination fetches the message when it meets any message holder for the first time, which can be either a home or a mobile node.

Wu et al. demonstrate that HS is optimal when the inter-meeting times between any two nodes and between a node and a community home follow exponential distributions.

6.4.4.3 Performance Analysis

Wu et al. adopt a continuous Markov chain to model the message dissemination and compute the expected delivery delay of HS. First they construct the *state transition graph* as follows:

$$G\langle S, \{\rho_{s,s'}(t)|s, s' \in S\}\rangle$$

s is a state $s = \{s_1, \ldots, s_h, \ldots, s_{h+n}\}$ satisfies:

$$\sum_{i=1}^{h+n} s_i = C,$$

where the i-th component s_i represents the number of message copies held by the i-th home (if $i \le h$) or node $i-h$ (if $i > h$). h is the number of homes in the network. And $\rho_{s,s'}(t)$ denotes the *probability density function* about the time t that it takes for the state transition from s to s'. An example of the state transition graph is shown in Figure 6.5.

Based on the state transition graph, it is derived that the expected delivery of the HS algorithm, denoted by D, satisfies:

$$D \le \begin{cases} \frac{1}{h\Lambda} + \frac{2}{\Lambda} + \frac{1}{C\Lambda} &, \quad C \le h \\ \frac{1}{h\Lambda} + \frac{2}{\Lambda} + \frac{1}{h\Lambda + (C-h)\lambda} &, \quad C > h. \end{cases}$$

where λ and Λ are the exponential parameters of the inter-meeting time between any

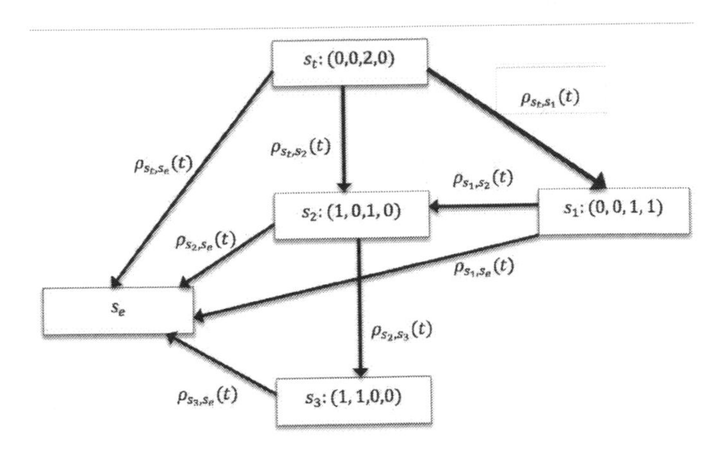

Figure 6.5: An example of the state transition graph (h = 2, n = 5).

two nodes and between a node and a community home, respectively. And C denotes the number of message copies.

In turn, given an arbitrary threshold $\Theta(\geq \frac{1}{h\Lambda}+\frac{2}{\Lambda})$ of the expected delivery delay of HS, it guarantees that $D \leq \Theta$ if we let C satisfy the following equation:

$$C = \begin{cases} \frac{1}{\Theta\Lambda-2-\frac{1}{h}} & , \quad \Theta \geq \frac{2}{h\Lambda}+\frac{2}{\Lambda} \\ \frac{\Lambda}{\lambda}\cdot(\frac{1}{\Theta\Lambda-2-\frac{1}{h}}+h) & , \quad \Theta < \frac{2}{h\Lambda}+\frac{2}{\Lambda}. \end{cases}$$

6.4.5 SOSIM

6.4.5.1 Basic Idea

The routing problem is converted to a process to resolve the social feature differences between a source and a destination [30] . However, [24] shows that merely distinguishing nodes by either the same or completely different social features is insufficient in reflecting the dynamic behaviors of the nodes. Reference [24] improves the social feature-based routing scheme and proposes a routing algorithm, namely SOSIM, using similarity metrics calculated based on the nodes' contact history to more accurately evaluate social similarities between node pairs. This routing algorithm of SOSIM is a delegation-based forwarding scheme.

6.4.5.2 Forwarding Algorithm

Suppose there are m social features for each node in the networks. Each individual node has a vector based on its social features. For example, a node X has a vector of $(x_1, x_2, \cdots, x_m)$, and another node Y has a vector $(y_1, y_2, \cdots, y_m)$. Metric $S(X,Y)$ is used to calculate social similarity between two nodes X and Y and it can be calculated in various ways. SOSIM provides three different ways to calculate $S(X,Y)$.

The forwarding scheme of SOSIM is based on delegation forwarding, where the quality of a node for a message is the social similarity between the node and a virtual node that are totally similar to the destination. Each message is associated with a threshold that records the highest quality it met in the history. Once the quality of the meeting node is higher than the message host node, the message is forwarded. Thus, the message is forwarded to the node that is more similar to the destination step by step and finally delivered to the destination.

6.4.5.3 Social Similarity Metrics

To evaluate the social similarity of two nodes more accurately, SOSIM looks at the nodes' past meeting ratios. In SOSIM, each node is associated with a vector. Node X has a vector $(x_1, x_2, \cdots, x_m)$, where $x_i = \frac{M_i}{M_{total}}$. That is,

$$(x_1, x_2, \cdots, x_m) = (\frac{M_1}{M_{total}}, \frac{M_1}{M_{total}}, \cdots, \frac{M_m}{M_{total}}), \tag{6.19}$$

where M_i is the number of meetings of X with nodes whose social feature F_i is the same as the destination's feature F_i, and M_{total} is the total number of meetings of X with any other node in the history we observe. Thus, $0 \leq x_i \leq 1$ for all $1 \leq i \leq m$. We suppose there is an ideal forwarder R for each message's destination. An ideal forwarder R is a theoretical node that meets people like destination D all the time [24]. Hence its vector consists of m 100%s: $(100\%, 100\%, \cdots, 100\%)$.

SOSIM provides three similarity metrics for its routing algorithm:

- *Tanimoto Similarity*: The Tanimoto coefficient to measure the similarity of X and Y is,

$$S(X,Y) = \frac{X \cdot Y}{X \cdot X + Y \cdot Y - X \cdot Y}. \tag{6.20}$$

  We can observe from Equation 6.20 that if X is very similar to Y, then the $S(X,Y)$ is approximate to 1.

- *Euclidean Similarity*: The Euclidean similarity of X to Y is defined as,

$$S(X,Y) = 1 - \frac{\sqrt{\sum_{i=1}^{m}(y_i - x_i)^2}}{\sqrt{m}}. \tag{6.21}$$

- *Weighted Euclidean Similarity*: To determine the weight of a social feature, Shannon entropy is used. The Shannon entropy for a given social feature is calculated as:

$$w_i = -\sum_{i=1}^{k} p(f_i) \cdot \log_2 p(f_i) \tag{6.22}$$

where w_i is the Shannon entropy for feature F_i, $(f_1, f_2, \cdots, f_k)$ are the possible values of feature F_i, and p denotes the probability mass function of F_i. The Weighted Euclidean Similarity [24] is as follows:

$$S(X,Y) = 1 - \frac{\sqrt{\sum_{i=1}^{m} w_i \cdot (y_i - x_i)^2}}{\sqrt{\sum_{i=1}^{m} w_i}}. \tag{6.23}$$

6.5 Multicast

6.5.1 Multicasting in Delay-Tolerant Networks

6.5.1.1 Basic Idea

While most social-based approaches mainly focus on forwarding data to a single destination, Gao et al. [9] study multicast in DTNs from the social network perspective. Although multicast is more effective in scenarios such as data dissemination and multi-party communication, it is more difficult to model and implement multicast in opportunistic DTNs.

In opportunistic networks, an essential difference between multicast and unicast is that data items have multiple destinations. Therefore, relay nodes should forward data items to as many destinations as possible. Gao et al. first focus on multicasting a single data item, and then generalize it to multicasting multiple data items. The two problems are formulated as follows:

- *Single-Data Multicast (SDM)*:
  $\{p, \mathbb{D}, T\}$: To deliver a data item to a set $\mathbb{D}$ of destinations.

- *Multiple-Data Multicast (MDM)*:
  $\{p, \mathbb{D}_1, \ldots, \mathbb{D}_n, s_1, \ldots, s_n, T\}$: To deliver a set of data items $d_1, d_2, \ldots, d_n$ with sizes $s_1, \ldots, s_n$ from a data source to destination sets $\mathbb{D}_1, \mathbb{D}_2, \ldots, \mathbb{D}_n$.

The challenge of the two problems is, given the average delivery ratio p and the time constraint T, how do we select a minimum number of relays?

The basic idea of this chapter is to develop social-based metrics, i.e., centrality and social communities, based on the probabilities of nodes forwarding data to their destinations. Based on the social-based metrics, the relay selections in SDM and MDM can be formulated uniformly as a knapsack problem:

$$min \sum_{k=1}^{n} x_k$$

$$s.t. \sum_{k=1}^{n} w_k x_k \geq W,$$

where $x_k \in \{0, 1\}$ indicates whether node N_k is selected as the relay, and the constraint indicates that the selected relays should satisfy the performance requirements in delivery ratio and delay.

Since the solution to this knapsack problem itself is trivial, this chapter mainly focuses on answering the following questions:

1) What are the appropriate social-based metrics for SDM and MDM, respectively?

2) How do we calculate the weights w_k of individual nodes?

3) How can the source calculate the total required weight W?

6.5.1.2 Single-Data Multicast

Gao et al. develop a centrality-based heuristic for the SDM problem based on the local knowledge of the data source. For SDM, relay selection can be done based on local knowledge of the data source about its contacted neighbors, because the data source only multicasts a single data item, and does not need to distinguish the data forwarding probabilities to different destinations for relay selections. The relays are selected among the contacted neighbors of the data source based on their centrality, to ensure that the required delivery ratio can be achieved within the time constraint.

This chapter proposes a new centrality metric based on the Poisson modeling of social networks. Suppose there are totally N nodes in the network, and the contact rate between any two nodes N_i and N_j is λ_{ij}. The cumulative contact probability (CCP) [9] of N_i is defined as:

$$C_i = 1 - \frac{1}{N-1} \sum_{j=1, j \neq i}^{N} e^{-\lambda_{ij} T}.$$

C_i indicates the average probability that a randomly chosen node in the network is contacted by N_i within time constraint T. Since all the nodes in DTNs can exchange their centrality values upon contact with each other, the data source knows the centrality values of all of its contacted neighbors when it select relays.

For relay R_i, which is in contact, the source node calculates the probability that R_i contacts a randomly chosen destination N_j within time T. For relay R_i, which is not in contact, the source node calculates the probability that the source contacts R_i, and then R_i contacts a randomly chosen destination N_j within time T. Since the average delivery ratio should be higher than p, the relay selection problem can be transformed to the unified knapsack formulation by taking logarithms on both sides of the inequality.

6.5.1.3 Multiple-Data Multicast

Gao et al. exploit a community-based approach to solve the MDM problem to which a localized heuristic is not applicable due to the node buffer constraints and the requirement of destination awareness. The "destination-awareness" capability is that, for MDM, the relays should be aware of their probabilities for forwarding each data item to the destinations. In this approach, a node maintains its destination-awareness about other nodes in the same social community. The data source selects relays among its contacted neighbors based on destination-awareness, and places appropriate data items on each relay. Gao et al. [9] use the concept of social forwarding path:

Definition 6.1 A k-hop *social forwarding path* $P_{AB} = (V_P, E_P)$ between two nodes A and B consists of a nodes set $V_P = \{A, N_1, N_2, \ldots, N_{k-1}, B\}$ and an edge set $E_P = \{e_1, e_2, \ldots, e_k\}$ with edge weights $\{\lambda_1, \lambda_2, \ldots, \lambda_k\}$. The *path weight* is the probability $p_{AB}(T)$ that a data item is forwarded from A to B along P_{AB} within time T.

Each node, using the community-based approach, maintains the "best" social forwarding path with the largest path weight to all the other nodes within the same community. If a node belongs to multiple communities, a separate table is maintained for each community. When two nodes contact, they exchange and update their social forwarding path tables.

Based on knowledge about the destinations, the data source S selects relays among its contacted neighbors. Suppose at S there are data items $d_1, \ldots, d_n$ with size $s_1, \ldots, s_n$ and destination sets $\mathbb{D}_1, \ldots, \mathbb{D}_n$, and S select relays among nodes $R_1, \ldots, R_m$ with buffer size $B_1, \ldots, B_m$ [9]. The relay selection problem is formulated at S as the following knapsack problem:

$$min|\{j|\sum_{i=1}^{n} x_{ij} > 0|$$

$$s.t. \begin{array}{ll} \sum_{i=1}^{n} x_{ij} s_i \le B_j, & for \quad j = 1, \ldots, m \\ \frac{1}{|\mathbb{D}_i|} \sum_{k \in D_i} \prod_{j=1}^{m} (1 - x_{ij} p_{jk}) \le (1-p), & for \quad i = 1, \ldots, n, \end{array}$$

where $x_{ij} \in \{0,1\}$ indicates that data item d_i is placed on relay R_j and p_{jk} is the probability for S to send data to destination k via R_j, in the form of the weight of the corresponding social forwarding path from S to k via R_j [9].

Due to the second set of constraints, such problem is NP-hard. Gao et al. propose an effective heuristic for relay selection consisting of two stages:

- First, based on the node buffer constraint of R_j, calculate the optimal data item selection for each R_j. Such optimal data item selection leads to the maximal average probability that a data item is forwarded to its destinations via R_j.
- Second, conduct the relay selection using the optimized data forwarding probabilities in the first phase as node weights.

6.5.2 SCOOP

In this section, we introduce another routing protocol, SCOOP, which is different from the previous works. SCOOP is a two-hop multi-cast routing protocol, whereas, those routing protocols discussed previously are multi-hop uni-cast routing protocols. Unlike the social-based routing scheme and the encounter-based routing scheme, [11] designs the routing protocol based on the simulation results that two-hop is enough for achieving an "optimal" performance, and then it proposes two-hop relaying strategies (referred to as SCOOP), which are fully decentralized and require only local information collected by individual devices.

6.5.2.1 Benefits of Two-Hop Relaying

Reference [11] chose four real-world data sets (shown as follows) to evaluate multi-hop relaying.

- Infocom trace [25]
- UCSD trace [18]
- DieselNet buses [3]
- SF Taxis trace [23]

The simulation results in [11] show that two hops yield nearly the same performance as using "any-hop" paths to disseminate information in the four real-world traces. This means that a two-hop relaying strategy can achieve delays close to the "optima." The designation of SCOOP is based on this result.

6.5.2.2 Positively Correlated Paths

Reference [11] shows that there are positive correlations across the two-hop relaying paths, and gives the statistics on the correlation coefficients between various paths. The results show that many routing paths are positive correlated. It implies that carefully selecting relays is crucial to optimize content distribution.

6.5.2.3 Relaying Strategy

Let $p_{i,u}(x)$ be the steady-state probability that a message of channel i is received within the deadline by user u under relaying strategy x. That is

$$p_{i,u}(x) = \mathbb{P}[A_{i,u} \leq t_i], \tag{6.24}$$

where $A_{i,u}$ is the age of a message of channel i when received by user u. We define the value of channel i to user u by $V_{i,u}(p_{i,u}(x))$, where $V_{i,u}$ is an increasing function [11]. The global system objective is to optimize the aggregate value of information channels across users in the system:

$$maximize \sum_{i \in I, u \in U} V_{i,u}(p_{i,u}(x)), \qquad x \in [0,1]^{|I| \times |U|}. \tag{6.25}$$

Assume that the cost for relay r to transmit and receive messages to be relayed at average rate a_r is captured through a cost function $C_r(a_r)$ assumed to be increasing, continuously differentiable, and convex [11]. This is accommodated by replacing the objective function in 6.25 by:

$$\sum_{i \in I, u \in U} V_{i,u}(p_{i,u}(x)) - \sum_{r \in U} C_r(a_r(x)). \tag{6.26}$$

Sub-gradient Algorithms:

SCOOP is based on the sub-gradient method, which amounts to updating the relay probabilities as follows, for every channel i and relay r,

$$\frac{d}{dt} x_{i,r} = \sum_{j \in I, u \in U} V'_{j,u}(p_{i,u}(x)) \frac{\partial}{\partial x_{i,r}} p_{j,u}(x). \tag{6.27}$$

The difficulty of this approach lies in evaluating the gradient in 6.27. SCOOP uses smoothed perturbation analysis (SPA) [5] and stochastic approximation to solve it. SCOOP provides a distributed relaying strategy; details can be found in [11].

References

[1] A. Balasubramanian, B. N. Levine, and A. Venkataramani. DTN routing as a resource allocation problem. In *Proc. ACM SIGCOMM*, 2007.

[2] A. Balasubramanian, B.N. Levine, and A. Venkataramani. How small labels create big improvements. In *Proceedings of ACM SIGCOMM*. ACM Press, 2007.

[3] A. Balasubramanian, B.N. Levine, and A. Venkataramani. Enabling interactive web applications in hybrid networks. In *Proceedings of ACM MobiCom*. ACM Press, 2008.

[4] J. Burgess, B. Gallagher, D. Jensen, and B. N. Levine. MaxProp: Routing for vehicle-based disruption-tolerant networking. In *Proc. of IEEE INFOCOM*, 2006.

[5] X.R. Cao. Perturbation analysis of discrete event systems: Concepts, algorithms, and applications. *European Journal of Operational Research*, pages 1–13, 1996.

[6] Elizabeth M. Daly and Mads Haahr. Social network analysis for routing in disconnected delay-tolerant manets. In *Proceedings of the 8th ACM International Symposium on Mobile Ad Hoc Networking and Computing*, pages 32–40, 2007.

[7] A. Doria, M. Udn, and D. P. Pandey. Providing connectivity to the Saami nomadic community. In *Proc. 2nd Int. Conf. on Open Collaborative Design for Sustainable Innovation*, Dec. 2002.

[8] H. Dubois-Ferriere, M. Grossglauser, and M. Vetterli. Age matters: Efficient route discovery in mobile ad hoc networks using encounter ages. In *Proc. of ACM MobiHoc*, 2003.

[9] Wei Gao, Qinghua Li, Bo Zhao, and Guohong Cao. Multicasting in delay tolerant networks: a social network perspective. In *Proceedings of the Tenth ACM International Symposium on Mobile Ad Hoc Networking and Computing*, pages 299–308, 2009.

[10] M. Grossglauser and M. Vetterli. Locating nodes with ease: Last encounter routing in ad hoc Networks through Mobility Diffusion. In *Proc. of IEEE INFOCOM*, 2003.

[11] D. Gunawardena, T. Karagiannis, Proutiere A., E. Santos-Neto, and M. Vojnovic. Scoop: Decentralized and opportunistic multicasting of information streams. In *Proceedings of ACM MobiCom*. ACM Press, 2011.

[12] P. Hui and J. Crowcroft. How small labels create big improvements. In *Proceedings of IEEE ICMAN* . IEEE Press, 2007.

[13] P. Hui, J. Crowcroft, and E. Yoneki. Bubble rap: Social-based forwarding in delay tolerant networks. In *Proceedings of ACM MobiHoc* . ACM Press, 2008.

[14] D.B. Johnson, D. A. Maltz, and J. Broch. DSR: the dynamic source routing protocol for multihop wireless ad hoc networks. In *Ad Hoc Networking*, Addison-Wesley, 2001.

[15] Yong Li, Yurong Jiang, Depeng Jin, Li Su, Lieguang Zeng, and D.O. Wu. Energy-efficient optimal opportunistic forwarding for delay-tolerant networks. *Vehicular Technology, IEEE Transactions on*, (9):4500–4512, 2010.

[16] A. Lindgren, A. Doria, and O. Schelen. Probabilistic routing in intermittently connected networks. *Lecture Notes in Computer Science*, 3126:239–254, August 2004.

[17] P.V. Marsden. Egocentric and sociocentric measures of network centrality. *Social Networks*, pages 407–422, October 2002.

[18] M. McNett and G.M. Voelker. Access and mobility of wireless PDA users. In *Mobile Computing Communications Review*, 2005.

[19] S.C. Nelson, M. Bakht, and R. Kravets. Encounter-based routing in DTNs. In *INFOCOM 2009, IEEE*, pages 846–854, 2009.

[20] M.E.J. Newman. Analysis of weighted networks. *Physical Review E*, 2004.

[21] G. Palla. Uncovering the overlapping community structure of complex networks in nature and society. *Nature*, pages 814–818, 2005.

[22] C. E. Perkins, E. M. Belding-Royer, and S. R. Das. Ad-hoc on-demand distance vector routing. In *IETF MANET DRAFT*, 2002.

[23] M. Piorkowski, N. Sarajanovic-Djukic, and M. Grossglauser. A parsimonious model of mobile partitioned networks with clustering. In *Proc. of COMNETS*, 2009.

[24] D. Rothfus, C. Dunning, and X. Chen. Social-similarity-based routing algorithm in delay tolerant networks. In *Proceedings of IEEE ICC*. IEEE Press, 2013.

[25] J. Scott, R. Gass, J. Crowcroft, P. Hui, C. Diot, and A. Chaintreau. CRAWDAD trace cambridge/haggle/imote/infocom. 2006.

[26] T. Spyropoulos, K. Psounis, and C. Raghavendra. Spray and focus: Efficient mobility-assisted routing for heterogeneous and correlated mobility. In *Proc. of IEEE PerCom*, 2007.

[27] Vikram Srinivasan, Mehul Motani, and Wei Tsang Ooi. Analysis and implications of student contact patterns derived from campus schedules. In *Proceedings of the 12th Annual International Conference on Mobile Computing and Networking*, pages 86–97, 2006.

[28] M. Y. Uddin, H. Ahmadi, T. Abdelzaher, and R. Kravets. A low-energy, multi-copy inter-contact routing protocol for disaster response networks. In *Proc. of IEEE SECON*, 2009.

[29] Amin Vahdat and David Becker. Epidemic routing for partially-connected ad hoc networks. In *Technical Report CS-200006*, Duke University, Apr. 2000.

[30] J. Wu and Yunsheng Wang. Social feature-based multi-path routing in delay tolerant networks. In *Proceedings of the 31th Annual IEEE International Conference on Computer Communications*. IEEE Press, 2012.

[31] Jie Wu, Mingjun Xiao, and Liusheng Huang. Homing spread: Community home-based multi-copy routing in mobile social networks. In *IEEE INFOCOM*, 2013.

[32] Q. Yuan, I. Cardei, and J. Wu. Predict and relay: An efficient routing in disruption-tolerant networks. In *Proceedings of the 10th ACM MobiHoc*. IEEE Press, 2009.

Chapter 7

Multicast in Opportunistic Networks

Yunsheng Wang
Kettering University
Flint, Michigan

Jie Wu
Temple University
Philadephia, Pennsylvania

CONTENTS

7.1 Introduction

In recent years, there has been much research activity regarding opportunistic networks, or delay-tolerant networks (DTNs) [9, 12], which are a type of wireless network for challenged environments. In opportunistic networks, end-to-end paths be-

Table 7.1 Multicast Routing Schemes in Opportunistic Networks

Single Node Model	Multiple Copies Model	Single Copy Model
[46], [41], [14]	[1], [44], [6], [40], [29], [27], [24], [37], [18]	[32], [10], [39], [15], [21]

tween some or all of the nodes in the network do not exist most of the time. DTNs are characterized by intermittent connectivity, unknown mobility patterns, and limited network capacity. These networks have a variety of applications, including crisis environments, such as emergency response and military battlefields [20, 26], vehicular DTN road communication [19, 4], deep-space communication [25, 3], connectivity of developing countries [28, 22], and social contact networks [23, 11].

While multicasting in the Internet and mobile ad hoc networks have been studied extensively, efficient multicasting in opportunistic networks is a considerably different and challenging problem, due to the probabilistic nature of contact among nodes. There also has been some work on multicast routing protocols in opportunistic networks [29, 43, 46, 10, 35, 14, 24, 21, 41, 39, 37, 1, 6, 15, 27, 18, 32, 44, 40]. In this chapter, we survey the most recent contributions on DTN multicast routings.

The existing research focuses on three models:

- ***Single node*** (also called ***ferry***) ***model***, in which one single node holds all destinations, and delivers them to each destination, as they contact one another, through movement.

- ***Multiple-copy model***, in which the destination set is replicated at a contact once a certain condition related to the quality of the encountered node is satisfied. Most of them are extensions of unicast routing protocols.

- ***Single-copy model***, where a single copy of each destination is maintained where destinations can be scattered at different nodes. Each destination is forwarded to an encountered node if it has a higher probability of reaching the corresponding destination.

The researchers have proved that network coding can potentially improve the performance of the routing protocols in opportunistic networks [45, 38, 13, 30]. However, this is out of the scope of this chapter.

7.2 Multicast

Multicast is the delivery of a message from a generated source node to a group of destinations. Compared to the unicast routing schemes, multicast routing schemes save bandwidth and achieve higher efficiency. In traditional Internet multicast routing, optimal distribution paths from the source to the destinations, and the routing, can be easily controlled by a tree. In delay mulitcast-tolerant networks (DTNs), the mobile nodes are constrained by the limitations of battery capacity, computation capability, and storage. At the same time, due to high mobility and intermittent connection, multicasting in DTNs becomes a challenging problem.

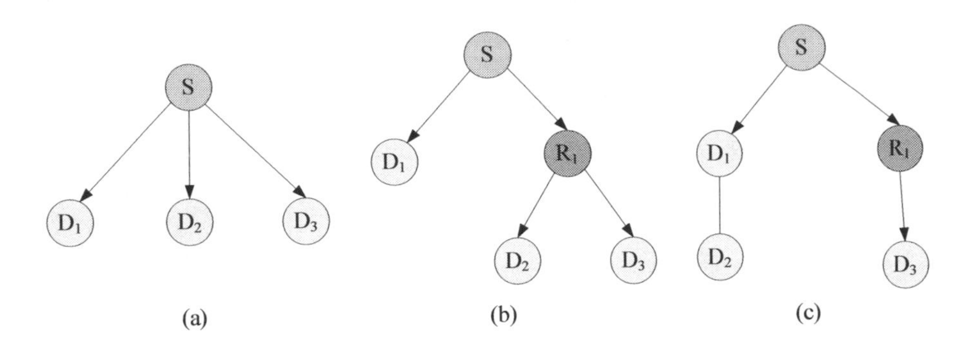

Figure 7.1: Illustration of a multicast tree.

There are different types of multicast trees, as seen in Figure 7.1. In Figure 7.1 (a), the source node will forward the message directly to the destination nodes. In Figure 7.1 (b), there are multicast relay nodes, which help the source forward the message. As shown in Figure 7.1 (c), the destination nodes can also act as the relay nodes, and can forward the message after they receive it.

Internet Protocol (IP) multicast is a bandwidth-conserving technology that reduces traffic by simultaneously delivering a single stream of information to thousands of corporate recipients and homes. Applications that take advantage of multicast include videoconferencing, corporate communications, distance learning, and the distribution of software, stock quotes, and news. IP multicast delivers source traffic to multiple receivers, without adding any additional burden to the source or the receivers, while using the least amount of network bandwidth of any competing technology.

A mobile ad hoc network (MANET) is composed of mobile nodes without any infrastructure. Mobile nodes self-organize to form a network over radio links. The goal of a MANET is to extend mobility into the realm of the autonomous, mobile and wireless domains, where a set of nodes form the network routing infrastructure in an ad hoc fashion. Multicasting can improve the efficiency of the wireless link when sending multiple copies of messages, by exploiting the inherent broadcast property of wireless transmission. Hence, reliable multicast routing plays a significant role in MANETs. In MANET multicast schemes, some construct multicast trees to reduce end-to-end latency, while others build mesh to ensure robustness. Some protocols create overlay networks, and use unicast routing to forward packets. Energy-aware multicast protocols optimize either the total energy consumption or system lifetime of the multicast tree.

DTN-based store-carry-forward communication is well suited to multicast and data sharing services, since bundle storage and forwarding to multiple contacts is an essential feature of DTN bundle routers. The goals for DTN multicast are: achieving a high total delivery rate for messages to interested receivers, achieving a fairly distributed delivery rate for messages from different sources, achieving minimum delivery delays, and optimizing resource utilization.

The multicast protocols in wired networks can be classified into two schemes: *shortest path multicast tree protocol* and *core-based tree multicast protocol* [5]. The single shortest path multicast tree can be constructed by applying Dijkstra's spanning tree algorithm [7]. In the core-based tree multicast protocol, a single node acts as the core of the tree, from which branches emanate. Several multicast routing protocols have been proposed for the Mobile Ad Hoc Network (MANET), which can be summarized as either tree-based or mesh-based, according to the kind of routes they create [42]. In the tree-based MANET multicast schemes, all the routes form a tree infrastructure with the source node as the root; thus, there is only one single path between each pair of sender and receiver. In the mesh-based MANET multicast schemes, a mesh infrastructure is maintained as the routing information, that is, more than one path between each sender and receiver pair exists, so it is more robust but less efficient. The single-copy model in DTNs is borrowed from the tree-based multicast protocols from wired networks and MANET, which had not constructed the multicast tree initially, because of the uncertainty property of the DTN environment. In some other DTN multicast schemes, the message will be forwarded to the nodes with higher activity levels, such as contact frequency, or higher centrality values. These nodes can be considered as the core of the tree. According to the challenge environment of DTNs, it is hard to construct a multicast tree from the source to the destinations. Researchers use the historical information or social network property, to predict the future contacts, in order to perform a better multicast process.

7.3 Single-Node Model

In the *single-node* (also called *ferry*) *model*, one single node holds all destinations, and delivers them to each destination as they contact one another, through movement. As shown In Figure 7.2, a ferry moves around, picks up the message from the source node, and drops it off to the destination nodes.

In [46], Zhao, Ammar, and Zegura proposed the basic single-node model, together with new semantics for DTN multicasting, which explicitly specify temporal constraints on group membership and message delivery. Reference [46] was the first study of multicasting in DTNs. The authors concluded that even with partial knowledge, multicast routing algorithms can perform efficiently in DTN environments. They also summarized that topology information is more important than group membership information in performing efficient multicasting in DTNs.

Yang and Chuah presented a two-stage single-node model in [41], where routes to destinations are first identified through a ferry, followed by the message delivery along the discovered routes. Their scheme included two stages: the *ferry-based interdomain multicast* and the *encounter-based intradomain multicast routing scheme with redundancy*. In the former stage, one or more message ferries were deployed to deliver interdomain traffic from one domain to another. The latter stage was used as the intradomain multicast routing scheme by a group leader, to deliver messages it received from a visiting ferry to its local multicast members.

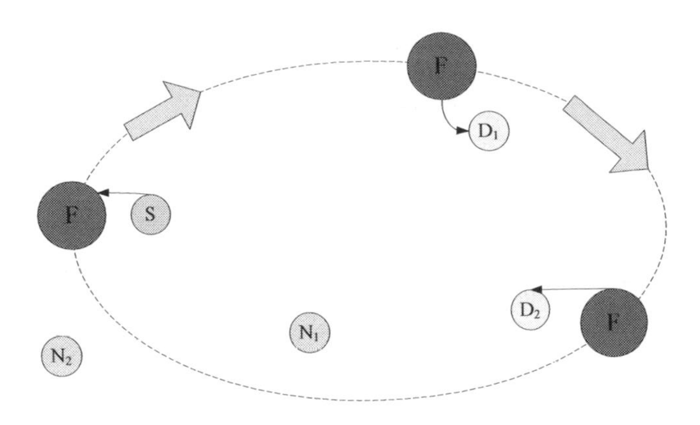

Figure 7.2: Illustration of a single-node model in DTN multicasting.

In [14], the authors extended the two-hop relay strategy in DTN multicasting. This is also considered to be a single-node model. The relay node is responsible for delivering messages directly to each multicast receiver. In this situation, the relay node is the ferry. The authors reported that their proposed multicast routing protocol can achieve the throughput upper bound $\Theta(n\lambda)$ for the case $n_s n_d = O(n)$, and $\Theta(\frac{n^2\lambda}{n_s n_d})$ for the case $n_s n_d = \omega(n)$, where n is the number of nodes, and the inter-contact time of an arbitrary pair of nodes follows an exponential distribution with rate λ. There are n_s sources, each of which is associated with n_d random destinations.

7.4 Multiple-Copy Model

In the multiple-copy model, the destination set is replicated at a contact once a certain condition related to the quality of the encountered node is satisfied. The number of replications, or copies, is hard to control in the DTN environment in the multiple-copy model, in which *ticket-based* or *token-based* schemes were proposed. There are different ways to measure the multicast ability of each node. The researchers used the inter-contact time, contact frequency, and contact duration of the node, with the destinations or all of the nodes as the evaluation metric. Recently, social network concepts have been introduced to measure the multicast ability of each node, such as centrality, similarity, and so on.

Epidemic routing [36], using a flooding-based scheme, is also suitable for DTN multicast. However, using epidemic routing will cause high overhead. In [1], Abdulla and Simon studied the effects of different controlled epidemic routing schemes using time-to-live (TTL), message expiration times, forwarding probabilities, and the number of copies to spread. The authors also analyzed the basic multicast routing scheme metrics such as message delivery ratio, message delay, and buffer occupancy.

In [44], the authors proposed a dynamic tree-based multicast approach in DTNs. A unique multicast tree is constructed for each bundle, and the tree is adjusted at each intermediate node, according to the current network conditions. When a node receives a bundle, it will dynamically adjust an initially constructed tree from the current knowledge of the network conditions. A newly discovered path will be quickly utilized by way of this adjustment. This multicast scheme also achieves better efficiency performance when the probability of link unavailability is high, and the duration of link downtime is large.

The approach discussed above relies on the opportunistic connectivities among nodes for delivery, without making use of additional information such as node location and node velocities. In 2009, Chuah and Yang introduced a context-aware multicast routing in [6], which is a node-density-based adaptive multicast scheme. Their scheme estimated the local node density and 2-hop neighbor contact probability to enhance the multicast routing scheme. This node-density-based adaptive multicast routing scheme can address the challenges of opportunistic link connectivity in DTNs, by using information like node locations and velocities. Hence, this scheme is flexible, and can adapt to different network environments.

The multicast routing schemes discussed above assumed a route discovery process that is similar to the mobile ad hoc network routing approaches. In [40], the authors introduced an encounter-based multicast scheme in DTNs, which was built on top of the PROPHET routing scheme [16]. The authors built a dynamic tree by referring to the contact probability between two nodes, rather than the shortest path route between two nodes. This scheme allowed the sender to have a replication capacity such that a partial mesh was built. This approach is an extension of the PROPHET [16] protocol. This scheme allows nodes to cache the data until a good next-hop node is found to relay the messages to the destinations, without considering the mobility pattern of each node.

Another multicast extension to the DTN's unicast PROPHET [16] protocol has been proposed in [29]. The authors introduced the use of information such as the node position and its moving direction, in addition to the estimation of the probability of contact between nodes, which can reduce the number of copies of the message to multicast. The authors demonstrated that if there are a minimum number of contacts between nodes, multicast works efficiently, minimizing the number of message replications done in the network. Additionally, they demonstrated that multicasting can improve DTNs' efficiency, for it permits the saving of resources as the number of message replications is minimized in the network. Fewer replications mean that less processing capacities are needed, more memory and bandwidth are available, and packets suffer less delay and loss.

Network coding is a mechanism in which nodes encode two or more incoming messages, and forward encoded messages instead of forwarding them as they are. In [24], Narmawala and Srivastava designed a network–coding–based multi-copy routing scheme for DTN multicast. The authors proposed different packet purging schemes to drain messages out of the network, which takes advantage of features of network coding, to increase buffer efficiency.

In [18], Lo and Luo extended their work [17] from unicast to multicast in DTNs, which takes advantage of the contact behavior in DTN environments to predict good message relay nodes that help to forward messages to their destinations. The approach dynamically adjusts the number of message copies generated, depending on different network situations. The authors also proposed a suitable buffer management system to determine the transmission order of these buffered messages when they have the opportunity to be forwarded or copied to other nodes. In their scheme, they can use the received time (entering time into a buffer) and the hop count (number of hops taken from the source) for buffer management. This control method can significantly reduce redundant message copies in the network.

In [27], the authors implemented the multicast routing protocol in vehicular delay-tolerant networks, which was a combination of geographic-based routing [34], PROPHET [16], and Spray-and-Wait [31] protocols. A differentiation between dense and sparse scenarios was used, according to an estimation of the number of nodes met. If a dense scenario was detected, the source is assigned a TTL to the created message, according to the message priority. Messages were scheduled according to priority and general buffer occupancy estimation. In sparse scenarios, better results were achieved with a weighted fair queuing (WFQ) [33] scheduling policy, which sorts the message buffer fairly. In their scheme, a dynamic assignment of the maximum number of MULE nodes, and a careful choice of MULE nodes, was made. The list of contacted nodes was always sorted according to their vehicle density estimation, thus it was possible to forward messages first to those contacts with more contact probabilities. Another implemented feature was the choice of MULE vehicles, according to their movement direction, which has to be at least 45° different from the movement direction of the node that carries the bundles; this contributes to a better spreading of the bundles.

In delegation forwarding (DF) multicast [37], the authors extended the delegation forwarding algorithm [8] in DTN multicast. The message holder for each destination would replicate the copy (for that destination) and forward it to an encountered node that has a higher quality than all previous nodes seen so far, with respect to that particular destination. The proposed multicast scheme was based on the dynamic multicast trees, as shown in Figure 7.3.

For any node i, the forwarding problem is a simple question: Upon contact with node j, should node i forward the message to node j? For many algorithms, the answer to the forwarding question is "forward the message if node j's quality is higher than that of node i." However, the cost of this approach can still be quite high. To reduce the cost, we seek to forward the message only to the highest-quality nodes that have previously met. Conceptually, we would like to forward less, and give the message to the nodes which are the best candidates for eventual delivery to the destinations. Thus, the forwarding question is whether node j is among the very highest-quality nodes.

The delegation forwarding multicast algorithm's main idea is to assign a quality value (which is static), and a level value (which is dynamic) for each node to each destination. Initially, the level value τ_{ia} for destination a of each node is equal to its quality value x_{ia} for destination a. During the routing process, a message holder i compares the quality x_{ja} of the node j it meets with its level value τ_{ia}. It only forwards the message to a node with a higher quality value than its own level value, and "asks"

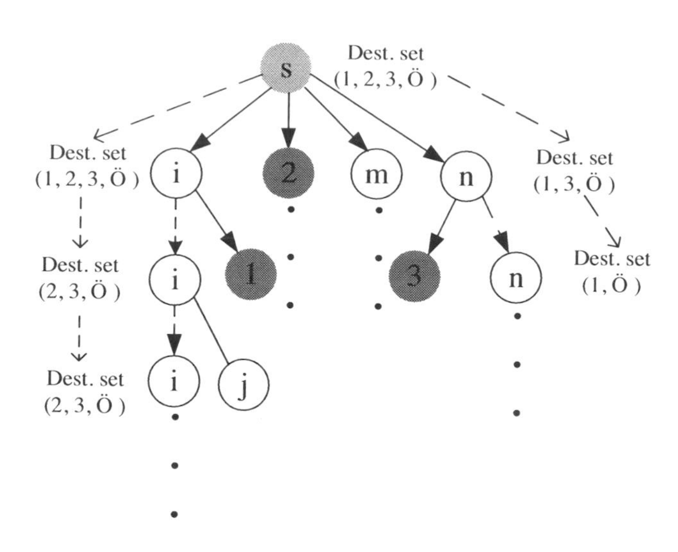

Figure 7.3: Illustration of a delegation forwarding multicast tree in DTNs.

this node to help forward the message to destination a. This approach does not need global knowledge. Each node decides whether it should or should not forward the message by itself. This is suitable for a distributed environment, such as a DTN. In addition, the message holder also raises its own level to meet the higher quality. If node j is one of the destinations, node i will forward a message to it, and will also use the strategy to determine whether node j is a good relay to forward the message.

The DF algorithm is shown in Figure 7.4 and Algorithm 11. The main difference from the previous two models is, in DF, the message is replicated and, after the

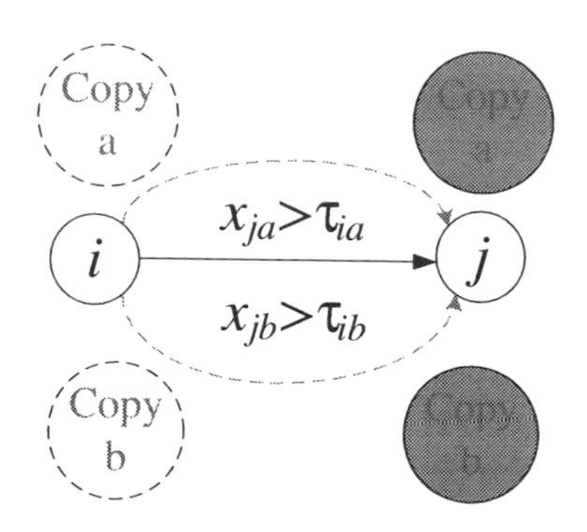

Figure 7.4: Delegation forwarding multicast in DTNs.

Algorithm 11: Delegation forwarding multicasting

```
1  There are N nodes in the network.
2  There are D destinations that need to be multicast.
3  Node n has quality x_nd and level τ_nd for destination d.
4  INITIALIZE ∀n,d : τ_nd ← x_nd.
5  On the contacts between node i, which is the message holder for destination
    a and node j:
6  if x_ja > τ_ia then
7     τ_ia ← x_ja
8     if node j does not have the message for destination a then
9        replicate a message to node j.
10    end if
11 else
12    if node j is the destination a then
13       replicate a message to node j.
14    end if
15 end if
```

forwarding process, initial message holder i and its relay node j will both have a copy of the message; therefore, there will be multiple nodes to seek the destinations. This means that DF can dramatically reduce the cost and delay.

The cost of DF in DTN unicast is given in [8]. Wang et al. gave a formal analysis of the cost of delegation forwarding multicast in [37]. For any node i maintaining a quality metric for destination a: x_{ia} (which lies between $(0,1]$) and a level value τ_{ia}, we focus on the gap $g_{ia} = 1 - \tau_{ia}$ between the current level and 1. The node that generates the message has a level value initially equal to its quality value, *i.e.*, $\tau_{ia} = x_{ia}$. The authors denote the initial gap $g_a = 1 - x_{ia}$.

Theorem 1 *Given the level value τ_{ia}, the expected number of forwardings in the delegation forwarding multicast model is*

$$E[F_{delegation}] \lesssim \frac{1}{2}\sqrt{N} + \frac{1}{3}D \cdot \sqrt{N},$$

where N is the number of nodes, and D is the number of destinations.

Proof 7.1 Suppose a node updates its gap value n times. The node's current gap is denoted as the random variable G_n. Since nodes meet according to rates that are independent of node quality, the node is equally likely to meet a node with any particular quality value. The next update of the gap of the nodes then occurs as soon as it meets a node with a higher quality value than G_n, and all values above this level are equally likely.

Hence,

$$G_{n+1} = G_n \cdot U, \tag{7.1}$$

where U is independent of G_n and follows a uniform distribution on $(0,1]$. According to [8], in the multicast scheme, the authors found:

$$E\left[G_{n+1}|G_n\right] = \frac{G_n}{2}, \text{ hence, } E\left[G_n\right] = \frac{\sum_{a=1}^{D} g_a}{2^n}.$$

Moreover, from Equation (7.1), we see that G_n approximately follows a lognormal distribution, with median $(\sum_{a=1}^{D} g_a)/e^n$. Hence, the distribution is highly skewed with most of the probability mass below the mean. So with a large probability, it has $G_n \leq (\sum_{a=1}^{D} g_a)/2^n$.

As in [8], the replication process can be described by a dynamic binary tree T, which contains all the nodes that have a copy of the message. The authors defined the set $B_a = \left\{i|x_{ia} \geq 1 - \frac{g_a}{\sqrt{N}}\right\}$, $a \in \{1,2,...,D\}$, which is called the *target set*. The authors identify a subtree of the tree T in which children are excluded for nodes having a threshold above $1 - \frac{g_a}{\sqrt{N}}$. All nodes in the subtree have a gap less than $\frac{g_a}{\sqrt{N}}$. This subtree is called the *target-stopped tree*.

According to [8], the essential observation is the following: if n is close to $log_2\left(\sqrt{N}\right)$, then except for a small probability, a node at generation n in the tree has a gap of at most $g_a/2^n \leq g_a/\sqrt{N}$. Hence, the authors safely assume that the target-stopped tree has a depth of at most n. Note that the total number of nodes appearing at generations $0,1,\ldots,n-1$ is at most $2^n = \sqrt{N}$.

In [8], Erramilli et al. offer the number of forwardings in the delegation forwarding unicast model. Hence, in the delegation forwarding multicast model of D destinations, the total size of this tree is at most:

$$C_{delegation} \lesssim \sqrt{N} + |\sum_{a=1}^{D} B_a| = (1 + \sqrt{\sum_{a=1}^{D} g_a}) \cdot \sqrt{N}.$$

Then, the total number of forwardings can be obtained:

$$F_{delegation} \lesssim \frac{1}{2}(1 + \sqrt{\sum_{a=1}^{D} g_a}) \cdot \sqrt{N},$$

since, $\sqrt{\sum_{a=1}^{D} g_a} \leq \sum_{a=1}^{D} \sqrt{g_a}$.

Hence,

$$\int_0^1 \sqrt{\sum_{a=1}^{D} g_a} dg_a \leq \sum_{a=1}^{D} \int_0^1 \sqrt{g_a} dg_a = \frac{2}{3} D.$$

Therefore,

$$E[F_{delegation}] = \int_0^1 \frac{1}{2}(1+\sqrt{\sum_{a=1}^{D} g_a})\cdot\sqrt{N}dg_a \lesssim \frac{1}{2}\sqrt{N}+\frac{1}{3}D\cdot\sqrt{N}. \quad (7.2)$$

7.5 Single-Copy Model

In order to reduce the overhead, the single-copy model is introduced in DTN multicast. There is only one copy for each destination, and destinations can be scattered at different nodes. Each destination is forwarded to an encountered node if it has a higher probability of reaching the corresponding destination.

Srinivasan and Ramanathan proposed a reliable anonymous multicast scheme in DTNs [32]: a scheme to ensure the reliable delivery of messages to all multicast receivers in DTNs, by making use of non-multicast nodes to reduce delivery latency. Instead of relying on a fixed multicast tree to guarantee anonymous message delivery, the authors utilized the changes in connectivity as they occur to deliver the messages. The anonymity part of this scheme ensures that, for a certain node i, no other node in the network can predict whether it is a multicast receiver, based on previous message interactions. Multicast receiver identities are kept secret from the custodial nodes that provide guarantees to such receivers. In addition, the provided anonymous authentication scheme enables a node i to authenticate a multicast group member j, while keeping the identities of both i and j hidden. This scheme also made controlled use of nonmulticast nodes to substantially improve the average latency of message delivery when there were no malicious nodes in the network.

In [10], Gao et al. developed a single-copy model where the forwarding metric is based on the social network perspective, which is a social-based approach, based on a notion of "ego-centric betweenness," in order to optimize multicast performance in DTNs. The authors formulated relay selections for multicast as a unified knapsack problem, by exploiting node centrality and social community structures. They assumed that individual contact pairs can be modeled as independent Poisson processes, with rates equal to the respective link weight. The authors validated this assumption on two popular real traces (INFOCOM and MIT), and find it to hold for a significant percentage of contact pairs. In this paper, two multicast problems were considered: single-data multicast and multiple-data multicast, whose goals were to deliver a data item or a set of data items to a set of destinations within the time constraint T. The additional optimization objective will minimize the number of relays used to achieve the average delivery ratio p. The authors also analyzed the exponential component in pair-wise inter-contact times.

The authors considered the node selfishness on DTN multicasting in [15]. They investigated how the selfish behaviors of nodes affect the performance of DTN multicast, in two-hop relaying and epidemic relaying schemes. Their work was based on theoretical analysis and experimental simulations, which reveal the following two characteristics of the impact of selfish behavior on the routing performance. First, the routing performance (i.e., delivery ratio, delivery cost, and delivery latency) is seriously degraded when a major portion of the nodes in the network are selfish. Second, the impact on the routing performance was related to the non-cooperative action of selfish behavior (i.e., not forwarding the messages, and dropping them). Specifically, the behavior of not forwarding the messages reduces the delivery cost, while the behavior of dropping the messages increases the delivery cost. However, both of them decrease the delivery ratio, and prolong the delivery latency, even if messages are eventually delivered. Their results also indicated that different selfish behaviors may have different impacts on different performance metrics. In addition, selfish behaviors influenced epidemic relaying more than two-hop relaying did.

In [21], the authors studied the DTN multicast problem from the graph indexing point of view. They solved the problem of minimizing the remote communication cost for multicast in DTNs. The authors analyzed this problem in the case of scheduled trajectories and known traffic demands, and proposed a solution based on a novel graph indexing system. Their system can solve the demand cover problem optimally on large real instances (data sets with millions of events and queries with thousands of nodes) in less than 10 seconds, in most cases.

In [39], Wu and Wang proposed a non-replication multicasting scheme in DTNs. The address of each destination was not replicated, but was assigned to a particular node based on its contact rate level and activity level. Their scheme was based on a dynamic multicast tree, where each leaf node corresponds to a destination. Each tree branch was generated at a contact based on the compare-split rule. The *compare* part determined when a new search branch was needed, and the *split* part decided how the destination set should be partitioned. Figure 7.5 illustrates the priority-based-split in [39].

In [39], the authors introduced two metrics to control the compare-split process: *contact rate level* and *active level.* They assume that there are N nodes in the whole network. The destination set of a multicast is represented as $D = \{1,2,...,n\}$. Each node a is associated with a contact rate vector $(f_1^a, f_2^a, ..., f_n^a)$, where f_i^a indicates the frequency that node a meets destination i in a given period T. f_i^a is also called the *contact rate level* for destination i in period T. We use $(c_1^a, c_2^a, ..., c_n^a)$ to indicate the number of times that node a meets destination i in a given period T. Hence, the contact rate level f_i^a can be presented as follows:

$$f_i^a = \frac{c_i^a}{T}. \tag{7.3}$$

In order to involve the effect of recent information, we also define T', which is considered as a recent period. $(c_1^{a'}, c_2^{a'}, ..., c_n^{a'})$ indicates the frequency that node a meets destination i in a given period T'. Hence, we define the contact rate level f_i^a as follows:

$$f_i^a = w \cdot \frac{c_i^{a'}}{T'} + (1-w) \cdot \frac{c_i^a - c_i^{a'}}{T - T'}, \tag{7.4}$$

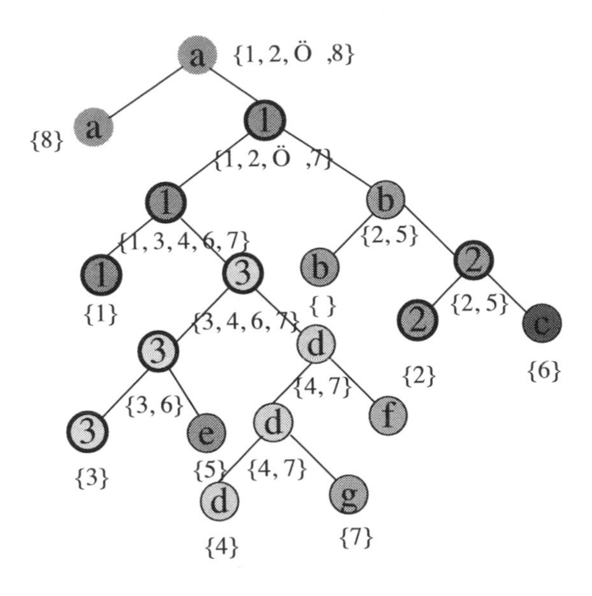

Figure 7.5: Illustration of the non-replication multicasting scheme in DTNs.

where w is the weight of the recent information for the contact rate level.

The *active level* of node a: A_a is the number of total contacts per time unit T that node a meets with all other nodes in the network.

$$A_a = \sum_{i=1}^{N} f_i^a \tag{7.5}$$

Figure 7.6 illustrates the system models considering the recent information.

The first step for the non-replication multicasting scheme [39] is *compare*. When node a, with a subset of destinations $D' \subseteq D$, had a contact with a new node b, without any destination subset, node a would first send D' information to node b. Nodes a, and b exchanged their contact rate vectors, $(f_1^a, f_2^a, ..., f_m^a)$ and $(f_1^b, f_2^b, ..., f_m^b)$, upon their contact, where f_i^a indicates the frequency that node a meets destination i in a

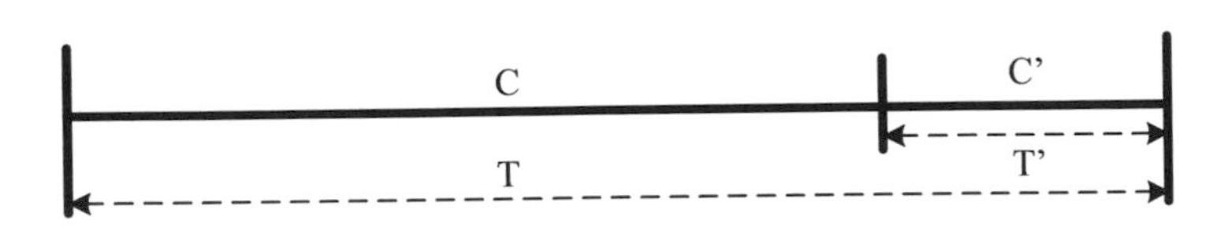

Figure 7.6: Illustration of the system models.

given period T. f_i^a is also called the *contact rate level* for destination i in period T. m is the size of the subset D' of destination set D. After comparing these two nodes' sum of the contact rate levels for all destinations, if $\sum_{i=1}^{m} f_i^b > \sum_{i=1}^{m} f_i^a$, then we advance to the next step, *split*.

Note that two rounds of exchanges are used. One round can be saved by exchanging $(f_1^a, f_2^a, ..., f_n^a)$ and $(f_1^b, f_2^b, ..., f_n^b)$. $(f_1^a, f_2^a, ..., f_m^a)$ and $(f_1^b, f_2^b, ..., f_m^b)$ can then be extracted locally.

The second step is to *split* the destination set. Suppose that $d_i = f_i^a - f_i^b$ is called the *contact rate difference* between nodes a and b for destination i. The activity levels A_a can be denoted by the total number of times that node a comes into contact with all other nodes.

$$A_a = \sum_{i=1}^{N} f_i^a \tag{7.6}$$

The destination set splitting is based on the ratio of two encountering nodes' activity levels. The ratio k can be denoted as:

$$k = \lceil \frac{A_a}{A_a + A_b} \times m \rceil. \tag{7.7}$$

1. Both a and b generate the contact rate difference vector $(d_1, d_2, ..., d_m)$. Find the kth largest element in $O(m)$ operations using a general *selection algorithm* [2].

2. Node a keeps k nodes that have higher values than, or equal values to, the kth largest element. In the case of a tie, when two contact rate differences are equal, the node ID is used to break the tie.

3. Node b keeps $m - k$ nodes that have lower values than, or equal values to, the kth largest element.

In step (1), the optimal linear solution is used to find the kth largest element. The major goal in using the non-replication multicasting scheme [39] in DTNs is to ensure that the delivery of multicast information was done over different paths. Each path had a relatively high contact frequency of reaching the corresponding destination subset quickly. Then, multiple holders for destination nodes can search for destinations in parallel. These solutions can reduce the multicast cost. The number of forwardings is a major metric to measure the cost of the multicasting process. Compare-split can also reduce the latency in DTN multicasting.

Suppose that D_a was the destination subset kept in node a, and that D_b is the destination subset assigned to b; the authors proposed to maximize the *combined contact rate* of a and b as follows:

$$max\{\sum_{i \in D_a} a_i + \sum_{j \in D_b} b_j\}. \tag{7.8}$$

Theorem 2 *Suppose that D_a and D_b are two subsets as a result of the kth element partition. $d_i = f_i^a - f_i^b$ is called the* contact rate difference *between nodes a and b for destination i. The maximum combined contact rate of visiting any of the destinations within a time period occurs when, for each $i \in D_a$ and $j \in D_b$, $d_i \geq d_j$.*

Proof 7.2 It is clear that any other partition (including the optimal one) can be generated through a sequence of swaps between two elements, one each from D_a and D_b. We show that each swap will deteriorate the combined contact rate level. Suppose i in D_a and j in D_b are swapped. Based on the condition $d_i \geq d_j$, we have $f_i^a - f_i^b \geq f_j^a - f_j^b$, or $f_i^a + f_j^b \geq f_i^b + f_j^a$.

Note that $f_i^a + f_j^b$ is the combined contact rate involving destinations i and j, whereas $f_j^b + f_i^a$ is the combined contact rate after the swap of i and j. The theorem follows.

This optimal split algorithm can partition the destinations to nodes with higher contact rate levels; hence, it can reduce the number of forwardings and latency in DTN multicasting.

7.6 Discussion and Future Work

It has been eight years since the first DTN multicast paper was published [46]. Since then, the field of DTN multicasting has attracted tremendous attention and research. With the growing amount of social data, ubiquitous systems, and mobile social media applications in everyday life, the analysis of social networks is receiving increased attention. While there is a large body of research concerning online social networks, important aspects of offline opportunistic mobile social networking, which is a new type of DTN, still remains largely unexplored. Mobile computing services are becoming highly personalized and influenced by user location, social behavior, and interests. Exploiting various aspects of the social behavior of mobile users can improve multicast performance in the opportunistic mobile social networks.

In the future of opportunistic network multicasting, researchers can use different types of technology to benefit the routing schemes, such as RFID, Bluetooth, and smartphones, to record the participants' contact information, geographic information, phone call or message information, and so on. There are two directions for opportunistic network multicasting:

- *Social-behavior-based multicasting*: using the knowledge of the participants' social information for multicasting guidance. This social information includes social features, community properties, strong and weak ties, centrality, influential, etc.
- *Mobility-based multicasting*: There are two important features of mobile networks. One is the mobility of the nodes, which results from the intrinsic nature of humans, compelling them to travel with their devices from one

location to another. Another important feature is that direct communication between any two devices is only possible when they are within transmission range of each other. These two features make such networks highly dynamic in terms of their connectivity, and strongly dependent on human mobility patterns. Exploiting the human mobility and relationship models in opportunistic mobile social networks can improve the performance of DTN multicasting.

7.7 Conclusion

In this chapter, we have reviewed the multicast routing schemes in opportunistic networks. We classified these schemes into three categories: single node, multiple-copy, and single-copy models. In the single-node model, one particular node will be responsible for multicasting the message to all destinations. Although the overhead and security can be controlled, the message delivery delay will increase dramatically. In the multiple-copy model, researchers usually design a multicast tree to forward the message. These routing protocols can achieve high delivery rates and small delivery delays, but it is still a challenging problem to effectively control the number of copies of the message within an opportunistic environment. In the single-copy model, each destination has only one copy. These schemes can control the communication overhead and buffer occupancy, and perform well, all at the same time.

References

[1] M. Abdulla and R. Simon. Controlled epidemic routing for multicasting in delay tolerant networks. In *Proc. of IEEE International Symposium on Modeling, Analysis and Simulation of Computers and Telecommunication Systems (MASCOTS)*, pages 1–10, 2008.

[2] M. Blum, R. W. Floyd, V. R. Pratt, R. L. Rivest, and R. Endre Tarjan. Time bounds for selection. *J. Comput. Syst. Sci.*, 7(4):448–461, 1973.

[3] S. Burleigh, A. Hooke, L. Torgerson, K. Fall, V. Cerf, B. Durst, K. Scott, and H. Weiss. Delay-tolerant networking: an approach to interplanetary Internet. *IEEE Communications Magazine*, 41(6):128–136, June 2003.

[4] D. Câmara, N. Frangiadakis, C. Bonnet, and F. Filali. *Vehicular delay tolerant networks*. Book chapter N23 in *Handbook of Research on Mobility and Computing: Evolving Technologies and Ubiquitous Impacts*, IGI Global, 2011, ISBN: 9781609600426, 2011.

[5] X. Chen and J. Wu. Multicasting techniques in mobile ad hoc networks, pages 25–40. *The Handbook of Ad Hoc Wireless Networks*. CRC Press, Inc., 2003.

[6] M. Chuah and P. Yang. Context aware multicast routing scheme for disruption tolerant networks. *Int. J. Ad Hoc Ubiquitous Comput.*, 4(5):269–281, July 2009.

[7] T. H. Cormen, C. E. Leiserson, R. L. Rivest, and C. Stein. *Introduction to Algorithms*. The MIT Press, 1997.

[8] V. Erramilli, M. Crovella, A. Chaintreau, and C. Diot. Delegation forwarding. In *Proc. of the ACM International Symposium on Mobile Ad Hoc Networking and Computing (MobiHoc)*, pages 251–260, 2008.

[9] K. Fall. A delay-tolerant network architecture for challenged internets. In *Proc. of the ACM Conference on Applications, Technologies, Architectures, and Protocols for Computer Communications (SIGCOMM)*, pages 27–34, 2003.

[10] W. Gao, Q. Li, B. Zhao, and G. Cao. Multicasting in delay tolerant networks: a social network perspective. In *Proc. of the Tenth ACM International Symposium on Mobile Ad Hoc Networking and Computing (MobiHoc)*, pages 299–308, 2009.

[11] P. Hui, A. Chaintreau, J. Scott, R. Gass, J. Crowcroft, and C. Diot. Pocket switched networks and human mobility in conference environments. In *Proc. of the ACM SIGCOMM Workshop on Delay-Tolerant Networking (WDTN)*, pages 244–251, 2005.

[12] S. Jain, K. Fall, and R. Patra. Routing in a delay tolerant network. In *Proc. of the ACM Conference on Applications, Technologies, Architectures, and Protocols for Computer Communications (SIGCOMM)*, pages 145–158, 2004.

[13] A. Khreishah, I. M. Khalil, and J. Wu. Distributed network coding-based opportunistic routing for multicast. In *Proc. of the ACM MobiHoc '12*, pages 115–124, 2012.

[14] U. Lee, S. Oh, K. Lee, and M. Gerla. Relaycast: scalable multicast routing in delay tolerant networks. In *Proc. of IEEE International Conference on Network Protocols (ICNP)*, pages 218 –227, 2008.

[15] Y. Li, G. Su, D.O. Wu, D. Jin, L. Su, and L. Zeng. The impact of node selfishness on multicasting in delay tolerant networks. *IEEE Transactions on Vehicular Technology*, 60(5):2224 –2238, June 2011.

[16] A. Lindgren, A. Doria, and O. Schelén. Probabilistic routing in intermittently connected networks. *SIGMOBILE Mob. Comput. Commun. Rev.*, 7(3):19–20, July 2003.

[17] S. Lo and W. Liou. Dynamic quota-based routing in delay-tolerant networks. In *Proc. of the IEEE Vehicular Technology Conference (VTC)*, pages 1 –5, 2012.

[18] S. Lo and N. Luo. Quota-based multicast routing in delay-tolerant networks. In *Proc. of International Symposium on Wireless Personal Multimedia Communications (WPMC)*, pages 544 –548, 2012.

[19] P. Luo, H. Huang, W. Shu, M. Li, and M. Wu. Performance evaluation of vehicular DTN routing under realistic mobility models. In *Proc. of the IEEE Wireless Communications and Networking Conference (WCNC)*, pages 2206 –2211, 2008.

[20] R. Malladi and D. P. Agrawal. Current and future applications of mobile and wireless networks. *Commun. ACM*, 45(10):144–146, October 2002.

[21] M. Mongiovi, A.K. Singh, X. Yan, B. Zong, and K. Psounis. Efficient multicasting for delay tolerant networks using graph indexing. In *Proc. of the IEEE INFOCOM*, pages 1386 –1394, 2012.

[22] M. Musolesi and C. Mascolo. A framework for multi-region delay tolerant networking. In *Proc. of the ACM Workshop on Wireless Networks and Systems for Developing Regions (WiNS-DR)*, pages 37–42, 2008.

[23] M. Musolesi and C. Mascolo. Car: Context-aware adaptive routing for delay-tolerant mobile networks. *IEEE Transactions on Mobile Computing*, 8(2):246–260, 2009.

[24] Z. Narmawala and S. Srivastava. Midtone: Multicast in delay tolerant networks. In *Proc. of the International Conference on Communications and Networking in China (ChinaCOM)*, pages 1 –8, 2009.

[25] NASA. NASA successfully tests first deep space Internet. `http://www.nasa.gov/home/hqnews/2008/nov/HQ_08-298_Deep_space_internet.html`, 2008.

[26] NASA. Disruption tolerant networking for space operations (DTN). `http://www.nasa.gov/mission_pages/station/research/experiments/DTN.html`, 2012.

[27] A. Palma, P.R. Pereira, A. Casaca, and null. Multicast routing protocol for vehicular delay-tolerant networks. In *Proc. of IEEE International Conference on Wireless and Mobile Computing, Networking and Communications (WiMob)*, pages 753 –760, October 2012.

[28] A. Pentland, R. Fletcher, and A. Hasson. Daknet: rethinking connectivity in developing nations. *Computer*, 37(1):78 – 83, January 2004.

[29] J. Santiago, A. Casaca, and P. Pereira. Multicast in delay tolerant networks using probabilities and mobility information. *Ad Hoc & Sensor Wireless Networks*, 7(1-2):51–68, 2009.

[30] L. Sassatelli and M. Medard. Inter-session network coding in delay-tolerant networks under Spray-and-Wait routing. In *Proc. of the 10th International Symposium on Modeling and Optimization in Mobile, Ad Hoc and Wireless Networks (WiOpt)*, pages 103–110, 2012.

[31] T. Spyropoulos, K. Psounis, and C. S. Raghavendra. Spray and Wait: an efficient routing scheme for intermittently connected mobile networks. In *Proc. of the 2005 ACM SIGCOMM Workshop on Delay-Tolerant Networking (WDTN)*, pages 252–259, 2005.

[32] K. Srinivasan and P. Ramanathan. Reliable anonymous multicasting in disruption tolerant networks. In *Proc. of the IEEE Global Telecommunications Conference (GLOBECOM)*, pages 1 –5, 2008.

[33] D. Stiliadis and A. Varma. Latency-rate servers: a general model for analysis of traffic scheduling algorithms. *IEEE/ACM Transactions on Networking*, 6(5):611 –624, October 1998.

[34] I. Stojmenovic. Position-based routing in ad hoc networks. *IEEE Communications Magazine*, 40(7):128 –134, July 2002.

[35] S. Symington, R. C. Durst, and K. Scott. Custodial multicast in delay tolerant networks. In *Proc. of IEEE Consumer Communications and Networking Conference (CCNC)*, pages 207 –211, 2007.

[36] A. Vahdat and D. Becker. Epidemic Routing for Partially Connected Ad Hoc Networks. Technical Report, Dept. of Computer Science, Duke University, 2000.

[37] Y. Wang, X. Li, and Jie Wu. Multicasting in delay tolerant networks: Delegation forwarding. In *Proc. of the IEEE Global Telecommunications Conference (GLOBECOM)*, pages 1 –5, 2010.

[38] J. Widmer and J. Le Boudec. Network coding for efficient communication in extreme networks. In *Proc. of the ACM SIGCOMM Workshop on Delay-Tolerant Networking (WDTN)*, pages 284–291, 2005.

[39] J. Wu and Y. Wang. A non-replication multicasting scheme in delay tolerant networks. In *Proc. of the IEEE International Conference on Mobile* Ad Hoc *and Sensor Systems (MASS)*, pages 89 –98, 2010.

[40] Y. Xi and M. Chuah. An encounter-based multicast scheme for disruption tolerant networks. *Computer Communications*, 32(16):1742–1756, 2009.

[41] P. Yang and M. Chuah. Efficient interdomain multicast delivery in disruption tolerant networks. In *Proc. of the International Conference on Mobile Ad-Hoc and Sensor Networks (MSN)*, pages 81 –88, 2008.

[42] S. Yang and J. Wu. New technologies of multicasting in MANET. *Design and Analysis of Wireless Networks*, 2005.

[43] Q. Ye, L. Cheng, M. Chuah, and B. D. Davison. Performance comparison of different multicast routing strategies in disruption tolerant networks. *Comput. Commun.*, 32(16):1731–1741, October 2009.

[44] Q. Ye, L. Cheng, M. Chuah, and B.D. Davison. Os-multicast: On-demand situation-aware multicasting in disruption tolerant networks. In *Proc. of the IEEE Vehicular Technology Conference (VTC)*, volume 1, pages 96 –100, 2006.

[45] X. Zhang, G. Neglia, J. Kurose, D. Towsley, and H. Wang. Benefits of network coding for unicast application in disruption-tolerant networks. *IEEE/ACM Transactions on Networking*, 2012.

[46] W. Zhao, M. Ammar, and E. Zegura. Multicasting in delay tolerant networks: semantic models and routing algorithms. In *Proc. of the ACM SIGCOMM WorkShop on Delay-Tolerant Networking (WDTN)*, pages 268–275, 2005.

Chapter 8

Interest-Based Data Dissemination in Opportunistic Mobile Networks: Design, Implementation, and Evaluation

Wei Gao
Department of Electrical Engineering and Computer Science
The University of Tennessee
Knoxville, Tennessee

Wenjie Hu
Department of Computer Science and Engineering
The Pennsylvania State University
University Park, Pennsylvania

Guohong Cao
Department of Computer Science and Engineering
The Pennsylvania State University
University Park, Pennsylvania

CONTENTS

This research focuses on providing pervasive data access to mobile users without support of cellular or Internet infrastructure. Two mobile users are able to communicate and share data with each other whenever they opportunistically move into the Bluetooth communication range of their smartphones. We designed and implemented our system on Android-based smartphones, and deployed our system to students at the Penn State University campus. Our system dynamically captures the interests of mobile users at runtime, and intelligently distribute to users the data that they are interested in. It also provides a unique research facility for investigating the interests and the data access patterns of users in various mobile environments.

8.1 Introduction

With recent technical advances and popularization of smartphones which are able to store, display, and transmit various types of media content, it is important to provide pervasive data access to mobile users, and distribute media content to these users promptly and efficiently. A straightforward solution is to provide such data access via the 3G cellular network infrastructure. However, the data access will create a huge amount of data traffic for the 3G network, which imposes immense pressure on the limited spectrum and the backend resources of 3G networks, and hence adversely affects the quality of service. This also motivates research on offloading 3G traffic to other communication networks such as WiFi [23, 1], but the coverage of WiFi accessibility is still limited nowadays as reported in [1], especially for mobile users which keep moving.

To address this challenge, in this chapter we develop a practical system which enables mobile users to autonomously transmit and share data with each other by exploiting the unused *opportunistic communication links* between the short-range radios (i.e., Bluetooth) of their smartphones. Two users are able to communicate when they opportunistically *contact* each other, i.e., move into the Bluetooth communication range of each other. The networks consisting of these users are called *opportunistic mobile networks*, which are also known as delay-tolerant networks (DTNs) [7] or Pocket Switched Networks (PSNs) [14].

The contributions of our work are two-fold. First, our system provides prompt and efficient access on categorized web data to mobile users, based on their interests in the data. More specifically, our system dynamically captures the interests of mobile users at runtime, and intelligently sends them their interested data. The power constraints of smartphones are also taken into account, and we propose solutions to trade-offs between power consumption and data availability of mobile users. To the best of our knowledge, this is the first work to implement opportunistic data sharing/access among smartphones. Previous work has focused on analyzing the contact or communication patterns of mobile users based on the collected user behavior records [19, 18, 1], and furthermore propose data sharing schemes based on the analysis results [19, 3, 24]. However, these schemes have never been implemented or evaluated with practical user involvement and realistic media content.

Second, our system is going to be deployed over a large number of smartphone users in mobile environments, and hence provides a unique research facility for investigating user interests and the data access patterns of mobile users. Comparatively, although a lot of experimental systems have been developed to monitor and record the behaviors of mobile users in various mobile environments, ranging from a university campus (Dartmouth [17], MIT Reality [6], UCSD [20]), conference sites (INFOCOM [5]), and urban areas (DieselNet [2]), they are limited to record and analyze the contact process among mobile users. The characteristics of users' interest and data access patterns are generally neglected.

The rest of this chapter is organized as follows. We motivate the proposed work with potential mobile environments and applications that benefit from our system in Section 8.2. The design and the implementation of our system are described in

Sections 8.3 and 8.4, respectively. The experiment results are presented in Section 8.5. Section 8.6 summarizes the related work. Section 8.7 discusses future research directions and concludes the chapter.

8.2 Motivation

The practical applicability of our system can generally be motivated by the following mobile environments.

Large-scale public commute systems, in which opportunistic communication links can be found among commuters on the bus, train, ferry, or subway. First, due to the mobility and physical constraint of public transportation, 3G or WiFi network coverage is usually unavailable to commuters. For example, smartphones do not have 3G signal in an underground subway, and commuters on a train may frequently lose WiFi connections when the train is moving. Second, the population density of mobile users is usually high in such environments. This high density, on one hand, ensures the existence of opportunistic communication links among commuters. On the other hand, it ensures that a sufficient amount of data is available among commuters, and many users may share common interests. For example, most of the commuters will be interested in daily news, weather reports, or traffic information.

Public events, in which opportunistic communication links can be found among mobile users at a stadium, museum, theater, or shopping mall. Although 3G or WiFi network coverage may be available at these places, the high population density and traffic demand during the event—sports game, concert, or holiday sale—may incur fierce competition for the limited channel bandwidth and significantly reduce data transmission rate. Particularly, most of the users will be interested in the same data during the event. For example, users may be interested in the live score of other concurrent matches during an NBA game, and they may be interested in coupon or discount information during a holiday sale at a shopping mall. In these cases, the efficiency of data access can be significantly improved when users share data among themselves.

8.3 System Design

Our system provides mobile users with access to categorized media news, which is dynamically retrieved by our backend server from well-known news websites, including CNN, *New York Times*, and BBC, on a hourly basis. The raw data retrieved from these websites is classified into a finite number of categories, and is re-formatted for easy processing at smartphones. It is important to notice that, the functionality of our system is not limited to media news, and it can be applied to any categorized media content ranging from photos at Flickr, video clips at YouTube to music at iTunes.

Our system is developed on smartphones which are distributed to voluntary students at the Pennsylvania State University. The process of data access among mobile users is illustrated in Figure 8.1. The categorized news is accessible to mobile users via a subset of smartphones which are employed as "super users." A super user re-

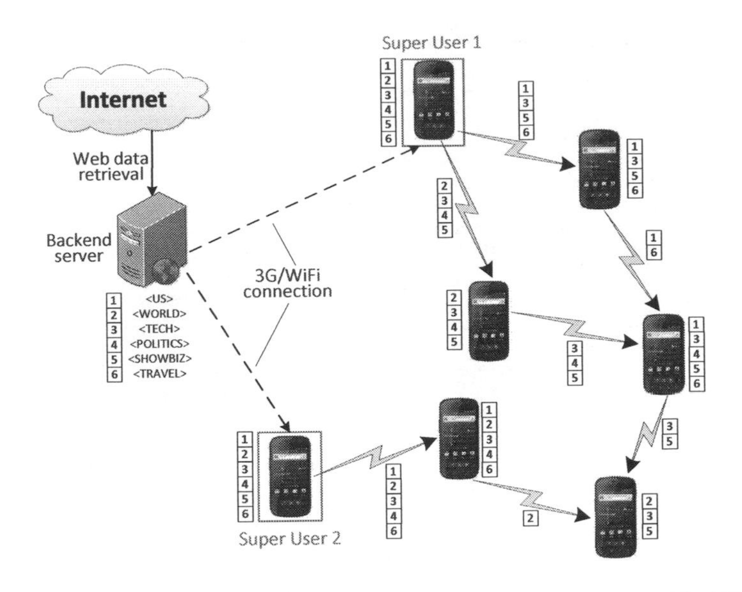

Figure 8.1: Overview.

trieves the up-to-date news from the backend server on a hourly basis, when it has 3G or WiFi connections available. Such news is then shared and transmitted among mobile users when they opportunistically contact each other, according to their interests in the news. In this way, we ensure that a user only receives the news that he is interested in. As illustrated in Figure 8.1, the amount of data available to a user will not decrease during the multi-hop data sharing process, because a user generally receives data from multiple users.

In general, the process of data sharing and transmission is transparent to the mobile users. A user only reads the news that he is interested in when his phone reports that new data is available.

8.3.1 Collecting Data from the Web

A web crawler is running at the backend server to collect news from the aforementioned websites periodically, in the form of HTML webpages. In general, these webpages are usually redundant and contain a large amount of stuff such as the CSS/frame specifications, outlinks, advertisements, Javascripts, etc., which are not related to the news content. As a result, these webpages are re-formatted by our system, so that the data size can be reduced significantly and the pieces of news are more convenient for the smartphones to process and display. The comparison between the original webpage and the re-formatted news page from CNN is illustrated in Figure 8.2.

The web crawler is responsible for re-formatting the webpages, which can be computationally expensive and hard to realize on smartphones. The important at-

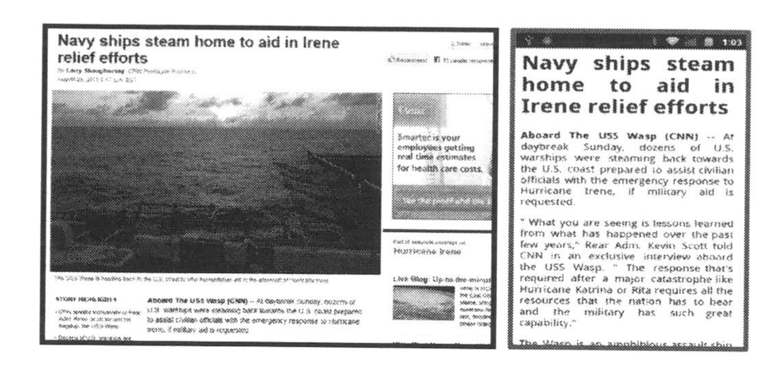

Figure 8.2: Comparison between the original webpage and the re-formatted news page from CNN.

tributes of the news, such as the publishing date, title, and category, are extracted from the webpage. These attributes, along with the main text of the news, is saved as a self-defined simple HTML file as shown in Figure 8.3(a) and transmitted to smartphones. Correspondingly, a data handler is implemented at each smartphone to manage the received HTML files. This management is realized by using a special XML file named `fileManifest.XML` as shown in Figure 8.3(b). Each HTML file (i.e., a piece of news) is mapped to a `<file>` element in `fileManifest.XML`, in which the attributes of the piece of news are also recorded.

To show the efficiency of such webpage re-formatting on reducing file sizes, we analyzed the news webpages that we collected from CNN during one day. The average size of original webpages is around 47KB. As shown in Figure 8.4, the average size of HTML files after re-formatting is only 7.4% of the original webpages.

```
<html>
  <head>
     <category>
     <pub date>
     <title>
  </head>
  <body>
     <description>
      <main body>
  </body>
</html>
```

(a) Self-defined HTML file format

```
<filelist>
  <file>
    <category>
    <pub date>
    <title>
    <description>
    <read date>
   <file path>
  </file>
  .....
</filelist>
```

(b) fileManifest.XML format

Figure 8.3: Format of HTML and XML files.

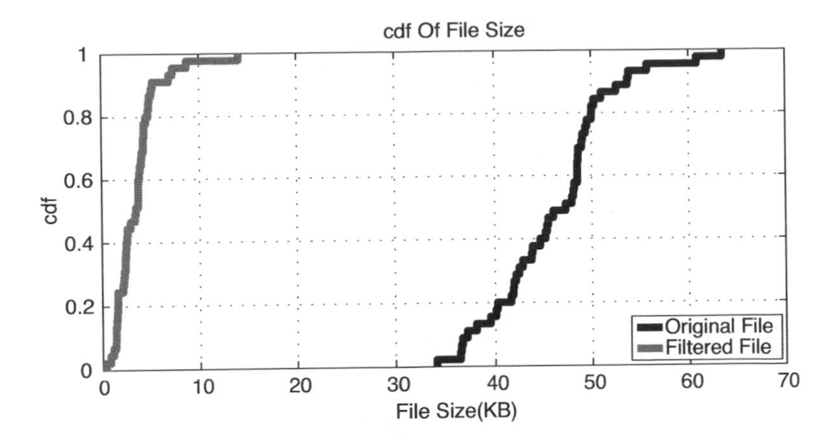

Figure 8.4: CDF of file size after filtering.

8.3.2 User Interest Profile

Each smartphone maintains a *User Interest Profile (UIP)*, which is manually initialized when the phone is first distributed to the user. Later on, the UIP of a user is dynamically updated by his phone whenever he reads the news that he is interested in. When two smartphones contact each other, they determine which news should be transmitted to each other based on their maintained UIPs.

The UIP of a mobile user is maintained with respect to the set of categories (denoted as $\mathcal{C}$) into which the media news is classified. As shown in Figure 8.5, each category $j \in \mathcal{C}$ is associated with an interest index I_j which ranges from 1 to 100. The values of interest indices specified by users are internally normalized by our system to ensure that $\sum_{j=1}^{|\mathcal{C}|} I_j = 100$ at any time.

In order to update the UIP, a smartphone maintains N_j as the number of times that the user reads the news in the category j and N_j is initialized as 1. Whenever a user reads a piece of news belonging to the category $k \in \mathcal{C}$, the UIP of the user is updated as follows. First, the corresponding interest index I_k is updated to $I_k \cdot \frac{N_k+1}{N_k}$. Afterward, the interest indices over all the categories are normalized to ensure that $\sum_{j=1}^{|\mathcal{C}|} I_j = 100$. More specifically, for each category $j \in \mathcal{C}$, we update I_j to $I_j \cdot \frac{100}{100+I_k \cdot \frac{1}{N_k}}$. Note that the value of I_k before update is used for this normalization.

It is easy to see that N_k monotonically increases as time elapses. As a result, the UIP of a user gradually becomes stable over time.

8.3.3 Data Transmission

8.3.3.1 Basic Approach

Due to mobility, two users may only be in contact for a limited period of time, and it is usually hard to predict such contact duration. Therefore, when a smartphone A

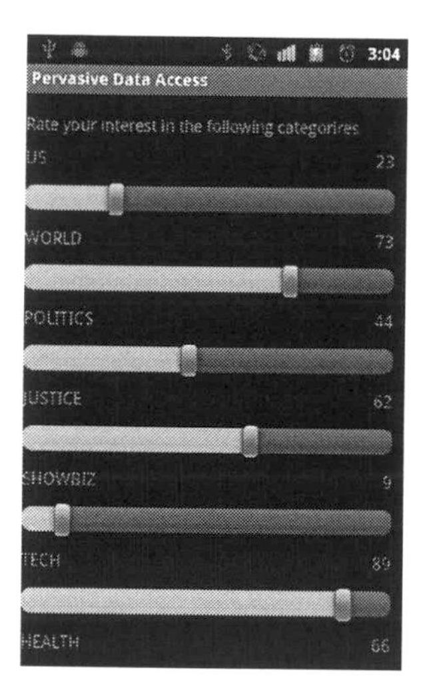

Figure 8.5: User Interest Profile (UIP).

contacts another phone B, A needs to determine an appropriate sequence for transmitting the pieces of news that B does not have to B, so as to ensure that the news B is interested in can be transmitted before the contact ends.

A straightforward solution is to determine such sequence based on B's UIP over different categories of news. However, B may only receive news of a few categories. Instead, we propose a probabilistic solution for determining such sequence over all the categories. Suppose that A has the set $\mathcal{D}$ of news that B does not have. The process of such data transmission is described by Figure 8.6, where $C(d_k)$ and $S(d_k)$ indicate the category and size of the piece of news d_k, respectively, and I_j indicates the interest index of B at category j. In general, A determines which piece of news in $\mathcal{D}$ to be transmitted to B one by one before the contact with B ends, and the news in category c has the probability $I_c/\sum_{j=1}^{|C'|} I_j$ to be transmitted next. After c is probabilistically determined, A transmits the piece of news in category c with the smallest size to B.

8.3.3.2 Considering Data Freshness

The freshness of different pieces of news is also considered in determining the data transmission sequence. For a piece of news d_i, which is in the category $c_i \in C$ and was generated at T_i, the freshness of d_i is calculated as:

$$F_i = \begin{cases} 1, & T_{now} - T_i < T_G \\ \frac{1}{\frac{T_{now}-T_i}{T_G}} = \frac{T_G}{T_{now}-T_i}, & other, \end{cases} \tag{8.1}$$

where T_{now} is the current time and T_G is the time granularity for calculating data freshness. For example, if we set T_G to one day, all pieces of news generated in the current day will have $F_i = 1$, while the news generated yesterday will have $F_i \in [1/2, 1)$.

Algorithm 12: Data transmission (A, B)

```
1  while contact is not over && 𝒟 ≠ ∅ do
2      p ← Uniform[0,1]                          // Random generator
3      C' ← ∅
4      for i = 1 to |𝒟| do
5          C' ← C' ∪ {C(d_i)}
6      for c = 1 to |C'| do
7          if p > (Σ_{j=1}^{c-1} I_j) / (Σ_{j=1}^{|C'|} I_j) && p ≤ (Σ_{j=1}^{c} I_j) / (Σ_{j=1}^{|C'|} I_j) then
8              Break                             // Transmit data in category c
9      A transmits data d_j to B, j = argmin_{d_k ∈ 𝒟, C(d_k)=c} {S(d_k)}
10     𝒟 ← 𝒟 \ {d_j}
```

In general, we prioritize to transmit the most important pieces of news first within the limited contact duration between mobile users. The importance of a piece of news takes both user interest and data freshness into account, and the importance P_i for a piece of news d_i is defined as

$$P_i = mI_i/100 + (1-m)F_i, \tag{8.2}$$

where m is used to adjust the weight between user interest and data freshness. As a result, each piece of news can be mapped into the square area shown in Figure 8.6. In data transmission, we introduce interest threshold S_I, freshness threshold S_F, and Priority Boundary Line (PBL) to realize the balance between user interest and data freshness. S_I and S_F are the lower bounds of user interest and data freshness. If a piece of news is too old, or attracts little interest from a user, it will fall into area A and not to be transmitted. Otherwise, the news corresponding to area B is fresher and attracts the interest of most users. These data are prioritized for being transmitted in the first place.

Determining transmission priority for pieces of news falling into the areas C and D is more complicated. We divide C and D into smaller areas based on PBL. We believe that users are still interested in some data in the areas $C1$ and $D1$. We set up the PBL to be across (S_I, S_F) and with a gradient determined by m. We will only send data whose priority is above the PBL.

8.3.4 Multi-Party Data Transmission

Furthermore, we explore the possibility of multi-party communication using Bluetooth radios of smartphones, i.e., a user transmits data with multiple peers simultaneously. Such multi-party communication is particularly challenging in Android-based smartphones due to the hardware limitation, which can be elaborated in the following three perspectives. First, current smartphones do not support reuse of the Bluetooth radio. In a TCP/IP connection, a user can use multiple threads in the OS to

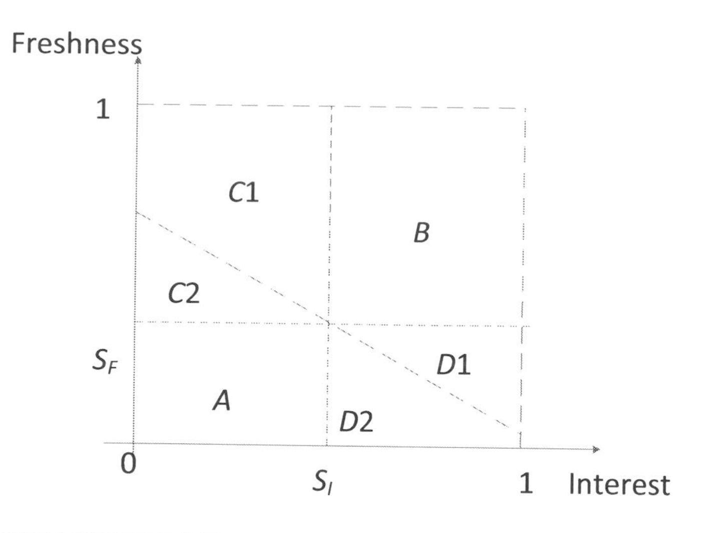

Figure 8.6: Data transmission priority considering user interests and data freshness simultaneously.

support multi-party communication with multiple counterparts simultaneously. However, such reusability is not applicable to Bluetooth communication. Due to the hardware limitation of Bluetooth radios in smartphones, it can only support one connection at one time, and will not respond to other connection requests given an existing Bluetooth connection. Second, Bluetooth radio cannot check the channel availability before transmission, like CSMA/CA does in WiFi. Third, broadcast is basically not supported by Bluetooth.

In this chapter, we propose three new schemes to support multi-party data transmission using Bluetooth radio of smartphones.

- *Sensing before sending:* When a smartphone wants to communicate with others via its Bluetooth radio, it will first check the status of Bluetooth radio. If the Bluetooth radio is available, it will wait for a specific time period and start the connection. If not, it will wait for the end of previous communication.
- *E2E back after failing:* If a smartphone initializes a Bluetooth connection but fails, it believes that the other phone is busy for another communication and then back off for a random time period $t_{back} \in [0, \mathcal{W}]$ before trying again. If it fails the n-th time, it will select a back time $t_{back} \in [0, 2^n\mathcal{W}]$. After the 5th try, the back window will remain unchanged and not be enlarged again.
- *Switching connection:* The Bluetooth connection between two smartphones will be terminated if there is no more file to transmit or after transmitting S_N number of files. After termination, the two smartphones will not connect to each other again during the next time period S_A, to give other smartphones the opportunity to communicate.

Algorithm 13: Load balancing $(\mathcal{S}, \mathcal{S}', \mathcal{U}, K)$

1 $n \leftarrow 0, \mathcal{U}' \leftarrow \emptyset$
2 **while** $n < K$ **do**
3     $j \leftarrow \arg\max_{k \in \mathcal{S}' \setminus \mathcal{U}} \{E(k)\}$
4   j is selected as the super user
5     $\mathcal{U} \leftarrow \mathcal{U} \bigcup \{j\}, \mathcal{U}' \leftarrow \mathcal{U}' \bigcup \{j\}, \mathcal{S}' \leftarrow \mathcal{S}' \setminus \{j\}$
6     $n \leftarrow n+1$
7   **if** $\mathcal{U} == \mathcal{S}$ **then**
8         $\mathcal{U} \leftarrow \mathcal{U}'$ // Every user has been selected, so reset $\mathcal{U}$

8.3.5 Power Constraint of Smartphones

Smartphones usually consume more power than traditional cell phones due to their advanced functionalities of running complicated applications and processing various media content. Hence, the battery power on smartphones is usually limited and should be taken into account during the system design. In our previous work [13, 11], we have made initial efforts on reducing the power consumption of smartphones by optimizing their contact detections. We aim to further address the power constraints of smartphones via workload balancing and adjustment of data access strategies.

8.3.5.1 Balancing the Workload of Super Users

As illustrated in Figure 8.1, the super users are responsible for retrieving the up-to-date news from the backend server on a hourly basis, and hence consume their battery power much faster. To balance the workload of super users, in our solution, all mobile users (denoted as set $\mathcal{S}$) in the network are scheduled to be super users in a round-robin manner, and the super users are changed daily.

Due to the difficulty of distributed coordination among mobile users, super users are selected at the backend server according to the status of the smartphones' battery power. Each smartphone reports its remaining battery power to the server at 3 a.m. every day via 3G or WiFi connections. The server selects K super users from the set $\mathcal{S}'$ including all the users having reported to the server. The selection process is described by Figure 8.8, where $E(k)$ indicates the remaining battery power of user k, and $\mathcal{U}$ indicates the set of users that have been selected as the super user recently and will not be the super user again in the near future. In general, the users with the maximum remaining battery power of their smartphones are selected as the super users.

The server is responsible for notifying each selected super user about the selection result. Since the execution time of Algorithm 13 is negligibly short, any selected super user can be notified. At the same time, the super users of the previous day automatically revoke their super user status at 3 a.m.

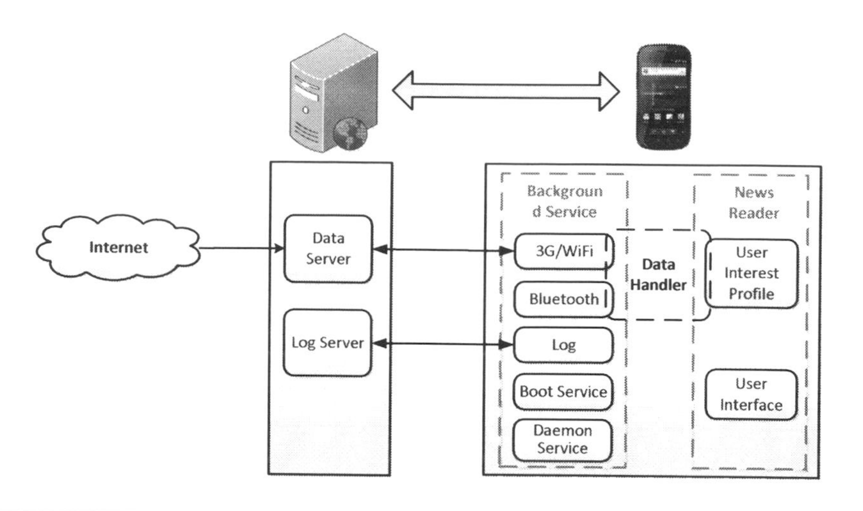

Figure 8.7: System implementation.

8.3.5.2 Trade-off between Power Consumption and Data Availability

Our system also enables the mobile user to flexibly trade-off between the power consumption of data transmission of their smartphones and the availability of media news. Such trade-off is realized by a user-controlled parameter λ valued in the range $[0,1]$. $\lambda = 0$ means that the user wants to conserve the maximum amount of battery power and does not want to receive any data, while $\lambda = 1$ means that the user wants all the available data regardless of the transmission overhead.

Such trade-off can be easily integrated with the process of determining the data transmission sequence described in Algorithm 12. More specifically, every time a piece of news d_j is picked, A only has probability λ_B to send this news to B, where λ_B is the trade-off parameter specified by B. In practice, a user can adjust λ at any time according to the power status of the smartphone.

8.4 System Implementation

The implementation of our system is illustrated by Figure 8.7. At the server side, the web crawler retrieves data from the Internet as described in Section 8.3 and transmits the re-formatted news to the super users. The log module is responsible for recording the contact and data access behaviors of mobile users. At the smartphone side, the handling of system execution and data transmission is implemented as background system services. These services then interact with the news reader. The news reader displays the received news to the user as shown in Figure 8.8 and maintains the UIP of user. Such interaction is realized via the data handler which manages the received news.

Figure 8.8: News reader.

8.4.1 Development Platform

The system was developed on the Samsung Nexus S phone, which is equipped with a 1-GHz processor to support efficient processing of media contents, and a 16-GB SD card is pre-installed on the phone to provide sufficient storage space. Both AT&T 3G and 802.11 b/g/n networks are supported on this phone.

Our system is implemented based on the Android 2.3.3 (Gingerbread) OS. It would be desirable if our system could be developed as a stand-alone application which is independent from the OS kernel, and could be downloaded and run on any smartphone. However, it is challenging to realize this in practice due to the following system requirements:

- Our system is required to be continuously running in the background for a long period of time without user involvement, so as to enable the smartphones to share the media news with each other. However, this capability is not supported by some mobile OSs for third-party applications. For example, iOS on the iPhone only enables the system applications to be running in the background, and other applications can only be "wakened" up by push notifications. Instead, we implemented this capability on the Android OS by developing specific system applications.

- A smartphone in our system should be able to periodically detect the existence of other smartphones within the communication range of its Bluetooth radio. This requires the phone's Bluetooth interface to be persistently set at the *discoverable* mode. However, the current Android OS limits the Bluetooth interface to be at discoverable mode for up to 120 sec. Active user involvement is required periodically to resume the discoverable mode. In our implementation, the Android OS kernel is modified to address this limitation.

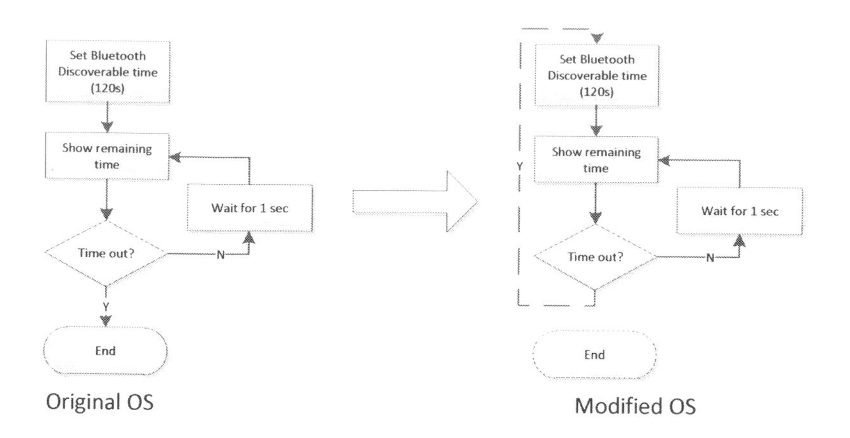

Figure 8.9: Extending Bluetooth discoverable time.

8.4.2 System Implementation

In this section, we describe how the aforementioned challenges are addressed to ensure that our system can run correctly on the Android OS.

8.4.2.1 Discoverable Mode of Bluetooth

We require the Bluetooth interface of a smartphone to be persistently set at the discoverable mode so that the phone can periodically detect other nearby peers. However, in Android OS, whenever the Bluetooth interface is set at the discoverable mode, a system timer of 120 sec is triggered, and the discoverable mode is set off when the timer expires. Our solution to solve this problem is shown in Figure 8.9. We remove this limit and keep monitoring the system timer. The timer is reset whenever it expires, so that the discoverable mode is resumed as soon as it has been set off.

This modification, by itself, cannot completely ensure that Bluetooth is always in discoverable mode, as users or other application processes can still turn Bluetooth off. We developed a specific daemon service to solve this problem, which is described in Section 8.4.2.3.

8.4.2.2 Removing the System Dialog Requesting Bluetooth Permission

In the default workflow of Android OS, each time when a user application requests to open the Bluetooth interface, the Android OS will pop up a dialog requesting user permission to do that, and hence user involvement is required for setting up the Bluetooth discoverable mode. To address this problem, we modified the kernel of the Android OS, so that each time a request is sent to the OS to set up the Bluetooth discoverable mode, a parameter T is specified to indicate the elapsed time we request for the discoverable mode. In particular, we set T to be a negative value, so that we

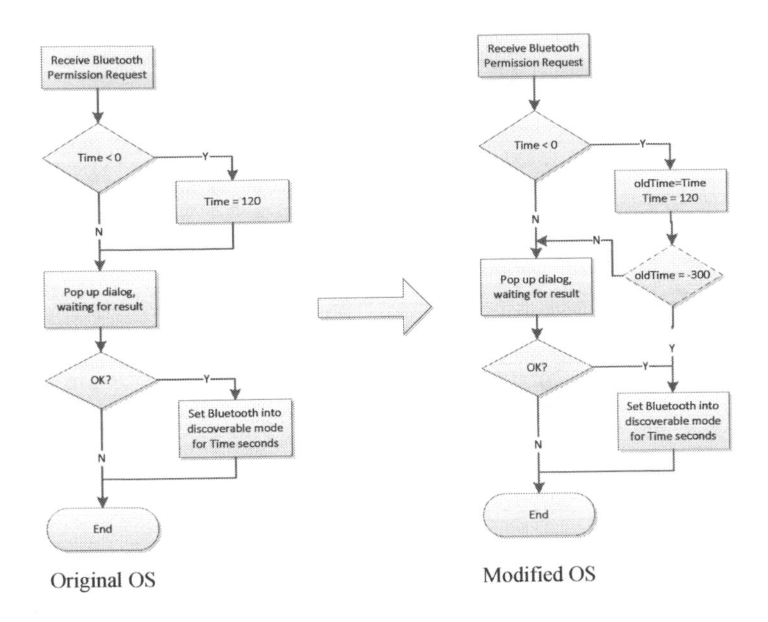

Figure 8.10: Remove Bluetooth permission request dialog.

are able to modify the system workflow and skip over the system dialog requesting user permission. The elapsed time for Bluetooth discoverable mode is then set as 120 seconds automatically, as shown in Figure 8.10.

8.4.2.3 *Boot and Daemon Services*

We developed boot service as a special system application running in the background, which is started with the Android OS and automatically starts our system afterward. A daemon service is developed to make sure that our system is resilient to unexpected failures of smartphones, such as power depletion, system crash or user mistakes. The boot service performs the following checking tasks every 30 seconds:

- *Checking Bluetooth discoverable mode:* If Bluetooth is not in the discoverable mode, it sends a request to enable it for -300 seconds. Thus, the system will set Bluetooth into discoverable mode without bothering the user.
- *Checking Bluetooth server status:* In order to start Bluetooth communication as soon as two devices contact, the Bluetooth server socket of the client is active all the time. If not, we will restart the Bluetooth server socket.
- *Checking neighbor:* The daemon service will also scan for neighbors. If there is one neighbor M available and the Bluetooth device is not connected to any others, it connects to M automatically.

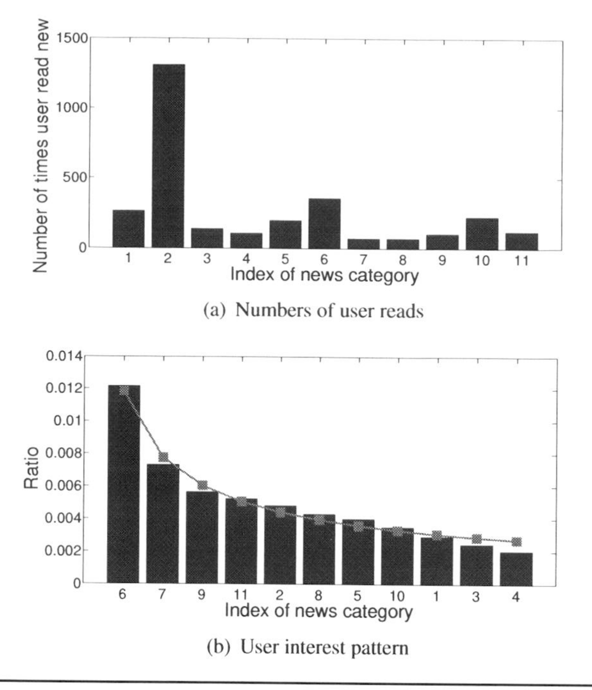

(a) Numbers of user reads

(b) User interest pattern

Figure 8.11: Behaviors of user data access over different news categories.

The system should work immediately after the phone is booted without user action. Thus, the boot service also listens to the specific signal that the system generates when being booted. Then, it triggers the daemon service to start, and our system starts to work properly.

8.5 Experimental Results

We deployed the system with 20 graduate students in the Departments of Computer Science and Engineering (CSE) and the Information Science and Technology (IST) at the Pennsylvania State University, to evaluate their user interests and data access patterns during the time period between September 2011 and March 2012. A total number of 45,807 pieces of media news in 11 categories were retrieved from the CNN news website and shared among mobile users. The amounts of news in different categories are listed in Table 8.1. We observe that over 40% of the news is found in the categories of "US" and "WORLD," and other news is basically evenly distributed among the remaining nine categories. During the experiment, these news items have been transmitted and shared 655,794 times among the 20 users, and have been read 2,970 times by the users.

8.5.1 Data Access Patterns of Mobile Users.

As shown in Figure 8.11(a), the news read by users is unevenly distributed over different categories, and this distribution is determined by the amount of news available

Table 8.1 Amount of News in Different Categories

Index	News categories	Amount of news
1	US	5500
2	WORLD	13755
3	POLITICS	3969
4	JUSTICE	3915
5	SHOWBIZ	3992
6	TECHNOLOGY	2754
7	HEALTH	1199
8	LIVING	1816
9	TRAVEL	1547
10	OPINION	3946
11	OTHER	3396

in different categories. Furthermore, we evaluate the user interest and the data access patterns of mobile users using the ratio of the number of times users read the news over the amount of news available in the network. This ratio over different news categories is shown in Figure 8.11(b). Surprisingly, we observe that variation of this ratio over different news categories can be accurately approximated by power-law curve fitting, which is consistent with the well-known result that the user access pattern over web contents exhibits Zipf characteristics [4].

To better interpret the data access patterns of mobile users, we also evaluate the amount of news that users receive during the experiment periods separately, and the evaluation results are shown in Figure 8.12. We observe that the distribution of the amount of news that users receive is also skewed, and consistent with the data access patterns shown in Figure 8.11(a). In comparison, the result in Figure 8.11(b) provides more accurate characterization of users' data access patterns.

We are also interested in the data access behaviors of mobile users during different time periods in a day. Figure 8.13(a) shows that more than 90% of the news is received by smartphones during the time period between 9 a.m. and 8 p.m, which indicates that most opportunistic contacts happen among mobile users when they are

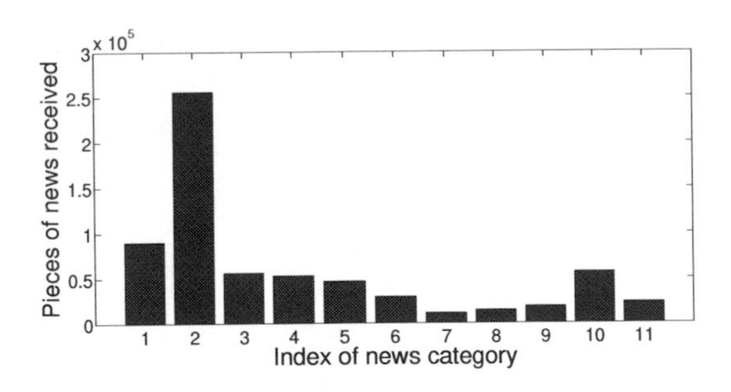

Figure 8.12: Amounts of news received by users.

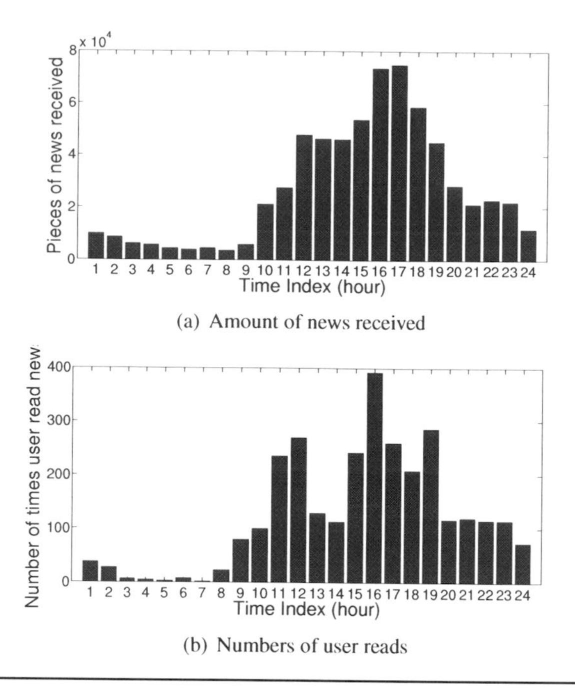

(a) Amount of news received

(b) Numbers of user reads

Figure 8.13: Temporal patterns of data access behaviors of mobile users.

on campus during the daytime. Moreover, Figure 8.13(b) shows that more than 45% of the news is read by users during two specific time periods, i.e., 10 a.m.–12 p.m. and 3 p.m.–7 p.m. This result highlights the possible heterogeneity of user interests over time, and motivates us to further explore the temporal differentiation of user interest profiles for more accurate identification of the data that users are interested in.

8.5.2 Characterization of Social Communities

Students involved in our experiment also exhibit noticeable social correlations with each other, when they stay in the CSE department building. As a result, we also explored the social community structure among these mobile users during the experiment period. Social community structure in mobile environments is characterized based on contact patterns of mobile users, such that a community consists of mobile users that frequently contact each other. Palla et al. [22] defines a k-clique community as a union of all k-cliques (complete subgraphs of size k) that can reach each other through a series of adjacent k-cliques. We employ a distributed method for detecting k-clique communities using the cumulative node contact duration as the threshold for community detection [15]. The result of community detection with $k = 3$ is shown in Figure 8.14. Surprisingly, we observe that students are naturally classified into two groups according to their affiliations with the CSE or IST departments. This community structure therefore indicates the consistency between users' social connectivity and their contact patterns in practice.

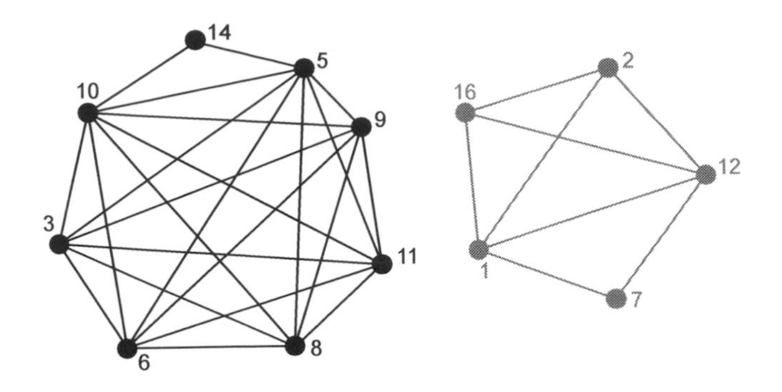

Figure 8.14: Social community structure among mobile users, 81730 sec is used as the threshold.

Moreover, we are also interested in the changes of such social community structures over time. As shown in Figure 8.15, we evaluated the social community structure among mobile users during the daytime and nighttime, respectively. Figure 8.15(a) shows that the social community structure during the daytime is similar with that in Figure 8.14, and indicates that users' interactions during the daytime are the main factor determining social community structure in the network. In contrast, Figure 8.15(b) shows distinct characteristics of the community structure. Basically, the two communities merge with each other during the nighttime. This indicates that social correlation among mobile users is more casual during the nighttime, so that all the mobile users can be grouped into a large community.

8.6 Related Work

Data access in opportunistic mobile networks can be provided in various ways. Data can be actively disseminated by the data source to appropriate users based on their interest profiles [8]. Publish/subscribe systems [25] are most commonly used for such data dissemination, and they usually exploit social community structures among mobile users to determine the brokers. Caching is another way to provide data access among mobile users. Distributed determination of caching policies or locations for minimizing data access delay has been studied [24, 16], but they generally rely on specific assumptions to simplify the network conditions. In [24], it is assumed that all the mobile devices contact each other with the same rate. In [16], mobile users are artificially partitioned into several classes such that users in the same class are statistically identical. Recent research efforts [9] developed more generic solutions by assuming the heterogeneity of users' contact and interest patterns, and further propose to cache data at specific network locations which can be easily accessed by other users in the network. The freshness of data cached at mobile users has also been taken care of in a fully distributed manner [10].

Recent research efforts have been focusing on exploring the realistic characteristics of mobile network environments and user behaviors for efficient design and

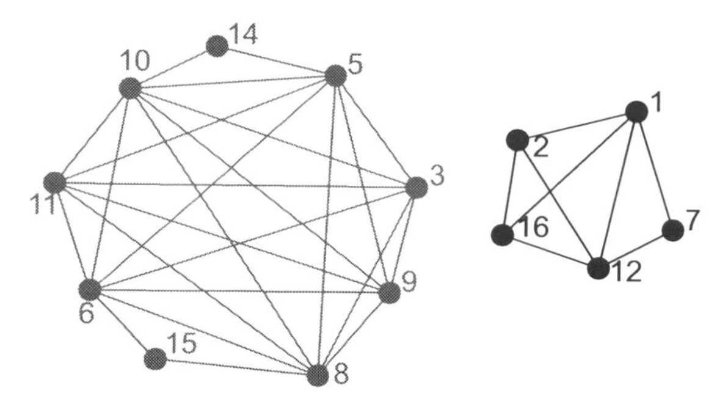

(a) User communities during daytime with 61880 secs as threshold

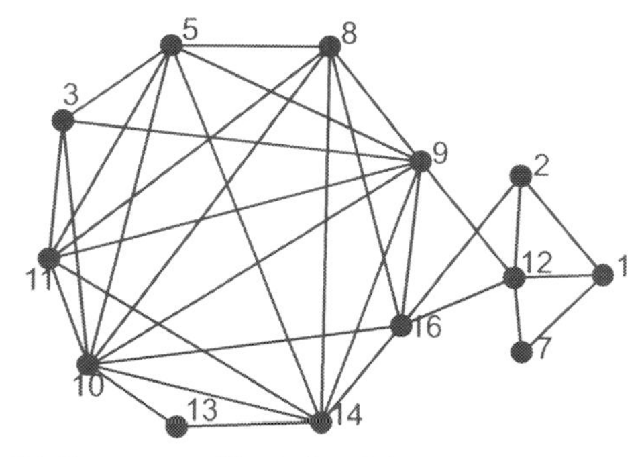

(b) User communities during nighttime with 5029 secs as threshold

Figure 8.15: User communities during different time periods.

implementation of data access systems [21, 12]. References [18, and 1] focus on investigating the coverage of WiFi networks in urban areas, and further improve data accessibility by transferring 3G traffic to WiFi. References [19] studied the traffic patterns of commuters in metropolitan subway systems, which is exploited for intentional media sharing. However, none of these systems has been implemented with realistic media content being transmitted among mobile users, and the efficiency of data access is only evaluated via trace-based simulations with synthetic network traffic patterns.

8.7 Conclusion and Future Directions

In this chapter, we focused on developing a practical system which provides pervasive data access to smartphone users without support of cellular or Internet infrastructure. We designed and implemented our system on Android-based smartphones, and deployed our system to a number of students at the Pennsylvania State University. Our system is able to efficiently share the media news that users are interested in when they opportunistically contact each other, and can also be used for capturing and analyzing the interest and data access patterns of mobile users.

In the near future, we plan to deploy our system on a large number of smartphones (more than 100), and distribute these phones to voluntary students from various departments of Penn State University. Being different from our system deployment described in Section 8.5, the increase of system scale may significantly change the characteristics of user interests and their data access patterns. We expect to have a large amount of traces which record the realistic data sharing and access behaviors of users in various mobile environments. These traces could provide unique research facility for mobile network researchers, and motivate future research on providing efficient data access with respect to realistic user interests.

We plan to improve the data access efficiency of our system by taking the temporal and spatial variations of user interests into account. In Section 8.3, our system determines the data that a user would be interested in based on his cumulative UIP over a long period of time. However, as shown in Figure 8.13, the interest of a mobile user may vary during different time periods in a day or at different places. For example, a student in Computer Science may be generally interested in information technology, but will more likely be interested in food during lunch time or when he is at a restaurant. The consideration of such temporal/spatial differentiations of user interest leads to more accurate identification of the data that users are interested in, and further improves the efficiency of data access.

We also plan to add security protection functionality into our system to protect mobile users during data access and sharing. We notice that the peer-to-peer data access can be utilized by malicious attackers to distribute malicious code, virus, or worms among mobile users, and can also facilitate spammers to spread advertisements or other unwanted information. Particularly, the speed of such spread can be significantly accelerated due to the high population density in the application scenarios described in Section 8.2. We plan to develop efficient distributed authentication methods and network trust architecture to ensure the authenticity and integrity of the shared web data. The development of these security protection methods will also take social interactions among mobile users into account.

Acknowledgments

This work was supported in part by Network Science CTA under grant W911NF-09-2-0053.

References

[1] A. Balasubramanian, R. Mahajan, and A. Venkataramani. Augmenting mobile 3G using WiFi. In *Proceedings of MobiSys*, pages 209–222, 2010.

[2] A. Balasubramanian, R. Mahajan, A. Venkataramani, B. Levine, and J. Zahorjan. Interactive WiFi connectivity for moving vehicles. In *Proceedings of ACM SIGCOMM*, pages 427–438, 2008.

[3] C. Boldrini, M. Conti, and A. Passarella. ContentPlace: Social-aware data dissemination in opportunistic networks. In *Proceedings of MSWiM*, pages 203–210, 2008.

[4] L. Breslau, P. Cao, L. Fan, G. Phillips, and S. Shenker. Web caching and Zipf-like distributions: Evidence and implications. In *Proceedings of INFOCOM*, volume 1, 1999.

[5] A. Chaintreau, P. Hui, J. Crowcroft, C. Diot, R. Gass, and J. Scott. Impact of human mobility on opportunistic forwarding algorithms. *IEEE Trans. on Mobile Computing*, 6(6):606–620, 2007.

[6] N. Eagle and A. Pentland. Reality mining: Sensing complex social systems. *Personal and Ubiquitous Computing*, 10(4):255–268, 2006.

[7] K. Fall. A delay-tolerant network architecture for challenged internets. *Proc. SIGCOMM*, pages 27–34, 2003.

[8] W. Gao and G. Cao. User-centric data dissemination in disruption tolerant networks. In *Proceedings of INFOCOM*, 2011.

[9] W. Gao, G. Cao, A. Iyengar, and M. Srivatsa. Supporting cooperative caching in disruption tolerant networks. In *Proceedings of the IEEE International Conference on Distributed Computing Systems (ICDCS)*, 2011.

[10] W. Gao, G. Cao, M. Srivatsa, and A. Iyengar. Distributed maintenance of cache freshness in opportunistic mobile networks. In *Proceedings of the 32nd IEEE International Conference on Distributed Computing Systems (ICDCS)*, pages 132–141, 2012.

[11] W. Gao and Q. Li. Wakeup scheduling for energy-efficient communication in opportunistic mobile networks. In *Proceedings of IEEE INFOCOM*, 2013.

[12] B. Han and A. Srinivasan. Your friends have more friends than you do: Identifying influential mobile users through random walks. In *Proceedings of the 13th ACM International Symposium on Mobile Ad Hoc Networking and Computing (MobiHoc)*, pages 5–14, 2012.

[13] W. Hu, G. Cao, S. Krishanamurthy, and P. Mohapatra. Mobility-assisted energy-aware user contact detection in mobile social networks. In *Proceedings of IEEE International Conference on Distributed Computing Systems (ICDCS)*, 2013.

[14] P. Hui, A. Chaintreau, J. Scott, R. Gass, J. Crowcroft, and C. Diot. Pocket switched networks and human mobility in conference environments. In *Proceedings of the 2005 ACM SIGCOMM Workshop on Delay-Tolerant Networking (WDTN)*, pages 244–251. ACM, 2005.

[15] P. Hui, E. Yoneki, S. Chan, and J. Crowcroft. Distributed community detection in delay tolerant networks. *Proc. MobiArch*, 2007.

[16] S. Ioannidis, L. Massoulie, and A. Chaintreau. Distributed caching over heterogeneous mobile networks. In *Proceedings of the ACM SIGMETRICS*, pages 311–322, 2010.

[17] M. Kim, D. Kotz, and S. Kim. Extracting a mobility model from real user traces. In *Proceedings of INFOCOM*, 2006.

[18] K. Lee, I. Rhee, J. Lee, S. Chong, and Y. Yi. Mobile data offloading: How much can WiFi deliver? In *Proceedings of ACM CoNEXT*, 2010.

[19] L. McNamara, C. Mascolo, and L. Capra. Media sharing based on colocation prediction in urban transport. In *Proceedings of MobiCom*, pages 58–69, 2008.

[20] M. McNett and G. Voelker. Access and mobility of wireless PDA users. *ACM SIGMOBILE CCR*, 9(2):40–55, 2005.

[21] N. P. Nguyen, T. N. Dinh, S. Tokala, and M. T. Thai. Overlapping communities in dynamic networks: Their detection and mobile applications. In *Proceedings of the 17th Annual International Conference on Mobile Computing and Networking (MobiCom)*, pages 85–96, 2011.

[22] G. Palla, I. Derényi, I. Farkas, and T. Vicsek. Uncovering the overlapping community structure of complex networks in nature and society. *Nature*, 435(7043):814–818, 2005.

[23] M. R. Ra, J. Paek, A. B. Sharma, R. Govindan, M. H. Krieger, and M. J. Neely. Energy-delay trade-offs in smartphone applications. In *Proceedings of MobiSys*, pages 255–270. ACM, 2010.

[24] J. Reich and A. Chaintreau. The age of impatience: Optimal replication schemes for opportunistic networks. In *Proceedings of ACM CoNEXT*, pages 85–96, 2009.

[25] E. Yoneki, P. Hui, S. Chan, and J. Crowcroft. A socio-aware overlay for publish/subscribe communication in delay tolerant networks. In *Proceedings of MSWiM*, pages 225–234, 2007.

Chapter 9

Exploiting Social Information in Opportunistic Mobile Communication

Abderrahmen Mtibaa
Electrical & Computer Engineering
Texas A&M University
Doha, Qatar

Khaled A. Harras
Computer Science Department
Carnegie Mellon University
Doha, Qatar

CONTENTS

9.1 Introduction

The growth of social interaction has evolved from the confined requirement of physical proximity, to telegraph and telephone networks, ultimately exploding to various forms of cyber-based interactions enabled by the ubiquity of the Internet. These social interactions, no longer requiring physical presence amongst participants, are enabled by many applications including email, chat, and more recently, online social network (OSN) services such as Facebook, Google+, MySpace, or LinkedIn. These applications create a virtual space through which users build social networks of their acquaintances, where they can freely interact, independently of where they are located. However, when people with similar interests or common acquaintances are within physical proximity of one another, beyond human sensory ranges, they have no automated way to identify potential relationships. The relationship between virtual social interactions and physical meetings remains largely unexplored. Exploiting high-granularity physical proximity and social information has been recently tackled in the domain of mobile opportunistic networks [22, 32].

Mobile opportunistic networking becomes an interesting and challenging problem because enabling it provides solutions to the intermittent aspect of wireless connections nowadays, especially in situations where the necessary infrastructure is unavailable, costly, or overloaded. We show that in such networks using social information to enable and enhance node cooperation is fundamental for the message delivery process. The lack of node cooperation in such cases, where a node may refuse to act as a relay and settle for sending and receiving its own data, causes considerable degradation in performance. To ensure this desired cooperation, we investigate leveraging social information in order to address four main challenges: (i) efficiently selecting the next hop toward a destination in a way that minimizes delay and maximizes delivery success rate, (ii) ensuring fair resource utilization among participating mobile devices, (iii) creating scalable solutions to accommodate the wide-scale adoption of such solutions, and (iv) enabling trustful communication between users.

In this chapter, we present the research we have conducted to address each of the challenges listed above. In order for this research to be more credible, we adopt a data-driven approach where our research, analysis, and evaluation, are based on multiple mobility traces collected from conferences, university campuses, and metropolitan cities. These traces model mobile opportunistic networks, including mobile device connection and disconnection times, as well as information about the social interaction, relationships, or common interests amongst participants. In the following sections, we begin by introducing the overarching methodology adopted in our work, along with the details of the traces used, followed by the solutions we propose to address the four challenges listed above.

9.2 Methodology: A Data-Driven Approach

An important aspect in mobile opportunistic network algorithm evaluation is the mobility pattern of the nodes, which determines the contact pattern as well as the contact duration. Therefore, the literature is ripe with mobility models for opportunistic networks that have been extensively surveyed [4, 18, 14]. According to these surveys, the mobility pattern can be influenced by physical factors, such as obstacles, speed limits, and boundaries defined by streets, as well as social factors, including preferred places, working or free times, and even the movement of friends. Thus, it is important to choose accurate mobility models that can reflect realistic scenarios in which the protocol is expected to be deployed.

The contact pattern can be modeled using two possible methods. The first method is to create a synthetic model that generates patterns according to certain rules or equations describing a class of users or applications. The second option is to gather, analyze, and build upon real mobility traces, collected from experimental deployments. Mobility traces that have been gathered by the research community can be mainly broken down to either contact traces or GPS footprint traces. The former provides a list of contacts, defining the two or more nodes participating in a contact, as well as the start and end times of that contact. GPS traces provide the location of a node within a given moment in time. Thus, GPS traces are more powerful since they allow for more complex studies, for example, varying the communication range of the nodes.

There are various trade-offs between adopting mobility traces versus synthetic mobility models. Mobility traces are attractive since they exhibit the real behavior of nodes in a certain scenario, as opposed to synthetic models, which can simplify or overlook some details of node mobility. However, traces can have shortcomings such as the existence of gaps and inconsistencies, and the low number of participating nodes as a result of the difficulty of running long experiments at a large scale. Synthetic models, on the other hand, are more flexible, since they frequently provide parameters that can be adjusted, allowing for different scenarios. For example, one test possible with synthetic models, which is quite hard to do in traces, is to change characteristics of the network such as the average number of nodes in a community, the average speed, or the number of evaluated nodes.

We focus our efforts in this chapter on utilizing real behavior mobility traces. This decision is in part because synthetic models have been extensively utilized in the past. Furthermore, we believe the realism of traces outweighs their shortcomings, which can be minimized by considering a variety of data sets representing different traces. In this section, we describe the experimental data traces we adopt throughout the chapter, followed by the overarching methodology utilized in our analysis and evaluation.

9.2.1 Experimental Data Traces

Our work, analysis, and evaluation throughout this chapter are based on six experimental data sets. CoNext07, Infocom06, CoNext08, Dartmouth01, and Hope08 are state-of-art human mobility traces publicly available on CRAWDAD.[1] However, such data sets are collected in relatively small areas such as conferences and campuses. Since it is generally costly to run an experiment in very large areas (e.g., city-wide), we *artificially* created the sixth data set, which uses the San Francisco taxi cab trace [30] coupled with three mobility traces that we utilize to represent three different sub-communities; Infocom06, CoNext07, and CoNext08. Table 9.1 summarizes the characteristics of the following experimental data sets.

[1] `crawdad.cs.dartmouth.edu/`

Table 9.1 Dataset Properties

	Hope08	Dartmouth01	SanFrancisco11		
			CoNext07	Infocom06	CoNext08
duration	3 days	3 days	3 days	3 days	3 days
mobility detection	WiFi	WiFi	Bluetooth	Bluetooth	Bluetooth
# nodes	414	100	27	47	22
median inter-contact	30mins	6 mins	10 mins	15 mins	12 mins
median contact time	90 sec	160 sec	240 sec	150 sec	120 sec

9.2.1.1 CoNext07 [22]

Visitors to a conference were asked to carry a smartphone device during 3 consecutive days with the MobiClique application installed [22]. Prior to the experiment start, each participant was asked to indicate other CoNext participants he knew or had a social connection with. During the experiment, our social networking application indicated when a contact, or a contact of a contact, was in Bluetooth range. This connection neighborhood was then displayed on the user's device; the user could then add new connections or delete existing connections based on the physical interaction consequent to the application notification. In addition, the devices also logged any other device that was detected; the scanning period was set to a scanning interval of 2 minutes. The CoNext07 data set was collected on 28 Windows Mobile devices that were given to a preselected set of participants the first day of the conference. Each device was used for an average of 2.2 days since people arrived and departed at different times.

9.2.1.2 Infocom06 [5]

This trace was collected with 78 participants during the IEEE Infocom 2006 conference. People were asked to carry an experimental device (i.e., an iMote) with them at all times. These devices logged all contacts between experimental devices (called internal contacts) using periodic scanning every 2 seconds. In addition, they log contacts with other external Bluetooth devices (e.g., cell phones, PDAs). We are presenting results for internal contacts only in this chapter. People were also asked to fill out questionnaires with their nationalities, languages, countries, cities, academic affiliations, and topics of interest. Based on this information, we consider in this chapter three different social graphs from this experiment; based on users (*i*) common topics of interest when two users shared *k* common topic, (*ii*) Facebook graph, and (*iii*) social profile (union of nationality, language, and city).

9.2.1.3 CoNext08 [29]

This experiment was performed at the CoNEXT 2008 conference using smartphones with the MobiClique application installed. The main difference compared to the CoNext07 experiment is in the parameterizations: we had 22 participants and the neighborhood discovery was randomized to be executed at intervals of 120 +/− 45 seconds. In addition, the social profile of MobiClique was initialized based on the user's Facebook profile. The initial list of interests contained user-selected Facebook groups and networks from each profile. As in the CoNext07 experiment, the social network evolved throughout the experiment as users could make new friends and discover or create new groups (i.e., interest topics) and leave others. For our work, we consider the collected contact trace and the final social graph of 22 devices since the remaining devices were not collecting data on each day of the experiment.

9.2.1.4 Dartmouth01 [16]

We used the WiFi access network of the Dartmouth campus [9]. This data set spans roughly 1300*x*1300 square meters and over 160 buildings, with about 550 802.11b access points deployed. Dartmouth College covers student residences, sports infrastructures, administrative buildings, and academic buildings. The data set contains logs of client MAC addresses, and SSIDs of access points as well as their positions. We assume that two nodes are able to communicate if they are connected simultaneously to the same access point. We use this trace to generate contacts between 100 nodes in order to simulate message propagation in a pure ad hoc manner. We note that the ping-pong effect in the Dartmouth trace [9] will not affect such assumption.

9.2.1.5 Hope08 [1]

The Hope08 dataset was collected during the 7th HOPE conference. This experiment had a large number of participants (around 770) to collect and exchange contact information (after an explicit connection setup using send/receive pings). The dataset contains the location of participants, every 30 seconds, as well as their topics of interest in the conference. This dataset is also publicly available in the CRAWDAD database.

We assume that contact opportunities are available when two people are tagged in the same place during a minimum period of time δt. In this chapter, we consider $\delta t = 2\,mins$ as a minimum contact duration in order to avoid using very short contact opportunities for forwarding. During the conference, participants were asked to select up to five interests they may share with other conference attendees. These interests are used to build a social graph connecting all participants. We note that there are only 414 nodes connected in the social graph as shown in Table 9.1. We believe that, nowadays, the world is socially connected: Facebook announced an average of 3.74 degrees of separation between its members in 2011.[2]

9.2.1.6 SanFrancisco11

Owing to the lack of *large-scale* experimental data sets, we *artificially created* SanFrancisco11. This data set uses the San Francisco taxi cab trace [31] coupled with three human mobility traces in order to represent three sub-communities of the San Francisco, area such as the airport, downtown, and the sunset areas. To the best of

[2] `bbc.com/news/technology-15844230`

our knowledge, this is the largest data set that captures human mobility contacts as well as human social properties in large-scale networks.

The San Francisco taxi cab trace contains GPS coordinates of approximately 500 taxis collected over 30 days. Each trace has the reported time and location for each taxi. We incorporate traces for the duration of 3 days, interpolate the movement of the cabs, and then generate the contacts between these taxis. We assume a contact has occurred when a taxi comes within 100 meters from another taxi. The contacts trace thus contains the starting timestamp of when the contact has occurred and the ending time stamp of when the contact has finished. It also contains the IDs of the two cabs that happened to be in contact with each other during that time.

We artificially incorporate three real human mobility traces in the three different areas of San Francisco city. Taxis are moving between these areas. Contacts between taxis and nodes within an area are added based on the same contact time and inter-contact time distribution [5] of the corresponding area. We utilize three existing data sets, Infocom06, CoNext07, and CoNext08, to represent three different sub-communities and emulate a large-scale "semi-realistic" data set. The resulting data set contains, therefore, 3 sub-communities in three different areas of San Francisco. Taxis are moving between these areas. Contacts between taxis and nodes within an area are added based on the same contact time and inter-contact time distribution [5] of the corresponding area.

9.2.2 Evaluation Methodology

In mobile opportunistic networks, we are generally interested in delivering data among a set of N mobile wireless nodes. Communication between two nodes is established when they are within radio range of each other. Data is forwarded from source to destination using these opportunistic *contacts*. We model the evolution of contacts in the network by a time varying graph $G(t) = (V, E(t))$ with $N = |V|$. We assume that the network starts at time t_0 and ends at time T (T can be infinite). We call this temporal network [15] the *contact graph*. Each $G(t)$ describes the contacts between nodes existing at time t. Such a time-varying graph model can be obtained from a mobility/contact trace or from a mobility model along with knowledge of radio properties (e.g., radio range).

In our evaluation, we compute the sequence of optimal paths found between any source and destination in the data set. From the sequence of delay-optimal paths we deduce the delay obtained by the optimal path at all times. We uniformly combine all the observations of a trace among all sources, destinations, and for every starting time (the time in seconds when the message m was generated by the source node S). We assume that contacts between any two nodes are long enough to successfully transfer a message. We present this aggregated sample of observations via its empirical CDF. The detailed computation process can be found in [5]. Compared with previous generalized Dijkstra's algorithm, this algorithm directly computes representation of paths for all starting times in our considered data sets.

We utilize the following metrics to evaluate a given forwarding algorithm f: (i) the *normalized success rate within time t*: the probability of f to successfully find a path within time t, when sources, destinations, and message generation times are uniformly chosen at random. If no path exists, we include an infinite value in the distribution. We then normalize by the CDF given by an epidemic forwarding algorithm, and (*ii*) the *normalized cost*: the fraction of contacts (i.e., number of replica copies) used by f normalized by the fraction of contacts used by epidemic forwarding algorithm (the most expensive).

9.3 PeopleRank: A Social Opportunistic Forwarding Algorithm

For many years, mobile scenarios focused on communication between mobile devices and fixed access points. With the emergence of a new generation of powerful mobile devices, novel communication paradigms were possible, namely, ad hoc data transfer between the mobile devices. In such ad hoc network settings, the devices are often disconnected from each other and use Bluetooth, Wifi, or any other wireless connectivity to exchange and forward data in an opportunistic hop-by-hop manner.

To forward messages in such mobile opportunistic networks, a device must decide to which device the message should be sent. If a message needs to be delivered to a specific destination, efficient forwarding is of key interest and is one of the most challenging problems in mobile opportunistic networks. In these networks, the proposed techniques span from greedy broadcasting to all neighboring devices to mech anisms where the message is forwarded to its destination only when this destination is met. A compromise between these two extremes is to forward messages to a subset of the encountered nodes. The challenge that arises is how to select the next hop toward the destination such that delivery delay is minimized while maximizing the chance of the original message to reaching the final destination; i.e., messages have to be forwarded to nodes that are likely to meet the destination.

Whilst contact-and learning-based forwarding schemes have been quite popular [13, 19, 3, 8, 35], there has been less work on social-based forwarding [22, 6, 27, 36]. We develop a systematic approach to the use of social interaction as a means to guide forwarding decisions in an opportunistic network. We find that social interaction information alone is not sufficient and needs to be augmented in some way with information about contact statistics. We therefore develop an approach that combines these two pieces of information. The main challenge in combining social and contact information to guide forwarding decisions stems from the significant structural differences between these two sources of information. We approach this problem by developing PeopleRank, where nodes are ranked using a tunable weighted combination of social and contact information [28]. Our technique gives higher weight to social information in cases where there is correlation between that information and the contact trace information. More specifically, we introduce an opportunistic forwarding algorithm using PeopleRank which ranks the importance of a node using a combination of social and contact-graph information. Our evaluation shows that PeopleRank manages to deliver messages with a near-optimal success rate (i.e., close to

Epidemic Routing) while reducing the number of message retransmissions by 50% compared to Epidemic Routing.

9.3.1 The PeopleRank Algorithm

Inspired by the success of the PageRank algorithm to measure the importance of a web page, we propose a smart modification of this algorithm to rank nodes in social opportunistic networks.

The PageRank algorithm performs a random walk on the World Wide Web graph, where the nodes are pages, and the edges are links among the pages. It gives the probability distribution used to represent the likelihood that a person randomly clicking on links will arrive at any particular page. The PageRank is given by the following equation:

$$PR(p_i) = (1-d) + d \sum_{p_j \in M(i)} \frac{PR(p_j)}{L(j)}, \tag{9.1}$$

where $p_1, p_2, ..., p_n$ are the pages, $M(i)$ is the set of pages that link to p_i, $L(j)$ is the number of outbound links on page p_j, and d is a damping factor which is defined as the probability, at any step, that the person will continue clicking on links. Various studies have tested different damping factors, but it is generally assumed that the damping factor will be set to around 0.85.

Google describes the PageRank as a vote from page A to page B if page A links to B. Moreover, the importance of the page is proportional to the volume of votes this page could give. We apply the same concept in our algorithm to tag people as "important" if they are linked (in a social sense) to many other "important" people. We assume that only friends could vote for each other because they are more likely to recommend each other.

In the same way web pages are hyperlinked, one could establish a social graph between persons linked through social relationships such as friendship, or common interests. We denote such a social graph $G = (V, E)$ as a finite undirected graph with a vertex set V and an edge set E. An edge $(u, v) \in E$ if, and only if, there is a social interaction between nodes u and v (i.e., u and v are friends or they are sharing k common interests).

Consequently, the PeopleRank value is given by the following equation:

$$PeR(N_i) = (1-d) + d \sum_{N_j \in F(N_i)} \frac{PeR(N_j)}{|F(N_j)|}, \tag{9.2}$$

where $N_1, N_2, ..., N_n$ are the nodes (devices), $F(i)$ is the set of friends that links to N_i, and d is the damping factor which is defined as the probability that the person will still be connected to the network. Note that in the following section, we will examine the role and impact of parameter d in greater detail.

The PeopleRank of a node can be calculated either in a centralized way (as done with PageRank by Google) or it could be determined in a distributed fashion. The straightforward method is to compute the PeopleRank of each node in a centralized way by crawling the entire social graph.

A more interesting approach in our setting is to determine the PeopleRank in a distributed fashion. Therefore, we propose an extension of the basic algorithm that allows us to compute the PeopleRank of each node in a distributed system. The proposed algorithm is given by Algorithm 14. The PeopleRank value is updated each time a node meets a neighbor in its social graph (e.g., friend). The two neighbors exchange their PeopleRank values as well as the number of their neighbors. Then, each node updates its PeopleRank value using the formula given in Equation 9.2.

As described, the algorithm gives more authority to frequently seen neighbors. For example, when two "friends" are in different countries and do not met very often, Algorithm 14 does not contribute to increasing the PeopleRank values of the two nodes.

Algorithm 14: PeopleRank(i)

```
1  Require |F(i)| ≥ 0
2     PeR(i) ⇐ 0
3     while 1 do
4        while i is in contact with j do
5           if j ∈ F(i) then
6              send(PeR(i), |F(i)|)
7              receive(PeR(j), |F(j)|)
8              update(PeR(i))
9           end if
10       end while
11 end while
```

9.3.2 PeopleRank Evaluation

The performance of a new forwarding algorithm like PeopleRank is always determined by two main factors: (i) the average message delivery delay; and (ii) the overhead induced by the forwarding mechanism, i.e., the number of replicas in the system. In the following, we present the performance of PeopleRank, and we compare our results with the performance achieved with other well-known forwarding schemes.

In Figure 9.1, we plot the PeopleRank delay and cost performances for the six data sets presented in Section 9.2. We compare the delay distribution (CDF) of PeopleRank, *Centrality*, and the degree-based algorithm with an epidemic approach performing the same number of retransmission (cost). The plot shows that our PeopleRank algorithm outperforms all the other forwarding rules on all six data sets. PeopleRank and the centrality-based algorithm perform with around 90% of the maximum success rate within the 10-minute timescale using MobiClique, and Infocom (Face-

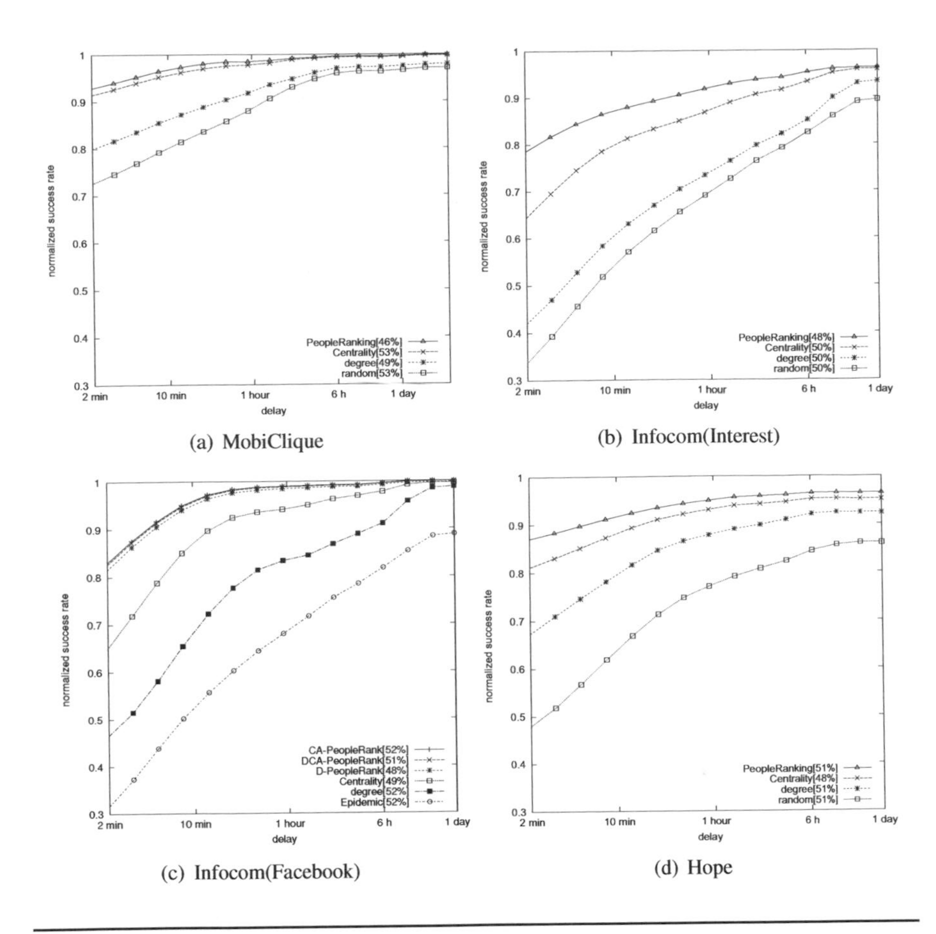

(a) MobiClique

(b) Infocom(Interest)

(c) Infocom(Facebook)

(d) Hope

Figure 9.1: *PeopleRank* vs. centrality vs. degree.

book) data sets. In these data sets, PeopleRank achieves the best success rates while using only 50% of contacts. Note again that the Centrality-based algorithm required centralized computation (as defined in [22]) and is way more complex to compute than the other rules under consideration.

In Figure 9.1, we plot the PeopleRank performance using the same contact trace between the preselected Infocom participants and three different social graphs connecting those participants (their Facebook profiles, common interests, and other social information). In the Infocom06 experiment, users were asked to fill their interest as a mandatory field in the questionnaire, additional fields such as name, nationality, affiliation, and city are optional. Using the friendship graph extracted from Facebook, the PeopleRank algorithm performs with a success rate above 95% normalized by the optimal flooding delay. It improves the performance by 45% of the random distribution using the same number of retransmissions. Indeed, Facebook is a explicitly defined social graph; a link is established if, and only if, the two users accept this friend relationship. Similar results were shown using the union social graph which is based on geographic location as defined by academic affiliation or cities. Geographic location helps users to socialize more often and meet with each other more frequently. The interest-based social graph uses common interests and common topics to link users. However, this information is not efficient to identify people that are likely to meet and socialize with each other.

Extensions of the PeopleRank algorithm, as well as investigating the impact of the damping factor for the performance of the PeopleRank algorithm, are detailed in [28].

9.4 Ensuring Fairness in Mobile Opportunistic Networking

Ensuring fairness is particularly important for mobile opportunistic networks since it acts as a major incentive for node cooperation. In fact, the lack of node cooperation, where a node refuses to act as a relay and settles for sending and receiving its own data, causes considerable delay degradation in the network. Previous studies have considered an absolute fair allocation of user resources. These studies have shown that while absolute fairness ensures a global sense of fairness among nodes, it deals with high end-to-end delivery delays [7, 2]. It is then a primary consideration whether there exists a trade-off relationship between mobile device resource consumption fairness and message delivery efficiency [23, 26].

9.4.1 The Efficiency Fairness Trade-off

Our discussion of efficiency and fairness highly depends on the definition of those two concepts. In this section, we let "*efficiency*" denote the successful delivery ratio of a given forwarding algorithm within t seconds. Higher efficiency means shorter end-to-end delay and more successful message delivery in the network.

To define fairness, one may consider it as a means to provide users with incentives to collaborate in mobile opportunistic communication—i.e., encourage them to contribute to forwarding messages and remain longer in the network. We therefore define "*fairness*" as a relative equality in the distribution of resource usage among neighboring nodes in the network—i.e., a forwarding algorithm is "fair" if the capacity assignment of a given node N in the network is equivalent to those of N's neighbors.

While rank-based forwarding techniques have demonstrated great efficiency in performance [6, 27, 3, 28], they do not address the rising concern of fairness amongst various nodes in the network. Higher-ranked nodes typically carry the largest burden in delivering messages, which creates a high potential for dissatisfaction amongst them. An absolute fair treatment of users, however, causes significant end-to-end delay and message delivery performance degradation [2, 17, 20]. Consequently, there is

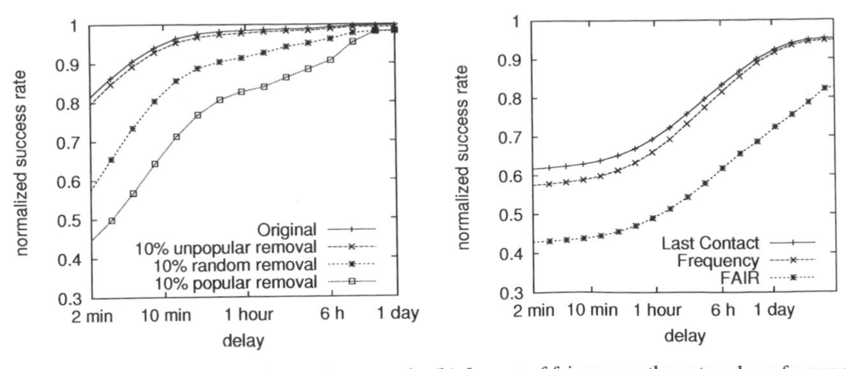

(a) PeopleRank performance using node removal (b) Impact of fairness on the network performance

Figure 9.2: Efficiency vs. fairness trade-off (Infocom06 data set).

no technique to date that ensures both fairness and efficiency. It is therefore important to consider whether there exists a trade-off relation between fairness and efficiency.

9.4.1.1 Absolute Efficiency

As discussed earlier, efficiency deals with high potential of dissatisfaction among popular nodes. Let us assume that dissatisfied nodes decide to boycott the forwarding process. We show the impact that these popular nodes have on network performance when they refrain from message forwarding; popular nodes may be relevant but not critically needed for a network.

Methodology: We use a technique, which we call "*node removal*," to exclude nodes from the forwarding process. We study the impact of excluding popular nodes on the overall forwarding performance by applying this technique to the mobility traces we have. We also studied the impact of removing the unpopular nodes. From the sequence of delay-optimal paths in our data sets, we deduce the delay obtained by the optimal path at all times. We uniformly combine all the observations of a trace among all sources, destinations, and for every starting time (the time in seconds when the message m was generated by the source node S). We present this aggregated sample of observations via its empirical CDF. We plot the success rate of forwarding algorithms normalized by the success rate of flooding within a given message delivery delay. This technique is a relevant metric used to analyze forwarding algorithms and is discussed in more detail in [5].

Figure 9.2-(a) plots the empirical CDF of the normalized delivery success rate of the PeopleRank forwarding algorithm (an example of an efficient rank-based forwarding algorithm according to [28]). We have observed similar results for LC and FR algorithms. Figure 9.2 shows the impact of excluding the most popular nodes on the successful delivery rate; 10% popular removal consists of removing the 10% most popular nodes from the original data set (here, popularity rank is given by the PeopleRank values [28]). We show that if only 10% of the most popular nodes boycott the forwarding process, PeopleRank's success rate performance degrades by roughly 20% within a 10-minute timescale. PeopleRank performance shows insignificant regress when we exclude 10% of the most unpopular nodes; only 1.5% less within a 10-minute timescale. This result is consistent with the fact that unpopular nodes do not contribute as much to the forwarding process as the popular nodes do. Therefore, it is indeed shown that popular nodes in the network are more suitable than others to deliver a given message to its destination.

As a summary, we observe that popular nodes in the network are more suitable than others to deliver a given message to its destination; selecting preferential relays in forwarding decisions gives better forwarding performance. It is indeed important to further satisfy popular nodes. Providing fairness is therefore crucial since the unfair treatment of users is considered as a disincentive to participation in the communication process.

9.4.1.2 Absolute Fairness

While fairness is our goal, absolute fairness amongst all nodes is not. In this section, we discuss the impact of absolute fairness on the overall network performance. Let us assume an absolutely fair allocation of resources across nodes in the network. We perform an offline study of the path availability in our data sets, where each node forwards the same number of messages compared to all other nodes in the network. We ensure fairness by balancing cost across nodes. We call this offline fair path establishment technique, the FAIR algorithm. FAIR provides, by definition, a uniform fair allocation of user resources among nodes in the network.

Figure 9.16(b) compares two contact-based ranking algorithms (i.e., LC and FR) to the FAIR algorithm using the Infocom06 data set. We show how an absolutely fair allocation of resources leads to significant performance regression. The probability of success within 10 minutes decreases from 72% to 43% for Last Contact algorithm and from 60% to 43% for the Frequency algorithm. The reason behind this regression is that an absolutely fair balancing of cost across nodes causes significant end-to-end delay and success rate performance degradation.

9.4.2 Real-Time Distributed Approach for Fairness-Based Forwarding

In this section, we quantitatively identify the trade-off relationship between fairness and efficiency. We first provide some intuition regarding the desired fairness we seek while introducing our satisfaction index metric.

9.4.2.1 Desired Fairness and Satisfaction Index

In a mobile opportunistic network, we can roughly divide nodes into three different categories with respect to their popularity ranking; *popular*, *semi-popular*, and *unpopular* nodes (as shown in Figure 9.3). While ranking nodes in the network may

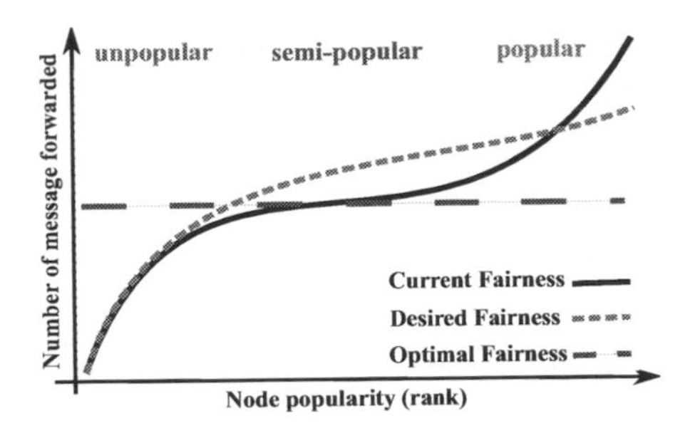

Figure 9.3: Node categories with respect of their popularity rank.

be considered to efficiently forward messages in DTN, it creates a high potential of dissatisfaction amongst nodes since they forward more messages compared to others (as shown by the solid line in Figure 9.3). On the other hand, an absolutely fair allocation of resources yields poor forwarding performance (dashed line in Figure 9.3). Inspired by the previous results shown in Figure 9.16, we believe that the desired fairness is different from an absolutely fair distribution of resources (optimal fairness). It is indeed important to increase the satisfaction of popular nodes by reducing their cost. Unpopular nodes, however, are unlikely to be impactful if they were involved in the forwarding process. Therefore, our goal is to further satisfy popular nodes by moving from a situation where popular nodes carry the largest burden in delivering messages to a "fair distribution" of this burden among *popular* and *semi-popular* nodes as shown by the desired fairness curve, dotted line, in Figure 9.3.

To reach this desired situation, we need a metric by which we can assess the level of fairness amongst the nodes. We define a *satisfaction index* (*SI*) as the difference between the message load distribution given by the forwarding process (current fairness) and uniform distribution among nodes (desired fairness). Our goal is to construct paths between any source and destination that verify the condition below:

$$\exists C \geq 0, \forall i \in 1..n,$$
$$SI(N_i) = load(N_i) - \sum_{j=1}^{n} load(N_j) \geq 0 - C, \tag{9.3}$$

where N_i represents a node i in the network, n is a total number of nodes, $load(N_i)$ is the current load at node N_i, and C is a positive constant integer. The goal is then to verify if there exists a relatively small positive integer C, such that we can construct paths between any source-destination pair and satisfy Equation 9.3.

We now propose a real-time distributed framework that helps rank-based forwarding algorithms utilize potential nodes to forward messages while satisfying the fairness property given by Equation 9.3. We call our systematic framework, *F*airness-based *O*pportunistic networkin*G* (*FOG*).

9.4.2.2 The FOG Framework

Whenever two nodes are within close proximity of each other, they separately run an update process. They first update their relative maximum burden *max* and relative minimum burden *min* per unit time. They then decide whether or not to forward a message *m* if the following condition is verified; (*i*) the encounter node N_j is the destination node of the message *m*, or (*ii*) N_j is higher ranked compared to N_i and its burden ($burden(N_j)$) did not exceed the relative average. Finally, the receiver of the message should update its actual burden value.

Algorithm 15: FOG(N_i)

---

1 $max \leftarrow 0$
2 $burden \leftarrow 0$
3 $time \leftarrow system_{t}ime$
4 $min \leftarrow time$
5 **while** True **do**
6    $max \leftarrow max(max, burden)$
7    $min \leftarrow min(min, burden)$
8   **while** N_i in contact with N_j **do**
9       $update(max, min)$
10   **if** $Rank(N_j) \geq Rank(N_i)$ AND $burden(N_j) < (max + min)/2$ OR $N_j = destination(m)$ **then** $Forward(m, N_j)$
11   **end if**
12   **if** $receiving(m, N_i)$ **then**
13          $burden \leftarrow (burden + 1)/\Delta T$
14   **end if**
15   **end while**
16 **end while**

FOG as a fully distributed framework, relies on local information at the node to estimate the distribution of the burden among neighboring nodes. We believe that nodes are satisfied by comparing themselves to their neighbors and acquaintances and require only relative equality amongst their neighbors in the network. We note that an offline technique to measure the satisfaction index *SI* among all nodes is presented in [26].

9.4.2.3 FOG Evaluation

In the following, we evaluate the performance of three state-of-the-art social forwarding algorithms (PeopleRank, LC, and FR) relying on the two previously described data sets (Infocom06, and CoNext07). We also use a data-driven artificial trace to evaluate the scalability of FOG in large-scale social networks.

In Figure 9.4, we compare the *SI* and the message distribution among nodes using FOG and the offline approach (described in detail in [26]). In addition, we compare the success rate of these forwarding algorithms. We show that FOG outperforms all the offline rules we have tested and achieves one of the best SI/success rate trade-

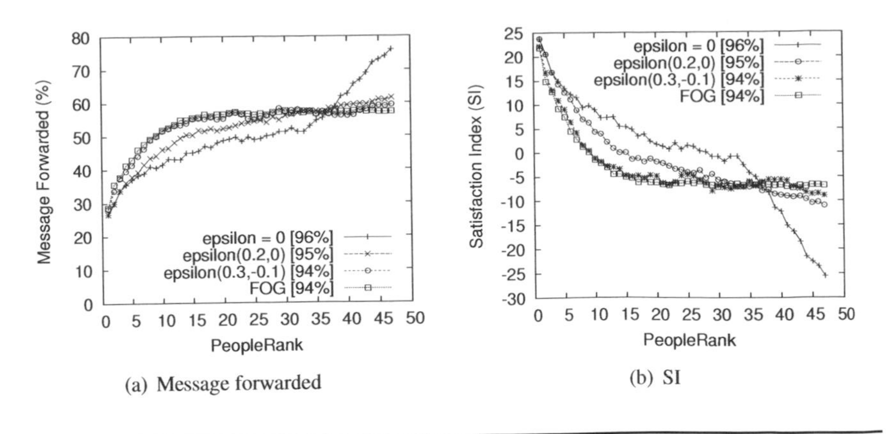

(a) Message forwarded

(b) SI

Figure 9.4: Comparison of FOG and the offline technique (Infocom06 data set).

offs. FOG also ensures a more fair distribution of the burden among the nodes; the SI of popular and semi-popular nodes remain close to zero and verify Equation 9.3 with a relatively small constant $C = 3$. FOG explicitly prohibits forwarding messages when the burden is not fairly distributed among neighbors; popular nodes are more likely to meet all other nodes as well as each other. The burden is therefore fairly distributed among them. We note that FOG is also efficient given that it keeps the success rate close to the optimal (2% less compared to the original PeopleRank success rate performance).

***Scalability of FOG*:** In order to study the scalability of our approach, we use the SanFrancisco11 trace in our evaluation. As described previously in Section 9.2, the data set contains 3 communities interconnected by very mobile nodes (the cabs).

We evaluate the performance of FOG integrated with each of the three rank-based forwarding algorithms in large-scale networks relying on the San Francisco modified data set. Figure 9.5 plots for the PeopleRank algorithm (*a*) the normalized number of messages forwarded and (*b*) the normalized success rate performance using the original algorithm and the FOG-based extension. We show that FOG-based extensions ensure better distribution of resources among the nodes while keeping the success rate performance close to optimal (original); we also show that FOG-PeopleRank gives a significantly high improvement compared to the FAIR algorithm (described in Section 9.4.1) in large scale networks: +20% in the success rate within a 10-minute timescale. In PeopleRank, socially well-ranked nodes carry the highest burden since they are more likely to forward a message to its destination. FOG-PeopleRank therefore ensures an acceptable trade-off between efficiency and fairness in large-scale networks.

***Cost of FOG*:** Cost, in addition to end-to-end delay and success rate metrics, is a very important evaluation metric in opportunistic networks. We define the cost of a

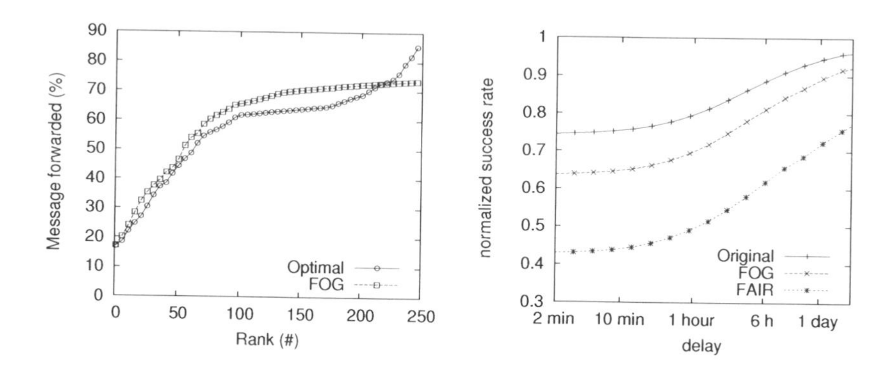

Figure 9.5: Comparison of FOG and offline approach performance (using the modified SF dataset).

forwarding algorithm as the fraction of contacts involved in the forwarding process. FOG is designed to fairly distribute the burden among semi-popular and popular nodes. In this section, we study the impact of the fair distribution of the burden given by FOG on the overall cost.

Figure 9.6 compares the cost of FOG-PeopleRank to FAIR (absolute fairness) and the original version of PeopleRank using three different data sets (Infocom06, Conext07, and modified SF data sets). The original version of PeopleRank performs well and achieves the best normalized cost in all scenarios we have tested; PeopleRank, as a social-based ranking algorithm, was designed to provide a fairly good cost/-efficiency trade-off. We note that rank-based forwarding algorithms can be fair and keep the cost and the performance close to the optimal. FOG-PeopleRank outperforms the FAIR algorithm while using only 2% to 6% more replicas of the message m compared to the cost of the original PeopleRank algorithm.

It is crucial to be aware that, while the total number of replicas in the network increases, the cost increase per node is not significant as shown in the previous plots. Moreover, as the number of nodes in the network grows, reflecting more realistic deployment environments, the increase in cost becomes significantly smaller.

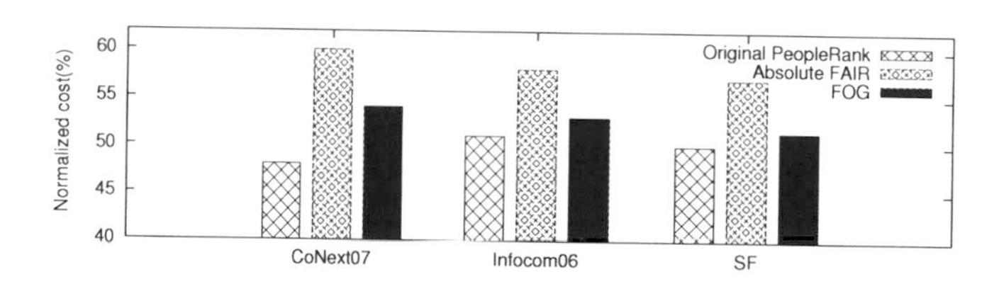

Figure 9.6: Normalized cost of extended PeopleRank versions.

9.5 Forwarding in Large-Scale Mobile Opportunistic Networks

This section contributes to a better understanding of the weaknesses of existing social forwarding algorithms in large-scale mobile opportunistic networks, and proposes insights and solutions to deal with these issues [24, 25]. We propose partitioning large-scale communities into multiple sub-communities based on various common social characteristics such as locality or social interests. We introduce CAF, a community-aware framework which can easily be integrated with most of the existing social forwarding algorithms, in order to improve their performance in large-scale networks. CAF uses particular nodes called MultiHomed nodes to disseminate messages across the sub-communities in the network. Original social forwarding algorithms can then behave normally within a local sub-community. Besides the simplicity of CAF, it causes negligible overhead compared to that induced by state-of-the-art algorithms, such as BubbleRap [11], which compute the global node ranking in the large-scale network. CAF remains a distributed forwarding algorithm and relies on a local social/contact information to estimate future transfer opportunities. We utilize the largest data set (to the best of our knowledge) that captures human mobility contacts in large-scale networks in addition to the corresponding social information. Our results show that we obtain a performance increase of around 40% compared to the state-of-the-art social forwarding algorithms, while incurring a marginal increase in cost. We also show that CAF outperforms BubbleRap and achieves 5% to 30% delivery rate improvement.

9.5.1 *Forwarding Drawbacks in Large-Scale Opportunistic Networks*

It has been shown that rank-based forwarding techniques are one of the most promising methods that provide a high success rate and reasonably low cost [22, 10, 27]. However, these studies focus on small data sets, and a small number of nodes. We believe that such techniques present serious limitations in large scale opportunistic networks.

Figure 9.7(a) illustrates a scenario where rank-based forwarding attempts to disseminate a message M generated by a source S to a destination D. Without any notion of communities, M is forwarded in the wrong direction relying on "globally popular nodes" in the network (i.e., the BubbleRap technique [12]). These particular nodes, although popular, may not be able to deliver the message to all the nodes in the network. In Education City (EC), a campus that includes 6 US university branches in Doha, Qatar, the founder of EC could be a very popular person in the whole campus (globally popular node), but is not likely suitable to relay the message M to a student in a particular university on campus. However, other nodes may be locally popular (e.g., within a university campus as shown in Figure 9.7(b)) and more suitable to deliver this message to its destination in a specific sub-community. Therefore, particular nodes, which we call MultiHomed nodes such as postmen or campus bus drivers are more suitable to disseminate the message M across sub-communities. This ap-

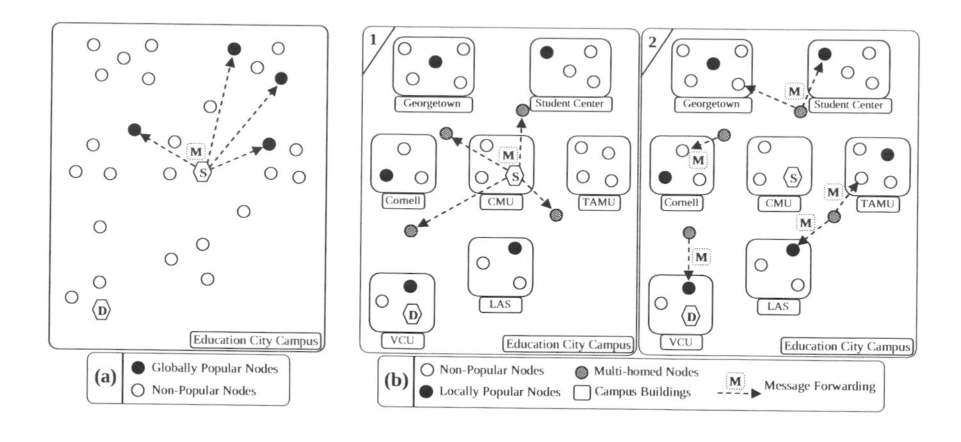

Figure 9.7: Example showing (a) the weaknesses of existing rank-based forwarding algorithms in a large-scale community, and (b) identifying sub-communities within the same large-scale one, and using MultiHomed nodes to disseminate messages to these sub-communities.

proach is opposed to the BubbleRap algorithm that uses globally popular nodes to disseminate the message to all communities. Therefore, the main idea (shown in Figure 9.7(b)) is to first break down the original large scale community into multiple sub-communities, then disseminate the message to these sub-communities. Afterward, locally popular nodes can then deliver the message M within its sub-community.

Figure 9.8(a) plots the normalized success rate of the 5 rank-based forwarding algorithms: PeopleRank, Degree-Based, Simbet, FRESH Total, and Greedy Total with respect to the SanFrancisco11 data set. Note that the value of the CDF (Cumulative Distribution Function) represents the probability of successfully finding a path within time t, when sources, destinations, and message generation times are chosen at random. If no path exists, we include an infinite value in the distribution. We then normalize by the CDF given by an epidemic forwarding algorithm.

Despite the fact that node popularity matches the contact properties of nodes, there are 25% to 55% losses compared to Epidemic forwarding, within a 10-minute timescale. In fact, in large-scale networks, rank-based forwarding algorithms lose many opportunities to reach destinations in optimal delays. Similar results are observed using the San Francisco data set. Similar results were shown for the Hope08, and Dartmouth data sets [25]. These preliminary results match the intuition presented in the EC example shown in Figure 9.7 and strongly motivate the need for agile solutions that take such situations into account.

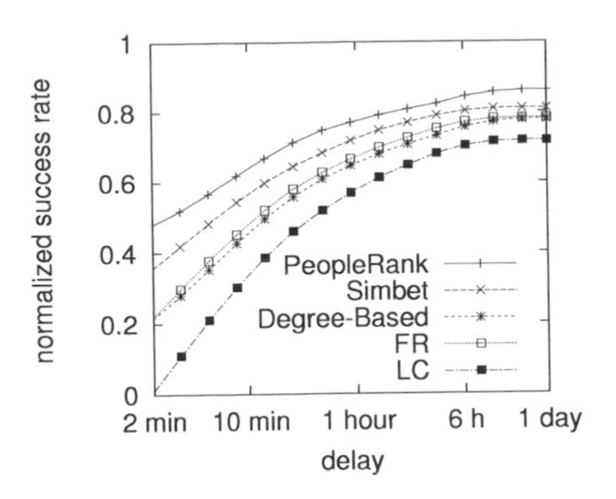

Figure 9.8: Scalability issues of rank-based algorithms relying on the SanFrancisco11 data set.

9.5.2 *Forwarding within Sub-Communities*

So far, we have shown that using node popularity in large-scale areas has serious drawbacks. Rank-based forwarding techniques, while showing promising results in small communities, present serious limitations in large-scale networks. Our main hypothesis is that in large-scale networks, where multiple sub-communities may exist, social and contact prediction has its limitations.

In this section, we introduce and compare the impact of different large-scale community classification techniques, and ensure that the state-of-the-art rank-based forwarding algorithms perform well within the resulting sub-communities. In the next section, we introduce CAF, a community-aware framework, that can be easily integrated with these algorithms to improve their performance in large-scale networks.

9.5.2.1 Classification and Forwarding in Sub-Communities

A common property of networks is cliques or communities; circles of friends or acquaintances in which every member knows every other member. In large-scale networks, people can be regrouped into sub-communities. Our experimental data set can be classified into multiple communities according to different classification techniques. The SanFrancisco11 data set is by default classified into three communities; airport, downtown, and sunset areas (geographic classification). We propose two community classification techniques for the Dartmouth01 and the Hope08 data sets:

Activity-Based Classification: The Dartmouth College campus has over 160 buildings. Usually people visiting the campus are interested in a few buildings. People can be classified based on their activity or interests. The intuition behind this classification is people who frequent more athletic buildings are more suitable to meet each other and socialize during their athletic activities. The campus contains more than a dozen athletic facilities and fields. We note that in the Dartmouth campus, activity buildings are not colocated.

We compute the contact duration between a particular device and any athletic building access point. If this duration is more than 3 hours a day, we consider the student carrying the device as an athletic student. Similarly, we define academic and residential communities.

Geographic Classification: In mobile opportunistic communication, physical proximity is fundamental for message dissemination. We propose to group people living in close proximity into a single geographic community. Such classification could be done using Facebook or other online social applications where users explicitly specify their current city or neighborhood. Using Hope08 and Dartmouth01 data sets, we classify users within physical proximity into a single sub-community.

The Dartmouth campus area is roughly 1300 x 1300 square meters, and people going to campus every day are mostly visiting the same places. Usually, these places are selected in a way that minimizes walking distance. To capture this classification, we split the Dartmouth campus into regions (Northwest *NW*, Northeast *NE*, Southwest *SW*, and Southeast *SE*).[3] A node i belongs to a region R if it has been connected to more access points belonging to this corresponding region as compared to the other regions.

The Hope08 dataset can also be grouped according to geographic classification; during the conference, only people in the mezzanine or the 18th floor were tracked. We therefore consider two geographic communities; *mezzanine* and *18th*.

Combination Classification: The combination classification is defined as an activity classification within a geographical area. For example, in the Dartmouth campus, given the three different building types and 4 different geographical areas, the combination will give us a maximum $3*4$ different sub-communities. For example, the athletic buildings are mostly concentrated in only two geographical areas in the Dartmouth campus.

9.5.2.2 Impact of Different Community Classifications on Forwarding Performance

After classifying the large-scale community, we show now rank-based forwarding algorithms performance within a single sub-community. We plot the normalized success rate of two rank-based forwarding algorithms, PeopleRank, and Frequency (FR), according to the two community classifications described above. Similar results have been obtained with other rank-based forwarding algorithms [25].

From Figure 9.9 we generally observe that geographic classification leads to better PeopleRank performance. PeopleRank achieves 92% to 97% of normalized success rate within 10-minute timescales according to the geographic classification in Figure 9.9(b), and 90% to 94% within the same timescale according to the activity-based classification in Figure 9.9(a). These results confirm that short distances (e.g., people living in the same neighborhood or region) typically lead to strong social ties, and relevant social classification.

[3] `http://www.dartmouth.edu/~maps/campus/close-ups/index.html`

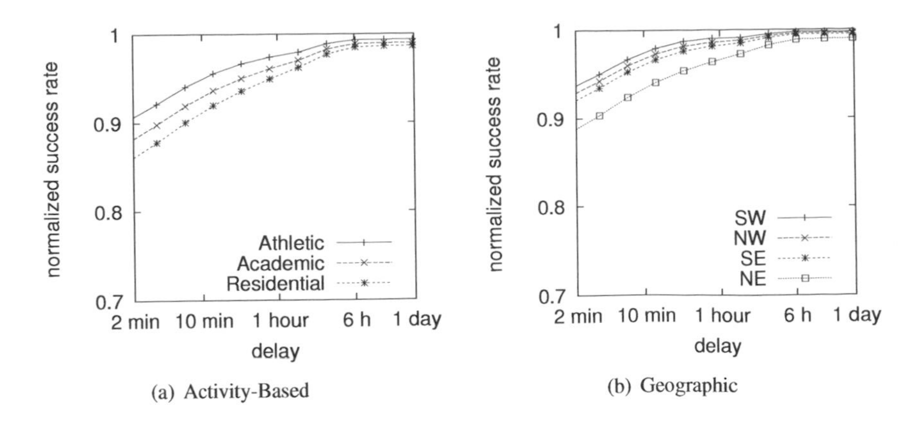

Figure 9.9: Normalized success rate distribution of PeopleRank relying on different community classification (Dartmouth01).

Moreover, we notice that, in Figure 9.9(a), PeopleRank achieves higher success rates among athletic users than others according to the activity-based classification. Relying on the athletic community, PeopleRank outperforms its own success rate performance by roughly 3% and 5% within 10-minute timescales compared to respectively, the academic and the residential communities. As described above, most of the athletic buildings are located in the southeast corner of the campus which leads to a combination of geographic and activity-based classification. While PeopleRank achieves poor performance across multiple communities, within a single sub-community, it performs fairly well, and achieves more than 80% success rate within 10 minutes. Users within a single sub-community are able to communicate efficiently using a rank-based forwarding algorithm such as PeopleRank. Similar results obtained for Hope08 and SanFrancisco11 data sets are shown in [25].

We also investigate the the performance of Greedy algorithm within the Dartmouth01 sub-communities. We plot the performance of Greedy according to both geographic (Figure 9.10(a)), and activity-based (Figure 9.10(b)) classifications. The FRESH forwarding algorithm achieves roughly an 80% success rate within most sub-Dartmouth01 communities. We also obtain similar results using Simbet and degree-based forwarding algorithms [25].

We note that this chapter does not investigate new classification techniques; we assume that people can be classified based on their social profile information [33], location, IP addresses, or phone numbers. Our goal is to confirm that rank-based forwarding algorithms achieve satisfactory performance within sub-communities regardless of the classification methods. However, it was shown in the previous section that they suffer in large-scale networks where multiple sub-communities may exist. We therefore propose a strategy to help existing rank-based forwarding algorithms deal with this issue and successfully forward messages across multiple sub-communities.

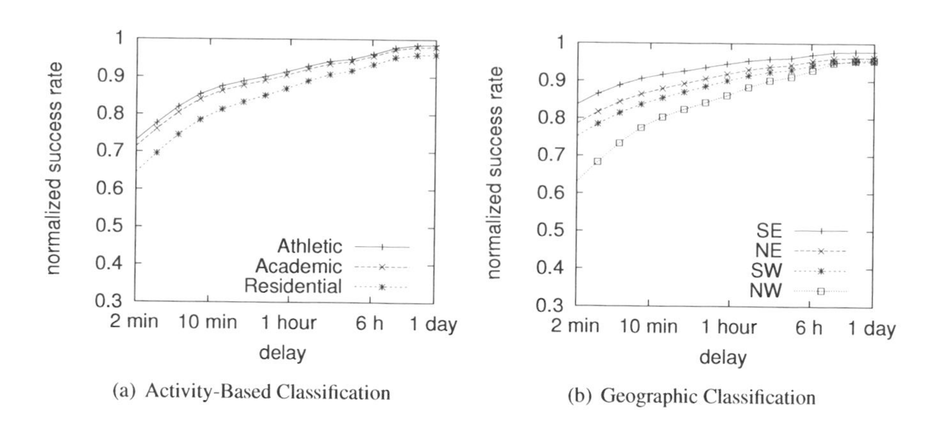

(a) Activity-Based Classification

(b) Geographic Classification

Figure 9.10: Normalized success rate distribution of Greedy relying on different community classification (Dartmouth01).

9.5.3 Forwarding across Sub-Communities: The Community-Aware Framework (CAF)

Motivated by the satisfactory performance of rank-based forwarding within single sub-communities, we propose a *community aware framework* (CAF). This framework can easily be integrated with most rank-based forwarding algorithms in order to deal with the weaknesses described above in large-scale networks.

The community-aware framework (CAF) relies on the fact that rank-based forwarding algorithms operate normally within the same sub-community *SC*. Indeed, messages will be forwarded relying on rank-based techniques toward nodes which belong to the same sub-community.

On the other hand, particular nodes will operate as an inter-community backbone and circulate messages to other sub-communities where they will then be forwarded according to a rank-based forwarding rule. We call these backbone-like nodes *MultiHomed* nodes (MH). MultiHomed nodes are characterized by their higher mobility and the fact that they belong to multiple sub-communities (i.e., MultiHomed nodes move from one sub-community to another). These nodes can be postmen, buses, cabs, etc. depending on the large community. We then rank these MultiHomed nodes (MH_{rank}) according to the number of sub-communities they belong to; i.e., $SC(i)$ is equal to the number of sub-communities node i belongs to. For example, if we consider the geographic classification in the Dartmouth01 data set, MH nodes belonging to four sub-communities are highly ranked compared to MH nodes belonging to only two or three sub-communities. Therefore, MH nodes carry a message forward to other MH nodes according to a non-decreasing MH_{rank}.

Algorithm 16 summarizes the additional operations on top of the current state-of-the-art rank-based forwarding algorithm which we will refer to as RBFA in Algorithm 16. Besides the simplicity of our proposed algorithm, we emphasize that the

Algorithm 16: CAF-RBFA(node i)

```
1  Require: Node i is running a Rank-Based Forwarding Algorithm RBFA
2  Require: RBFA(i) denotes the rank of node i according to RBFA
3  Ensure: MH^i_rank ← SC(i)
4    while (1) do
5      while (i is in contact with j) do
6        update(RBFA(i), RBFA(j))
7        while (∃ m ∈ buffer(i)) do
8          if
           [SC(i) == SC(j) & RBFA(j) ≥ RBFA(i)]
           OR [j = destination(m)]
           OR [MH^j_rank ≥ MH^i_rank] then
9            Forward(m, j)
10         end if
11       end while
12     end while
13   end while
```

overhead is relatively negligible compared to the overhead induced by BubbleRap to compute the global node ranking in the whole system (in a large-scale network). Our proposed algorithm remains a distributed forwarding algorithm and relies on local social/contact information to estimate future transfer opportunities. Next, we evaluate the CAF-extended version of three state-of-the-art rank-based forwarding algorithms that integrate this proposed framework.

9.5.3.1 The Impact of Community Classification on CAF-Enabled Rank-Based Forwarding Algorithms

We investigate the impact of different community classification (described in the previous section) on the performance of CAF. We show a representative set of results for the CAF-PeopleRank algorithm. Similar results have been observed for CAF-Simbet and CAF-Degree-based, but are not shown due to space limitation.

Figure 9.11 plots the normalized success rate of the extended PeopleRank algorithm, with different classification techniques, in the Dartmouth01 data set. We notice that extended PeopleRank outperforms the original PeopleRank for all timescales (5% to 30% of success rate improvement). Furthermore, such improvement differs from one community classification to another. Geographic classification gives better performance than activity-based classification; indeed, the activity-based classification in the Dartmouth campus groups people belonging to specific buildings. However, these buildings are not always geographically close to each other, and therefore messages sent from a specific building can take a long time to reach other members of the same sub-community. We finally note that, combining two community classification approaches leads to better success rates with more than a 30% improvement compared to the activity-based classification.

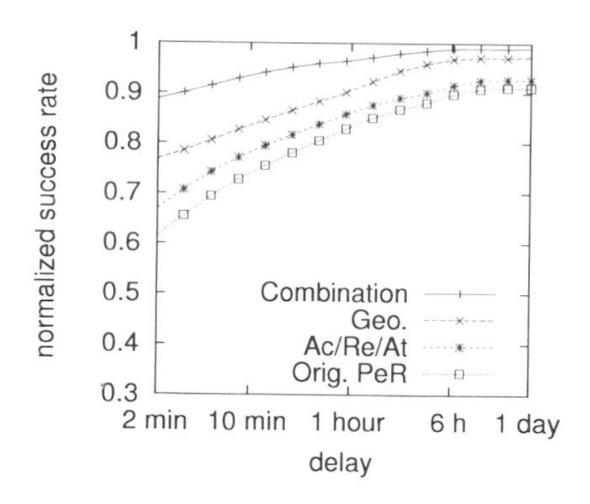

Figure 9.11: Impact of community classification on CAF-PeopleRank success rate (Dartmouth01 data set).

9.5.3.2 *CAF vs. BubbleRap*

Initially, we consider the SanFrancisco11 data set using only 5% of the total cabs to represent the MultiHomed nodes. We evaluate the performance of our proposed framework (CAF) and compare it against: (*i*) the original rank-based forwarding scheme (without CAF), and (*ii*) the BubbleRap algorithm. Later, we analyze the impact of different values of cabs fraction on the performance and justify our 5% choice in the evaluation.

We apply CAF to state-of-the-art social-based forwarding algorithms: PeopleRank, Simbet, and Degree-Based forwarding. Figure 9.12 compares the performance of the extended versions of these three algorithms against the original versions (without CAF) and the BubbleRap algorithm using the SanFrancisco11 data set. We first show that in the three plots, the CAF extended algorithms outperform the corresponding original algorithms for all timescales; for a 10-minute timescales CAF-PeopleRank outperforms PeopleRank by roughly 40% more delivery success rate, while CAF-Simbet and CAF-Degree-Based achieve higher respectively, 30% and 25% better success rates compared to their original algorithms. For larger timescales, CAF performance remains superior, however the improvement is less significant since the proposed framework is designed for a better dissemination of the message in order to reach the destination in fewer delays.

Moreover, we show that CAF outperforms BubbleRap especially for PeopleRank and Simbet; the probability of successfully delivering the message using CAF-PeopleRank or CAF-Simbet is 5% to 30% higher than the success rate probability achieved by BubbleRap for all timescales. The reason behind this result is that BubbleRap uses node degree to estimate social centrality. However, it was shown that node degree cannot be considered to efficiently estimate future contacts [22]. Therefore, BubbleRap performs poorly compared to centrality-based forwarding al-

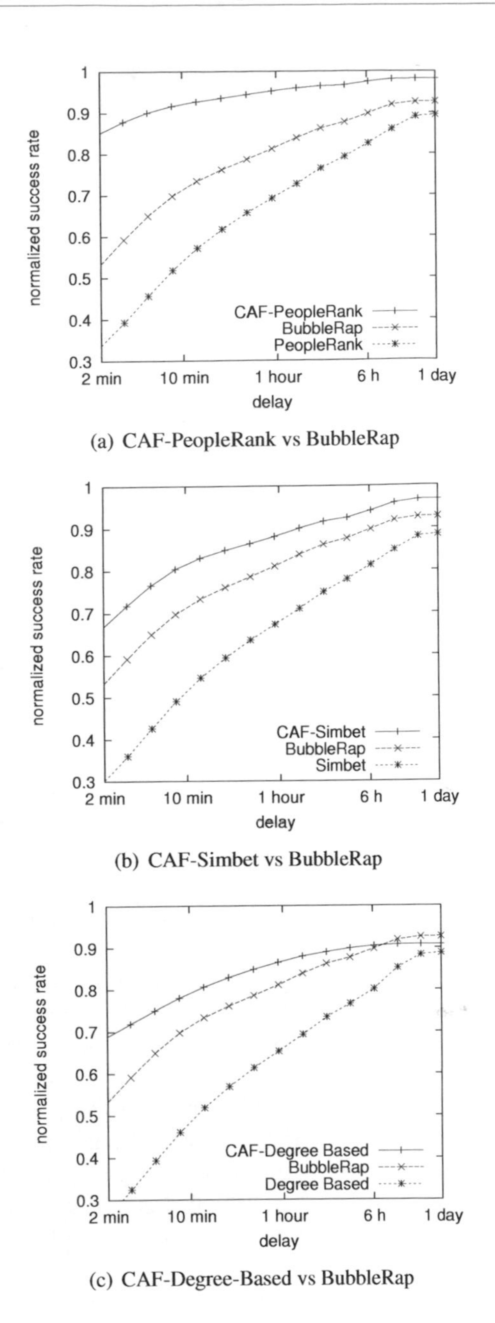

(a) CAF-PeopleRank vs BubbleRap

(b) CAF-Simbet vs BubbleRap

(c) CAF-Degree-Based vs BubbleRap

Figure 9.12: Comparison of CAF-PeopleRank, CAF-Degree-based, and CAF-Simbet with BubbleRap (SanFrancisco11 data set using only 5% cabs).

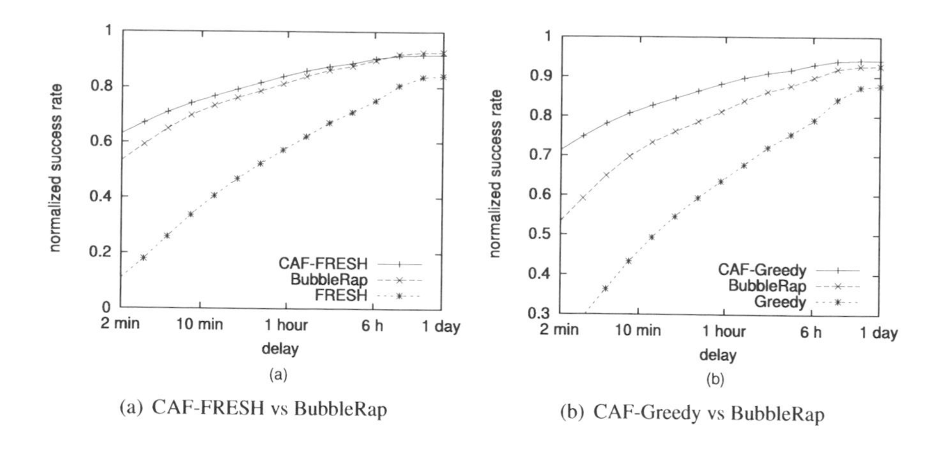

(a) CAF-FRESH vs BubbleRap

(b) CAF-Greedy vs BubbleRap

Figure 9.13: Comparison of CAF-FRESH and CAF-Greedy with BubbleRap (SanFrancisco11 data set using only 5% cabs).

gorithms (CAF-Simbet and PeopleRank) and performs better compared to CAF-Degree-Based only in very large timescales.

We also apply CAF to state-of-the-art contact-based forwarding algorithms: FR, and FRESH. Figure 9.12 shows the performance improvement introduced by CAF compared to the original performance given by each of these algorithms. CAF applied to FRESH and Greedy achieves from 30% to 45% success rate improvement compared to the original FRESH (34% success rate) and Greedy (43% success rate) algorithms.

We finally compare CAF-FRESH and CAF-Greedy to BubbleRap. CAF achieves slightly better performance than BubbleRap; for instance, CAF-Greedy achieves 8% better success rate compared to BubbleRap. BubbleRap uses contact properties to compute social communities. However, CAF uses explicit social data to infer sub-communities and forward accordingly. It has been shown in [10, 27] that when social data does not present a strong correlation with the mobility data, this translates to social forwarding performance degradation. This may explain the better performance given by this algorithm in large timescales compared to CAF-Degree-Based algorithm. However, in shorter time-scales CAF-Degree-Based achieves a 7% higher success delivery rate than BubbleRap. This can be explained by the use of explicit MultiHomed nodes instead of globally popular nodes as shown in the motivating example (Figure 9.7).

9.5.3.3 The Impact of MultiHomed Nodes

The number of MultiHomed nodes (MH) needed to make CAF efficient may depend on different factors such as the number of communities, distance between communities, MH mobility, etc. Obviously, the more MultiHomed node that are available for the system, the more successful it will be. Our goal in this section is to under-

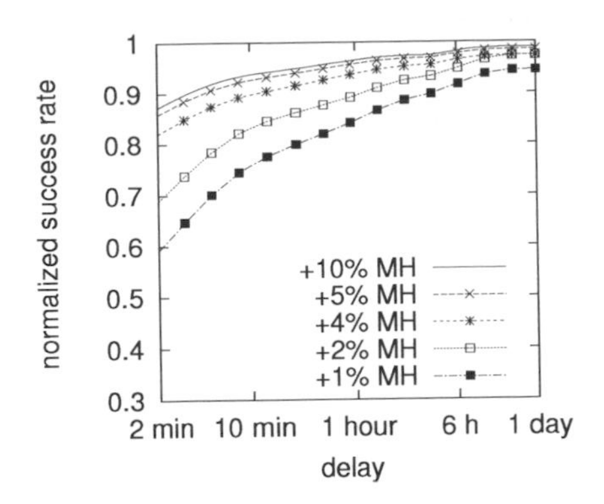

Figure 9.14: Normalized success rate distribution of CAF-PeopleRank across multiple communities (SanFrancisco11).

stand the impact of the number of MH nodes on the system's performance. We use the SanFransisco11 data set and vary the fraction of cabs used in the trace (we randomly pick $x\%$ of the total number of cabs); cabs in this data set connect the three disconnected areas of the city of San Fransisco, and may operate as MH nodes.

In Figure 9.14, we plot the normalized success rate of the CAF-PeopleRank algorithm with different fractions of MultiHomed nodes in the SanFrancisco11 data set. Obviously, the more MH nodes used the better performance CAF can achieve. We show that the improvement is significant for the first MH added; the improvement from 1% to 2% of MH is roughly 10% of the success rate, however it is only 0.7% from 5% to 10% of MH nodes. We show that with only 5% of MultiHomed nodes (only 10 cabs in the data set) the CAF-PeopleRank algorithm achieves a near-optimal performance; it performs at more than a 90% success rate compared to epidemic forwarding (within a 10-minute timescale) which represents only 0.7% less than the optimal given by 10% of MH nodes (20% of MH gives no significant improvement compared to 10% of MH). These results are very promising since they do not require a large number of participants to be involved in order for CAF to be successful, and hence, minimizes the deployment barrier for such solutions.

An important observation to share from the figure is that adding MultiHomed nodes such as taxis in the SanFrancisco11 data set, is also beneficial for shorter delays. This might appear strange since one might expect that taxis may need non-negligible time to drive from one community to another, and so, only large time delays would be affected. However, this effect can simply be explained by the fact that taxis are also used to improve the performance within a single sub-community; e.g., within downtown, taxis could be considered as a better relay to efficiently disseminate the message within such sub-community.

Figure 9.15 plots a comparison of CAF-Simbet, CAF-BubbleRap, and CAF-PeopleRank success rate performance using 2% and 5% MultiHomed nodes. We

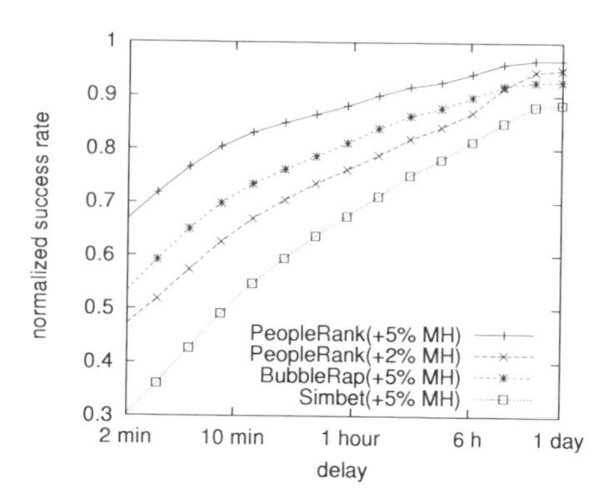

Figure 9.15: CAF-Simbet vs CAF-BubbleRap vs CAF-PeopleRank (SanFrancisco11).

show that CAF-PeopleRank outperforms all other schemes while using only 2% compared to 5% MultiHomed nodes used by the other scheme. This result was verified for short timescales where CAF-PeopleRank was able to identify and disseminate messages to the more suitable relays to reach more destinations.

9.5.3.4 The Cost of CAF

One may claim that CAF can be costly. The cost of a forwarding algorithm, defined as the fraction of contacts involved in the forwarding process, is very important in opportunistic networks. It is obvious that CAF uses more contacts than the original version since it disseminates the message first to all other sub-communities, and then proceeds normally within each sub-community. We therefore quantitatively compare the cost of each rank-based forwarding algorithm and study more options in order to reduce the cost of the extended version of rank-based forwarding algorithms.

In addition to the basic CAF we have presented, we propose and evaluate the cost of a community destination-aware framework (CDAF). Such framework assumes a priori knowledge of the destination's sub-community. Such a simplistic assumption, while surrealistic, is widely used in the literature [21]. We assume that the source node will be able to search, retrieve, and append the destination's community to the message; e.g., it can use an IPv6 address or simply append the city where the destination lives (e.g., fetched from Facebook) to the original message. Knowing the destination's community C_m, CDAF will then be able to forward the message m to the MultiHomed node that belongs to C_m. As a result, CDAF will narrow down the set of nodes able to receive the message m by focusing more on the destination's area, and then the overall overhead generated by the forwarding process.

Algorithm 17 summarizes the additional operations on top of the CAF-RBFA forwarding algorithm running at a given node i. We assume that node i knows the list of sub-communities it belongs to. Nodes i MH^{i}_{rank} will be simply the number of communities it belongs to. Algorithm 17 will then operate similarly to Algorithm 16 with one additional condition at line 5 where $CDAF(i)$ checks if the encounter node's MH belongs to the message's destination sub-communities(C_m) or not. We use this framework as a benchmark and compare the overhead of CDAFs, CAFs, to the overhead of the original rank-based forwarding algorithms.

Algorithm 17: CDAF(node i)

```
1  Require: Node i is running CAF-RBFA(i)
2  Ensure: C_list(i) ← {C_j | i ∈ C_j}
3  Ensure: MH^i_rank ← ||C_list(i)||
4      while (1) do
5          while (i is in contact with j) do
6              update(CAF-RBFA(i),CAF-RBFA(j))
7              while (∃ m(D@C_m) ∈ buffer(i)) do
8                  if (C_m ∈ C_list(j) ) then
9                      if
10                     [(C_list(i) ∩ C_list(j)) AND (CAF(j) ≥ CAF(i))]
11                     OR [j = destination(m)]
12                     OR [MH^j_rank ≥ MH^i_rank] then
13                         Forward(m, j)
14                     end if
15                 end if
16             end while
17         end while
18     end while
```

Figure 9.16 compares the normalized cost of the different schemes using (a) the SanFrancisco11 data set, and (b) the Hope08 data set. The forwarding process of our CAF framework disseminates additional replicas of the same message to reach the destination in shorter delays; CAF uses 2% to 10% more replicas as compared to the original forwarding. However, while incurring such marginal increase in cost, CAF shows a success delivery rate increase of around 10% to 45% compared to state-of-the-art rank-based forwarding algorithms. Similar results are found using different data sets and using other forwarding algorithms as benchmarks [24, 25].

We also show that CAF is almost as costly as BubbleRap and uses slightly fewer contacts than BubbleRap in PeopleRank and Simbet cases (cost decreases by 2% to 5%). BubbleRap uses 4% less contacts compared to CAF-Degree-Based. However, we would like to emphasize the fact that the cost metric does not include the overhead generated by the K-CLIQUE community detection algorithm used by BubbleRap [12]. This is partly explained by the slightly better performance of BubbleRap compared to the CAF-Degree-Based algorithm.

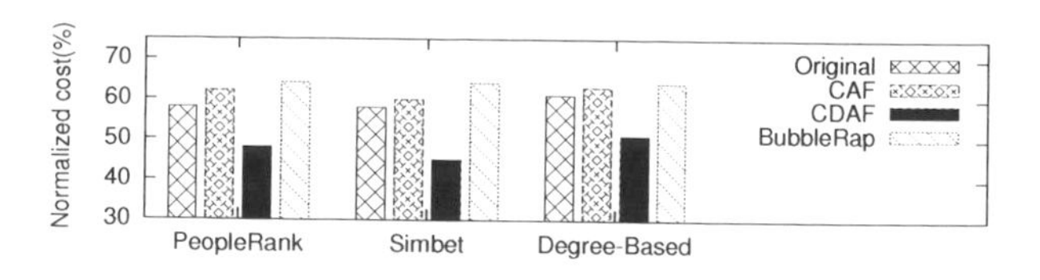

Figure 9.16: Normalized cost of different social forwarding schemes vs. BubbleRap (Hope08 dataset).

We also investigate the cost of our proposed frameworks CAF and CDAF when applied to contact-based forwarding algorithms. Figure 9.17 compares the normalized cost of CAF, CDAF, BubbleRap, and the original algorithms FRESH and Greedy using the two data sets SanFrancisco11, and Hope08. Obviously, contact-based algorithms are more costly than social-based algorithms. This is partly explained by the larger delivery delay incurred by these algorithms, which increases the number of message replicas in the network. We also show that CDAF decreases the cost considerably by 10% to 20%.

9.6 Social-Based Trust in Mobile Opportunistic Networks

Since users may not want to forward messages in opportunistic networks without incentives, we introduce a set of trust-based filters to provide the user with an option of choosing trustworthy nodes in coordination with personal preferences, location priorities, contextual information, or encounter-based keys.

In this section, we focus on trust establishment in opportunistic communication. The establishment of trust is more challenging in infrastructure-less networks such as opportunistic networks where no centralized mechanisms can be easily deployed. Figure 9.18 shows an opportunistic communication scenario between Alice and Bob.

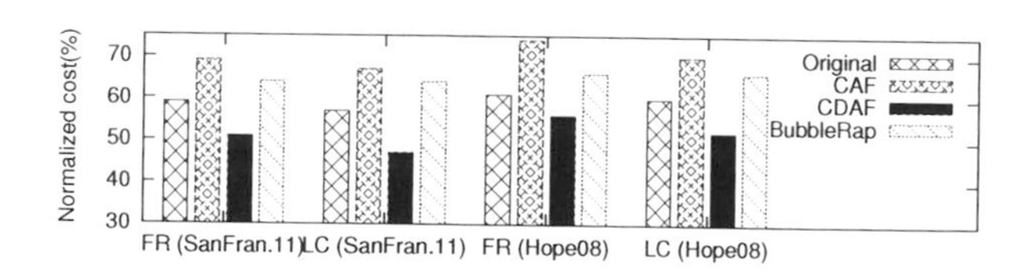

Figure 9.17: Normalized cost of contact-based forwarding schemes (Greedy, and FRESH) vs. BubbleRap.

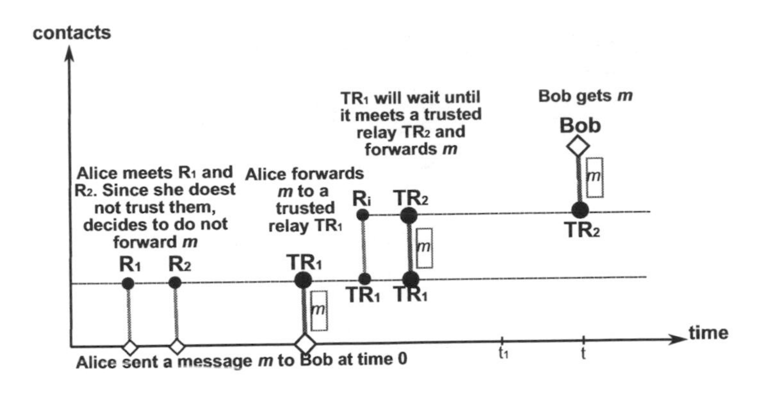

Figure 9.18: Trust establishment in opportunistic communication.

In the absence of trust, Alice will try to maximize the probability of reaching Bob by sending her message m to all other relay nodes R_i in the network. To avoid unwanted communication and establish a trusted environment that increases node cooperation, Alice will only forward her message m to relay nodes that she can trust (trusted relay TR_i). However, this trust-based communication environment may introduce additional delay by filtering out unwanted communication opportunities (e.g., Bob could be reached at time $t_1 \ll t$ through R_2, but since Alice did not trust R_2, she will miss an opportunity to reach Bob with shorter delay).

We leverage social relationships between users to facilitate and enable trustful communication between users. We propose and study a set of social trust filters to identify the subset of contacts between nodes that are allowed in the forwarding path. We utilize explicit social information coupled with real human mobility trace to establish trustworthy communication between a particular node and: (i) a relay (Relay-to-Relay trust), or (ii) the source node (Source-to-Relay trust). We couple these two approaches with three social-based trust filters (common friends, common interests, the distance in the social graph) to introduce and study six trustworthy forwarding techniques to enable trustworthy communication. Our results depict that simple social relationships between users can be utilized to ensure trust in opportunistic networks. While these results show a cost incurred for this trust manifested in additional end-to-end delays, this cost is justified by the fact that in the absence of trust, users may not cooperate and the performance then drops significantly. Our trust filters, therefore, yield a fair trade-off between trust and success rate by achieving more than 35% success rate compared to an untrusted environment where only 10% of the nodes refuse to cooperate in absence of trust.

We investigate the trade-off between trust and efficiency in opportunistic networks. We propose and study a set of social trust filters. These filters use explicit social relationships between nodes to ensure a trust communication in mobile opportunistic networks. We first motivate the use of trust filters in opportunistic networks.

Table 9.2 Social-Based Trust Filters in Opportunistic Networks

Trusted Entity	d-Distance	Common Friends	Common Interests
Relay-to-Relay	R2R: d-distance	R2R: CF	R2R: CI
Source-to-Relay	S2R: d-distance	S2R: CF	S2R: CI

We then evaluate, based on real mobility traces, the performance of a trustful communication in opportunistic networks using different sets of social trust filters.

9.6.1 Social-Based Trust Filters

We present a set of trust filters that uses social relationships between users to establish trustful communication between these users. Two users may trust each other if they share i common interests, have f common friends, are friends, or if they are both friends of the message's sender user, etc.

We introduce, in Table 9.2, two major techniques for trust establishment: (*i*) Relay-to-Relay trust, and (*ii*) Source-to-Relay trust. Relay-to-Relay trust uses a transitive trust approach to establish a trusted path between a source node S and a destination node D relying on trust between every two successive relay nodes on this path. Source-to-Relay trust requires pre-establishment of trust communication between the source and all relay nodes used in the path. We couple these two techniques with three social estimators of trust based on the distance in the social graph, common interests, and common friends.

We utilize two main metrics to evaluate the proposed trust filters: (*i*) the *normalized success rate within time t*: the probability of successfully delivering the message to its destination within time t normalized by the same probability given by epidemic routing [34] (ideal scenario with no trust: optimal success rate within the same time t), and (*ii*) the *normalized cost*: the fraction of contacts (i.e., number of replica copies) normalized by the fraction of contacts used by epidemic forwarding (the most expensive).

9.6.2 Relay-to-Relay Trust

The basic filters to estimate trust communication between two nodes i and j utilize simple social properties between these two nodes. We now propose a set of social estimation of trusted communication between these two nodes.

The distance in the social graph between two nodes can be a good trust estimation metric. The shorter the social distance they are from each other, the more they trust each other; friends ($dist = 1$) trust each others more than friends-of-friends ($dist = 2$), where $dist(i, j)$ measures the shortest distance between two nodes in a graph.

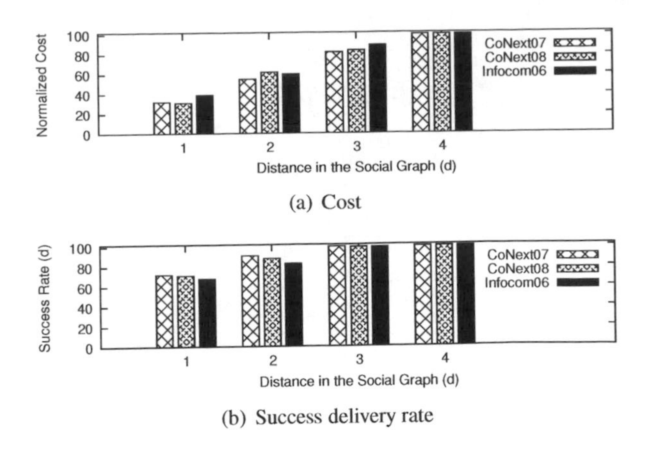

Figure 9.19: Performance evaluation of R2R: d-distance.

We introduce the *d-distance trust filter* such that only the contacts $< i, j >$ are allowed in the forwarding process satisfying $d\text{–}distance(i, j) = dist(i, j) \leq d$, where $d = 1..D$ is an integer no greater than the diameter D of the social graph.

We plot in Figure 9.19 the success delivery rate and the cost of *d–distance* filter approach. Our results show that trusted communication may not be efficient; the more trust a node demands, the lower success delivery rate it achieves. For example, if we only allow direct friends to be involved in the forwarding process (i.e., $d = 1$), we reduce the cost to less than 40% compared to flooding. However the success rate within a 10-minute timescale decreases to roughly 70%. Moreover, if we relax such assumption and allow friends-of-friends to be involved in the forwarding process we increase the success rate to more than 80% while using only 50% of the total contacts used in Epidemic forwarding.

Common interests between two nodes has largely been considered as a good approximation of social similarities [22, 6]. People sharing common interests tend to go to similar locations, events, etc. Note that using k common interests as an approximation of social similarity between two users does not require an explicit confirmation from both users. We consider the common interest-based filter ci such that $< i, j >$ contacts are allowed in the forwarding process $\iff ci(i, j) = \sum_{I} 1_{i,j \in s(I)} \geq k$, where $s(I)$ is the set of users that subscribe to the interest I.

Figure 9.20 evaluates the common interest-based filter performance. All the data sets show that the more common interests the filter requires, the worse the success rates the system achieves; with roughly 50% cost (no less than 3 common interests) the overall success rate is no more than 80%. Moreover, if only users that share no less than 5 interests participate, the performance drops below 30%. Besides the fact that this filter is using implicit social information where there is no explicit so-

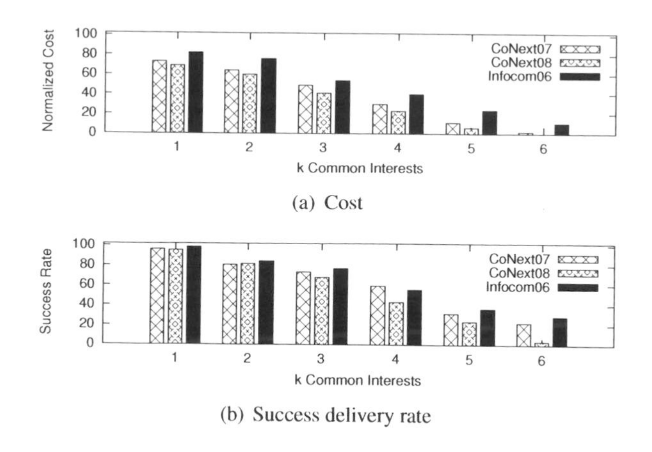

(a) Cost

(b) Success delivery rate

Figure 9.20: Performance evaluation of R2R: common interests.

cial connection between the end users, this filter gives poor results compared to the friendship filter (d–$distance$).

Friendship is usually considered as a good approximation of a strong social bond between people. However, a typical friendship graph may contain strong relationships such as family members or best friends, and also "wayward" friends or family members that someone may not communicate with at all. In social network theory, social relationships are viewed in terms of nodes and ties. Researchers differentiate then between weak and strong ties in social networks. In this section, we focus more on strong social ties since people may not trust friends if they do not communicate often. Common friends are largely used by online social networks such as Facebook and LinkedIn in their content recommendation systems.

We introduce the common friends-based filter which allows only $< i, j >$ contacts in the forwarding process $\iff cf(i, j) = \sum_n 1_{n \in F(i) \cap F(j)} \geq k$, where $F(i)$ is the set of node i's friends.

We show in Figure 9.21 that by only allowing nodes that share no less than 2 friends we reduce the cost by a factor of 3 and we achieve roughly 80% of the optimal success rate. Moreover, if $k = 3$, the normalized success rate is no less than 60% with no more than 20% of the total number of contacts used in Epidemic forwarding.

Inspired by the previous results that show that friendship based filters give better cost/success rate performance compared to implicit social based filters (i.e., common interests), we propose a combination of them. Note that $< i, j >$ contacts selected by the common friends filter implicitly verify the following $dist(i, j) \leq 2$. We propose a combination of 1-distance and common friends-based filters.

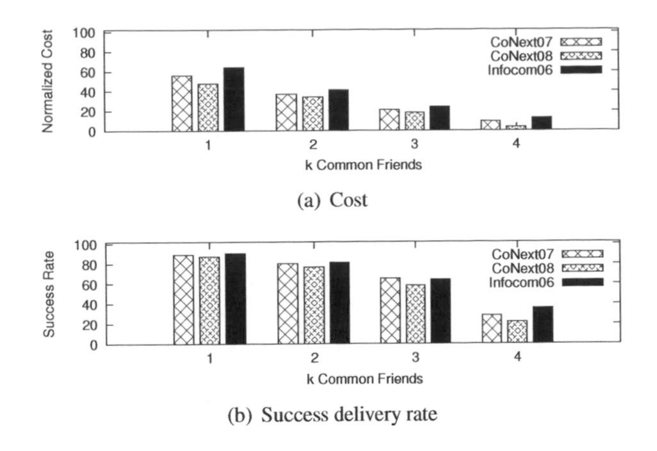

(a) Cost

(b) Success delivery rate

Figure 9.21: Performance evaluation of R2R: common friends.

In order to compare the combination filter to all previous approaches, we aggregate all results from all data sets in only one plot (Figure 9.22-(a)). The x-axis is the normalized cost and on the y-axis we have the corresponding normalized success rate. We observe that the closer the dots are to the top left corner, the better the performance achieved; only a few contacts were efficiently used to forward messages to all destinations. We show that the success rate of Relay-to-Relay filters exponentially increase with cost; a near-optimal performance is achieved within a cost of 70%. This implies that this technique, while filtering out many communication opportunities, is able to deliver a message with a near-optimal success rate performance. We also show that the combination filter outperforms all other filters proposed and achieves the best cost success rate trade-off.

In the figure, we use "x% untrusted environment" to denote the network configuration where x% of the most popular nodes are dissatisfied and decide to not participate in the forwarding process in an epidemic setting. We compare the performance of our filters to the performance of 5% and 10% untrusted environments. We observe that it is possible to ensure trust with almost an optimal success delivery rate within a 10-minute timescale, and outperform the 10% untrusted environment by roughly 35% with the same cost. Moreover, our filters achieve the same success rate given by 10% untrusted environment with 50% fewer contacts.

9.6.3 *Source-to-Relay Trust*

The Relay-to-Relay trust technique implicitly assumes that contacts are independent from each other, and a node j should only trust node i from which it will receive

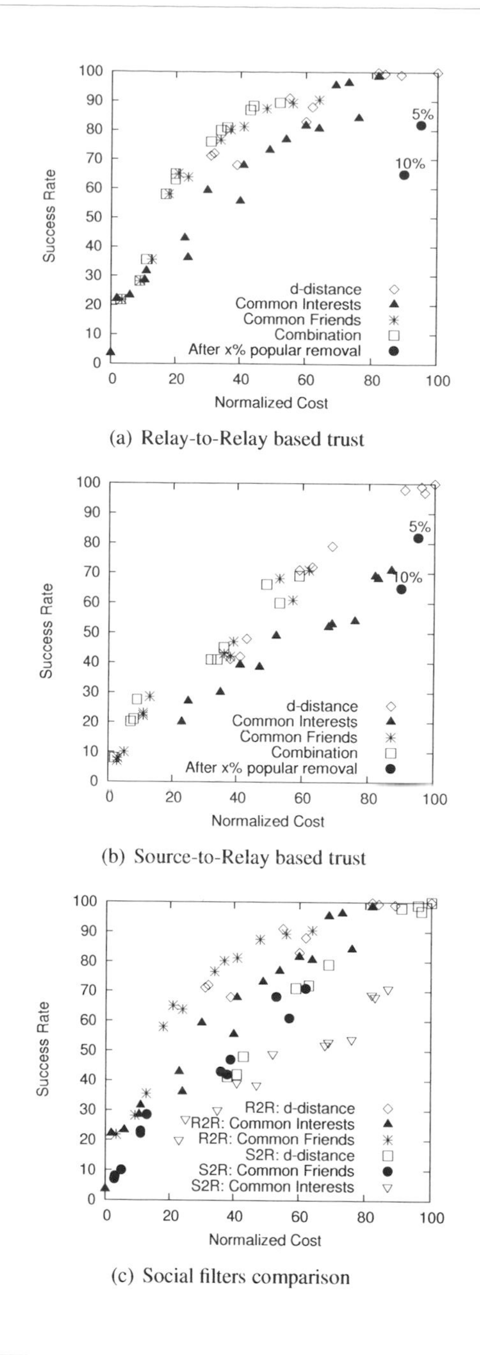

(a) Relay-to-Relay based trust

(b) Source-to-Relay based trust

(c) Social filters comparison

Figure 9.22: Comparison of social filters-based trust techniques.

the message. Opposed to Relay-to-Relay trust, which uses hop-by-hop trust establishment, source-based filters use an "end-to-end" trust approach. It estimates trust between the actual node and the source S. We compare again the three trust filters; distance in the social graph, common friends, and common interests as shown in Table 9.2. While receiving a message, each R_i filters those messages generated by an untrusted source node S.

We compare in Figure 9.22-(b) the distribution of both cost and success rate of the Source-to-Relay filters. Overall, we observe that Source-to-Relay filters yield a poor performance; as opposed to the Relay-to-Relay technique, Source-to-Relay success rate increase linearly with cost. The distance filter achieves no more than a 42% success rate when d=1 (i.e., only friends of the source node S are allowed to forward messages). Moreover, the common friends technique outperforms the two other techniques. However, it achieves only 50% success rate with 50% of the contacts. Finally, we also show that our filters still outperform the 5% and 10% untrusted environment performance by 5% to 12%.

In order to compare our two proposed trust approaches, we aggregate all results from all data sets in only one plot (Figure 9.22-(c)). We show that the common friends-based filter outperforms all the other studied approaches. It achieves one of the best cost/success rate trade-offs with more than 60% of success rate using no more than 30% of the contacts. Moreover, the S2R: d-distance gives the worst performance and achieves only 50% of the optimal success rate using roughly 50% of the contacts. However, since it is hard to quantitatively measure trust, this plot cannot show the best trust estimation approach. Finally, we show that Source-to-Relay trust, which can be considered as a good approximation of trust, leads to a poor cost/success rate performance.

9.7 Conclusion

The emergence of online social network applications together with the arrival of a new generation of powerful mobile devices makes social information between people easily accessible and exploitable. This chapter develops an approach for the use of social interaction information as a means to guide forwarding decisions, establish trustworthy paths, and ensure fair utilization of resources in mobile opportunistic networks. Our results indicate that social information can indeed be leveraged to further improve the performance of existing forwarding schemes that use only contact information or mobility prediction techniques.

In this chapter, we have introduced a social opportunistic forwarding algorithm that uses the PeopleRank metric to rank the relative "importance" of a node using a combination of social graph and contact graph information. We have shown that PeopleRank achieves an end-to-end delay and a success rate close to those obtained by flooding while reducing the number of retransmissions to less than 50% of the that induced by flooding. We have also shown that despite the fact that social information is used as a good predictor for human mobility, social forwarding algorithms in general, and PeopleRank in particular, suffer in large-scale networks where the social graph consists of different communities. We have therefore proposed CAF, a community-aware framework where social information can be considered to guide and improve forwarding decisions within a sub-community. Across multiple sub-communities, CAF helps existing social forwarding algorithms improve their performance by roughly 40%, and achieves a 5% to 30% better success delivery rate than BubbleRap, with negligible incurred cost of no more than 10%.

We have also addressed the rising concern of fairness amongst various nodes in mobile-based opportunistic networks. Rank-based forwarding algorithms, particularly social-based algorithms, are typically designed to reduce the number of data replicas in the network to conserve bandwidth and reduce end-to-end delay. In such algorithms, higher ranked nodes carry a much heavier burden in delivering messages, which can create high levels of dissatisfaction amongst them. We have proposed FOG, a real-time distributed framework that maximizes the overall user satisfaction using existing state-of-the-art rank-based forwarding algorithms. FOG helps rank-based forwarding algorithms utilize potential nodes to forward messages while satisfying fairness properties. It also implements different opportunistic fairness algorithms and selects the suitable one regarding given network properties. In this chapter, we have introduced two real-time distributed algorithms: the Proximity Fairness Algorithm (PFA) and the Message Context Fairness Algorithm (MCFA). Our evaluation shows that FOG uses both fairness algorithms PFA and MCFA to ensure relative equality in the distribution of resource usage among neighbor nodes while maintaining a high delivery success rate, and a close-to-optimal cost performance.

We have finally used social information as a means to study trust establishment in opportunistic networks. Our work highlights the trust/efficiency trade-off in mobile opportunistic networks where social relationships between people can be considered to establish trust in opportunistic networks. We have proposed multiple social-based trust filters, and shown that we can ensure trusted communication in opportunistic networks, but this may introduce additional end-to-end delay. Our trust filters, however, yield a fair trade-off between trust and success rate.

References

[1] Aestetix and C. Petro. CRAWDAD data set hope/amd (v. 2008-08-07). Downloaded from http://crawdad.cs.dartmouth.edu/hope/amd, Aug. 2008.

[2] A. Balasubramanian, B. Levine, and A. Venkataramani. DTN routing as a resource allocation problem. *SIGCOMM Comput. Commun. Rev.*, 37(4):373–384, 2007.

[3] J. Burgess, B. Gallagher, D. Jensen, and B. N. Levine. Maxprop: Routing for vehicle-based disruption-tolerant networking. In *Proceedings of IEEE Infocom*, 2006.

[4] T. Camp, J. Boleng, and V. Davies. A survey of mobility models for ad hoc network research. *Wireless Communications and Mobile Computing: Special issue*

on Mobile Ad Hoc Networking: Research, Trends and Applications, 2:483–502, 2002.

[5] A. Chaintreau, A. Mtibaa, L. Massoulié, and C. Diot. The diameter of opportunistic mobile networks. In *Proceedings of ACM CoNext*, 2007.

[6] E. M. Daly and M. Haahr. Social network analysis for routing in disconnected delay-tolerant manets. In *MobiHoc '07*, pages 32–40, New York, NY, USA, 2007. ACM.

[7] S. Guo and S. Keshav. Fair and efficient scheduling in data ferrying networks. In *Proceedings of the 2007 ACM CoNEXT conference*, CoNEXT '07, pages 13:1–13:12, New York, NY, USA, 2007. ACM.

[8] K. A. Harras, K. C. Almeroth, and E. M. Belding-Royer. Delay tolerant mobile networks (dtmns): Controlled flooding in sparse mobile networks. In *NETWORKING 2005. Networking Technologies, Services, and Protocols; Performance of Computer and Communication Networks; Mobile and Wireless Communications Systems*, pages 1180–1192. Springer, 2005.

[9] T. Henderson, D. Kotz, and I. Abyzov. The changing usage of a mature campus-wide wireless network. In *MobiCom '04: Proceedings of the 10th Annual International Conference on Mobile Computing and Networking*, pages 187–201, 2004.

[10] T. Hossmann, T. Spyropoulos, and F. Legendre. Know thy neighbor: Towards optimal mapping of contacts to social graphs for DTN routing. In *Proceedings of IEEE INFOCOM*. IEEE, 2010.

[11] P. Hui, J. Crowcroft, and E. Yoneki. Bubble rap: Social-based forwarding in delay tolerant networks. In *Proceedings of ACM MobiHoc '08*, New York, NY, USA, 2008. ACM. (also appeared as Cambridge University TR: UCAM-CL-TR-684).

[12] P. Hui, J. Crowcroft, and E. Yoneki. Bubble rap: Social-based forwarding in delay tolerant networks. *IEEE Transactions on Mobile Computing*, 99(PrePrints), 2010.

[13] S. Jain, K. Fall, and R. Patra. Routing in a delay tolerant network. In *Proceedings of ACM SIGCOMM*, pages 145–158, 2004.

[14] D. Karamshuk, C. Boldrini, M. Conti, and A. Passarella. Human mobility models for opportunistic networks. *Communications Magazine, IEEE*, 49(12):157–165, 2011.

[15] D. Kempe, J. Kleinberg, and A. Kumar. Connectivity and inference problems for temporal networks. In *Proceedings of the ACM International Symposium on the Theory of Computing*, 2000.

[16] D. Kotz, T. Henderson, I. Abyzov, and J. Yeo. CRAWDAD data set dartmouth/campus (v. 2009-09-09). Downloaded from http://crawdad.cs.dartmouth.edu/dartmouth/campus, Sept. 2009.

[17] S. S. Kulkarni and C. Rosenberg. Opportunistic scheduling for wireless systems with multiple interfaces and multiple constraints. In *Proceedings of the 6th ACM International Workshop on Modeling Analysis and Simulation of Wireless and Mobile Systems*, MSWIM '03, pages 11–19, New York, NY, USA, 2003. ACM.

[18] G. Lin, G. Noubir, and R. Rajaraman. Mobility models for ad hoc network simulation. In *INFOCOM 2004. Twenty-Third Annual Joint Conference of the IEEE Computer and Communications Societies*, volume 1. IEEE, 2004.

[19] A. Lindgren, A. Doria, and O. Schelén. Probabilistic routing in intermittently connected networks. *SIGMOBILE Mob. Comput. Commun. Rev.*, 7(3):19–20, 2003.

[20] Y. Liu and E. Knightly. Opportunistic fair scheduling over multiple wireless channels. In *INFOCOM 2003. Twenty-Second Annual Joint Conference of the IEEE Computer and Communications. IEEE Societies*, volume 2, pages 1106 – 1115 vol. 2, April 2003.

[21] M. Mauve, A. Widmer, and H. Hartenstein. A survey on position-based routing in mobile ad hoc networks. *Network, IEEE*, 15(6):30 –39, Nov/Dec 2001.

[22] A. Mtibaa, A. Chaintreau, J. LeBrun, E. Oliver, A.-K. Pietilainen, and C. Diot. Are you moved by your social network application? In *WOSN'08: Proceedings of the First Workshop on Online Social Networks*, pages 67–72, New York, NY, USA, 2008. ACM.

[23] A. Mtibaa and K. Harras. Fog: Fairness in mobile opportunistic networking. In *Proceedings of IEEE SECON*, 2011.

[24] A. Mtibaa and K. Harras. Social forwarding in large scale networks: Insights based on real trace analysis. In *Proceedings of IEEE ICCCN*, 2011.

[25] A. Mtibaa and K. A. Harras. CAF: Community aware framework for large scale mobile opportunistic networks. *Computer Communications*, 36(2):180–190, 2013.

[26] A. Mtibaa and K. A. Harras. Fairness-related challenges in mobile opportunistic networking. *Computer Networks*, 57(1):228–242, 2013.

[27] A. Mtibaa, M. May, and M. Ammar. On the relevance of social information to opportunistic forwarding. *International Symposium on Modeling, Analysis, and Simulation of Computer Systems*, 0:141–150, 2010.

[28] A. Mtibaa, M. May, M. Ammar, and C. Diot. Peoplerank: Social opportunistic forwarding. In *Proceedings of IEEE INFOCOM*. IEEE, 2010.

[29] A.-K. Pietiläinen, E. Oliver, J. LeBrun, G. Varghese, and C. Diot. MobiClique: Middleware for mobile social networking. In *WOSN'09: Proceedings of ACM SIGCOMM Workshop on Online Social Networks*, August 2009.

[30] M. Piorkowski, N. Sarafijanovic-Djukic, and M. Grossglauser. CRAWDAD data set epfl/mobility (v. 2009-02-24). Downloaded from http://crawdad.cs.dartmouth.edu/epfl/mobility, Feb. 2009.

[31] M. Piorkowski, N. Sarafijanovoc-Djukic, and M. Grossglauser. A parsimonious model of mobile partitioned networks with clustering. In *The First International Conference on COMmunication Systems and NETworkS (COMSNETS)*, January 2009.

[32] D. Rabbou, A. Mtibaa, and K. Harras. Scout: Social context-aware ubiquitous system. In *2012 5th International Conference on New Technologies, Mobility and Security (NTMS)*, pages 1–6, May 2012.

[33] C. Tantipathananandh, T. Berger-Wolf, and D. Kempe. A framework for community identification in dynamic social networks. In *KDD '07: Proceedings of the 13th ACM SIGKDD International Conference on Knowledge Discovery and Data Mining*, pages 717–726, New York, NY, USA, 2007. ACM.

[34] A. Vahdat and D. Becker. *Epidemic Routing for Partially-Connected Ad Hoc Networks*. Technical Report CS-2000-06, University of California San Diego, Jul 2000.

[35] A. V. Vasilakos, Y. Zhang, and T. Spyropoulos. *Delay Tolerant Networks: Protocols and Applications*. CRC Press, 2012.

[36] J. Wu and Y. Wang. Social feature-based multi-path routing in delay tolerant networks. In *INFOCOM, 2012 Proceedings IEEE*, pages 1368–1376. IEEE, 2012.

Chapter 10

Exploiting Private Profile Matching for Efficient Packet Forwarding in Mobile Social Networks

Kuan Zhang
Department of Electrical & Computer Engineering
University of Waterloo
Waterloo, Canada

Xiaohui Liang
Department of Electrical & Computer Engineering
University of Waterloo
Waterloo, Canada

Rongxing Lu
Department of Electrical & Computer Engineering
University of Waterloo
Waterloo, Canada

Xuemin (Sherman) Shen
Department of Electrical & Computer Engineering
University of Waterloo
Waterloo, Canada

CONTENTS

10.1 Introduction

Mobile Social Network (MSN), as one of the promising social networking platforms, have become progressively popular and brought numerous benefits in various fields due to the portability and pervasiveness of smartphones [1, 2, 3]. With MSNs, users can discover friends with similar interests or preferences in the local proximity, exchange videos or images, and share traffic, shopping, or health information with others. MSNs are usually designed in the hybrid architecture as shown in Figure 10.1, where users often need to send data to their friends. In MSNs, while some users can directly access online social networks via the Internet or cellular networks, other users may not be able to access the Internet at some time in a mobile environment. If the Internet is not available, opportunistic contacts can be an alternative method to help users to stay connected with their friends. Since mobile devices (smartphones and tablets, etc.) are usually taken by humans, the dynamic and unpredictable human mobility results in the intermittent connectivity and dramatically impacts their communications.

Recently, some researchers [4, 5] studied social features and made use of such intermittent connections and human mobility to improve communication performance. One of the applied social features [6] is that users with common interests or preferences probably exist in the same social community, and likely encounter each other

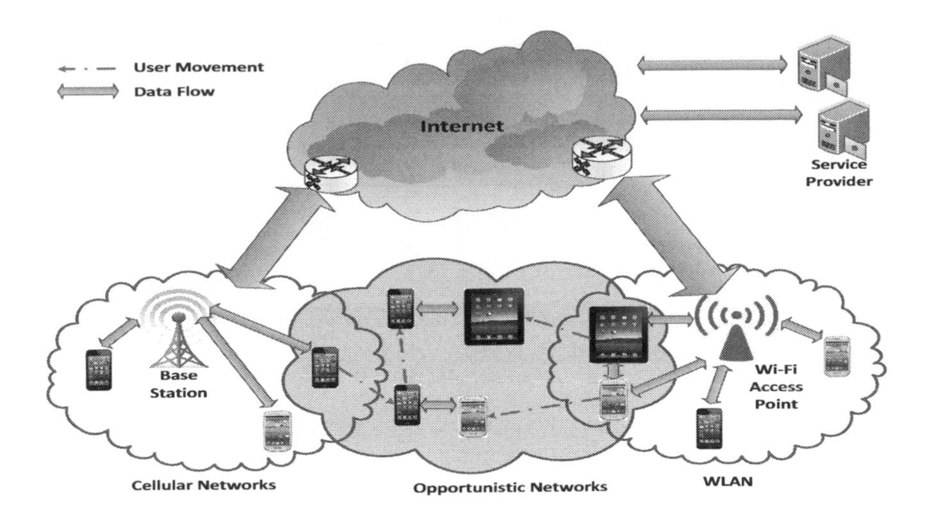

Figure 10.1: Mobile social networks.

in physical locations. For example, at a conference, George and Bob are graduate students who participate in the same session on computer networks. George has encountered Bob in the first day's session, and they have already known of each other's participation. This indicates that George and Bob may belong to a common community. When Alice is looking for George and wants to send a message to him, she needs to first discover her neighbors, for instance, Bob and Charlie in Figure 10.2. Then,

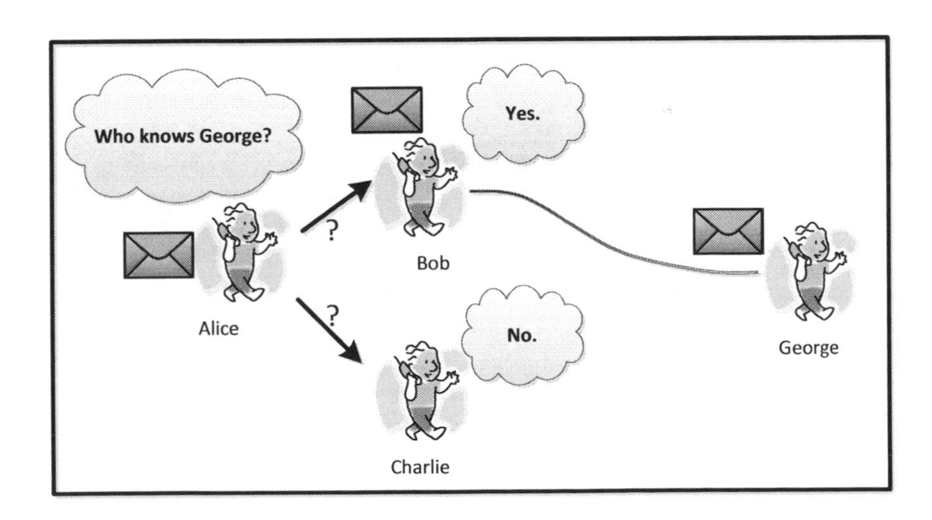

Figure 10.2: Packet forwarding in MSNs.

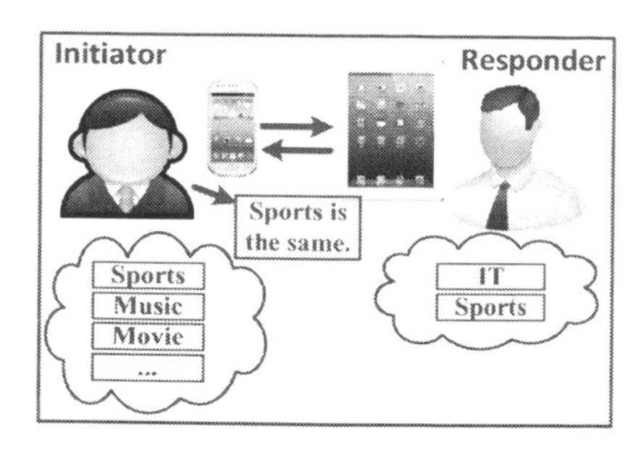

(a) Community profile matching

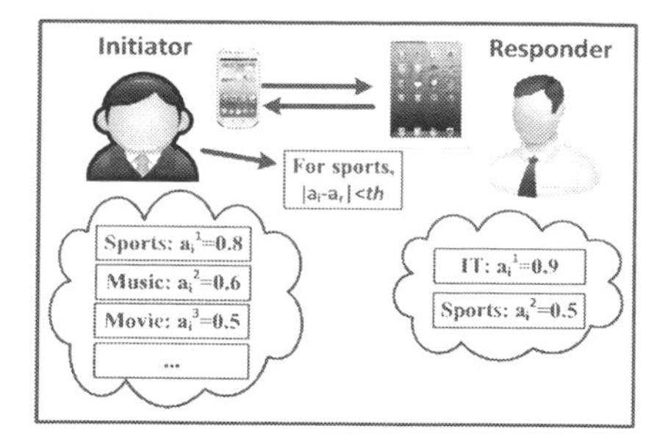

(b) Active degree profile matching

Figure 10.3: Private profile matching overview.

she broadcasts the forwarding request to her neighbors. Because Bob and George possibly belong to the same community, i.e., the computer network session, they likely encounter each other in the future. On the other hand, Charlie does not have such a relationship to George. Therefore, Alice prefers to forward the packet to Bob, who has a larger possibility to successfully forward the packet to George. From the above example, we intend to study the features of social community and explore an efficient packet forwarding protocol in MSNs.

On the other hand, the existing packet forwarding protocols [4, 7] disclose user's social information to enable users to learn more about the relay candidates and select the optimal one among them. Such approaches raise serious privacy concerns and might disclose user's private information. In fact, from the perspective of users, they may not be willing to disclose their preference and history activities to persons with who they are not familiar. In other words, they would like to merely share profiles with others having common interests or events. In the above example, when Bob and George encounter each other, they would secretly compare and match their profiles to determine whether the other belongs to some common communities or not, as shown in Figure 10.3(a). Only if both Bob and George belong to the same communities will they share such information with each other; otherwise, neither Bob nor George discloses any community information to the other. In this way, the user's privacy can be preserved.

In this chapter, we will apply privacy-preserving techniques with packet forwarding to enhance communication performance and protect user's sensitive information from disclosure. Specifically, we propose POMP, a Private prOfile Matching aided efficient Packet forwarding protocol for MSNs. The proposed POMP protocol is characterized by optimal relay selection and private profile matching. With the private matching results, POMP utilizes the community feature to forward packets and preserve user privacy during both profile matching and packet forwarding phases. Our major contributions are three-fold.

• First, we propose the POMP protocol, which exploits the common community feature and selects the relays having the most common communities with the destination to help forward packets. POMP also includes two private profile matching protocols to enable community matching and active degree matching without disclosing the uncommon information between the participating users. According to the private profile matching results, the forwarding capability, reflected by the common communities between the relay and the destination, can be calculated for each neighbor of the source. For users in the same community, we also utilize active degree profile matching in order to stimulate users to forward packets to users with similar active degrees.

• Second, we present an enhanced POMP protocol (ePOMP) where the optimal relay is selected with the consideration of the multi-hop neighbors. Furthermore, with a recommendation mechanism, users recommend the encountered users with common communities to others and help the source obtain more knowledge of relays and improve the forwarding efficiency.

• Finally, we evaluate the performance of POMP and ePOMP through extensive trace-based simulations. Our simulation results validate the efficiency of the proposed packet forwarding protocols in terms of delivery ratio, average delay and communication overheads. In addition, the privacy analysis demonstrates that POMP and ePOMP preserve user privacy by minimizing the disclosure of users' sensitive information.

The remainder of this chapter is organized as follows: In Section 10.2, we investigate the related work on packet forwarding protocols and private profile matching in MSNs. Network model and design goals are presented in Section 10.3. The details and homomorphic properties of the Paillier cryptography system are introduced in Section 10.4. In Section 10.5, we present the details of our proposed POMP protocol, followed by the privacy analysis and the performance evaluation in Sections 10.6 and 10.7, respectively. Finally, Section 10.8 concludes the chapter.

10.2 Related Work

10.2.1 Packet Forwarding

Packet forwarding is of great importance in MSNs, and draws considerable attention from the research field. The original epidemic routing [8] was proposed to maximize the packet delivery ratio, while minimizing the average delay. However, it consumes considerable communication overhead. Since then, there have been many attempts are trying to maintain forwarding performance with reasonable resource consumption. Spyropoulos et al. [9] address the high communication overhead of flooding-based packet forwarding protocols, and propose-Spray-and-Wait routing scheme. Spray-and-Wait first disseminates a number of copies into the network, and then keeps the copy-holders waiting until the packet is successfully forwarded to the destination. In addition, Yuan et al. [10] investigate routing based on the prediction of future contacts and make use of users' mobility to facilitate delay-tolerant network (DTN) routing. Two observations are made: 1) users usually move around several frequently visited locations instead of randomly walking; 2) some users' behaviors are semi-deterministic and could be predicted according to their past mobility. In addition, Lindgren et al. [11] propose a delivery predictability metric which indicates the probability of a user meeting another certain user. By using this metric, an optimal relay is selected to forward the packet.

Despite the above fundamental forwarding protocols, recent attempts study and exploit social characteristics, such as centrality [7], community [12], betweenness [13], and social proximity [14] in order to fully make use of the intermittent connection among users and efficiently forward packets. Hui et al. [7] study centrality and community features, and propose a human impact-based routing protocol to purposely forward packets to users with some social relationship with the destination. Wu et al. [4] investigate a social feature-based multi-path routing in MSNs, and utilize entropy to analyze user's social features for efficient routing. In [5], Bulut et al. introduce friendship to measure the degree of users' direct and indirect connections in order to improve forwarding efficiency. Gao et al. [15] study transient contact distribution, transient connectivity, and transient community structure, and develop an efficient data forwarding protocol with these novel metrics. Unfortunately, most of these works have not paid adequate attention to security and privacy problems. They require a large amount of social or contact information exchange so that an increasing portion of private and sensitive information is disclosed. The trade-off between privacy and forwarding accuracy is also studied in [16]. In addition, Cabaniss et al. [17] investigate the characteristics of social groups and propose a DSG-N^2 routing algorithm based on a probability scheme. However, the DSG-N^2 only relies on mobile users to forward packets without any stationary store devices so that the performance requirements cannot be satisfied in specific scenarios. Link et al. [18] propose a geographic routing scheme (GeoDTN) based on the mobility of participant nodes in disruption-tolerant networks. Similar to DSG-N^2, GeoDTN relies on social group, in which all the nodes are not static, and cannot significantly improve the delivery ratio.

At the same time, Aviv et al. [19] investigate human mobility patterns and propose a Return-to-Home protocol, which keeps most packets in the social spots to improve the forwarding efficiency. They, unfortunately, cannot achieve reliability, location privacy, and anonymity simultaneously. According to this principle, Lu et al. [20] propose SPRING, a privacy-preserving forwarding protocol, and enable users communicate with each other more reliably by deploying store-and-forward roadside facilities at intersections with high social ties. In [21], Lin et al. propose a social tie assisted packet forwarding protocol (STAP) aiming to preserve the receiver's location privacy. STAP utilizes a pseudonym-changing strategy in order to enhance the location privacy for the receivers. Furthermore, the assistance of RSUs improves the transmission efficiency. Lu et al. [22] utilize the idea of "Sacrificing the Plum Tree for the Peach Tree," which is one of the Thirty-Six Strategies of Ancient China, to protect the receiver's location privacy, and propose an SPF protocol to enable users to disclose insensitive locations but preserve the sensitive location. In both SPRING and STAP, a large number of RSUs have to store all packets being transmitted in the network, which also produces more communication overhead for the relay vehicles passing by. All of the spot-based forwarding protocols require the pre-set fixed devices to store-and-forward packets to mobile users, and the number of such social spots exponentially increases with the size of the network region. As such, they are not suitable for the short-term or large area scenarios due to the scalability issues.

10.2.2 Private Profile Matching

Private profile matching protocols for MSNs [23] have been extensively investigated. For example, Yang et al. [24] develop a fully distributed MSN platform, called E-SmallTalker, that exchanges common attributes and matches user's profile without a pre-established Bluetooth connection. For ehealthcare applications, Lu et al. [25] propose a private symptom matching protocol which preserves user's sensitive information when discovering users with similar symptoms. Li et al. [26] introduce privacy levels during profile matching in MSNs. The privacy levels can be customized by users with different amounts of personal information disclosed in different privacy levels. Zhang et al. [1] design a family of fine-grained private matching protocols in order to enable two users to match without disclosing private information. The family of protocols achieve diverse privacy levels based on a wide range of matching metrics. In addition, Persona [27], an online social network with user-defined privacy, blinds user data by using attribute-based encryption and applies fine-grained profile matching. Recently, Liang et al. [28] investigate k-anonymity features and introduce anonymity risk level in profile matching. Finally, they propose fully anonymous profile matching protocols for MSNs to enable users to anonymously execute profile matching without disclosing any private information, even though the matching protocols are run in multiple rounds by several users.

We focus on privacy preservation not only in profile matching but also during packet forwarding, and improve the communication efficiency at the same time.

10.3 Problem Definition

In this section, we first formulate the network model and threat model. Then we identify our design goals including the efficiency and privacy goals.

10.3.1 Network Model

We consider a homogenous MSN composed of a trust authority (TA) and N mobile users.

1. *Trust Authority (TA)* is a trustable, powerful, and storage-rich entity. In the initialization phase, TA bootstraps the entire network, and is not involved in profile matching and packet forwarding. Once bootstrapping, TA is able to produce secure seed keys, which are used to generate the session keys, for all legitimate users with the consideration of security and privacy. Regarding the profiles of users, TA verifies and issues each user a certificate for each

community. Here, we define a community in which a certain number of users participate.

2. *Mobile users* can be denoted by $\mathbb{U} = \{u_1, u_2, ..., u_n\}$, where N is the total number of the users in the network. All mobile users are equipped with portable communication devices that have equal communication range and can communicate with each other bi-directionally. Considering the real-world scenario, the power and storage of the communication devices are constrained toward mobile users. When registering to TA, each user is able to legitimately obtain a unique identity and key materials which should be securely kept. During profile matching and packet forwarding phases, mobile users are able to generate their own signatures and simultaneously verify the identity of others.

3. *Communities* indicate that every pair of users in a specific community have a high social tie with each other. In other words, users with similar interests probably encounter each other, i.e., in sessions during a conference, campus activities, or sports club events, etc. Let Λ be the total number of communities in the whole network. Each user u_i, who joins community λ, is assigned $a_i^\lambda = 1$, where $a_i^\lambda \in \{0,1\}$, and a unique integer $A_i^\lambda \in [1, \rho]$ as the active degree value indicating the activeness in this community. If $A_i^\lambda = \rho$, user u_i is the most active one in community λ. For example, user u_i who joins community 1, 3, 4, $\cdots$, i, securely keeps its own community sets $\tilde{a}_i = \{a_i^1, a_i^3, a_i^4, \cdots, a_i^\Lambda\}$ and $\tilde{A}_i = \{A_i^1, A_i^3, A_i^4, \cdots, A_i^\Lambda\}$.

4. *Profile Matching*: The user starting the profile matching is defined as the initiator, while the other user who receives the profile matching request and replies with the matching results is defined as the responder. In Figure 10.3(a), the initiator u_i and the responder u_r acquire the common communities that both users belong to, which is called community profile matching. In Figure 10.3(b), u_i and u_r want to know whether the difference in their active degree values is within a threshold γ or not. This type of profile matching is called active degree profile matching. We define the risk of privacy leakage $PL = \mathcal{PL}(x, y)$, where x and y denotes the amount of the common and uncommon information between two users u_i and u_j, respectively. The larger gap of the community and active degree results in the higher risk level of privacy leakage.

10.3.2 Threat Model

Malicious users exist in the network, and get involved in profile matching and/or packet forwarding. Generally, most users are honest but curious about other users' personal profiles, especially the communities and the corresponding active degree value that others have. Note that such users can honestly follow the protocols. When profile matching, a user's information may be disclosed to others.

10.3.3 Design Goals

Our design goal is to develop an efficient and privacy preserving packet forwarding protocol in MSNs. We identify the efficiency and privacy goals, respectively.

1. *Efficiency Goal*: In MSNs, end-to-end connection cannot be guaranteed all the time. Furthermore, strangers may not get adequate knowledge about others in MSN scenarios so that the complete information about relays or destinations is difficult to obtain for packet forwarding. In this chapter, we aim to design an efficient packet forwarding protocol in which communities and active degrees are exploited to facilitate forwarding. Well-matched users should be able to exchange with their own friends with the same communities according to community profile matching results so that others may know their potential friends even though they have not met yet.

2. *Privacy Goal*: The main privacy goal is to preserve user's private communities and active degrees from being disclosed to others outside these communities. In both profile matching and packet forwarding, a specific community of an individual user can be merely disclosed to others who also belong to the same community. In other words, only the users with the same communities can know that they belong to these communities. If a user does not belong to a specific community, he/she cannot learn anything about who belongs to this specific community at all. We intend to explore an effective scheme to privately match user's community information without disclosing any other uncommon information. Furthermore, to match users' active degree values, we will design the profile matching protocol in a fine-grained fashion.

10.4 Preliminaries

In this section, we introduce a classical homomorphic encryption scheme, Paillier cryptography [29, 30], which is the basis of our proposed protocol. Paillier cryptography system consists of three steps:

1. **Key Generation:** An entity chooses two large primes p and q, and computes $N = pq$. The base $\mathcal{G}$ is randomly selected, where $\mathcal{G} \in \mathbb{Z}^*_{N^2}$ and $\gcd(\mathsf{L}(\mathcal{G}^{\kappa} \bmod N^2), N) = 1$. Here, $\mathsf{L}(x)$ is defined as $(x-1)/N$, while κ is the least common multiple between $p-1$ and $q-1$, $\kappa = lcm(p-1, q-1)$. The public key is $\langle N, \mathcal{G} \rangle$ and the private key is κ.

2. **Encryption:** Let $x \in \mathbb{Z}_N$ be the plaintext and $r \in \mathbb{Z}_N$ be a random number. The ciphertext can be calculated by

$$\mathsf{Enc}(x \bmod N, r \bmod N) = \mathcal{G}^x r^N \bmod N^2, \tag{10.1}$$

where $\mathsf{Enc}(\cdot)$ is the Paillier encryption operations on two integers modulo N.

3. **Decryption:** Given a ciphertext $c \in \mathbb{Z}_{N^2}^*$, the corresponding plaintext can be derived by

$$\mathsf{Dec}(c) = \frac{\mathsf{L}(c^{\kappa} \bmod N^2)}{\mathsf{L}(\mathcal{G}^{\kappa} \bmod N^2)} \bmod N, \tag{10.2}$$

where $\mathsf{Dec}(\cdot)$ denotes the decryption operation.

The Paillier cryptography system has two significant properties which are especially suitable for multi-party private computing, since it enables addition and multiplication over ciphertexts.

1. **Homomorphic:** For any $x_1, x_2, r_1, r_2 \in \mathbb{Z}_N^*$, we have

$$\mathsf{Enc}(x_1, r_1)\mathsf{Enc}(x_2, r_2) \equiv \mathsf{Enc}(x_1 + x_2, r_1 r_2) \bmod N^2 \tag{10.3}$$

$$\mathsf{Enc}^{x_2}(x_1, r_1) \equiv \mathsf{Enc}(x_1 x_2, r_1^{x_2}) \bmod N^2. \tag{10.4}$$

2. **Self-blinding:**

$$\mathsf{Enc}(x_1, r_1) r_2^{x_2} \bmod N^2 \equiv \mathsf{Enc}(x_1, r_1 r_2). \tag{10.5}$$

Therefore, we take advantage of the Paillier cryptography system in order to support diverse operations like addition and multiplication directly over ciphertexts and preserve user's private information when profile matching.

10.5 Proposed POMP Protocols

In this section, we present the details of the POMP protocol. First, we briefly describe POMP. Then, we theoretically analyze the forwarding procedures and propose POMP with detailed steps. Second, we propose community and active degree profile matching protocols (CPM and APM), which enable diverse profile matching with privacy preservation. Finally, we extend POMP to an enhanced POMP protocol with multi-hop relay selection and a recommendation mechanism to further improve the forwarding efficiency.

10.5.1 Overview of POMP

As shown in Figure 10.4, when a source u_s wants to send a packet to a destination u_d, u_s detects whether there is any user in its proximity. Once finding a neighboring user u_i, u_s first matches the community information with u_i. If the amount of the common communities between u_s and u_i is larger than a pre-set threshold TH, this neighbor user u_i is selected as a relay candidate. Secondly, u_s and u_i match their active degree in the common communities. If u_i's active degree is similar u_s's (i.e., the difference of the two active degree values is smaller than a threshold), the packet will be forwarded to u_i. Since some previous works [7] only consider forwarding within one community, we explore common communities to evaluate the relay capability and provide more opportunities to successfully forward packets.

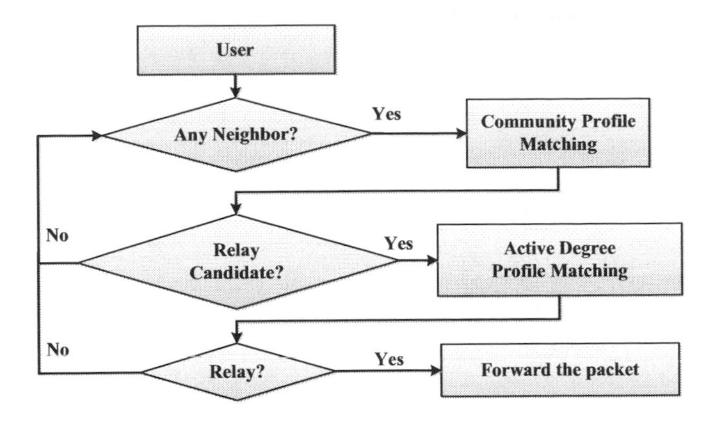

Figure 10.4: Overview of POMP.

10.5.2 POMP Protocol

First, we theoretically analyze the contact probability, since the forwarding in MSNs is based on users' contacts. The contact between every two mobile users u_i and u_j forms a Poisson process [31, 32] with the pairwise contact rate $\lambda_{i,j}$. The packet forwarding procedures can be formalized as follows.

Denote $S_{i,j}$ as a binary random variable following

$$S_{i,j} = \begin{cases} 1, & \text{if } u_i \text{ and } u_j \text{ contact within period } T; \\ 0, & \text{otherwise.} \end{cases}$$

Hence, $S_{i,j}$ follows a Bernoulli distribution as shown in Equation 10.6.

$$\overline{S_{i,j}} = 0 \cdot \int_T^{\infty} \lambda_i e^{-\lambda_i t} dt + 1 \cdot \int_0^T \lambda_i e^{-\lambda_i t} dt \tag{10.6}$$

Since the contacts between every pair of users are independent, the probability that u_i meets at least one user within the period T is shown in Equation 10.7.

$$\begin{aligned} P_i(t \leqslant T) &= 1 - \prod_{\substack{u_j \in \mathrm{U} \\ j \neq i}} (1 - \overline{S_{i,j}}) \\ &= 1 - e^{-\sum_{\substack{u_j \in \mathrm{U} \\ j \neq i}} \lambda_{i,j} T} = 1 - e^{-\lambda_i T} \end{aligned} \tag{10.7}$$

Therefore, t follows the power-law distribution with the probability distribution function (pdf) $f_i(t) = \lambda_i e^{-\lambda_i t}$, for any $t \geqslant 0$. The average contact interval is

$$\overline{E_i(t)} = \int_0^{\infty} t f_i(t) dt = \int_0^{\infty} t \lambda_i e^{-\lambda_i t} dt = \frac{1}{\lambda_i}. \tag{10.8}$$

If both the source u_s and the destination u_d are in the same community C_p, u_s can either directly forward the packet to u_d; or select another user u_r also in C_p to store, carry, and forward this packet to u_d in such an indirect fashion. Let t_1 be the time for u_s to discover u_r, and t_2 be the time for u_r to find u_d. Therefore, the probability $P^r_{s,d}(t = t_1 + t_2 \leqslant T)$ that u_s successfully forwards the packet to u_d via u_r is

$$\begin{aligned} P^r_{s,d}(t \leqslant T) &= \int_0^{t_1} \lambda_{s,r} e^{-\lambda_{s,r} t} dt \cdot \int_{t_1}^{T} \lambda_{r,d} e^{-\lambda_{r,d} t} dt \\ &= \int_0^T f_{s,r}(t) \otimes f_{r,d}(t) dt \\ &= \int_{t=0}^{T} \left(\int_{\tau=0}^{t} f_{s,r}(\tau) \cdot f_{r,d}(t-\tau) d\tau \right) dt, \end{aligned} \tag{10.9}$$

where $\otimes$ denotes the convolution. Since u_r knows the exact time $t_{s,r}$, we have Equation 10.10 and derive Equation 10.9 as shown in Equation 10.11.

$$P^r_{s,d}(t = t_1 + t_2 \leqslant T) \geqslant P(t_1 \leqslant t_{s,r}) \cdot P(t_2 \leqslant t_{s,r}) \tag{10.10}$$

$$\begin{aligned} P^r_{s,d}(t \leqslant T) &= \int_{t=0}^{T} \left(\int_{\tau=0}^{t} f_{s,r}(\tau) \cdot f_{r,d}(t-\tau) d\tau \right) dt \\ &\geqslant \int_{\tau_1=0}^{t_{s,r}} f_{s,r}(\tau_1) d\tau_1 \cdot \int_{\tau_2=0}^{T-t_{s,r}} f_{r,d}(\tau_2) d\tau_2 \\ &= \left(1 - e^{-\lambda_{s,r} t_{s,r}}\right) \cdot \left(1 - e^{-\lambda_{r,d}(T - t_{s,r})}\right) \end{aligned} \tag{10.11}$$

Considering both direct and indirect contacts between u_s and u_d, the probability that a packet forwards from u_s to u_d can be calculated as

$$\begin{aligned} p_{s,d}(t \leqslant T) &= 1 - (1 - P_{s,d}(t \leqslant T)) \cdot \prod_{\substack{u_r \in \mathbb{U} \\ r \neq s,d}} \left(1 - P^r_{s,d}(t \leqslant T)\right) \\ &\geqslant 1 - e^{-\lambda_{s,d} T} \cdot \prod_{\substack{u_r \in \mathbb{U} \\ r \neq s,d}} \left(1 - p^r_{s,d}\right), \end{aligned} \tag{10.12}$$

where $p^r_{s,d} = \left(1 - e^{-\lambda_{s,r} t_{s,r}}\right) \cdot \left(1 - e^{-\lambda_{r,d}(T - t_{s,r})}\right)$. Since $0 \leqslant 1 - p^r_{s,d} \leqslant 1$ where $u_r \in \mathbb{U}$ and $r \neq s, d$, $p_{s,d}$ will be even smaller after multiplying more items like $1 - p^r_{s,d}$. In practice, it means that more relays selected provide more opportunities for successful forwarding.

If more relays are selected to forward the packets, the probability of establishing such a multi-hop connection on the selected path within period T can be calculated as

$$P^{r \cdots r'}_{s,d}(t \leqslant T) = \int_0^T f_{s,r}(t) \otimes \cdots \otimes f_{r',d}(t) dt. \tag{10.13}$$

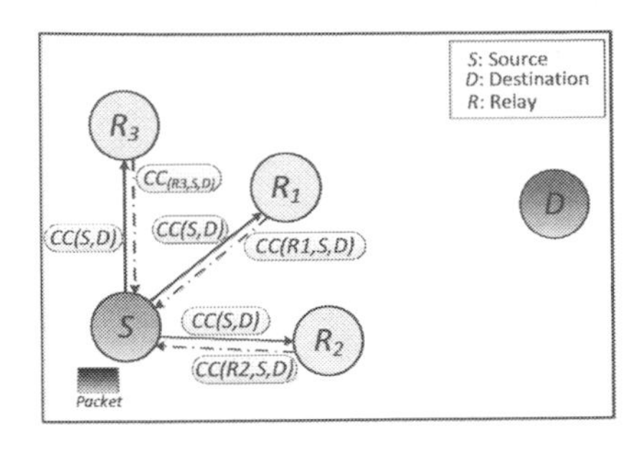

(a) One-hop profile matching

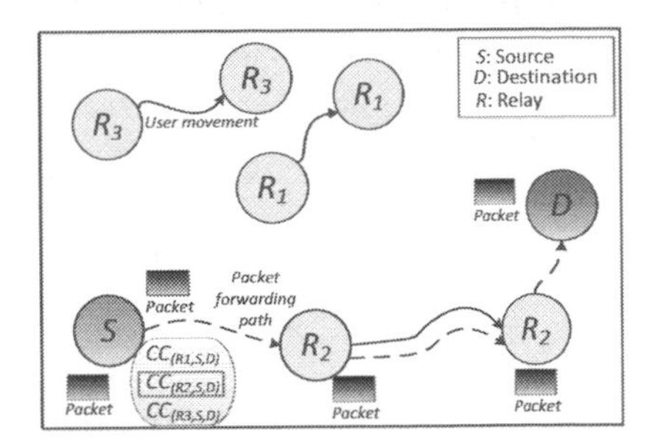

(b) Packet forwarding

Figure 10.5: Relay selection of POMP.

With the consideration of multi-communities, the probability that u_s forwards the packet to u_d is calculated by Equation 10.14. This probability is greater than the probability that u_s forwards the packet only in one community.

$$\begin{aligned}\mathcal{P}_{s,d}(t \leqslant T) &= 1 - \prod_{i \in \mathbb{CC}_{s,d}} (1 - p_{s,d}(t \leqslant T, i)) \\ &\geqslant \max_{i \in \mathbb{CC}_{s,d}} \{p_{s,d}(t \leqslant T, i)\}\end{aligned} \tag{10.14}$$

The more common communities that u_s and u_d have, the higher probability that they successfully forward packets. Therefore, the POMP protocol exploits such features to improve the forwarding efficiency. The detailed POMP protocol steps are illustrated as follows.

Step 1: When the source u_s wants to forward a packet to the destination u_d, the source maintains a common community set $\mathcal{CC}_{(s,d)}$ and first constructs a full community set $\mathcal{C}_s$ with the length Λ. Then, u_s encrypts this community set by u_s's public key $\mathcal{K}_{pub,u_s}$, and sends the 4-tuple $(ID_s||\mathsf{Cert}_{ID_s}||\mathcal{C}_s||\mathsf{Sign}_{\mathcal{K}_{pri,s}}(\mathcal{C}_s))$ to its one-hop relay u_r for profile matching.

Step 2: Running the community profile matching, u_s obtains the one-hop profile matching results at first. In Figure 10.5(a), the results are denoted by $\mathcal{CC}_{s,r} = \{CC_1, CC_2, \cdots, CC_r\}$, where $u_1, u_2, \cdots, u_r$ are one-hop neighboring users of u_s. If $max(\mathcal{CC}_{s,r}) \geqslant TH$, u_r, having the most common communities, is selected as the relay candidate. Then, u_s and u_r match their active degree in common communities. If the total difference $D_{s,r} \leqslant |\mathcal{CC}_{s,r}| \times th$, u_r is selected as a relay.

Step 3: After finding a relay u_r, u_s encrypts the packet to generate the ciphertext $\mathcal{C} = \mathsf{Enc}_{\mathcal{K}_{pub,d}}(\mathcal{P}) = \mathcal{G}^{\mathcal{P}} \cdot r^N \bmod N^2$. Then, u_s signs on this packet by using its private key. Here, u_s calculates $s_1 = \frac{\mathsf{L}(h(\mathcal{P})^{\kappa} \bmod N^2)}{\mathsf{L}(\mathcal{G}^{\kappa} \bmod N^2)} \bmod N$ and $s_2 = (h(\mathcal{P})(\mathcal{G}^{-1})^{s_1})s_2^N \bmod N^2$. Then, u_s sends $(ID_{u_s}||\mathcal{C}||ID_{u_d}||s_1||s_2)$ to u_r.

Algorithm 18: Forwarding of POMP

1 **Procedure**: Forwarding of the POMP
2 u_s wants to forward a packet to u_d
3 **Phase 1**: Relay Selection
4 **if** u_s discovers a one-hop neighbor u_x **then**
5 u_s encrypts $\mathcal{C}_s$ with $\mathcal{K}_{pub,u_s}$, and sends $(ID_s||\mathsf{Cert}_{ID_s}||\mathcal{C}_s||\mathsf{Sign}_{\mathcal{K}_{pri,s}}(\mathcal{C}_s))$ to u_x
6 u_s and u_x run profile matching, and u_s obtains the common community set $\mathcal{C}_{s,x} = \{a_1, a_2, \cdots, a_x\}$
7 $\mathcal{C}_{s,x}$ is added to $\mathcal{CC}_{s,r}$
8 **if** $max(\mathcal{CC}_{s,r}) \geqslant TH$ **then**
9 u_s selects u_r, whose $\mathcal{C}_{s,r} = max(\mathcal{CC}_{s,r})$
10 u_s and u_r matches their active degree values
11 **if** $D_{s,r} \leqslant max(\mathcal{CC}_{s,r}) \times th$
12 u_s selects u_r as the one-hop relay to forward the packet to u_d
13 **end if**
14 **end if**
15 **end if**
16 **Phase 2**: Forwarding
17 u_s encrypts the packet P and obtains the ciphertext $\mathcal{C} = \mathsf{Enc}_{\mathcal{K}_{pub,d}}(\mathcal{P}) = \mathcal{G}^{\mathcal{P}} \cdot r^N \bmod N^2$
18 u_s calculates $s_1 = \frac{\mathsf{L}(h(\mathcal{P})^{\kappa} \bmod N^2)}{\mathsf{L}(\mathcal{G}^{\kappa} \bmod N^2)} \bmod N$ and $s_2 = (h(\mathcal{P})(\mathcal{G}^{-1})^{s_1})s_2^N \bmod N^2$
19 u_s sends $(ID_{u_s}||\mathcal{C}||ID_{u_d}||s_1||s_2)$ to u_r.
20 **if** $h(\mathcal{P}) = \mathcal{G}^{s_1} \cdot s_2^N \bmod N^2$ **then**
21 u_r stores this packet
22 **end if**
23 **if** u_r meets the destination u_d **then**
24 **if** u_r and u_d mutually authenticate each other **then**
25 u_r forwards the packet to u_d
26 With $\mathcal{K}_{pri,d}$, u_d decrypts the plaintext as $\mathcal{P} = \frac{\mathsf{L}(\mathcal{C}^{\kappa} \bmod N^2)}{\mathsf{L}(\mathcal{G}^{\kappa} \bmod N^2)} \bmod N$.
27 **end if**
28 **end if**
29 **End Procedure**

Step 4: Receiving the packet from u_s, u_r first verifies $h(\mathcal{P}) \stackrel{?}{=} \mathcal{G}^{s_1} \cdot s_2^N \bmod N^2$. If valid, u_r stores this packet. In addition, u_r can also follow steps 1–3 to find another relay to help forward this packet.

Step 5: When the relay u_r meets the destination u_d, they first finish mutual authentication to verify the identities of both users. Then, u_r forwards the packet to u_d. With its own private key $\mathcal{K}_{pri,d}$, u_d is able to decrypt the plaintext $\mathcal{P} = \frac{\mathsf{L}(\mathcal{C}^{\kappa} \bmod N^2)}{\mathsf{L}(\mathcal{G}^{\kappa} \bmod N^2)} \bmod N$. The detailed procedures can be found in Algorithm 18.

10.5.3 Profile Matching

According to the two types of data (community and active degree), we design two profile matching protocols: CPM and APM, respectively.

1. *Community Profile Matching (CPM)*:

   Consider that user u_i discovers m neighbors u_{i_1}, u_{i_2}, ..., u_{i_m} in its proximity.

ID_i is denoted as the unique identity of user u_i, while a_i^λ denotes whether user u_i belongs to community λ or not, where $a_i^\lambda \in \{0,1\}$ is the community indicator. If u_i belongs to community λ, $a_i^\lambda = 1$. During the initialization phase, all the users within community λ are assigned 3-tuple $(h_{Root}^\lambda, h_{Left}^\lambda, h_{Right}^\lambda)$ by TA. We have $h_{Root}^\lambda = h_{Left}^\lambda + h_{Right}^\lambda$, where h_{Root}^λ, h_{Left}^λ and h_{Right}^λ are random numbers selected by TA. Users outside the community cannot possess the valid authenticated 3-tuple. Each user securely keeps its own community set $\mathcal{A}_i = \{a_i^1, a_i^2, \cdots, a_i^\lambda, \cdots, a_i^\Lambda\}$. The goal of community profile matching is to obtain the common community set $\mathcal{CC}_{i,j} = \mathcal{A}_i \cap \mathcal{A}_j$ between two users.

Step 1: When two users u_i and u_j want to match their profiles, the initiator u_i first picks h_{Left}^λ in each community λ. A new set is computed as $\mathcal{AH}_{Left,i} = \{h_{Left,i}^1, h_{Left,i}^2, \cdots, h_{Left,i}^\Lambda\}$, where

$$h_{Left,i}^\lambda = \begin{cases} h_{Left}^\lambda, & \text{if } u_i \text{ belongs to community } \lambda; \\ \mathsf{hash}(t_c), & \text{otherwise.} \end{cases}$$

Note that t_c is the current system time. When u_i does not belong to some communities, u_i calculates $\mathsf{hash}(t_c)$ instead. Afterward, u_i encrypts its community hash set $\mathcal{AH}_i$ with its own public key $\mathcal{K}_{pub,i}$ and obtains the ciphertext $c_i = \mathsf{Enc}_{\mathcal{K}_{pub,i}}(\mathcal{AH}_{Left,i})$. The signature for the encrypted ciphertext is generated by using u_i's private key $\mathcal{K}_{pri,i}$, and denoted as $\mathsf{Sign}_{\mathcal{K}_{pri,i}}(c_i)$. The ciphertext c_i is then concatenated with u_i's certificate Cert_i, including the identity and community certificates. Eventually, the 4-tuple $(ID_i||\mathsf{Cert}_i||c_i||\mathsf{Sign}_{\mathcal{K}_{pri,i}}(c_i))$ is sent to u_j.

Step 2: Receiving the 4-tuple, u_j first extracts u_i's public key and the signature of its identity ID_i. u_j then verifies u_i's certificate Cert_i. If invalid, u_j immediately terminates the protocol and reports this invalid user to TA for revocation. After verifying the authenticity of u_i's identity and community information, u_j picks h_{Right}^λ in each community λ that u_j belongs to, and produces

$$h_{Right,j}^\lambda = \begin{cases} h_{Right}^\lambda, & \text{if } u_j \text{ belongs to community } \lambda; \\ \mathsf{hash}(t_c), & \text{otherwise.} \end{cases}$$

If u_j does not belong to some community, u_j calculates $\mathsf{hash}(t_c)$, the hash value of the current system time t_c. The corresponding community set is obtained as $\mathcal{AH}_{Right,j} = \{h_{Right,j}^1, h_{Right,j}^2, \cdots, h_{Right,j}^\Lambda\}$.

Then, u_j encrypts $\mathcal{AH}_{Right,j}$ by using u_i's public key $\mathcal{K}_{pub,i}$ and produces $\mathsf{Enc}_{\mathcal{K}_{pub,i}}(\mathcal{AH}_{Right,j})$.

In addition, u_j calculates $\mathcal{A}_{Root,j} = \{h_{Root}^1, h_{Root}^2, \cdots, h_{Root}^\Lambda$, where h_{Root}^λ is the root value for community λ. u_j calculates the ciphertext as $c_j = \mathsf{Enc}_{\mathcal{K}_{pub,i}}(\vartheta(\mathcal{AH}_{Left,i} + \mathcal{AH}_{Right,j} - \mathcal{A}_{Root}))$, where $\vartheta \in (\frac{N}{2}, N)$ is a random integer. Note that N is the parameter of Paillier cryptography, and N is a large prime. Eventually, the 4-tuple $(ID_j||\mathsf{Cert}_j||c_j||\mathsf{Sign}_{\mathcal{K}_{pri,j}}(c_j))$ is sent back to user u_i.

Step 3: With the replied 4-tuple, u_i first checks the certificate of u_j. Verifying u_j's identity and community authenticity, u_i decrypts c_j and checks whether the plaintext equals 0 or not. If so, it indicates $h^{\lambda}_{left,i} + h^{\lambda}_{right,j} = h^{\lambda}_{Root}$. Therefore, the common community set $CC_{i,j} = C_i \cap C_j$ is obtained without disclosing any private information.

2. *Active Degree Profile Matching (APM)*:

We define A_i^{λ} as the active degree value for user u_i in community λ. A_i^{λ} is an integer $\in [1,\rho]$, where ρ is the maximum value for the specific community. Let γ denote the threshold of active degree value difference between two users determined by the initiator. Suppose that u_i and u_j have completed CPM and learn their common communities.

Step 1: When two users u_i and u_j want to match their active degree values with each other, the initiator u_i first encrypts its active degree value and corresponding threshold with its own public key $K_{pub,i}$. The ciphertexts are $c_i = \mathsf{Enc}_{K_{pub,i}}(A_i^{\lambda} + \gamma)$, $c_i' = \mathsf{Enc}_{K_{pub,i}}(A_i^{\lambda} - \gamma)$. The signature $\mathsf{Sign}_{K_{pri,i}}(\lambda, c_i, c_i')$ for the encrypted ciphertexts is generated with u_i's private key $K_{pri,i}$. The ciphertexts c_i and c_i' are concatenated with u_i's unique identity, certificate for the identity, community information, and the corresponding active degree values. Eventually, the 6-tuple $(ID_i||\mathsf{Cert}_i||\lambda||c_i||c_i'||\mathsf{Sign}_{K_{pri,i}}(\lambda, c_i, c_i'))$ is sent to u_j.

Step 2: Receiving the 6-tuple, u_j first extracts u_i's public key and the certificates. u_j then verifies Cert_i and u_i's community signature $\mathsf{Sign}_{K_{pri,i}}(\lambda, c_i))$. If invalid during the verification, u_j reports u_i to TA to draw ID_i into the revocation list and immediately terminates the protocol. If all valid, u_j encrypts its own active degree value A_j^{λ} by using u_i's public key $K_{pub,i}$. The ciphertexts are $c_j = \mathsf{Enc}_{K_{pub,i}}(\Psi)$ and $c_j' = \mathsf{Enc}_{K_{pub,i}}(\Psi')$, where $\Psi = \vartheta(A_i^{\lambda} - A_j^{\lambda} + \gamma)$ and $\Psi' = \vartheta'(A_i^{\lambda} - A_j^{\lambda} - \gamma)$, respectively. Both ϑ and $\vartheta' \in (1,\rho)$ are random integers. Eventually, the 6-tuple $(ID_j||\mathsf{Cert}_j||\lambda||c_j||c_j'||\mathsf{Sign}_{K_{pri,j}}(\lambda, c_j, c_j'))$ is sent back.

Step 3: With the replied 6-tuple, u_i first checks the certificates for identity, community, and active degree A_j^{λ}. Afterward, u_i calculates the product of Ψ and Ψ' as shown in Equation 10.15.

$$\Psi \cdot \Psi' = \vartheta(A_i^{\lambda} - A_j^{\lambda} + \gamma) \times \vartheta'(A_i^{\lambda} - A_j^{\lambda} - \gamma) \tag{10.15}$$

If $\Psi \cdot \Psi' < 0$, the difference between A_i^{λ} and A_j^{λ} is beyond the scope of the preset threshold γ; otherwise, the active degree value difference between u_i and u_j is within this threshold. The latter indicates that u_i and u_j are well matched based on the threshold γ.

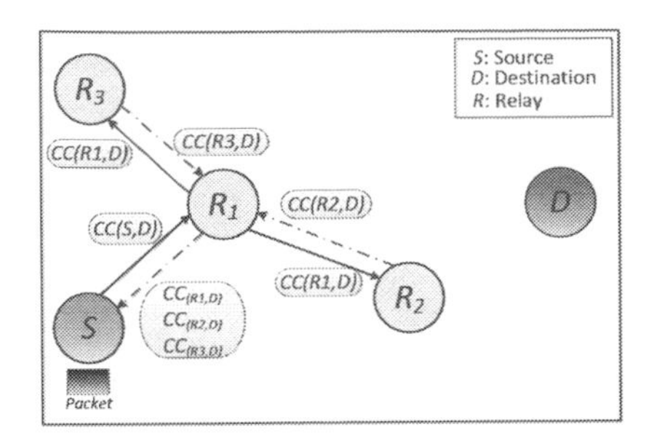

(a) Multi-hop profile matching

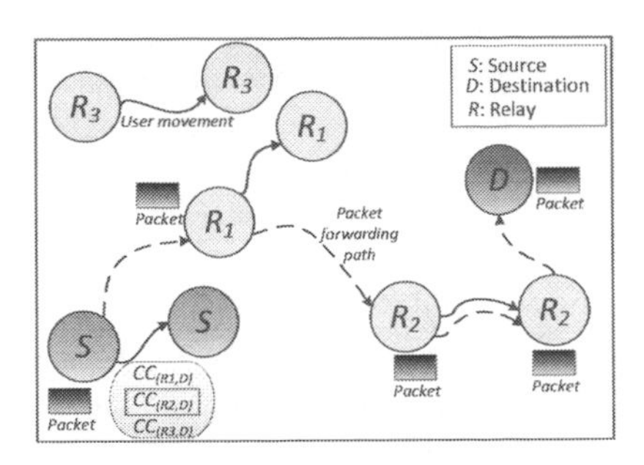

(b) Packet forwarding

Figure 10.6: Multi-hop relay selection of ePOMP.

10.5.4 Enhanced POMP (ePOMP) Protocol

In this subsection, we propose an enhanced POMP (ePOMP) protocol, where we exploit the multi-hop relay selection and recommendation mechanism to further improve the forwarding efficiency under the reasonable privacy level.

In the original POMP, only the one-hop neighbor to the source is considered as a relay to forward the packet. As shown in Figure 10.6, if there is a potential relay R_2 with two hops to the source, the source probably misses this better relay without knowing it at all.

To this end, we update the POMP protocol with multi-hop relay selection. First, the multi-hop neighboring users to the source u_s match their profiles to u_s. Second, the user, including one-hop and multi-hop neighboring users, having the most common communities can be selected as the optimal relay. The detailed steps of the ePOMP protocol are as follows:

Step 1: When the source u_s wants to forward a packet to the destination u_d, the source maintains a common community set $\mathcal{CC}_{(S,D)}$ and first constructs a full community set $\mathcal{A}_s$ with the length of Λ. Then, u_s encrypts this community set with u_s's public key $\mathcal{K}_{pub,u_s}$, and sends the 4-tuple $(ID_s||\text{Cert}_{ID_s}||c_s||\text{Sign}_{\mathcal{K}_{pri,s}}(c_s))$ to its one-hop relay u_r for to request a relay to the destination u_d.

Step 2: Running the CPM in a multi-hop manner, u_s obtains the multi-hop profile matching results at first. As shown in Figure 10.6(a), the results are denoted by $\mathcal{CC}_{r,d} = \{CC_r, CC_{r_1}, CC_{r_2}, CC_{r_x}\}$, where r_x is a two-hop user relayed by u_r. The forwarding capability for the neighboring user u_r is defined as $fc_r = (max\{CC_r, CC_{r_1} CC_{r_2}, \cdots, CC_{r_x}\}, r_i)$ where u_{r_i} is the two-hop user with the most common communities. If fc_r is greater than a preset threshold TH, u_r could be selected as the one-hop relay no matter whether u_r is likely to meet u_d or not. u_{r_i} is selected as the relay candidate. Then, with the multi-hop APM, u_s also obtains the active degree differences between every multi-hop relay. If the total difference $D_{s,r} \leqslant max(\mathcal{CC}_{s,r}) \times th$, u_r is selected as a relay.

Steps 3–5 finish the packet forwarding procedures, which are similar to POMP.

The forwarding capability of the one-hop neighboring user depends on the strongest user in the multi-hop path to the source. Since the multi-hop connection is already established near the source, the packet transmission delay of this kind of multi-hop communication is much lower than is produced by store-carry-and-forward procedures under intermittent connections in the conventional DTNs.

In addition, to enhance the relay's knowledge of the destination, we exploit a recommendation mechanism, where both users in the current contact exchange their encountered users having some common communities with the opposite. It consists of the following two phases: the common community user recommendation phase and the packet forwarding phase.

1. *Common Community Recommendation Phase*:

   When any two users are in proximity, they can well match their profiles by using CPM and APM with privacy preservation. If they belong to common communities, they can exchange some users also belonging to these common communities. Here, we define a friend as a user who belongs to the same community to which the two matched users belong. In addition, each user should be able to verify whether the recommended friends from the other are authenticated or not. Then, the friend information can be maintained in a weighted friendship graph $\mathcal{FG} = (FV, FE)$, where FV is the vertex, and FE is the edge weighting with the number of common communities between the connected two vertexes.

   Step 1: Having the well-matched profiles between two users u_i and u_j, u_i first selects its most k familiar friends $\mathcal{FR}_i = \{fr_{i_1}, fr_{i_2}, \cdots, fr_{i_k}\}$, where fr_{i_k} shares the most common communities with u_j to the best of u_i's knowledge.

   Step 2: Having the recommended friends, u_j updates its own friendship graph with these recommended friends.

   Step 3: u_j then feeds back its own most k familiar friends $\mathcal{FR}_j = \{fr_{j_1}, fr_{j_2}, \cdots, fr_{j_k}\}$ to u_i, where $\mathcal{FR}_j \cap \mathcal{FR}_i = \emptyset$ so that u_j does not return the same friends as u_i recommended.

   Step 4: Once u_i receives the feedback friendship set from u_j, u_i verifies the validity and updates its own friendship graph as depicted in *Step 2*. The detailed algorithm is illustrated in Algorithm 19.

2. *Packet forwarding phase*:

   Given the social friendship graph, we explore the ePOMP protocol with privacy preservation. Details are as follows.

   Step 1: When the source u_s wants to transmit a packet $\mathcal{P}$ to the destination u_d, u_s encrypts the packet with u_d's public key $\mathcal{K}_{pub,d}$ to generate the ciphertext $\mathcal{C} = \mathsf{Enc}_{\mathcal{K}_{pub,d}}(\mathcal{P})$. Afterward, u_s broadcasts the request to forward the packet to u_d.

   Step 2: If a neighbor u_r receives this forwarding request from u_s, u_r firstly queries its friendship graph to find $|\mathcal{CA}_{r,d}|$, which is the total number of common communities between u_r and u_d. Then, u_r continues to check whether

Algorithm 19: Common Community Recommendation

```
1  Procedure: Common Community Recommendation
2  u_i and u_j encounter
3  u_i selects most k familiar friends FR_i = {fr_i1, fr_i2, ..., fr_ik}, and sends FR_i to u_j
4  if FR_i are verified by u_j then
5      u_j updates its own friendship graph FG_j = FG_j ∩ FR_i
6      u_j sends FR_j = {fr_j1, fr_j2, ..., fr_jk} to u_i, where FR_j ∩ FR_i = ∅
7      if FR_j are verified then
8          u_i updates its own friendship graph FG_i = FG_i ∩ FR_j
9      else
10         u_i revokes u_j, and terminates
11     end if
12     u_j revokes u_i, and terminates
13 end if
14 end procedure
```

any of u_r's friends in its friendship graph belongs to more than TH common communities between u_r's friends and u_d. If so, u_r's friends can be recorded as potential relays r_p. If u_r has more than R potential friends including itself, u_r is eligible to forward the packet $\mathcal{P}$. Eventually, u_r feeds back to u_s for accepting the relay request.

Step 3: If u_s receives the acceptance of forwarding packet, u_s packages the ciphertext $\mathcal{C}$ with the signature, certificate, and destination information. The details of forwarding are similar to POMP in Section 10.5.2.

10.6 Privacy Analysis

In this section, we analyze privacy properties of our proposed POMP protocol. Our analysis focuses on how the POMP preserves the user's private information during profile matching and packet forwarding.

During profile matching, both CPM and APM protocols can preserve user privacy and well match the common communities without disclosing any uncommon community. Due to the properties of Paillier cryptography, the encrypted ciphertext are additive and multiplicative so that the matching results can be processed in the ciphertext. According to the security of Paillier cryptography, the responder cannot intercept the plaintext and obtain the initiator's profile at all. On the other hand, the responder injects the random number ϑ, which is the blind factor, in the matching results. Even though the initiator can decrypt the ciphertext by using its own private key, it can learn nothing except the common communities. If the initiator does not belong to the community, $\mathsf{hash}(t_c)$ is calculated to represent the attribute of the specific community. Since the system time t_c changes at a distinct time, the hash values generated during any two different profile matchings are not equal. The initiator cannot analyze the useful information $\mathcal{AH}_{Left,j}$ and $\mathcal{AH}_{Right,j}$ according to the statistics gathered from multiple profile matchings of different users.

The privacy characteristics of the CPM can apply for the APM as well. With the APM, since the two users know that the other one belongs to the specific community, the private information that should be preserved is the difference between the active degree values which reflect the user's activeness. According to the pre-set threshold th, each involved user can only know whether the opposite is similar to itself. A large, equal, or small relation cannot be inferred based on the APM results.

When packet forwarding, only the users with common communities are involved in relay selection and known to the users who also belong to such specific communities. The users without such common communities cannot know who belongs to the specific community through any attack, for example, the packet analysis attack and packet tracing attack presented in [20, 21].

In addition, TA operates in an off line pattern. Only in the initialization phase, TA generates the key materials and community information including the active degree values according to the users' history contacts. Afterward, mobile users can operate without any support from TA except the revocation of the detected malicious users. Moreover, the secure channel can be established by using Paillier cryptography to resist eavesdropping from outside attackers. Therefore, from the above privacy analysis, the proposed protocols can preserve user privacy during both profile matching and packet forwarding phases.

10.7 Performance Evaluation

To evaluate the effectiveness of the proposed protocols, we simulate them in a real-world human trace, and evaluate performance in terms of packet delivery ratio, average delay, and the number of copies.

10.7.1 Simulation Setup

Our simulation is based on the Infocom06 trace [33], which is a real human trace with 78 mobile users attending a conference in four days. Each mobile user is equipped with a dedicated Bluetooth device. Once two mobile users are in the proximity to each other, their attached Bluetooth devices can detect neighbors due to the Bluetooth discovery program running every 120 seconds. Totally, 128,979 contacts are logged with each contact being recorded including encountered users, and the start and end time. We separate the entire contact data into two parts: training set and simulation set. One third of the data are marked as the training set to produce user profiles, such as communities and corresponding active degree values, while the remaining data are used for simulation.

We assign the communities based on sociology theory [6]. Let $NC_{i,j}$ denote the total number of contacts between u_i and u_j. A complete graph of users can be established with each edge $E(u_i, u_j)$ weighted by $NC_{i,j}$. Removing the edges that are weighted less than 100, we obtain a graph G with 78 vertices and 2863 edges. Using the Bron-Kerbosch algorithm [34], we further discover all maximal cliques in G. Here, a clique is a complete subgraph where every two vertices have the high-weighted edge. Furthermore, a maximal clique is the clique that cannot be extended by including one more adjacent vertex. Eventually, we extract 7550 maximal cliques denoted as $CL_1, \cdots, CL_{7550}$ that each contains at least 15 users, and sort them in the descending order of their total weights.

We construct communities through the following steps. First, we scan the sequence of cliques from CL_1 to CL_{7550}. For a scanned clique CL_i, where $i = 1, 2, \cdots, 7550$, we try to find a clique CL_j that has been previously scanned and identified as a *core clique* which contains more than 80% vertices of CL_i. If there are multiple such cliques, the one with largest weight is selected as CL_j. If CL_j is found, CL_i is assigned with the same community as CL_j; otherwise, a new community could be generated and assigned to CL_i which is determined as a core clique. After community generation and assignment, the cliques with the same community are collected into one community. A community contains multiple users, while each user might belong to multiple communities. Totally, 349 communities corresponding to 349 communities are produced from the training data set. Finally, we select the first generated 100 communities and their corresponding communities for the simulation. Each of these selected communities contains at least 28 users with each user belonging to 38 communities on average.

Based on these 100 communities, we assign active degree values to each user in G. First, in community a, the corresponding community is denoted as $\mathcal{CM}_a$. For each user in $\mathcal{CM}_a$, we compute the weight sum of its incidental edges in $\mathcal{CM}_a$; for each vertex outside $\mathcal{CM}_a$, we compute the weight sum of its incident edges to the vertices in $\mathcal{CM}_a$. Second, we sort all users in the increasing order of their weight sums, and assign their ranks as their active degree values A_i^a for user u_i in community a. A larger weight sum indicates that this user is more likely to communicate with the users in $\mathcal{CM}_a$. Therefore, the community value A_i^a can reflect the active degree in the community.

10.7.2 Simulation Results

We evaluate the performance of the proposed POMP and ePOMP protocols compared with Epidemic, PROPHET [11], and BUBBLE Rap [7] protocols. Each user generates a packet to a randomly selected destination. The performance evaluations are as follows.

Figure 10.7 shows the performance comparisons among Epidemic, PROPHET, BUBBLE Rap, and POMP and ePOMP in terms of delivery ratio, average delay and number of copies. Although the Epidemic packet forwarding protocol performs the highest delivery ratio and lowest average delay, it costs a huge number of copies, and we cannot afford this level consumption in some applications. Simultaneously, PROPHET consumes the lowest communication overhead, but cannot satisfy the performance requirements at all. BUBBLE Rap is a popular forwarding protocol and consumes the reasonable resources with better performance. Compared with these sophisticated protocols, POMP and ePOMP create further trade-off between the communication overhead and the performance. Furthermore, ePOMP performs better than POMP does since ePOMP selects relays with consideration of multi-hop

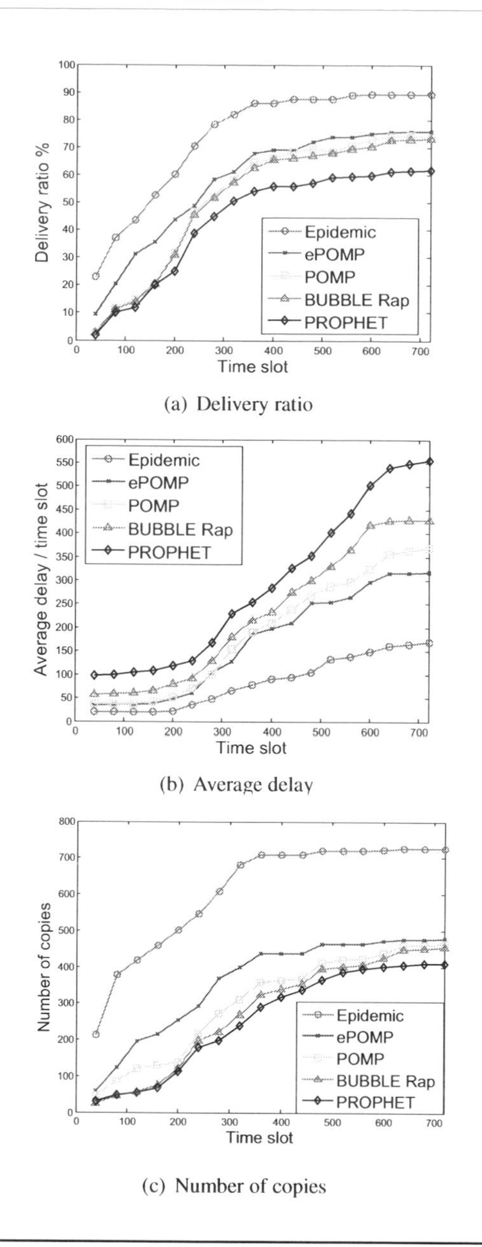

(a) Delivery ratio

(b) Average delay

(c) Number of copies

Figure 10.7: Performance comparison of different protocols.

neighbors and obtains benefits from the recommendation mechanism that enhances the user's knowledge.

Figure 10.8 shows the impact of TH on the performance of ePOMP. As shown in Figure 10.8(a), during the first 300 time slots, the delivery ratio of ePOMP increases quickly. The growth slows after the 400th time slot. As TH increases, the delivery

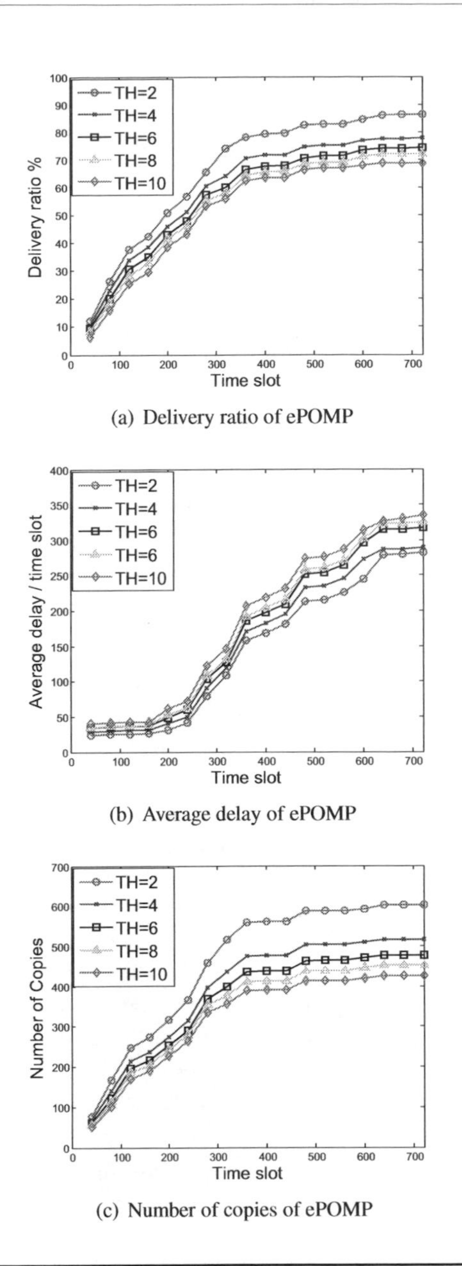

(a) Delivery ratio of ePOMP

(b) Average delay of ePOMP

(c) Number of copies of ePOMP

Figure 10.8: Impact of TH on performance of ePOMP.

ratio correspondingly decreases. This is because the increment of TH causes the sources to ignore the users that own fewer common communities with the destination. Therefore, some opportunities of forwarding are missed. Figure 10.8(b) shows the impact of TH on the average delay of ePOMP. Before the 300th time slot, the average

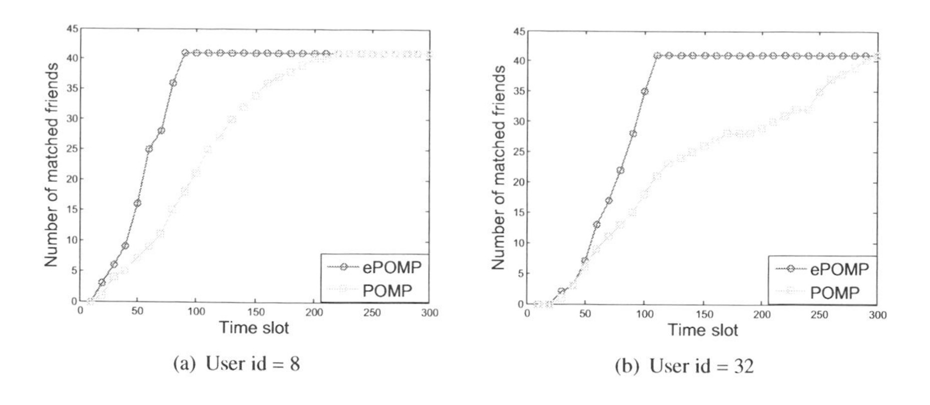

Figure 10.9: Comparison of the number of matched users.

delay changes slightly since ePOMP requires users to recommend their friends after profile matching. In this period, users dynamically update their friendship graph. As the increment of time slot, the average delay increases. Furthermore, the varying threshold impact the relay selection to some extent. A larger threshold results in a higher delay. As the threshold decreases, the number of copies correspondingly increases as shown in Figure 10.8(c), since more users are involved in the packet forwarding due to a lower threshold.

Figure 10.9 shows the comparison of the number of matched users that use ePOMP or not. We take two users in community 1 with the rank of 8th and 32nd, for example. As shown in Figure 10.9(a), for the 8th ranked user, it is active in community 1. Therefore, it can know many users in a shorter period than the 32nd ranked user shown in Figure 10.9(b). By using ePOMP, both users can directly or indirectly match more users with the common community. The reason is that users recommend friends after profile matching so that they can know the users in the same community even though they have not met yet. Based on this knowledge, ePOMP can achieve higher efficiency.

Regarding privacy loss among different protocols, we first define privacy loss metric as

$$\mathcal{PL}(CC,UC) = \frac{\alpha \times CC + \beta \times UC}{TC} \tag{10.16}$$

where CC denotes the common communities between the two users, UC denotes the uncommon communities, TC is the total number of the compared communities. Note that α and β are the privacy risk factors of the common and uncommon information, respectively. If the common communities are disclosed to the other users also having the same ones, the privacy risk level decreases correspondingly since users are encouraged to share such common information. However, if specific community information of a user is disclosed to another user who does not participate in the community, the privacy risk level dramatically increases. In our simulation, we

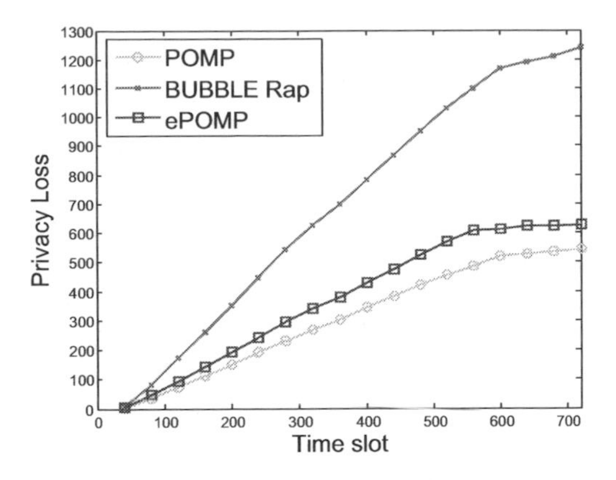

Figure 10.10: Comparison of privacy loss among different protocols ($\alpha = 0.5, \beta = 1$).

set $\alpha = 0.5$ and $\beta = 1$. The privacy loss reaches the maximum value 1 when all the information exchanged is uncommon to each other, and achieves the minimum value 0.5 when all that is common. We discover the privacy loss trends among BUBBLE Rap and the POMP and ePOMP. As shown in Figure 10.10, the POMP and ePOMP face the lower privacy loss, which validates the privacy effectiveness of the POMP and ePOMP. Furthermore, the ePOMP results in a little more privacy loss compared with the POMP, because the ePOMP exchanges more information among users with multi-hop relay selection and recommendation mechanism.

10.7.3 Computational Cost

In our protocols, the main computational cost is from encryption and decryption during the profile matching and forwarding. In the forwarding procedures, the computational cost is similar to the Paillier cryptograph. In this section, we mainly provide the numerical results of the computational cost of the profile matching protocols.

For the CPM, the initiator u_i requires 2 exponential operations and 1 modulation for the encryption of his profiles. The responder u_r needs 4 exponential operations and 2 modulations to add and blind his private profile in the ciphertext. Finally, u_i decrypts the ciphertext from u_r with 2 exponential operations and 3 modulations. Regarding the APM, the initiator u_i uses 4 exponential operations and 2 modulations to generate two ciphertexts. The responder needs 8 exponential operations and 4 modulations to finish the privacy-preserving profile matching. u_i decrypts the ciphertext with 2 exponential operations and 3 modulations.

10.8 Conclusions

In this chapter, we have proposed an efficient packet forwarding protocol (POMP) with private profile matching in MSNs to improve the forwarding efficiency and preserve user privacy. Based on analysis of social relationships between users, we have introduced common communities to facilitate the forwarding efficiency. Regarding privacy preservation, we have investigated the private profile matching protocols CPM and APM to match users' community and active degree information without disclosing any uncommon information. With these private profile matching results, POMP evaluates the forwarding capability of potential relays and selects the optimal one-hop relay to forward packets and protect user's private community information from being disclosed to strangers. In addition, we have proposed an enhanced POMP protocol by multi-hop relay selection and recommendation mechanism to further improve the forwarding efficiency. The privacy analysis demonstrates that user's private profile can be protected with CPM and APM protocols, while the extensive simulation results indicate that the POMP and ePOMP can achieve high efficiency and low delay with acceptable communication overhead. In our future work, we will investigate the impact of selfishness on cooperative profile matching and packet forwarding, and study how to encourage selfish users to participate.

10.9 Acknowledgment

This research has been supported by a research grant from the Natural Science and Engineering Research Council (NSERC).

References

[1] R. Zhang, Y. Zhang, J. Sun, and G. Yan, "Fine-grained private matching for proximity-based mobile social networking," in *Proc. of IEEE INFOCOM*, pages 1969–1977, 2012.

[2] X. Liang, X. Li, R. Lu, X. Lin, and X. Shen, "SEER: A secure and efficient service review system for service-oriented mobile social networks," in *Proc. of ICDCS*, pages 647–656, 2012.

[3] N. Kayastha, D. Niyato, P. Wang, and E. Hossain, "Applications, architectures, and protocol design issues for mobile social networks: A survey," *Proceedings of the IEEE*, 99(12): 2130–2158, 2011.

[4] J. Wu and Y. Wang, "Social feature-based multi-path routing in delay tolerant networks," in *Proc. of IEEE INFOCOM*, pages 1368–1376, 2012.

[5] E. Bulut and B. K. Szymanski, "Exploiting friendship relations for efficient routing in mobile social networks," *IEEE Transactions on Parallel and Distributed Systems*, 23(12): 2254–2265, 2012.

[6] D. J. Watts, “Small worlds: The dynamics of networks between order and randomness,” *Journal of Artificial Societies and Social Simulation*, 6(2), 2003.

[7] P. Hui, J. Crowcroft, and E. Yoneki, “BUBBLE Rap: Social-based forwarding in delay-tolerant networks,” *IEEE Transactions on Mobile Computing*, 10(11): 1576–1589, 2011.

[8] A. Vahdat and D. Becker, *Epidemic Routing for Partially-Connected Ad Hoc Networks*, Technical Report, Duke University, 2000.

[9] T. Spyropoulos, K. Psounis, and C. Raghavendra, “Spray and Focus: Efficient mobility-assisted routing for heterogeneous and correlated mobility,” in *Proc. of IEEE PerCom Workshops*, pages 79–85, 2007.

[10] Q. Yuan, I. Cardei, and J. Wu, “Predict and relay: An efficient routing in disruption-tolerant networks,” in *Proc. of ACM MobiHoc*, pages 95–104, 2009.

[11] A. Lindgren, A. Doria, and O. Schelén, “Probabilistic routing in intermittently connected networks,” *Proc. of ACM SIGMOBILE Mobile Computing and Communications Review*, 7(3), July 2004.

[12] J. Fan, J. Chen, Y. Du, W. Gao, J. Wu, and Y. Sun, “Geo-community-based broadcasting for data dissemination in mobile social networks,” *IEEE Transactions on Parallel and Distributed Systems*, 24(4): pp. 734–743, 2013.

[13] E. M. Daly and M. Haahr, “Social network analysis for information flow in disconnected delay-tolerant MANETs,” *IEEE Transactions on Mobile Computing*, 8(5): 606–621, 2009.

[14] T. Abdelkader, K. Naik, A. Nayak, N. Goel, and V. Srivastava, “SGBR: A routing protocol for delay tolerant networks using social grouping,” *IEEE Transactions on Parallel and Distributed Systems*, 24(12): 2472–2481, 2013.

[15] W. Gao, G. Cao, T. L. Porta, and J. Han, “On exploiting transient social contact patterns for data forwarding in delay-tolerant networks,” *IEEE Transactions on Mobile Computing*, 12(1): 151–165, 2013.

[16] G. Costantino, F. Martinelli, and P. Santi, “Investigating the privacy vs. forwarding accuracy trade-off in opportunistic interest-casting,” *IEEE Transactions on Mobile Computing*, 13(4): 824–837, 2014.

[17] R. Cabaniss, J. M. Bridges, A. Wilson, and S. Madria, “DSG-N^2: A group-based social routing algorithm,” in *Proc. of IEEE WCNC*, pages 504–509, 2011.

[18] J. Á. B. Link, D. Schmitz, and K. Wehrle, “GeoDTN: Geographic routing in disruption tolerant networks,” in *Proc. of IEEE GLOBECOM*, pages 1–5, 2011.

[19] A. J. Aviv, M. Sherr, M. Blaze, and J. M. Smith, “Evading cellular data monitoring with human movement networks,” in *USENIX Workshop on Hot Topics in Security (HotSec)*, pages 1–6, 2010.

[20] R. Lu, X. Lin, and X. Shen, "SPRING: A social-based privacy-preserving packet forwarding protocol for vehicular delay tolerant networks," in *Proc. of IEEE INFOCOM*, pages 632–640, 2010.

[21] X. Lin, R. Lu, X. Liang, and X. Shen, "STAP: A social-tier-assisted packet forwarding protocol for achieving receiver-location privacy preservation in VANETs," in *Proc. of IEEE INFOCOM*, pages 2147–2155, 2011.

[22] R. Lu, X. Lin, X. Liang, and X. Shen, "Sacrificing the plum tree for the peach tree: A socialspot tactic for protecting receiver-location privacy in VANET," in *Proc. of IEEE GLOBECOM*, pages 1–5, 2010.

[23] W. Dong, V. Dave, L. Qiu, and Y. Zhang, "Secure friend discovery in mobile social networks," in *Proc. of IEEE INFOCOM*, pages 1647–1655, 2011.

[24] Z. Yang, B. Zhang, J. Dai, A. C. Champion, D. Xuan, and D. Li, "E-smalltalker: A distributed mobile system for social networking in physical proximity," in *Proc. of ICDCS*, pages 468–477, 2010.

[25] R. Lu, X. Lin, X. Liang, and X. Shen, "A secure handshake scheme with symptoms-matching for mhealthcare social network," *MONET*, 16(6): 683–694, 2011.

[26] M. Li, N. Cao, S. Yu, and W. Lou, "FindU: Privacy-preserving personal profile matching in mobile social networks," in *Proc. of IEEE INFOCOM*, pages 2435–2443, 2011.

[27] R. Baden, A. Bender, N. Spring, B. Bhattacharjee, and D. Starin, "Persona: An online social network with user-defined privacy," in *Proc. of ACM SIGCOMM*, pages 135–146, 2009.

[28] X. Liang, X. Li, K. Zhang, R. Lu, X. Lin, and X. S. Shen, "Fully anonymous profile matching in mobile social networks," *IEEE Journal on Selected Areas in Communications*, 31(9): 641–655, 2013.

[29] P. Paillier, "Public-key cryptosystems based on composite degree residuosity classes," in *Proc. of EUROCRYPT*, pages 223–238, 1999.

[30] Paillier and Pointcheval, "Efficient public-key cryptosystems provably secure against active adversaries," in *Proc. ASIACRYPT: Advances in Cryptology*, 1999.

[31] W. Gao, Q. Li, B. Zhao, and G. Cao, "Social-aware multicast in disruption-tolerant networks," *IEEE/ACM Transactions on Networking*, 20(5): 1553–1566, 2012.

[32] A. Balasubramanian, B. N. Levine, and A. Venkataramani, "Replication routing in DTNs: A resource allocation approach," *IEEE/ACM Transactions on Networking*, 18(2): 596–609, 2010.

[33] J. Scott, R. Gass, J. Crowcroft, P. Hui, C. Diot, and A. Chaintreau, "CRAWDAD trace cambridge/haggle/imote/infocom (v. 2006-01-31)," Jan. 2006.

[34] C. Bron and J. Kerbosch, "Finding all cliques of an undirected graph (algorithm 457)," *Commun. ACM*, 16(9): 575–576, 1973.

Chapter 11

Privacy-Preserving Opportunistic Networking

Gianpiero Costantino
Istituto di Informatica e Telematica del CNR
Pisa, Italy

Fabio Martinelli
Istituto di Informatica e Telematica del CNR
Pisa, Italy

Paolo Santi
Istituto di Informatica e Telematica del CNR
Pisa, Italy

CONTENTS

11.1 Introduction

Opportunistic networks have recently attracted significant attention in the research community. This increasing interest in opportunistic networking is motivated by the fact that, given the growing popularity of smartphones and tablet PCs with peer-to-peer communication capabilities, opportunistic exchange of information between individuals is deemed an emerging communication paradigm.

Opportunistic networking raises new and challenging privacy concerns. In fact, the most promising routing and information dissemination approaches for opportunistic networks are based on a notion of *user profile*. Depending on the specific protocol, user profiles can be used to compactly represent user interests [17], mobility behavior [14], social contacts [6, 12], etc. In a typical opportunistic networking protocol, when two users meet they exchange their respective user profiles, which are used to decide whether a message exchange should be initiated, and which messages stored in the local buffers the users should exchange. Given that user profiles contain very sensible information—such as a user's interests, mobility pattern, list of friends, etc.—exchanging profiles in cleartext is likely to be considered an unacceptable privacy leak by the average user. For this reason, it has been recently acknowledged within the research community that minimizing privacy leakage while exchanging user profiles will be the key for achieving widespread user acceptance of forthcoming applications based on opportunistic networking.

In this chapter, we will survey a collection of approaches that have been recently proposed in the literature to address the need for minimizing privacy leakage during opportunistic user profile exchange. The rest of the chapter is organized as follows. In the next section, we will present a set of criteria according to which existing privacy-preserving opportunistic protocols can be categorized. In Section 11.3, we will describe the attacker models typically used in the privacy-preserving opportunistic networking literature. In Section 11.4, we will review existing cryptography-based approaches for privacy-preserving opportunistic networks, while Section 11.5 will be devoted to cryptography-free solutions. Finally, Section 11.6 concludes this chapter.

11.2 Taxonomy of Privacy-Preserving Opportunistic Protocols

Existing privacy-preserving opportunistic protocols can be categorized according to different criteria. A first, important criteria, which is also the main criteria used in this chapter, is whether the proposed solution makes use of cryptographic tools. Cryptography-based approaches challenge the limited computational and/or storage resources available on existing mobile devices; yet, they typically provide stronger privacy preservation guarantees than cryptography-free approaches. Cryptography-free approaches typically use randomization techniques to provide some form of privacy preservation while minimally impacting protocol performance.

The second criteria according to which existing approaches can be categorized is the application that they implement. While most of the approaches are only concerned with the first phase of an opportunistic networking application (the phase in which users exchange and compare their profiles, called *friend discovery* in the following), some approaches also consider a specific opportunistic networking application such as geo-routing, interest-casting, etc.

Finally, while in most cases the presented privacy-preserving solution is tailored to a specific application scenario, in a few cases more general privacy-preserving opportunistic networking frameworks are presented. In contrast to customized solutions, these frameworks allow user-friendly development of privacy-preserving protocols by means of a high-level programming language.

The main features of the privacy-preserving opportunistic networking protocols presented in this chapter are summarized in Table 11.1. Most of the proposed approaches are cryptography-based solutions, encompassing applications such as friend discovery, friend recommendation, and interest-casting. For cryptography-based approaches, the table also reports the cryptographic tools on which the protocols are based. MightBeEvil [11] and MobileFairPlay [5] are privacy-preserving programming frameworks for mobile devices; the application mentioned in the corresponding column is the application that was implemented by the respective authors to show the functionality of the framework. MightBeEvil, MobileFairPlay, and PCD [7] have

Table 11.1: Main Features of Privacy-Preserving Opportunistic Protocols

Name	Crypto	Tool	Fram.	Application	Impl.
MightBeEvil [11]	Yes	Garbled Circuits	Yes	Friend discovery	Yes
PCD [7]	Yes	Rsa, IBME	No	Friend discovery	Yes
MobileFairPlay [5]	Yes	Garbled Circuits	Yes	Interest-casting	Yes
Baglioni et al. [3]	Yes	Sketches	No	Friend recomm.	No
Zhang et al. [23]	Yes	Paillier cryptosystem	No	Friend discovery	No
FindU [15]	Yes	Homomorphic encrypt.	No	Friend discovery	No
PPBR [2]	No	–	–	Geo-routing	No
Hide-and-Lie [8]	No	–	–	Interest-casting	No

also been implemented and tested on mobile platforms. For this reason, these three cryptography-based approaches will be described in greater detail in Section 11.4. PPBR [2] and Hide-and-Lie [8] are non-cryptographic approaches for geo-routing and interest-casting, respectively. Since PPBR and Hide-and-Lie do not require cryptographic primitives, implementation issues for these protocols are less important than in cases involving cryptography-based solutions. PPBR and Hide-and-Lie will be described in Section 11.5.

It is important to observe that all the approaches reported in this book chapter are fully distributed, and do *not* require existence of a trusted third party (TTP), not even in an initial, boot-strapping phase of the protocols. Implementation and operation of a TTP in opportunistic networks is highly challenging and costly, and it is therefore considered hardly feasible in practice.

11.3 Attacker Models

Most existing approaches adopt relatively weak attacker models, which are acceptable in opportunistic network scenarios where the exchanged information is not likely to be very critical or valuable. Indeed, privacy-preserving opportunistic networking approaches are not concerned with the security of the messages exchanged between users which, if needed, can be secured with standard approaches. Instead, the objective is to preserve the information stored in user profiles, whose exchange is the pre-requisite for the (possible) subsequent message exchange. Observe that the information contained in user profiles is likely to be the most valuable information exchanged within opportunistic networks, thus justifying the goal of keeping this information as private as possible.

Typically, two attacker models are considered when analyzing the privacy-preservation properties of an opportunistic networking protocol: the *semi-honest* and the *malicious* attacker model.

In the former model, sometimes called *honest-but-curious* model, the attacker follows all the protocol steps as per specifications, but she can try to learn additional information about the other party, with the purpose of acquiring at least part of her private profile.

In the malicious model, the attacker is more powerful: she can deviate from the protocol specifications, e.g., by modifying or dropping messages, using fake user profiles during the interactions with other users, etc. Similar to the semi-honest model, the goal of the attacker is acquiring the information contained in the user profile of the attacked user.

Attacker models specific to particular privacy-preserving opportunistic networking protocols will be described in the respective sections.

11.4 Cryptography-Based Protocols

The use of cryptographic primitives in any type of network protocol allows a sender and a destination party to complete transactions while keeping information hidden through the main flow. However, this comes at the price of a more complicated communication protocol. In fact, the use of cryptography often requires execution of very complex and computationally intensive operations in order to obtain the desired level of protection. So, computer processors might be overloaded depending on the degree of protocol complexity.

The issue of computational complexity is even more critical in mobile devices, such as smartphones or tablet PCs. Mobile devices are built to be portable and energy efficient, and as a consequence they are equipped with much less powerful hardware than that equipping traditional PCs. For this reason, most cryptographic protocols are deemed not suitable to be run on smartphones, tablet PCs, and similar portable devices.

Recently, a few cryptography-based approaches to preserve privacy in opportunistic networks have been designed, implemented, and tested on mobile platforms, thus showing feasibility of cryptographic protocols on today's portable devices. In the following, we will describe in detail these approaches, namely, the Private Contact Discovery approach by De Cristofaro et al. [7], the MightBeEvil framework by Huang et al. [11], and the MobileFairPlay framework by Constantino et al. [5]. We will then conclude this section by briefly describing other cryptography-based opportunistic networking protocols, which have not been implemented on real mobile devices.

11.4.1 Private Contact Discovery

The work [7] of De Cristofaro et al. introduce a *private contact discovery* primitive according to which two users can discover their common friends in a privacy-preserving manner. The motivation for discovering common people in the respective contact lists comes from the observation made in social networking literature that new social contacts are more likely to arise between individuals sharing common friends.

The straightforward solution for performing this task would be the adoption of a trusted third party (*TTP*), seen as a trusted central server to find and output the common friends while not disclosing a party's contact list to the other party. However, as discussed in Section 11.2, use of a TTP is deemed not feasible in opportunistic networks.

In [7], the authors introduce the private contact discovery *(PCD)* primitive, which is a construct geared to preserve user privacy within applications that make use of contact lists. Also, PCD prevents users from falsely claiming friendships by introducing the notion of *contact certification.* For instance, Alice needs to obtain a certificate from Carol, attesting their friendship, to include Carol in her contact list. Thus, when Alice runs the contact list matching phase with Bob, she is able to keep hidden the possession of corresponding certificates with respect to non-common friends (and vice versa).

The cryptographic tool on which PCD is based is Index-Based Message Encoding *(IBME).* It is an encoding technique that combines a set of indexed input messages $m_1, m_2, \ldots, m_n \in M$, where M is a message space, into a single data structure S. Any message can be individually recovered from S using its index, which is chosen from an index space I specified, a priori, at encoding time.

IBME satisfies a security property called Index-Hiding Message Encoding *(IHME)*, which guarantees that no adversary, by observing the IBME structure *S* that encodes random messages, is able to learn any useful information about the deployed indices, even if the adversary knows some of the indices and/or messages.

A contact discovery scheme (*CDS*), at the heart of PCD, is defined as a tuple of four algorithm and protocols:

- $Init(1^k)$ is the parameter initialization for a generic user U;
- $AddContact(U \leftrightarrow V)$ is run by users U and V, when user V wishes to become a contact in user U's list;
- $RevokeContact(U,V)$ is a function, in which the identity of V is revoked from U's contact list;
- $Discover(V \leftrightarrow V')$ is the most important algorithm executed by V and V′ to discover whether they have common friends. Each user has a private input, which is coded as a triple $(role, CL, partner)$. The *role* field specifies the user function in the session as either *initializer* or *responder*, field *CL* is the contact list, and field *partner* is the name/ID of the supposed protocol partner. At the end of the algorithm, both users know whether they have common friends, while not disclosing the private contact list to the other party.

The pseudo-code of the *Discover* algorithm is reported in Figure 11.1. The algorithm uses Okamoto's technique described in [18] for RSA-based identity-based key agreement. In particular, to speed up the process, several instances of the Okamoto protocol are run at the same time. Then, all transferred messages are IMHE-encoded into a single structure before transmission. When IMHE messages are received, each of them is decoded and the probabilistic padding applied before is removed (line 17). In lines 23–24, a new round of message exchanges encoded with the IHME is performed. Finally, in line 29 each common user found is added to the SCL list, and the algorithm ends with either *"accept"* or *"reject"* depending on whether the size of the SCL list is larger than zero.

Experimental Analysis. De Cristofaro et al. tested their algorithm both with laptop and mobile devices. In the first case, experiments were conducted using an Intel XEON 2.6-GHZ CPU and an AMD NEO 1.6-GHz processor, while experiments on mobile devices were performed on an ARMv7 600-MHz CPU.

The results of the experiments performed in [7] show that the running time of the *Discover* algorithm on a laptop is less than one second, even when the size of the contact lists is larger than 100. When the algorithm is executed on the mobile ARM processor, it requires about 5 seconds to complete. This is still a reasonable time, although it is not clear whether the reported time refers to a complete, two-party implementation of the primitive, including the time requested to exchange data through a short range radio interface (WiFi or Bluetooth).

The private contact discovery protocol has not been made public by the authors in the form of a downloadable application, so it is not possible to download and test the application on a laptop or mobile device.

	V ON INPUT (init, CL, partner):		V' ON INPUT (resp, CL$'$, partner$'$):
1	$\mathcal{P} \leftarrow \emptyset,\ \mathcal{T} \leftarrow \emptyset$		$\mathcal{P} \leftarrow \emptyset,\ \mathcal{T} \leftarrow \emptyset$
2	FOR ALL $(U, \mathsf{cc}_{U \to V}) \in \mathsf{CL}$:		FOR ALL $(U, \mathsf{cc}_{U \to V'}) \in \mathsf{CL}'$:
3	PARSE $\mathsf{cc}_{U \to V}$ AS (N, e, g, σ_V)		PARSE $\mathsf{cc}_{U \to V'}$ AS $(N, e, g, \sigma_{V'})$
4	$(b, t) \leftarrow_R \mathbb{Z}_2 \times \mathbb{Z}_{N/2}$		$(b, t) \leftarrow_R \mathbb{Z}_2 \times \mathbb{Z}_{N/2}$
5	$\vartheta \leftarrow (-1)^b g^t \sigma_V \bmod N$		$\vartheta \leftarrow (-1)^b g^t \sigma_{V'} \bmod N$
6	$k \leftarrow_R [0, \lfloor p/N \rfloor - 1]$		$k \leftarrow_R [0, \lfloor p/N \rfloor - 1]$
7	$\theta \leftarrow \vartheta + kN$		$\theta \leftarrow \vartheta + kN$
8	$\mathcal{P} \leftarrow \mathcal{P} \cup \{(N, \theta)\}$		$\mathcal{P} \leftarrow \mathcal{P} \cup \{(N, \theta)\}$
9	$\mathcal{T} \leftarrow \mathcal{T} \cup \{(U, N, e, t)\}$		$\mathcal{T} \leftarrow \mathcal{T} \cup \{(U, N, e, t)\}$
10	$\mathcal{M}_V \leftarrow \mathsf{iEncode}(\mathcal{P})$	$\xrightarrow{\mathcal{M}_V}$	$\mathcal{M}_{V'} \leftarrow \mathsf{iEncode}(\mathcal{P})$
11		$\xleftarrow{\mathcal{M}_{V'}}$	
12	$\mathsf{sid} \leftarrow \mathcal{M}_V \parallel \mathcal{M}_{V'}$		$\mathsf{sid} \leftarrow \mathcal{M}_V \parallel \mathcal{M}_{V'}$
13	$\mathcal{P}' \leftarrow \emptyset,\ \mathcal{T}' \leftarrow \emptyset$		$\mathcal{P}' \leftarrow \emptyset,\ \mathcal{T}' \leftarrow \emptyset$
14	FOR ALL $(U, N, e, t) \in \mathcal{T}$:		FOR ALL $(U, N, e, t) \in \mathcal{T}$:
15	IF partner $\notin U.\mathsf{crl}$:		IF partner$'$ $\notin U.\mathsf{crl}$:
16	$\theta \leftarrow \mathsf{iDecode}(\mathcal{M}_{V'}, N)$		$\theta \leftarrow \mathsf{iDecode}(\mathcal{M}_V, N)$
17	$\vartheta \leftarrow \theta \bmod N$		$\vartheta \leftarrow \theta \bmod N$
18	$r \leftarrow (\vartheta^e / H_N(\mathsf{partner}))^{2t} \bmod N$		$r \leftarrow (\vartheta^e / H_N(\mathsf{partner}'))^{2t} \bmod N$
19	$c_0 \leftarrow H(\mathsf{sid} \parallel r \parallel 0)$		$c_0 \leftarrow H(\mathsf{sid} \parallel r \parallel 0)$
20	$c_1 \leftarrow H(\mathsf{sid} \parallel r \parallel 1)$		$c_1 \leftarrow H(\mathsf{sid} \parallel r \parallel 1)$
21	$\mathcal{T}' \leftarrow \mathcal{T}' \cup \{(U, N, c_1)\}$		$\mathcal{T}' \leftarrow \mathcal{T}' \cup \{(U, N, c_0)\}$
22	ELSE: $c_0 \leftarrow_R [0, p - 1]$		ELSE: $c_1 \leftarrow_R [0, p - 1]$
23	$\mathcal{P}' \leftarrow \mathcal{P}' \cup \{(N, c_0)\}$		$\mathcal{P}' \leftarrow \mathcal{P}' \cup \{(N, c_1)\}$
24	$\mathcal{M}'_V \leftarrow \mathsf{iEncode}(\mathcal{P}')$		$\mathcal{M}'_{V'} \leftarrow \mathsf{iEncode}(\mathcal{P}')$
25		$\xrightarrow{\mathcal{M}'_V}$	
26	$\mathsf{SCL} \leftarrow \emptyset$	$\xleftarrow{\mathcal{M}'_{V'}}$	$\mathsf{SCL} \leftarrow \emptyset$
27	FOR ALL $(U, N, c_1) \in \mathcal{T}'$:		FOR ALL $(U, N, c_0) \in \mathcal{T}'$:
28	IF $c_1 = \mathsf{iDecode}(\mathcal{M}'_{V'}, N)$:		IF $c_0 = \mathsf{iDecode}(\mathcal{M}'_V, N)$:
29	$\mathsf{SCL} \leftarrow \mathsf{SCL} \cup \{U\}$		$\mathsf{SCL} \leftarrow \mathsf{SCL} \cup \{U\}$
30			
31	IF $\mathsf{SCL} \neq \emptyset$ THEN		IF $\mathsf{SCL} \neq \emptyset$ THEN
32	TERMINATE WITH "ACCEPT"		TERMINATE WITH "ACCEPT"
33	ELSE		ELSE
34	TERMINATE WITH "REJECT"		TERMINATE WITH "REJECT"

Figure 11.1: The Discover algorithm from [7].

Attacker Model. The attacker model considered in [7] is half-way between the semi-honest and malicious attackers as described in Section 11.3. More specifically, the attacker can deviate from prescribed behavior, but only to perform a limited set of actions. The adversary is modeled as a PPT machine interacting with the protocol participants and having access to queries like *Discovery*, *Revoke*, and others. The property that the authors propose to protect users from disclosing non-matching contacts to other participants is the *Contact-Hiding* security. The latter one is modeled as a game where the goal of the adversary is to decide which of two contact lists: CL_0^*, CL_1^* is used by some challenge session π^*.

The authors show in the paper that a CH-adversary cannot distinguish between the contact lists with non-negligible probability. Hence, the Contact Discovery protocol is *Contact-Hiding* secure under the RSA assumption on safe moduli, in a Random Oracle model.

11.4.2 The MightBeEvil Framework

MightBeEvil has been introduced by Chapman et al. in [11] as a framework to run secure-two party computation (STC) in a mobile environment. In a secure-two party computation, there are two parties involved in the protocol: Alice and Bob, each holding some private data x and y, respectively. The goal of STC is to jointly compute the outcome of a function $f(x,y)$, without disclosing one party's input to the other party: at the end of the execution, both Alice and Bob know the outcome of $f(x,y)$, but they do not know the value of the other party's input.

The first implementation of secure-two party computation was made by Yao in the 1980s [1], presenting a solution for the well-known "Millionaire Problem." In the Millionaire Problem, two users want to know which of them is richer, without revealing to the other party his/her own amount of money. In particular, the problem is one of computing whether condition $x < y$ holds, where Alice knows only "x," and Bob knows only "y." At the end of the protocol execution, Alice knows only the outcome of the evaluation of condition $x < y$, without knowing y (similarly for Bob).

The general idea of the MightBeEvil framework is to allow people to easily write functions that can be run in a secure way. In fact, while a user implements the function with a high-level code, the framework translates it into Boolean garbled circuits, which implement the secure two party computation. The main idea is to create a sort of "encrypted circuit" for the function f, and then to obliviously compute the output of the circuit without learning any intermediate value.

One of the improvements brought by MightBeEvil regards the memory required to store the total garbled circuits in memory. The framework adopts a strategy to process all the gates in a pipeline in order to avoid saving the entire circuit. During each evaluation, the circuit generator, i.e. the party that prepares an "encrypted" version of a circuit for a function f, and the circuit evaluator, i.e. the party that obliviously computes the output of the circuit without learning from the computation, instantiate the circuit structure. Subsequently, the garbled gates generated are sent to the other participant in the same order defined in the circuit structure. Once the client receives the gates, it associates them with the corresponding gate of the circuit. Finally, the evaluator determines which gate to evaluate next based on the available output values and tables. Once a gate has been evaluated, it is discarded in order to reduce the number of truth tables stored in the memory.

The MightBeEvil framework has been published online,[1] to allow people to download and create Android applications using secure-two party computation. To show the functionality and efficiency of the developed framework, Chapman et al. developed a friend discovery application. Similar to [7], the goal of the application is to discover common friends in the users' contact lists. Compared to the PCD primitive by De Cristofaro et al., the authors of MightBeEvil claim to have developed a more flexible system. In fact, the friend discovery application presented in [10] is just one application that can be implemented using the MightBeEvil framework, while the work of De Cristofaro et al. requires new security proofs that are hardly adaptable to other applications. However, MightBeEvil flexibility is paid in terms of

[1] The MightBeEvil Framework online: `http://www.mightbeevil.org/`.

Table 11.2: Total Running Time Time for Discovering Common Friends with MightBeEvil [10]

Set Size	32	64	128	256
Time (seconds)	68	124	285	598

computational performance, as evident from the results of the tests performed by the authors of [10] and reported below.

The friend discovery application based on the MightBeEvil framework has been designed for a scenario in which the two interacting users are registered with the same WiFi network. The unique identifier used in the contact list can be either the email address or the phone number, and is hashed to a 24-bit value. Finding common friends in the respective lists then amounts to computing the intersection of the sets of hash values generated from the contact lists.

Experimental analysis. The authors of [10] have implemented and tested the friend discovery application on Android Nexus One phones equipped with a 1-GHz Scorpion processor. The application running time for different sizes of contact lists are reported in Table 11.2. As seen from the table, running time increases considerably with the contact list size, ranging from about $1\,min$ with 32-number contact lists to about $10\,min$ with 256-number lists. Comparing MightBeEvil running times with those reported in [7], the price to pay for flexibility in terms of running time is notable: DeCristofaro et al. report a running time of about $5\,sec$ with 100-number contact lists, which should be compared with a reported running time of 4 to $5min$ with MightBeEvil. The comparison is even more penalizing for MightBeEvil if we consider that the mobile platform used in [10] is more powerful than that used in [7] ($1-GHz$ vs. $600-MHz$ processors).

Despite the relatively long running times reported in [10], MightBeEvil is very interesting especially considering that the framework can be successfully run on smartphones where the computational power is less performant compared to desktop or server CPUs. In fact, the authors claim that devices are roughly 1000 times slower than typical desktops. For instance, running on Android Nexus One phones, protocols execute at a speed of approximately 100 non-free gates per second compared to 10,000 gates per second for desktop.

Attacker model. The authors of MightBeEvil adopted for their work the semi-honest threat model. As reminder, a semi-honest attacker is assumed to follow the protocol but she may attempt to learn additional information about the other party's input. The authors say that it is possible to achieve security of their framework against *malicious* adversaries by adopting an oblivious-transfer protocol that is secure against malicious adversaries. However, this comes at the expense of increased protocol complexity.

Another aspect to highlight is that a secure-two party protocol does not prevent a participant from being dishonest during the protocol execution. In fact, these protocols provide privacy regarding the participants' input, but do not guarantee that a malicious user could construct a circuit that produces an incorrect result without decision, or, moreover, that she uses a fake input value.

11.4.3 The MobileFairPlay Framework

Similar to MightBeEvil, the MobileFairPlay framework proposed by Constantino et al. in [5] is an implementation of secure two-party computation on a mobile platform. More specifically, the authors of [5] present a mobile implementation of the FairPlay[16] framework for secure two-party computation. FairPlay is a very powerful STC framework that allows the protocol designer to write high-level procedures with a language called Secure Function Definition Language *(SFDL)*. SFDL programs are then compiled by FairPlay into optimized garbled Boolean circuits. The FairPlay framework was later extended to deal with multi-party computation in [4]. A nice feature of FairPlay is that secure functions written using the SFDL language are guaranteed to be secure against both semi-honest and malicious attackers (recall Section 11.3).

As described in [5], two main issues have to be addressed for porting FairPlay to a mobile environment such as Android. First, the Java object computed as the outcome of the FairPlay compilation phase has to be made compatible with the JavaVM used in the mobile phone, which is usually quite different from the standard one. In particular, Android phones use the DalvikVM. The first step in the porting process has then been translating a Java object as produced by FairPlay into a `.dex` file executable on the DalvikVM. Second, FairPlay uses TCP/IP for communication between parties, which is not suitable for setting up and operating a direct radio communication between two smartphones. In MobileFairPlay, the TCP/IP connection has been replaced with a Bluetooth connection, as customary in opportunistic networking.

11.4.3.1 Application to Interest-Cast

To show the operation and efficiency of MobileFairPlay, the authors of [5] present a privacy-preserving implementation of interest-casting. More specifically, the implemented protocol performs the following steps: (1) find people in a user's neighborhood through a Bluetooth scan operation, (2) connect to another user and discover whether the two users have similar interest profiles without disclosing sensitive information, and 3) share messages between users' devices only if their profiles are similar.

We now briefly describe the interest-cast model used in [5], which is borrowed from [17]. According to Mei et al. [17], each user in an opportunistic network is characterized by an *interest profile*, which is used to drive the information dissemination process within the network. Notice that similar forwarding protocols based on a notion of a user's "social profile" have been recently introduced in the literature [9, 22].

User interests are modeled as an m-dimensional vector in a common m-dimensional *interest space*, where the number m of interests is typically much smaller than the number n of nodes in the network. More formally, the *interest profile* of user

A is defined as:

$$I_A = (a_1, \ldots, a_m) ,$$

where $a_i \in [1, max]$ is an integer representing A's interest in the i-th topic of the interest space. Note that interests are expressed as integers in the range $[1, max]$, with 1 representing no interest and *max* (an arbitrary integer > 1) representing maximum interest.

Let S be a user denoted as the message *source*. According to the definition of interest-casting, the message M generated by S (which can be thought of as a piece of information node S wants to share within the network) should be delivered to all nodes in the set $\mathcal{D}(S, \gamma)$, where

$$\mathcal{D}(S, \gamma) = \{U \in \mathcal{N} | sim(U, S) \geq \gamma\} , \tag{11.1}$$

where $sim(U, S)$ is a similarity metric used to express similarity between a node U and S's interest profiles, with relatively higher similarity values representing relatively more similar interests, and γ is the *relevance threshold* (set by S). Set $\mathcal{D}(S, \gamma)$ is called the set of *relevant destinations*, and in principle it is not known in advance to node S. Instead, set $\mathcal{D}(S, \gamma)$ is implicitly defined by S's interest profile, and by the relevance threshold γ.

In the interest-cast implementation presented in [5], the authors use a similarity metric called the vector-component-wise (*vcw*) metric, which is different from the cosine similarity metric used in [17]. Usage of a different similarity metric is motivated by the fact that the *vcw* computation can be easily translated into an SFDL program, which is not the case for cosine similarity.

Formally, the *vcw* similarity metric is defined as follows. Let $S = (s_1, \ldots, s_m)$ and $U = (u_1, \ldots, u_m)$ be the interest profiles of users S and U, respectively. We have:

$$vcw(U, S, \lambda) = \begin{cases} 1 & \text{if } \forall i \in \{1, \ldots, m\},\ |u_i - s_i| \leq \lambda \\ 0 & \text{otherwise} \end{cases} ,$$

where $\lambda \in [0, max]$ is an integer parameter used to narrow/widen the scope of the interest-cast. More specifically, by setting $\gamma = 1$ in definition (11.1), we have that $\mathcal{D}(S, 1)$ corresponds to the set of all nodes in the network if $\lambda = max$, while $\mathcal{D}(S, 1) = \{S\}$ if $\lambda = 0$.

In the interest-cast application presented in [5], a user is able to:

1. set up her own profile regarding her degree of interest in different topics, expressed as an integer value in the $[1, 100]$ interval;

2. start a new connection with another user and check whether they have similar interests without disclosing its own profile;

3. wait for incoming connections;

4. exchange files with another user in case of successful interest profile matching; the application supports exchange of different file types including pdf, txt, jpg, etc.

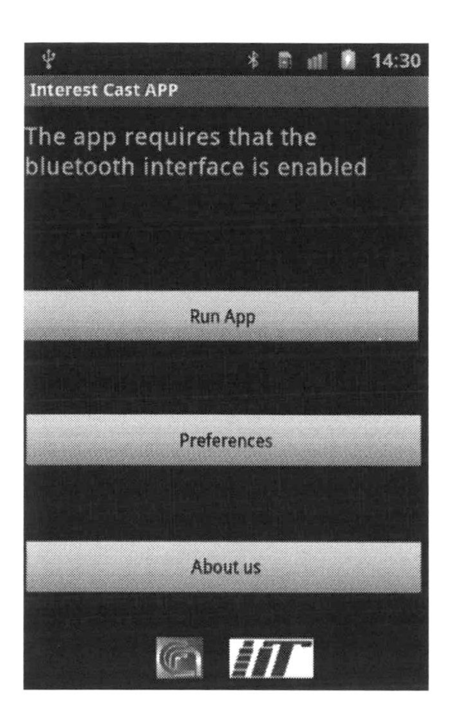

Figure 11.2: App main window.

Figure 11.2 displays the main window of the interest-cast application application.

User profile. The application has a window in which a user can set up her own profile; see Figure 11.3. The possible values that can be inserted are between 1 and 100, where 1 means no interest, and 100 the maximum possible interest for a topic. Examples of topics in the profile are: *cars, books, movies, sports, television, games,* and others. Finally, the user sets the value of λ required for computing the *vcw* similarity metric.

Privacy-preserving hand-shaking. Figure 11.4 represents the main phases of the application. Two are the participants involved in the protocol: Alice and Bob. The scenario considers that Alice is walking and meets another user (Bob). She tries a connection toward Bob. Once the Bluetooth connection has been established, they can start running the interest-cast match.

Secure computation of the forwarding condition. Once Bob gets an incoming connection from Alice, he randomly selects a number of topics from his profile to be used to compute the *vcw* metric. The number of tested topics is set to four in [5]. The tested topics are sent in cleartext to Alice. Once Alice receives the topics, the Interest Cast function implemented with secure two-party computation starts. The function represents the kernel of the application, and the corresponding *SFDL* code is shown in Listing 11.1.

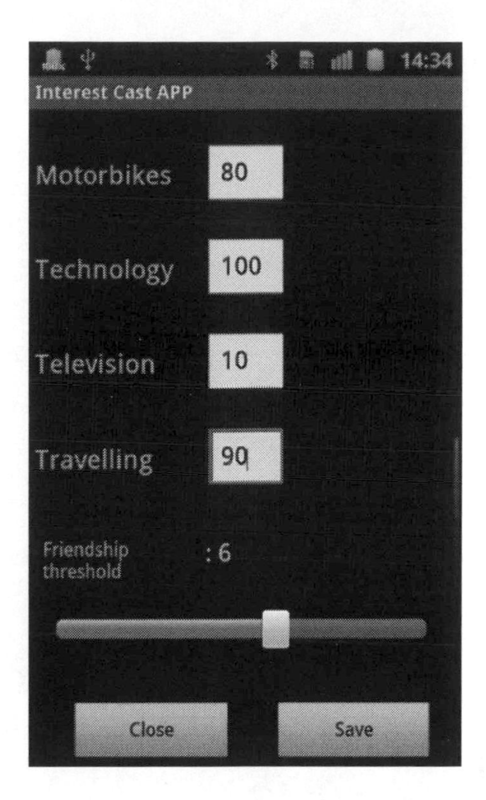

Figure 11.3: User's profile window.

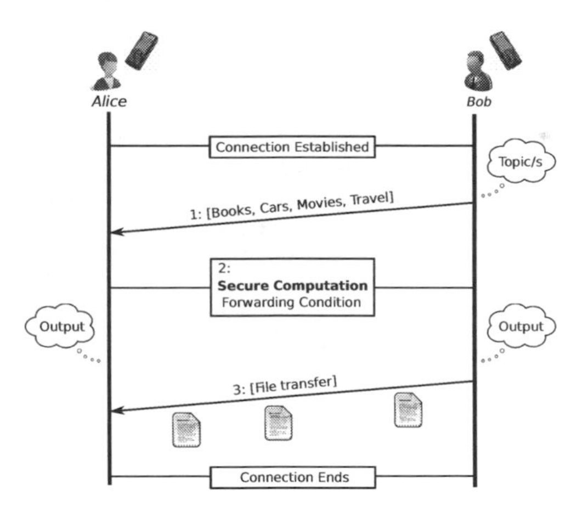

Figure 11.4: Protocol steps.

```
1  program InterestCast {
2  // Number of topics
3  const int = 4;
4  // 4-bit integer as type
5  type Key = Int<4>;
6  // 8-bit integer as type
7  type int = Int<8>;
8  // Alice's array
9  type AliceInput = int[4];
10 // Bob's array, one position is for the threshold
11 type BobInput = int[5];
12 // Alice's output-array. One position for each topic comparison
13 type AliceOutput = Boolean[4];
14 // Bob's output-array. One position for each topic comparison
15 type BobOutput = Boolean[4];
16
17 type Output = struct {AliceOutput alice, BobOutput bob};
18 type Input = struct {AliceInput alice, BobInput bob};
19
20 //The function that implements the secure forwarding condition
21 Function Output output(Input input)
22 {
23 var int tmp;
24 var int threshold;
25 var Key i;
26
27 // The threshold is initialised
28 threshold = input.bob[4];
29
30 // For all topics
31 for (i = 0 to int-1)
32 {
33 tmp = (input.bob[i] - input.alice[i]);
34
35 //If tmp is less than 0, we negate tmp in order to make it positive
36 if (tmp < 0)
37 {
38      tmp = ~tmp;
39 }
40
41 // if the difference is less than the threshold,
42 // the forwarding condition for the i-th topic is verified
43 if (tmp <= threshold)
44 {
45      output.alice[i] = 1;
46      output.bob[i] = 1;
47 }
48 else
49 {
50      output.alice[i] = 0;
51      output.bob[i] = 0;
52 }
53    }}}
```

Listing 11.1: Secure forwarding condition written in SFDL code.

The code in SFDL is quite simple to understand. Lines 2 to 18 list all variables and data structures needed in the function. The most important function must be named as *output* to be correctly compiled, and it contains the topic matching of Alice and Bob. The first input of Bob represents the threshold used to compare the difference of degree of interest in each topic (line 28). In the next **for** structure, all topics are processed sequentially. Then, the difference between Bob's and Alice's interests is done (line 33). If the difference is less than zero, then the opposite is taken (line 38). Instead, if it is positive it is compared with the threshold (line 43). Finally, if the difference is less than the threshold the output is positive, otherwise negative.

If all tested topics are successfully matched, the common interest for those topics has been discovered and both participants can start the last phase of the protocol.

File transfer. Once the hand-shaking phase has established that Alice and Bob have similar interests, Bob sends his files to Alice. The application allows exchanging files of any kind and any extension; in fact, raw bytes are exchanged during the file exchange phase, allowing transfer of files of arbitrary format like text files (`.txt`), pdf files, image files (`.jpg`), etc. For instance, a file can represent a movie advertisement regarding all cinemas in a city, an advertisement of a rock music concert, or any other kind of information.

11.4.3.2 Interest-Cast Execution Time

Compiling phase. Each time that the application is run, it requires some seconds to compile to SFDL code into Boolean garbled circuits. The speed of this process depends on the hardware. Smartphones are slower to compile compared to traditional PCs. Table 11.3 shows the time needed to compile the function described in Listing 11.1 with one tested topic, for some current Android smartphones. Table 11.4 reports compilation time with four tested topics. As seen from the tables, compilation times are reasonable, even for the more complex case of four-topic testing.

Running time. The most important performance parameter is represented by the time required to run the Interest-Cast function after compilation.

Figure 11.5 shows the time needed to run the secure function of the *vcw* metric in the case of one topic comparison. This time includes the time needed to exchange the garbled Boolean circuits through the Bluetooth interface, and to compute the output. Results are obtained considering the case in which the role of Alice is kept

Table 11.3 Compilation Time Required for a One-Topic Function

Smartphone	Time (ms)
Samsung Galaxy S2	382
Samsung Galaxy S-Plus	446
Samsung Galaxy S	492
Lg Optimus Dual	449
Htc Desire	489

Table 11.4 Compilation Time Required for Four-Topics Function

Smartphone	Time (ms)
Samsung Galaxy S2	4471
Samsung Galaxy S-Plus	5347
Samsung Galaxy S	6605
Lg Optimus Dual	5352
Htc Desire	6512

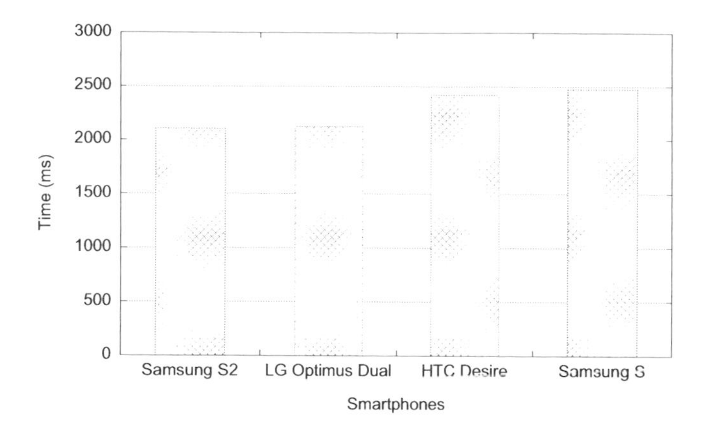

Figure 11.5: One topic: Bob's running time of the secure function.

fixed. In fact, while Bob is run four times with four different smartphones, Alice is instead always run on a Samsung Galaxy S-Plus. Figure 11.6 reports the running times obtained with four topics used to compute the *vcw* metric. Even in this case, the behavior is similar to the previous case. As expected, the best result is obtained with the most powerful device, namely the Samsung S2 phone, which has a dual-core 1.2-GHz processor. It is interesting to observe that, as opposed to the compilation time, which increases super-linearly with the number of tested topics, the duration of the secure function computation (hand-shaking) is *sub-linear* with the number of topics used to compute the *vcw* metric. For instance, considering the case of Alice running on Samsung Galaxy S-Plus and Bob running on Samsung Galaxy S2, the running time is $2.1\,sec$ with one topic, and only $3.7\,sec$ with four topics. Observe also that running times of the hand-shaking phase were acceptable ($< 5sec$) independent of the specific mobile phone used to run the secure function.

Attacker model. MobileFairPlay inherits the threat model from the FairPlay framework, since the "secure" part of the framework uses the SFDL language. In [16], it is shown that FairPlay (and, hence, MobileFairPlay) has strong security properties in the context of two-party computation. The framework is shown to be secure against a malicious attacker. In particular: *i*) a malicious attacker cannot learn

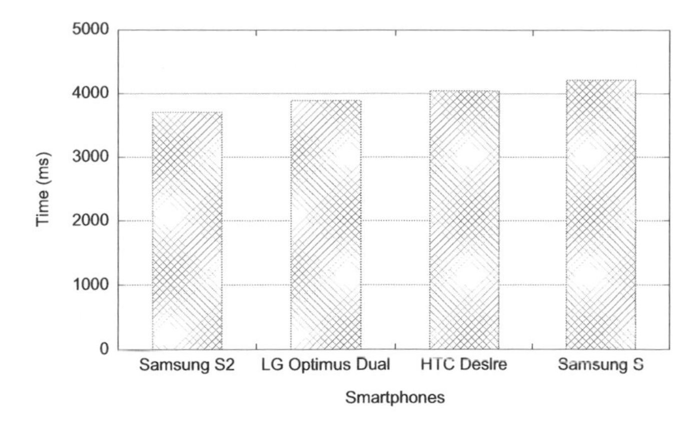

Figure 11.6: Four topics: Bob's running time of the secure function.

more information about the other party's input than it can learn from a TTP that computes the function; and *ii*) a malicious attacker cannot change the output of the computed function. Notice that, as customary in secure two-party computation, there is an asymmetry on the provided security guarantees: in particular, there is no way to prevent Alice from terminating the protocol prematurely, and not sending the outcome of the computation to Bob. This situation can be detected by Bob, but cannot be reversed.

11.4.4 Other Cryptography-Based Protocols

Other cryptography-based opportunistic networking protocols have been proposed in the literature, and although unlike the ones presented in this section, they have not been implemented in a mobile platform.

In [3], the authors present a privacy-preserving protocol for friend recommendation in mobile social networks. The goal is to discover whether the contact list of two users A and B are sufficiently similar, in which case user B is recommended as a possible friend to user A. Contact list similarity is verified by means of Jaccard's index computation, which is performed in a privacy-preserving way by means of suitably defined hash functions, called *sketches*. Sketches are exchanged between users and used to compute the Jaccard's index instead of the original contact lists.

In [15], the authors present a privacy-preserving protocol for friend discovery. The considered scenario is a bit different from a typical opportunistic networking scenario: there are n users, each holding a portable device and in direct contact with each other. An initiator user launches the friend discovery process, with the goal of identifying the user with a profile more similar to the user's own profile amongst the $n-1$ candidate friends. User profiles are defined in terms of a set of attributes A_i, which a user can or cannot have. Profile matching is performed based on number of common attributes between users. Thus, the initiator's goal is to identify the user with the largest number of common attributes amongst the $n-1$ candidate friends. To accomplish this task in a privacy-preserving manner, the authors define a scheme based on the Shamir's Secret Sharing Scheme [21] and on additive homomorphic encryption.

In [23], the authors present a set of privacy-preserving protocols for friend discovery. User profiles are defined as an interest vector with integer values, similar to that proposed in [17]. Different metrics are defined to determine whether two profiles are similar, including the *vcw* metric used in [5]. The presented set of privacy-preserving protocols (one for each metric) is based on Paillier cryptosystems [19].

11.5 Cryptography-Free Protocols

As described in the previous section, the use of cryptographic primitives when executing communication protocols poses significant computational challenges to today's mobile devices. While the computational power of mobile devices is expected to increase constantly in years to come, thus lessening computational issues related to the execution of cryptographic protocols, it is still true that cryptography-free communication protocols are much more efficient and apt for mobile devices than cryptography-based protocols. In this section, we present two approaches that have been recently proposed in the literature to provide some form of privacy preservation without using cryptographic tools.

11.5.1 Probabilistic Profile-Based Routing

Probabilistic Profile-Based Routing (PPBR) has been proposed in [2] as a protocol for privacy-preserving geo-routing in opportunistic networks. The goal of a geo-routing protocol is to deliver a message to all users located in a specific place, such as a building, a shopping center, a square, etc. Thus, the "address" of a message is not a specific user ID as in unicast communication, but the identifier of a location within a digital map.

The PPBR design goal is to geo-route messages while disclosing minimal information about users and their mobility patterns. In particular, the protocol should not leak the sender's identity, nor should it reveal information about previous location of users involved in the protocol.

PPBR uses user mobility profiles to estimate individuals who are good candidates to deliver a message toward the targeted location. To build the mobility profile, the authors of [2] suggest dividing the area of interest into square cells, and to periodically save a user's GPS coordinates, i.e. longitude and latitude. The mobile device periodically records its GPS location, maps it to a square cell, and determines the fraction of time spent within each square cell. More specifically, the authors introduce the following formula to establish a user profile according to her past locations:

$$p[\langle x,y\rangle] = \begin{cases} \frac{|W_{\langle x,y\rangle}|}{|W|} & \text{if } \langle x,y\rangle \in W \\ 0 & \text{otherwise} \end{cases}, \tag{11.2}$$

where W is a list of location entries already mapped to a square cell, $W_{\langle x,y \rangle}$ is the sub-list of W containing location entries occurring within the cell $\langle x,y \rangle$, $p[.]$ is the function returning the visiting frequency for cell $\langle x,y \rangle$, and $|.|$ indicates the length of the list.

In order to improve privacy preservation (see below for details), PPBR is based on the principle that when a device A carrying a message encounters another device B, it delivers the message to B without a preliminary "friend discovery" phase typical of many opportunistic networking protocols. It is device B that, upon receiving the message from A, performs a local computation based on its own mobility profile and the information contained in the message, with the purpose of estimating its "suitability" of acting as message carrier. In case the estimated "suitability" is below a pre-defined threshold, the message is removed from the buffer and dropped. Otherwise, it remains stored in device B's buffer.

Summarizing, in PPBR a device (node) that receives a message has two possible choices: 1) to become the next carrier of the packet in case it is a good candidate, 2) to discard the packet otherwise. However, to avoid flooding the network with multiple copies of the same packet, the authors adopt the strategy of allowing a node to forward only k copies of the message, where k is a protocol parameter. Furthermore, the threshold used to determine whether a receiver node should carry or drop the message is set in such a way that, on the average, only 1 out the k nodes receiving a copy of the message from node A will hold the message and continue forwarding.

The above described phase in which a node forwards k copies of the message to candidate carriers is called the *passing phase*. After the passing phase is over, the node simply keeps a copy of the message in the buffer until it either reaches the target location, or a *TimeToLive* message timer expires. This phase is called the *holding phase* of the protocol. Notice that, in order to improve privacy preservation, a node becoming carrier is allowed to start the passing phase only when it has moved at distance at least d away from the current position, where d is another protocol parameter. This heuristic prevents an eavesdropper from observing a successful interaction between a message carrier and a new forwarder.

11.5.1.1 Forwarding Packet Condition

Each node calculates its own location profile with the expression (11.2). Then, a node estimates whether it is the best candidate amongst the k message recipients by computing a similarity metric between its own mobility profile and the message's destination. The similarity metric is calculated considering the square cell identified as the message destination, and neighboring cells. So, even if a node was not in the target cell in the past, it may become the next carrier if it has frequently visited nearby cells.

Formally, the similarity metric between a node n with mobility profile p and a message m addressed to a_m is computed as follows:

$$sim(p, a_m) = p[a_m] + \sum_{\substack{a_p \in p \\ a_p \neq a_m}} \frac{p[a_p]}{dist(a_p, a_m)^2}, \tag{11.3}$$

where $dist(a_p, a_m)$ is the Euclidean distance between grid-points a_p and a_m.

The threshold used to decide whether a node should forward or drop the packet is determined based on a notion of a *general node profile*, which is intended to model the mobility pattern of an "average user." This "average user" profile can be obtained by, e.g., analyzing large real-world data traces of people moving in the area of interest. Denoting by p_n and p_g the mobility profile of node n and of the average user, respectively, the formula below is used to normalize a user's similarity with respect to the "average user" profile:

$$\sigma = \frac{sim(p_n, a_m)}{sim(p_g, a_m)}. \tag{11.4}$$

In the formula (11.5), σ represents the *marginal similarity* of user n. Higher values of σ indicate the node would be a good message carrier, while lower values indicate a poor message carrier.

The last step is determining the threshold τ used to determine whether a message should be forwarded or dropped. The threshold is calculated locally and in a private way by each node. First, a node computes σ for every grid square in p_g.

$$\bar{\sigma} = \left\langle \frac{sim(p_n, a)}{sim(p_g, a)} \middle| \forall a \in p_g \right\rangle \tag{11.5}$$

Then, the different values of $\bar{\sigma}$ are sorted in order to have $\bar{\sigma}_i < \bar{\sigma}_j$ if $i < j$. To allow a node to accept a message for about $1/k$ of destination addresses (cells), a node chooses τ such that $1/k$ values in $\bar{\sigma}$ are greater than τ; more precisely, $\tau = \bar{\sigma}_i$ and $i = \lfloor |\bar{\sigma}| * (k-1)/k \rfloor$, where $|.|$ denotes the length function.

11.5.1.2 Performance Evaluation

PPBR performance has been evaluated by means of simulation, considering three main aspects: 1) message delivery rate, 2) message delivery latency and 3) network load. Square cell dimension adopted in the simulations is of $200m \times 200m$, and the value of k is fixed at 10. Two data sets are used for the simulations: the first is the *Cabspotting Dataset* [20] containing GPS coordinates and timestamps of 536 taxicabs in the San Francisco area, where each cab represented a user with her own smartphone. The second data set is obtained running the *Self-Similar Least Action Walk* [13] mobility model, which is known to produce human-like trips.

In the simulations, PPBR has been compared with two probabilistic protocols: 1) *probabilistic random walk* and 2) *probabilistic flooding*. In the former protocol, a user accepts a carrier's message with a probability of $1/k$, i.e, 10%, independent of her mobility profile. In the latter protocol, whenever a carrier meets a new node, the message is duplicated with probability 0.1.

Regarding the *delivery rate* and *latency*, results show that the flooding protocol has the best latency and delivery rates, even if it creates a high network load. PPBR, instead, outperforms random walk both for median latency and delivery rate.

The *network load* is calculated as the average number of message duplicates in the system across all simulations. By design, PPBR does not guarantee that only a single message copy is present in the network. In fact, when a carrier meets other k nodes, she first sends the message to all k users, even if only one will get the message. However, the authors says that potentially PPBR could have zero duplicates, or more than one duplicate. Results show that probabilistic flooding has the highest network load, random walking probabilistic the lowest network load, and PPBR has an intermediate load.

Attacker model. PPBR has been designed to counter both passive and active adversaries. A passive adversary is able to observe message exchanges at a fixed number of locations during the protocol routing phase, but she cannot take part in protocol operation. On the contrary, an active adversary can also be a participant in the exchange phase, or even generate fake messages and/or drop them, and can then be considered a malicious attacker. However, adversaries who are able to physically follow a carrier and track her are currently not considered in PPBR, since this attack is deemed too costly in an opportunistic network scenario.

During the message exchange phase a malicious user is not able to understand whether the receiver is really getting the message. In fact, since the acceptance of a message is not acknowledged, an observer is not able to understand the outcome of the passing step, unless the attacker follows the new carrier for at least distance d to see if the node forwards the message or delivers it to the destination cell.

PPBR also guarantees the *de-anonymization* of the packet sender. Messages do not have trace of the sender node, and even if an attacker can observe the message exchange phase, she is not able to identify the originator of the message. However, this situation does not cover the case in which an attacker is able to observe the first step of the message delivery. In this case, an adversary can easily trace the originator.

11.5.2 Hide-and-Lie

The authors of Hide-and-Lie [8] propose a solution for increasing the users' privacy in an interest-casting application. Similar to [17] and [5], each user has an interest profile *IP* that defines the degree of interest for different topics with a probability ε, with $0 < \varepsilon \leq 1$. A message belonging to a category (topic) k is called primary for a node if $IP[k] = 1$, i.e., if a node has an interest in the message topic; otherwise, the message is called secondary. Messages are generated by special nodes, called message generator nodes, and are created with a fixed average rate of q messages per time step.

The structure of a message is the following:

$$M = [ID|CAT|data] , \tag{11.6}$$

where the *ID* is the unique identifier of *M*, *CAT* is the identifier of the message category, and *data* is the message content. The number of messages that a node can store is limited by the node buffer space. In particular, it is assumed that nodes prefer to download messages that are not already stored in their memory. This operation helps nodes to avoid wasting space, but adds a computational cost due to the need to filter messages.

To evaluate performance of their protocol, the authors introduce the notion of gain $(G_u(t))$ at a particular instant t, which is defined as follows. The gain of a node u at time t is defined as:

$$G_u(t) = \frac{O_u(t)}{A_u(t)} , \tag{11.7}$$

where $O_u(t)$ is the number of primary messages received by node u by time t, and $A_u(t)$ is the total number of messages generated by time t that are primary for u.

Attacker model. The attacker goal is tracking the target node in order to break her privacy. In particular, the attacker is assumed to learn the interest profile of a target user at instant t_0. At a later time t_1, the attacker observes the interest profiles of all nodes in the network, with the goal of identifying the profile corresponding to the target user. To this purpose, the attacker uses a similarity metric to compare the interest profiles observed at time t_1 with the profile of the target node observed at time t_0. The attack is successful if the attacker selects the interest profile of the target node amongst the profiles observed at time t_1, thus tracking the target user.

The attacker is assumed to estimate user profile $UP(u)$ of a node u as follows:

$$UP_u(t) = (EIP_u(t), CHM_u(t), IDL_u(t)) , \tag{11.8}$$

where the estimated interest profile *(EIP)* is a binary vector of k positions, with the ith position equal to 1, meaning that u has an interest in the i-th topic. The category histogram messages *(CHM)* shows, for each category, how many messages in the ID list belong to that category. Finally, *(IDL)* is the ID list of offered messages.

Attacker behavior. The attacker modeled by the authors performs her attacks in the following way:

1. The attacker identifies the target node (u_t) as the node to attack.

2. At the instant t_0, the attacker reads the interest profile of the target node $(UP_{u_T}(t_0))$.

3. After τ instants, i.e., at time $t_1 = t_0 + \tau$, the attacker reads the $UP_{u_i}(t_1)$ with $i \in [1..N]$ of each node and calculates a similarity metric between u_i and u_t.

4. Lastly, the attacker chooses the node most similar to the target node. If more than one node has the maximal similarity value, she chooses randomly among them. If the attacker selects node u_t, the attack is successful.

Attacker functions. The attacker can perform the attack in the following ways:

- **Prefiltered ID-based attack**: In this attack, it is assumed that nodes show their real interest profiles which, however, can change with time due to arrival/deletion of messages. In this attack, nodes show their own *EIP* at the instant t_1, then the attacker considers only those nodes whose EIP at time t_1 equals the target node's EIP at time t_0. Amongst these nodes, the attacker selects the one whose ID list has more elements in common with the ID list of the target node at time t_0.

- **Unfiltered ID-based attacker**: This function is similar to the previous one, but this time the attacker does not filter the *EIPs*: she selects the node whose ID list at time t_1 has the largest intersection with the ID list of the target node at time t_0.
- **Category histogram-based attacker**: In this attack, the attacker selects the node u whose $CHM_u(t_1)$ is the most similar to the $CHM_{u_t}(t_0)$. The similarity function is calculated using the χ^2-test.
- **Significant category-based attacker**: This function assumes that the interested categories are overrepresented in the ID-list, while the uninterested categories are underrepresented. To find the interested categories, the C categories must be classified into two clusters: the significant categories and the remaining ones. The clustering is done using the *k-means* algorithm on the CHMs. The result of the clustering algorithm is a binary vector of length C, where ones correspond to significant categories. Similarity of two binary vectors is defined as the Hamming distance of the vectors.

Defense mechanism. Observe that, in general, the more $UP_u(t_1)$ is different from $UP_u(t_0)$, the less likely it is that the attacker can link both profiles to the same user. Thus, a possible strategy to defend against an attacker is to purposely change the user's profile (*obfuscation*), using, e.g., randomization techniques. On the other hand, profile obfuscation might have detrimental effects on the interest-casting application performance.

The authors propose the obfuscation strategy called *Hide-and-Lie*. The term *Hide* refers to the fact that users can hide interesting categories, showing them as uninteresting. The term *Lie* refers to the fact that users can falsely claim uninteresting categories as interesting ones.

To run the *Hide-and-Lie* strategy, each node generates its obfuscated *EIP* from its real interest profile by inverting each category with a given probability λ. Inverting means that an uninteresting category can become interesting and vice versa. The λ parameter is used in *Hide-and-Lie* to tune the level of obfuscation in interest profiles. In particular, λ is a parameter ranging between 0 and 1. By setting $\lambda = 0$, nodes do not change their interest profiles (no obfuscation), while setting $\lambda = 1$ they completely invert their own profiles. Finally, an interesting is when $\lambda = 0.5$, which brings a totally randomized interest profile. However, to avoid having an interest profile completely different from the real one, which would impair interest-casting performance, the authors propose to set λ in the range between 0 and 0.5.

11.5.2.1 Simulations and Results

Results presented by the authors are obtained with a simulator written in C++, and two types of mobility models: the random walk *(RW)* and the restricted random walk *(RRW)* models. The setup used by the authors consider a fixed space where users can move in. When nodes arrive in specified meeting points, they run the message exchange phase, hence they are able to download messages from others. After a while, nodes move toward other meeting points.

The number of meeting points and mobility area depend on the mobility model adopted. In the RW model, the field size is a 15×15 unit with 30 meeting points; while in the RRW, the meeting points are 225 with a field area of 20×20 unit. The number of categories is fixed to 30, and the attacker delay τ is set to a value in $\{1, 50, 250, 500, 100\}$.

In the simulations, the authors analyze how the different attacker functions behave by changing the λ and τ parameters within the *Hide-and-Lie* strategy. The *prefiltered ID-based* attacker function assumes that nodes do not apply any privacy-enhancing technique. So, when λ is equal to zero, the attacker can easily link the target node interest profile with those recently collected. Instead, with $\lambda \neq 0$, the attacker can be deceived and as a consequence she will not distinguish the target node from the others.

As expected, the *unfiltered ID-based* attacker obtains a lower success probability with $\lambda = 0$ compared to the *prefiltered ID-based* function. Similarly to the *prefiltered ID-based* attack and to all the other attacks, the success probability decreases considerably with higher values of λ. This is because the number of the combinations of messages give enough variety to the attacker to identify nodes with higher probability when λ is low. However, when λ increases, the users collect messages from larger sets and they can hide more messages. Finally, by increasing the τ value, the probability of a successful attack decreases since nodes delete more messages, and for the attacker it becomes more and more difficult to match the new ID lists with the one of the target node.

The *category histogram-based* attacker function results are less sensitive to the value of τ. For low values of τ, it turns out to be less efficient than the ID-based attacker functions. This is because the *Hide-and-Lie* strategy causes substantial differences in the *EIP* values when τ is low, which are intolerable for the χ^2-test.

The *significant category-based* attacker function results or less sensitive to the value or τ. The advantage of this function comes from the fact the it tries to reveal the real node interest profile. However, by keeping a higher value of λ, for instance, close to 0.5, the success probability of this attack can be substantially reduced.

Generally, the authors claim that a good strategy of the *Hide-and-Lie* mechanism is to apply a high value of λ (close to 0.5). In fact, by following this approach, no attacker is able to better distinguish a node, independently from the τ value, than a naive attacker does by randomly choosing a node.

11.6 Conclusions

In this chapter, we discussed the need for preserving privacy in opportunistic networking, and surveyed existing approaches to privacy-preserving message exchange within opportunistic networks. As extensively discussed in this chapter, some cryptography-based solutions have been recently shown to perform efficiently on mobile devices, although the experienced running time depends heavily both on the data to be concealed (interest profile, contact list, etc.), as well as on specific protocol parameters. Cryptography-free solutions, while lightweight and easily executable in mobile environments, provide weaker privacy guarantees as compared to their cryptography-based counterparts. Given that privacy-preserving opportunistic networking has been addressed in the research literature only very recently, and the strong motivations for privacy support in opportunistic applications, we expect to see a flourishing of this young research field in the next few years, with more and stronger privacy-preserving solutions introduced by researchers.

References

[1] C. Andrew and C. Yao. Protocols for secure computations. In *23rd IEEE Symposium on FOCS*, pages 160 –164, 1982.

[2] A. J. Aviv, M. Sherr, M. Blaze, and J. M. Smith. Privacy-aware message exchanges for geographically routed human movement networks. In Sara Foresti, Moti Yung, and Fabio Martinelli, editors, *Computer Security—ESORICS 2012*, volume 7459 of Lecture Notes in Computer Science, pages 181–198. Springer Berlin Heidelberg, 2012.

[3] E. Baglioni, L. Becchetti, L. Bergamini, U.M. Colesanti, L. Filipponi, A. Vitaletti, and G. Persiano. A lightweight privacy preserving SMS-based recommendation system for mobile users. In *ACM RecSys*, 2010.

[4] A. Ben-David, N. Nisan, and B. Pinkas. FairplayMP: a system for secure multiparty computation. In *Proceedings of the 15th ACM Conference on Computer and Communications Security*, CCS '08, pages 257–266, New York, NY, USA, 2008. ACM.

[5] G. Costantino, F. Martinelli, P. Santi, and D. Amoruso. An implementation of secure two-party computation for smartphones with application to privacy-preserving interest-cast. In *Proceedings of the 18th Annual International Conference on Mobile Computing and Networking*, Mobicom '12, pages 447–450, New York, NY, USA, 2012. ACM.

[6] E. Daly and M. Haahr. Social network analysis for routing in disconnected delay-tolerant manets. In *ACM MobiHoc*, 2007.

[7] E. De Cristofaro, M. Manulis, and B. Poettering. Private discovery of common social contacts. In *Proceedings of the 9th International Conference on Applied Cryptography and Network Security*, ACNS'11, pages 147–165, Berlin, Heidelberg, 2011. Springer-Verlag.

[8] L. Dóra and T. Holczer. Hide-and-lie: enhancing application-level privacy in opportunistic networks. In *Proceedings of the Second International Workshop on Mobile Opportunistic Networking*, MobiOpp '10, pages 135–142, New York, NY, USA, 2010. ACM.

[9] D. Eppstein, M.T. Goodrich, M. Loffler, D. Strash, and L. Trott. Category-based routing in social networks: Membership dimension and the small-world phenomenon. In *IEEE Conf. on Computational Aspects in Social Networks (CASoN)*, 2011.

[10] Y. Huang, P. Chapman, and D. Evans. Privacy-preserving applications on smartphones. In *Proceedings of the 6th USENIX Conference on Hot Topics in Security*, HotSec'11, Berkeley, CA, USA, 2011. USENIX Association.

[11] Y. Huang, D. Evans, J. Katz, and L. Malka. Faster secure two-party computation using garbled circuits. In *Proceedings of the 20th USENIX Conference on Security*, SEC'11, Berkeley, CA, USA, 2011. USENIX Association.

[12] P. Hui, J. Crowcroft, and E. Yoneki. Bubble rap: Social-based forwarding in delay tolerant networks. In *ACM MobiHoc*, 2008.

[13] K. Lee, S. Hong, S.J. Kim, I. Rhee, and S. Chong. Slaw: A new mobility model for human walks. In *INFOCOM 2009, IEEE*, pages 855–863, 2009.

[14] J. Leguay, T. Friedman, and V. Conan. Evaluating mobility pattern space routing for DTNs. In *IEEE Infocom*, 2006.

[15] M. Li, N. Cao, S. Yu, and W. Lou. Findu: Privacy-preserving personal profile matching in mobile social networks. In *IEEE Infocom*, 2011.

[16] D. Malkhi, N. Nisan, B. Pinkas, and Y. Sella. Fairplay: A secure two-party computation system. In *Proceedings of the 13th conference on USENIX Security Symposium—Volume 13*, SSYM'04, Berkeley, CA, USA, 2004. USENIX Association.

[17] A. Mei, G. Morabito, P. Santi, and J. Stefa. Social-aware stateless forwarding in pocket switched networks. In *IEEE Infocom*, 2011.

[18] E. Okamoto and K. Tanaka. Key distribution system based on identification information. *IEEE J.Sel. A. Commun.*, 7(4):481–485, May 1989.

[19] P. Paillier. Public-key cryptosystems based on composite degree residuosity classes. In *Proc. EuroCrypt*, 1999.

[20] M. Piorkowski, N. Sarafijanovic-Djukic, and M. Grossglauser. A parsimonious model of mobile partitioned networks with clustering. In *Communication Systems and Networks and Workshops, 2009. COMSNETS 2009*, pages 1–10, 2009.

[21] A. Shamir. How to share a secret. *Communications of the ACM*, 22(11):612–613, 1979.

[22] J. Wu and Y. Wang. Social-feature based multi-path routing in delay tolerant networks. In *IEEE Infocom*, 2012.

[23] R. Zhang, Y. Zhang, J. Sun, and G. Yan. Fine-grained private matching for proximity-based mobile social networking. In *IEEE Infocom*, 2012.

Chapter 12

Incentivizing Participatory Sensing via Auction Mechanisms

Buster O. Holzbauer
Rensselaer Polytechnic Institute
Troy, New York

Boleslaw K. Szymanski
Rensselaer Polytechnic Institute
Troy, New York

Eyuphan Bulut
Cisco Systems
Richardson, Texas

CONTENTS

With the increasing availability of smartphones and other mobile devices with sensing capabilities, a class of problems known as participatory sensing is becoming increasingly popular. Participatory sensing refers to a sensing system in which humans voluntarily participate in the system, actively contributing to sensing, either by passively carrying devices or actively engaging in sensing. With respect to opportunistic networks, humans make mobility and decision making difficult to implement in dedicated sensor networks. Participatory sensing is also a more scalable solution since the costs are far lower than the costs associated with deploying a dedicated larger or computationally intensive sensor network. However, by involving humans, many issues must be addressed in designing a participatory sensing system. In this chapter we discuss several issues in participatory sensing system design, then introduce several papers on existing systems and study lessons that can be applied to designing a new system. Finally, we conclude by presenting our own participatory sensing system and results from its simulations.[1]

12.1 Introduction

Over the past decade, our ability to gather information about our world has drastically improved. Technology has allowed for cheaper sensors, better communication, hardware that can be powered longer, and increasingly mobile sensor networks [18]. This has led to a type of sensor network application often referred to as participatory sensing. Specifically, participatory sensing refers to a sensing system in which humans are carriers of sensing platforms known as "nodes" and voluntarily participate in the system. Various definitions with varying specifics exist in literature, but the general consensus is that the problem involves humans directly contributing to sensing, either by passively carrying devices or actively engaging in sensing.

In this chapter we discuss participatory sensing by first introducing its definitions in Section 12.2, then identifying some of the challenges that designers of such systems face in Section 12.3. Since the primary concern in running such systems is maintaining participation, we introduce economics concepts to help formalize the idea of incentives for rewarding long-term participation in Section 12.4. Having introduced foundations and challenges, we shift our focus in Section 12.5 to privacy, which is an important issue in participatory sensing. We then describe two more participatory sensing systems, one of which is our own system, in Section 12.6. Finally, we end the chapter with concluding remarks in Section 12.7.

12.2 Problem Definition

In order to monitor the environment, sensors have been used in a variety of situations ranging from static deployments such as personal weather monitoring, to mobile swarms of nodes designed to locate and track phenomena [1, 2, 11]. In sensor networks, range and lifetime of systems are limited by available power. In traditional mobile networks, sensors must dedicate some of their limited energy to movement, which detracts from the amount of energy they can use to sense, process, and communicate. By using mobility of living organisms, such as animals in ZebraNet [20], device energy does not have to be used for movement. Carriers will always go to areas of their interests, whether or not the application is suited to their lives. In ZebraNet, where the goal is to track a zebra population, the application is inherent to mobility of the carriers, which are the zebras themselves.

As mentioned previously, participatory sensing can be viewed as the problem of using voluntarily contributions from humans and their devices to collect measurements about a particular phenomenon. More discussion on how a system can leverage human involvement follows in Section 12.3.

12.3 Issues in Participatory Sensing

Participatory sensing is not without its challenges. Since it is still a type of sensor network, problems relating to hardware, communication, and application design found in traditional sensor networks still apply. Mobility is achieved through human movement [36], but integrating humans into the system introduces several new types of challenges as well. Before examining a paper on one of these challenges, we outline some of the key issues and their impact on designing participatory sensing systems.

12.3.1 Data

The purpose of a participatory sensing system is to perform some variety of sensing tasks. One of the most obvious design decisions is determining what measurements are required to achieve the system's goal, and what hardware is required to support these measurements. Beyond the basic type of measurements, additional requirements may be needed such as resolution of data, how often samples are provided, geographic coverage, etc.

Once data has been acquired it must be put to use, or stored so it can be used later. Ignoring privacy for the moment, a design decision still needs to be made about if the system will have a central repository, if data only exists on nodes, or if a third-party entity owns the data. Additionally, designers must anticipate what kinds of queries and reporting should be run on data, and ensure that these extractions are facilitated by the system's information flow and processing capabilities.

As an example, a myopic design for wildlife detection might be a system that runs a vision algorithm on collected images, and reports only if a particular species is in the image or not. Later, a query of interest might be "What is the population distribution of a particular species across the area of interest?" This question cannot be answered with the data stored (a binary "yes" or "no"), but could have been answered if the reports extracted from the vision algorithm had included a count of features corresponding to wildlife. Once a system is deployed it can be difficult to change design details. It may be difficult to communicate with users, access user-owned nodes and push updates. Additionally, doing so may incur costs in the form of inconvenience to users.

12.3.2 Coordination

An important decision, independent of the types of measurements taken, is who takes samples and how often. Assuming that privacy is not an issue, there are still several factors to consider. A simple sensor network deployment design is to have static sensors that sense and report on a fixed schedule. Without incurring communication overhead or adding a central controller, such a system cannot react to dynamic changes or events that cannot be perceived without a global view. If the system is synchronous, and run by one or more central controllers that dictate which nodes take measurements and which types of samples to take, these issues can be avoided. However, this "client-server" model adds communication overhead, and depending on the network conditions, can result in significant lag. Either the controller risks running on an outdated view of the sensors, or sensors may spend time idling while states are updated and while waiting for instructions from the controller.

Alternatively, decisions could be made by the human operators. To continue with the aforementioned wildlife detection example, in a synchronous system the controller might decide to request samples when nodes were known to be close to local areas of interest, or areas where no information was recorded. Without such a controller, the nodes might resort to simply taking an image periodically and submitting a measurement. When considering resource usage, it might be better if users took a picture when they spotted wildlife or were at a location known to require additional sampling. Such a user-driven approach is one alternative to having a central coordinator. Upon examination, the idea of users knowing which locations need additional data came up. This ties into the challenge in Section 12.3.1, since seeing something like a report of the number of measurements by location is yet another query that might not be anticipated by a designer who only considered the system's end goal. Such reporting also brings up software engineering challenges in APIs and usability.

A third option is to use a peer-to-peer, or ad hoc, network. In this case there is little to no structure, and events or queries drive the behavior of the network. A discussion of ad hoc techniques is outside the scope of this chapter. While in traditional sensor networks, work has been done on peer-to-peer setups [7], applying the paradigm of participatory sensing to such networks is an open problem [17].

12.3.3 Privacy and Security

By involving humans in the sensing process, the data that comes out of the system may reflect information about the participants. In general, this information is not part of the goals of the sensing task, but is an issue nonetheless. For example, many sensing tasks involve spatio-temporal context (a time and a place). If the identity of the user is not adequately protected, then access to reports could reveal exactly where someone was at a particular time. This constitutes a breach of privacy, and is undesirable [17].

To solve this issue, a designer might implement security measures to provide authentication and encryption so only authorized users could see records. This could be combined with an anonymization technique to let participants be credited for their contributions, but not directly exposed [36]. Having a secure repository or identity authority adds a point of failure to the system, and depending on the sensitivity of user data may not be sufficient. For example, consider a participatory sensing system in which users detect chemicals known to be byproducts of improvised explosive devices (IEDs). Such devices have been used in areas of unrest. Privacy violations could result in an individual being at risk of harm or suffering social losses.

A problem with anonymizing identifiers is that they are not sufficient to protect against cases where a potential adversary has prior knowledge about its victims. In the above example, if access to the reports was gained, the adversary could simply target locations that corresponded to residences to discourage participation. Here, the identity of the individual is not revealed directly through the report's identifier, but the location information is sufficient to cause privacy violation. This example illustrates that in order to provide privacy, attributes other than identification need to be considered. This challenge is addressed in several of the papers that we discuss throughout the chapter. Furthermore, this example illustrates that a solution in which nodes handle real-time queries with no central data repository is still at risk of privacy violations.

One solution to the problem of particular locations being compromising is to allow participants to disable participation in specific locations. However, if participants opt to send in very few measurements which are at locations and times where other participants also report, then they are less unique and thus harder to identify based on attribute analysis. This suggests a trade-off between exposure and privacy, as well as privacy and coverage of a system. Regulating the diversity of measurements ties into the previous issue of coordination since a controller can leverage its knowledge about the current data repository and participant privacy demands to minimize privacy loss when selecting which participants should sense. Note that not every application is well-suited to using a controller to make decisions about measurements.

Privacy is also a concern in anticipating data uses [12, 18]. If bandwidth and storage are not issues, one approach to prevent the system from being too short-sighted about data collection would be to collect as much data on as many types of measurements as possible, and append the richest metadata to all reports. However, the more information provided in reports, the easier it is to leverage prior knowledge and violate user privacy. Thus a balance must be struck between how much information needs to be stored to enable using data, and how limited the information should be to provide privacy protection. An example where privacy invasion may be overlooked is in the case of browser-based e-mail, which may collect a user's location data, but the primary functionality may blind users to the privacy risks [23].

12.3.4 Human Concerns

In participatory sensing, the device which humans use to sense is often their cellular device [17, 33]. This may be a feature phone or a smartphone, possibly with additional sensors interfacing through technologies such as Bluetooth. In all of these cases, the user has everyday uses for their devices, such as phone calls, messaging, and a wide variety of apps. These uses require energy, as does running a participatory sensing system [28]. Thus, supporting the participatory sensing task has a tangible cost. Furthermore, this cost may not be considered a renewable resource depending on the timescale, since there is no guarantee about when a user can charge their device next.

Communication also imposes upon the user. Bandwidth used for participatory sensing may cause performance degradation in other activities the user would normally do on their phones. In addition, the users may have a limited amount of data they are allowed to transmit and receive based on service provider restrictions.

Convincing users to go "off the beaten path" and perform sensing tasks or go places that are outside of their normal behavior is something the system should be designed to do if necessary. As a simple example, a sensing campaign to assess the levels of background noise in a city may require users to go to areas that are considered less desirable for reasons such as safety. While some users may normally traverse these areas, in order to get sufficient sampling, the system may need more users to travel through those regions. While human mobility is convenient to use when it coincidences with the needs of a sensing system, designers must be aware of its limitations.

If the sensing task is not passive, then users spend time and effort to contribute to it. Users may have a feeling that their time has worth, and justifying the use of their time for participatory sensing should be something designers are prepared to do. In the case of subjective and qualitative metadata added by users, additional resources may have to be spent to verify user input and manage users.

12.3.5 Participants

The salient feature of participatory sensing is the voluntary participation of users. Without users, the system cannot survive. Furthermore, humans can act in a variety of ways that do not benefit the system, such as never contributing, falsifying results, or providing information that does not satisfy requirements. While we do not discuss issues in reviewing data or recruitment of users in depth, research exists in literature on the topic [35].

Another issue is that since user participation is voluntary; they can stop participating at any time for any reason. Whether motivation is intrinsic (such as users participating in a sensing campaign that will better their community) or extrinsic (such as the system providing compensation for user participation), the system designers should be aware of what is needed to provide and reinforce motivation [33]. Additionally, system designers must be aware that participants can start or stop participating at any time. In Section 12.4 we cover terms and develop ideas about incentives to support the idea of extrinsic motivation to encourage participation.

12.4 Applying Market Mechanisms

Earlier in the chapter, it was established that persistent participation is essential to participatory sensing, and that this required potential users to be motivated to use the system. In Section 12.5, a scheme for participatory privacy regulation will be discussed. One of the advantages of the approach described in that paper is that by having users be involved in design, research, and regulation, they are intrinsically motivated to continue participating. Unfortunately, this kind of involvement is not always a realizable option. Even in cases where users are involved, incentive that is extrinsic to the sensing application can help reinforce participation. When users do not have any personal reason to join a system, or when they avoid a system because of perceived inconvenience incurred through participation, incentive is a useful tool. To discuss incentives, we refer to economic and market theory, which is a well-studied subject [22].

In economics, a market is a system with one or more goods. These goods, which are produced by sellers, have a price associated with them. Buyers purchase goods from the seller in exchange for currency. In this chapter we will consider the currency to be "incentive" and do not specify whether it is a monetary or otherwise tangible reward, or some sort of intangible reward such as points for ranking on a virtual leaderboard. One way to model participatory sensing with a central controller is to view it as a buyer of goods which are produced when participants perform sensing tasks. This makes participants the seller, meaning they place a price on the imaginary good produced by performing a sensing task. This price may be affected by factors about the human behind it, such as perceived sensitivity or valuation of their time and resources. Much like in markets with tangible goods, multiple producers, and supply exceeding demand, sellers (participants) are forced to compete with each other to try and sell their goods to a buyer. If participation is sufficient, this supply and demand relationship can be met and competition affects prices. Otherwise, the only limiting factor is the budget of buyers (sensing applications).

Deciding how to set prices under competition is not a trivial problem; however, before addressing this issue we examine another problem. How sellers and buyers interact must be defined to have any idea what sorts of strategies a buyer or seller might take to try and maximize their utility. Utility for a seller is the price paid less the cost of producing the good. For a buyer, the utility is their valuation of the good they wish to buy less the price they pay. Since buyers and sellers in a participatory sensing campaign have an interest in the same types of goods, we can apply the concept of an auction. In an auction, an auctioneer requests bids, and provides a good in exchange for payment from a winner selected by the auctioneer. In a reverse auction, the auctioneer still collects bids and selects a winner. However, the auctioneer gives the winner a payment and receives a good in exchange. While reverse auctions are not the only way to leverage incentive, they are the way our approach (discussed at the end of Section 12.6) manages incentive to address participation, inspired by Lee and Hoh's work described next [23]. The auction model considers buyers and sellers interacting directly, however other successful incentive mechanisms have been used, such as recursive incentive [30].

An auction mechanism defines the rules which determine how bids are submitted and how winners are determined from the bids. A simple auction mechanism is the first-price sealed bid auction. In this type of auction, bidders submit their prices to the auctioneer. All prices are secret, so participants do not learn each other's bids. The auctioneer then selects the winner with the best bid (highest if a forward auction, and lowest if a reverse auction). The advantages of this mechanism are that it is very easy for the auctioneer to run, and very easy for bidders to understand. However, the first-price sealed bid auction does not lend itself well to the recurring reverse auction scenario of participatory sensing. Recurring auctions simply mean that there are multiple rounds. Each round is like a single auction, however bidders and the auctioneer can learn over time from repeated rounds.

To illustrate why first price is not a good choice, we provide a simple example. Suppose that there are N bidders and in each round of the auction, M winners will be selected by the auctioneer. Further suppose that for any bidder i, there is some true valuation v_i^t, which is the lowest price bidder i is willing to offer. The auctioneer will select the M lowest bidders each time, by the rules of the auction. If the bids of the participants are static and participants are sorted so that for bidder i and j, with bids b_i and b_j, respectively, $\forall i, j : b_i < b_j$, then participants $1...M$ always win, and $M+1...N$ always lose. The auctioneer has no reason to change which participants it picks as winners, since despite the fact that $N-M$ participants never win, its expense is minimized.

However, Lee and Szymanski discovered the so-called "bidder drop phenomenon" [24, 25], which results from participants motivated by the belief that they should receive incentive at least some of the time. When expectations are not met, users stop participating by dropping out of the auction. A decrease in competition does not seem inherently bad—the remaining M nodes can satisfy the system, at least for some time. Even ignoring the eventuality of battery depletion, there is a problem with this situation. The assumption was that bids remain the same over time. However, suppose one or more nodes occasionally probe the market by increasing their bid slightly to b_i' when they win, and reducing their bid back to their original b_i if they lose. To examine the impact of this with the least complication, consider what happens with this exploration of price when only M participants remain. Any participant can increase its bid by an arbitrary amount and still win the next auction. As a result, the only limiting factor on how high the bids can go is the system's budget. This is an undesirable scenario, and can be prevented by keeping competition alive, which is done by maintaining participation. Thus, participation is important and cannot be sustained by first-price auctions, even assuming that participants are honest and do not collude.

Now that we have defined reverse auctions and suggested how they might be used in a participatory sensing campaign, we explore a paper that shares our beliefs and presents an auction mechanism [23]. In addition to presenting an auction mechanism,

Reverse **A**uction **D**ynamic **P**rice with **V**irtual **P**articipation **C**redit (RADP-VPC), and ways to measure its performance, the authors provide a formula for return on investment (ROI), which is a formal way to discuss participant tolerance to losses.

The authors postulate the use of reverse auction for distributing incentives to participants. A problem in modeling a reverse auction is that the true valuations of participants must be decided. The authors suggest that true valuation encapsulates all aspects of the "user's investment"—power consumption, resources, privacy, etc.—but this is a dynamic valuation. Depending on the location, time, campaign, resources available, etc., the true valuation of a particular participant may vary. This observation is in line with other research which suggests that certain locations may be sensitive and thus reflect a higher true valuation, or that changing social pressures might alter the risk of social costs associated with participating.

A fixed price mechanism is a simple solution, where all goods are viewed as equal and the auctioneer pays the same amount for any given measurement. However, due to the heterogeneous nature of prices, as well as the dynamic nature of valuations, fixed price incentive is not an optimal solution. Either the mechanism risks dispensing far too much incentive to retain participation, or selects a fixed price that leads to significant numbers of participants dropping out due to their expectations about winning not being satisfied.

Expectations about winning are formally viewed as whether or not a participant's ROI value is above a threshold or not. If ROI for a participant falls below a threshold, which is 0.5 in the paper, they stop participating. The authors define ROI as follows:

Definition 12.1 Let us assume that participant i at round r with ROI S_r^i, has participated in p_i^r rounds prior to r, with true valuation t_i, and tolerance to loss β_i. Then

$$S_i^r = \frac{e_i^r + \beta_i}{p_i^r \cdot t_i + \beta_i}.$$

□

As discussed in Section 12.4, first price auctions run the risk of prices growing out of control. The authors describe the same scenario, referring to the M nodes that always win as a "winning class," and the remaining nodes as a "losing class." They call the unchecked growth of prices "incentive cost explosion." To prevent this from happening, they add the concept of VPC—virtual credits that are rewarded to participants when they bid but do not win an auction round. The idea is that virtual credit will keep the cost from growing out of control by sustaining competition. RADP-VPC has a parameter α, which represents the amount of credit awarded for each consecutive round in which a participant loses. If participant i bids b_i^j in round j, and has lost k consecutive rounds, the auctioneer treats their bid as $b_i^j - k\alpha$. If the participant wins, k is reset to 0, and they are paid b_i^j. If a participant's true valuation is higher, they can eventually win through VPC and not drop off (as long as they have enough ROI tolerance). Lower true valuations still win when there is not enough VPC artificially pushing down the perceived bids of higher true valuation nodes. This allows all but exceedingly intolerant participants to win, thus keeping ROI values above threshold.

The application that the authors consider is a sensing task which requires a set number of measurements which are collected by a service provider. These measurements are collected through mobile devices and are all the same type of measurement. Collection is facilitated by a system that is designed to adapt to the changes in users' valuations, minimize the total expenses (the amount of incentive dispensed), and maintain quality of service which includes measurement precision, age, and geographic coverage. Age is important because in some cases, data is "perishable," meaning the usefulness of the data is affected by how long ago the data was sampled. Since the system is designed for perishable data, periodically new samples must be requested. This makes a system having recurring requests, which justifies designing with recurring auctions in mind.

Recruiting participants is another competition maintenance strategy. Since participants can drop out, recruiting former participants is a useful technique. If a participant dropped out, it means that the market conditions were not yielding incentive that matched their expectations. However, since the environment and prices are dynamic, it is possible that at a future point in time the distribution of bids will have changed such that a participant could rejoin and start winning. To facilitate recruiting former participants, any participant that is no longer participating is shown the highest price that won in the most recent round. If a participant sees that its true valuation is less than or equal to this revealed price, it should rejoin. The authors assume that only participants who have dropped out will receive this information. Suggested methods of delivery are e-mail and SMS. When a participant has dropped out, it needs to decide whether or not rejoining will benefit it. To do this it needs to calculate the expected ROI of rejoining. The authors use the following definition:

Definition 12.2 Let participant i at round r have participated in p_i^r rounds prior to r, with true valuation t_i, tolerance to loss β_i, and receive revealed price φ_r. Then the expected ROI, ES_i^{r+1} is:

$$ES_i^r = \frac{e_i^r + \varphi_r + \beta_i}{(p_i^r + 1) \cdot t_i + \beta_i}.$$

□

To evaluate the performance of RADP-VPC, the authors compare it to the **R**andom **S**election-based **F**ixed **P**rice (RSFP) mechanism, which randomly chooses participants until quality of service is met for the round, and pays them each a fixed price. Conceptually, the authors believe that RADP is better than RSFP from the service provider's view. This is because participants make the decisions about their prices based on their knowledge about current valuations, as opposed to RSFP where the mechanism is responsible for selecting a value that will satisfy the users' price expectations. To study the behavior of the mechanisms, strategies must be defined for the users (otherwise known as agents) who participate. The authors assume risk-neutral agents which react to winning or losing by modifying their bids. The authors formulate the utility U_i for participant i as follows:

Definition 12.3 Let U_i be the utility for participant i, bidding b_i^r in round r with credit for winning $c_i(b_i^r)$, true valuation t_i, and probability $g_i(b_i^r)$ of winning by bidding b_i^r. Then

$$U_i(b_i^r) = (c_i(b_i^r) - t_i) \cdot g_i(b_i^r).$$

□

Furthermore, the attribute of "risk-neutral" is important. When considering agent behavior, a designer must take into account how they view risk. Using the standard auction terminology, the authors distinguish the three risk attitudes and their corresponding objectives, namely:

1. Risk preference: Maximize $(c_i(b_i^r) - t_i)$.
2. Risk neutrality: Maximize $U_i(b_i^r)$.
3. Risk aversion: Maximize $g_i(b_i^r)$.

As a user's bid goes up, the gain from winning a round increases but probability of winning decreases. Since the goal is to optimize $U_i(b_i^r)$, participants must be aware of this trade-off as they set their bids. This leads to a simple bidding strategy which we adopt in the simulation discussed later in Section 12.6. If a participant loses in round r, then $b_i^{r+1} \leq b_i^r$ so $g_i(b_i^{r+1}) \geq g_i(b_i^r)$—either the bid and probability do not change, or the bid is lowered so the probability of winning might increase. Symmetrically, if a participant wins in round r, then $b_i^{r+1} \geq b_i^r$ so that $g_i(b_i^{r+1}) \leq g_i(b_i^r)$. This adaptive behavior is bounded by the constraint that for any given round r, $U_i^r > 0$ and $b_i^r > t_i$.

The descriptions of the experiments conducted can be found in the original paper. The results show that for a variety of distributions of t_i RADP-VPC results in lower total cost than RSFP. This is because in RADP-VPC, lower bid prices are favored. Because VPC prevents price explosion from happening, RADP-VPC results in a more efficient use of budget, which is reflected in the total cost being lower. The authors do not discuss how to tune the parameter α, but observe that while initially increasing, it results in a higher number of active participants, after a certain α the number of participants starts to decrease. This suggests that there is an optimal value for α. If the parameter is set too low, then the addition of recruitment can still result in lower total cost, since the dynamic nature of market is advertised to participants which can then rejoin based on their ES_i^r.

The authors claim that because of VPC, if there is correlation between geographic location and true valuation, RADP-VPC will retain additional participants and thus create a more geographically balanced set of data than a mechanism which does not account for participation. However, there will still be a bias toward lower true valuations, since data is time sensitive and $g_i(b_i^r)$ for a participant i is higher if t_i is lower.

This correlation may correspond to socio-economic factors that are geographic, such as economic disparity between neighborhoods.

Since the system is supposed to provide sufficient geographic coverage, additional design considerations can be taken. Considering arbitrarily defined regions, an auction can be run in each region. This removes dependency between regions, creating several markets or auctions. Since separate auctions are run, if each auction corresponds to a group of similar true valuations, the markets are less stratified and the bias caused by having a lot of participants with significantly cheaper true valuations is diminished. An important feature in the paper's experiments is that participants do not move between regions. Depending on the definition of regions, this can be an unrealistic assumption. Alternatively, if the regions are large enough to guarantee that mobile users do not move from one region to another, the auctions may be so large that no destratification will occur.

On the topic of privacy, the authors note that a limitation of the mechanism is that participant locations are revealed whether they win or lose a round. Since true valuation encapsulates several costs including privacy and resources, losers are penalized by bidding in the round, expending effort, and providing location, but not receiving incentive. The authors suggest encrypting data until winners are decided to get around this, but approximate location (regions, in the case of multiple smaller auctions) is still provided, and encryption precludes any sort of data quality enforcement. Another option is to use RADP-VPC with a "data broker." This effectively shifts the problem downstream to a third party who can manage security and privacy. However, such an entity may be able to enforce more specific policies that better cater to a participant, and allow the participant the opportunity to participate in auctions across service providers.

Finally the authors mention several real-world challenges that face RADP-VPC. In asynchronous systems, such as the one that is described by the next paper we examine, the concept of an auction round is difficult to define. The length of time that data is useful before perishing can help tune this value, since delaying decisions longer than the data lifespan means by the time the auctioneer selects winners, the data may no longer be useful. Calibration of rounds may also be considered based on supply and demand conditions. For example, if a newspaper is looking for photographs of an individual, and only one person has a photograph, that individual can dictate whatever price they want. The newspaper can either wait for additional photographs to become available (extending the auction round), or accept that there is a limited supply of measurements (images in this example), which may result in a higher total cost for the system.

Systems may be heterogeneous, meaning multiple types of measurements are required. Formulating a mechanism and selecting winners from an auction becomes a more difficult problem because measurements may not be equal in value, the system may not require the same number of each type of measurement, and users may be at risk of having resources depleted faster than expected if they are selected for many types of measurements. One possible solution is to run a separate auction for each type of measurement, but such auctions would be unable to determine if they were overutilizing a given participant. The participant has the option of changing their t_i

to reflect personal resources becoming increasingly scarce, but how this value should change is not clear and might not be a quantity that users could easily determine. This leads to another consideration, which is having a tool to automate bidding or assist in adjusting bids. Like other systems which have interfaces, a major software engineering concern is making systems easy to use and understand while still providing necessary information.

An important design consideration in any mechanism is to make sure it is robust against collusions. Colluding is the act of one or more participants working together and behaving in ways that may result in additional gains for the colluding parties at the expense of other bidders. As a simple example of the recruitment vulnerability, consider two participants i and j that decide to work together to maximize their profit by splitting i's profit. j participates in the first round and then quits. They then receive information about the highest winning bid every round since the system tried to attract it back. If frequent participation is required to avoid unnecessary disclosure, j can consistently bid a value that is very large, or learn a less suspicious value that guarantees losing auction rounds through adaptively increasingly their bid over time. j then provides i with the highest winning bid, and i can use this information to make winning bids much closer to the disclosed bid than they would otherwise be willing to make based on only $g_i(b_i^r)$. This results in a decrease in the efficiency of the overall system. The authors do not discuss collusion protection techniques in this paper.

The authors mention a paper that attempted to develop an understanding of true valuations. The study was done at a university with a limited demographic, so the value at which users were willing to sell their data (25 cents) is not necessarily applicable to all situations. However the insight that compensation can allow participation despite privacy concerns, and that valuation is situational and a multi-disciplinary problem, is still of value [14].

Privacy and security are not discussed beyond the need to keep recruitment messages personal, and data integrity being important, with the authors suggesting trust management [13]. The paper shows challenges in mechanism design, and one solution for an auction mechanism that is oriented toward retention of participants. The framework is general, and can be applied to any homogeneous client-server sensing system. Considerations included whether or not it could apply to asynchronous systems or heterogeneous systems, and that true valuation is dynamic and reflects all user valuation including cost of resources, privacy costs, and worth of a measurement. In addition, the paper provided a formal way to look at incentive and tolerance to losses through ROI.

The next paper we examine describes a system designed to study recycling practices at a university [34]. The measurements consisted of photographs of locations where trash and recycling were deposited (such as "waste bins") and optional tags that participants could input prior to submitting the photographs. This system is of particular interest to the chapter both because it is an example of asynchronous participatory sensing, and because it explores the effect of different incentive schemes applied to the task. While incentive does not necessarily involve application of the concepts introduced earlier regarding markets, the COMPETE mechanism, which is described in the following discussion, creates incentive-driven competition with the goal of promoting participation.

The authors define participatory sensing based on three requirements:

1. Users are involved in decisions about what will be collected. This is the same belief as expressed while discussing participatory privacy [36] at the beginning of Section 12.5, which has authors in common with this paper.

2. Users contribute data collected during daily routines. The mention of daily routines is important since it suggests that participatory sensing systems are tied to the patterns in human behavior.

3. Users are connected to the context/purpose of the tasks they perform.

Five incentive schemes were considered, four of which were micro-payment schemes. Micro-payments are incentives rewarded on a smaller, more frequent level. The incentive use discussed previously in this chapter has all been micro-payments since incentive is awarded based on single actions or measurements. To make a fair comparison between the five schemes, the maximum payout from each scheme was the same, namely the MACRO amount.

- MACRO: One large payment for joining the experiment
- HIGHμ: 50 cents/valid measurement
- MEDIUMμ: 20 cents/valid measurement
- LOWμ: 5 cents/valid measurement
- COMPETEμ: Between 1 and 22 centers/valid measurements, based on how many samples taken compared to other peers. Rankings were public, which differs from the "sealed bid" approach of competitive incentive mechanisms such as RADP-VPC.

Experiments were run using 55 Android phones. The authors found that COMPETE resulted in the highest number of samples, but competition motivated some users while others were indifferent or performed worse because of the competitive aspect. Micro-payment options did better than MACRO since there was a system-imposed sense of worth, where as MACRO users complained they were unsure what a measurement was worth. This shows that the campaign and mechanism influence how participants set their true valuations. Additionally, there was no incentive gain for a MACRO user for submitting a photo, so the payment scheme inherently did not encourage measurements. Looking at measurements over time, MACRO and COMPETE users became less motivated as the campaign continued. MACRO users cited a loss of novelty over time. COMPETE users "burned out" over time, with it becoming less important if they held a higher rank. HIGHμ, MEDIUMμ, and LOWμ users tended to ration out the number of measurements so that they would receive maximum incentive by the end of the campaign span. As users fell behind in quota, they would compensate later, causing slight increases. In the case of LOWμ, there is a sharper increase near the middle of the campaign, since many more measurements needed to be taken to reach the maximum reward compared to MEDIUMμ or HIGHμ.

The quality was not strongly affected by the payment scheme used—in all cases the percentage of invalid pictures is low. However, in the case of COMPETEμ, 10% or less of submissions had optional tags, while in all other schemes, an average of 50% or higher tags could be observed, with significant variation in the percentage.

Coverage was highest with COMPETEμ, where users would alter their routine to seek additional measurement opportunities. This conflicts with the participatory sensing definition suggesting data is collected during participants' daily routines. MACRO users did not alter behavior at all. The participants on the remaining micro-payment schemes would alter their behavior by going to measurement sites that they could see, but would not necessarily have measured if not on the micro-payment scheme.

From the above results, it is evident that the incentive scheme influences behavior of participants. Furthermore, the scheme was made known to the user at the beginning of the campaign, which supports the emphasis throughout the chapter on transparency and ensuring that users understand mechanisms in the system. Authors acknowledge it is an initial small-scale study. It is unclear if longer periods would have resulted in participants burning out regardless of payment scheme, and if the percentage of measurements with tags would have changed. Fixed micropayments appeared to perform the best—effectively the payment scheme translated into a goal that was easy for participants to conceptualize. Authors suggest that if a mechanism could be added to decrease "participant fatigue," then COMPETEμ might perform better. The issue of "participant fatigue" is important to consider in system design, since this means a mechanism with the purpose of maintaining participation must look at long-term behavior. One way this could be done is by using techniques discussed earlier in the chapter involving the ROI model and altering β over time. ROI does not measure a cost in "interest in participating" based on the potential toll on users of participating, however modeling such fatigue might be done in a manner similar to the ROI equations. Lastly, this paper demonstrates that having a payment scheme can improve coverage spatially and temporally, but not all designs result in an improvement. Deciding if coverage is a critical factor and addressing it is a challenge that participatory sensing system designers must consider, and this paper agrees with our identification of coverage as a challenge in Section 12.3.

A sensing system that was not a participatory sensing system, but made use of incentive to solve the problem of limited resources, was **S**elf-**O**rganizing **R**esource **A**llocation (SORA) [28]. The motivation for the system was that sensor networks are comprised of low-power devices able to compute and communicate. This is also the case with devices carried by humans in participatory sensing systems. The need to minimize energy used by the system and thus allow more energy to be used by the participant for normal tasks makes efficient allocation important. As discussed earlier in the chapter, the environment and human factors are dynamic, so the adaptive nature of SORA is also valuable to examine.

The nodes in the system are modeled as self-interested agents with the goal of maximizing "profit." In this paper, profit is virtual and is exchanged for virtual goods which are produced by performing actions. While this is not directly useful for participation, it illustrates a very different approach from reverse auctions to applying market mechanisms. Since this paper was published in 2005, deployments would have consisted of dedicated nodes instead of nodes carried by humans. Excluding the aspect of participation however, traditional distributed sensor networks share many of the other attributes and challenges of participatory sensing systems.

The actions that the authors describe are taking samples, aggregating stored measurements, listening for messages to forward, and transmitting messages. These same basic functions can be applied toward participatory sensing networks, though due to the use of mobile devices already connected to a service provider or wireless AP, communication tends to be less of a concern. However, as we will discuss when exploring CarTel in Section 12.6, aggregation and delivery concerns are still applicable to deployed participatory sensing systems.

The authors describe the adaptive behavior they expect from nodes as "dynamic," a term that was mentioned independently in Sections 12.3 and 12.4. The recurring use of "dynamic" highlights an important detail of designing for participatory sensing and in many traditional sensor network designs. Even if the system is assumed to be static, its environment is likely changing in ways that necessitate system adaption to it. By adding humans to participatory systems, the number of possible changes to which the system must react increases. The authors support this observation by indicating that a static schedule of actions, or a dynamic mix of actions on a fixed energy budget, will ultimately result in potential energy waste. This happens because different nodes are in different situations as defined by factors such as network topology and proximity to phenomena of interest. As the network and environment change, the optimal actions that any given node should take also change. The impact of a system being dynamic is significant enough that the authors credit existing work in market-oriented programming [8], but assert that SORA differs in that it solves a real-time allocation problem, whereas Wellman's work [42] only solves a static-allocation problem.

SORA applies reinforcement learning [38] by incorporating an exponentially weighted moving average (EWMA), a well-known filter. Each node computes utility $u(a)$ for an action a based on the probability of payment β_a and the price of the action's good, p_a. β_a represents the effect of the learning, and is adjusted using an EWMA filter with sensitivity α. We omit the equations here, but they are described by Mainland, Parkes, and Welsh in the original paper. In this way, nodes learn actions based on what benefits them. To influence the nodes, the system globally advertises a vector of prices that specifies how much the system is willing to pay for particular action-produced goods. The process of deciding which action to take is based on the current global prices and the current state of the node. The state of the node is the current energy budget. Note that goods are only purchased by the system if they are useful (submitting/aggregating an interesting measurement, or routing an interesting measurement toward the base station).

The system as described so far has not addressed how to incorporate energy. As mentioned before, a fixed energy budget poses problems for resource allocation, because nodes may consume energy too quickly. To rectify this, the authors use a "token bucket model" in which the bucket can hold a maximum amount of energy, (C), and the bucket fills at a rate of (ρ). The bucket size represents the largest amount of energy that the node is allowed to use at once. If C is set to the node's entire battery, then as in the fixed rate case, it is possible to deplete the battery rapidly, assuming ρ is not relatively large. ρ is a gradual "recharge over time" rate which may not represent physical charging, but rather it could be used to model the fact that in the case of user-operated nodes, users periodically recharge their devices. In our discussion of this paper, we only consider the original design, which is that ρ is designed to limit frequent bursts, while C controls the maximum burst of energy consumption allowed at once. It is also worth noting that here, energy is a constraint and modeled as a separate budget with a separate currency. This is unlike the previously mentioned RADP-VPC, where true valuation encapsulated all perceived costs, including usage of resources such as energy, and it used the same currency as the incentive.

The application that the authors consider is tracking vehicles through use of magnetometer measurements. Such an application would be hard to consider as a participatory sensing task. Yet, if the task could be accomplished by user's devices taking pictures and running image processing, and vehicles were differentiable, the task could be framed as a participatory campaign. Additionally, the authors state that SORA is not specifically designed for vehicle tracking, so the design lessons are general and can apply to participatory design. Specifically, SORA can used for other systems as long as the actions (and resulting goods) are defined, and any dependencies are explicitly stated. Since the nodes run a simple program, they cannot make assessments independently to determine dependencies, and rely on knowing that an action can or cannot be completed based on current goods (completed actions) explicitly.

In addition to being able to adapt and let different nodes express their circumstantial differences, the authors add a design goal of allowing control. The system operator should be able to control node behavior, and this is done simply through the global price vector. Any change made to this vector is propagated to the nodes. This incurs some overhead, but the authors mention that any of several existing "efficient gossip or controlled-flooding protocols" can be used. Still, the authors suggest that price vector updates are done infrequently. Unlike adapting at the individual level, control is important because some changes may require a global view to perceive and respond to them. The control is not absolute, since nodes must react to the changed price vector through the EWMA-based learning mentioned above. In experiments, the authors found that without large changes in the global price vector, the effect was hard to observe.

In order to adapt to changes, the nodes need to periodically try actions that are not the most profitable. This risk-taking behavior is implemented by an ε-greedy algorithm, where ε is a risk-taking factor (the authors use $\varepsilon = 0.05$). The nodes behave as expected and take the action that currently is believed to maximize their profit with probability $1 - \varepsilon$. The rest of the time, an action is chosen from all possible actions, with uniform probability of choosing any given action. By having $\varepsilon > 0$, nodes can never be completely blocked from learning about an action, regardless of the EWMA α chosen. An interesting design lesson is that nodes must be given the chance to explore the system, and this exploration can lead to local adaptivity. Whether a participatory system designer tries to anticipate participants deviating from "rational" behaviors or not, humans are liable to do so. For example, it is this deviation that leads to the incentive cost explosion in the case of reverse auctions. If a designer assumes participants always act according to the expected algorithm, the system cannot be designed to be robust against such behaviors.

The authors compare their algorithm against a static action schedule, a dynamic action schedule that adjusts based on the current energy budget, and a "Hood tracker" [43] to compare against a published system. Aggregation-based methods perform worse with respect to error, however this is due to error being measured based on where the target vehicle was when the base station received a given measurement. Thus the additional time spent collecting measurements and processing them during aggregation introduced time lag, which in turn increased the distance the vehicle moved before the base station received the measurement. Any actions taken that do not result in a measurement eventually arriving at the base station contribute to wasted energy. The authors note that "In a perfect system, with a priori knowledge... there would be no wasted energy." The difference in energy efficiency between SORA and the static or dynamic methods are about 40% once $C > 1500$. This is due to SORA's learning approach and shows that the reinforcement learning method results in much higher energy efficiency with small costs in accuracy.

Through experiments, the authors examine the effects of ε and α. We do not summarize those results here, however, what the authors do find is that the two parameters serve as a way to tune behavior prior to the experiment, while the global price vector allows for control during the experiment. In the case of participatory sensing, ε and α would be parameters the user could change, while the global price vector would be an example of something the system operator would change. While in both cases, parameters affect behavior, in the case of participatory sensing, the participants also express control through parameters. To prevent the two groups from working against each other, design should be oriented toward making a system easy to understand, transparent, and having operators and participants cooperate. This is in line with the philosophy suggested in the first paper in Section 12.5. The other design lesson is that sensor network applications require addressing "extreme resource limitation of nodes" and the fact that the environment or universe is not fully known and over time it changes.

12.5 Privacy-Oriented Approaches

While the designs so far have primarily focused on incentives and maintaining participation, we now shift focus to systems that were designed with a primary goal being privacy. The first paper selected serves as a transition from focusing on participation to focusing on privacy, by involving participants in the design of policies related to privacy. We then discuss two systems that are designed with privacy involved, but do not directly involve users in high-level decisions about information flow.

Privacy of participants and ethics regarding information collected by a participatory sensing system are certainly a concern. We begin by summarizing and discussing a paper that addresses these topics. For simplicity, we refer to "participatory urban sensing" as "participatory sensing" [36].

Shilton et al. state that designers of participatory sensing systems need to *proactively* take steps address to the needs and requests of users, which may be quite diverse. In addition, they bring up the idea of "social trust" by stating that users must be "significantly involved in the design process" in order to attain such trust. A definition of social trust is not provided in the paper, however the general idea is that participants in a system should be able to trust that the system will not misuse data they provide. The authors agree that user participation is an important challenge, and that addressing privacy through participation ("participatory privacy regulation," not to be confused with participatory sensing) is a way to use participants. This is an application of human involvement unlike those considered earlier in this chapter. Despite the lack of quantitative measurements, incorporating a participatory model into privacy offers insight into the complexities of privacy and participation, and is an approach that can be applied in design of such sensing systems.

In prior research that the Shilton's group did on sensing projects, they found that privacy concerns arose. These "serious privacy concerns" were identified when tasks included location tracking and image capture, which are both examples of sensing tasks that we independently suggested earlier in this chapter as uses for participatory sensing. According to the authors, issues about privacy were "one of the first ethical challenges." This supports our belief that in participatory sensing system design, privacy is an issue and must be addressed.

The authors note that sensing systems can be installed in which participation is passive and achieved simply by being in the same space as the sensor system. The passive nature of being "in" a participatory sensing system is backed up by other literature [5]. However, Shilton et al. suggest that participants must engage "with" the system in order to collect data that is not only useful, but ethical. Without participants being involved in design and usage, the authors indicate that data sampling may be invasive. As we will show both in the remainder of this section and in Section 12.6, in other systems, the participants are rarely viewed as designers of the system, and are instead presented with a fully developed system which may have no controls or limited controls through a set of designer-selected parameters.

In the paper, a list of privacy and security techniques are briefly discussed. We provide a short list of these techniques, but exclude references, which can be found in the original paper. The value of such a list is that it illustrates a wide array of tools that exist for designers considering participatory sensing design. While we do not examine the application of these principles in other works, several appear in the selected few works that we choose to review.

- Systems that provide warnings, notification, and/or feedback about privacy
- Ways to identify vulnerabilities in information systems (which could lead to unintended data access)

- Systems that allow users to choose what data they want to submit
- Identity management
- Selective retention / "forgetting" data
- Encryption of data
- Statistical anonymization

Personal and social variables dictate how a participant shares information. For example, not revealing the location of one's home, or appearing in a particular social role such as a manager [29]. The environment affects these decisions by affecting what a participant is comfortable with. Social norms, situational pressure, and personal relationships are just a few factors that can play into individual decisions about privacy. Understanding of information flow, or beliefs about flow affects the willingness of participants to share information. If there is belief that the flow of information is very limited, the privacy risk is low. If there is an incomplete understanding of where information can go, the potential privacy risk is higher and participants may be more reluctant to contribute. Understanding information flow involves beliefs about who has access to the information, how those entities spread information, and to whom information is spread.

The paper introduces participatory privacy regulation, which is designed to allow decisions at both the individual level and in groups. These decisions develop policies about how the sensing system can collect, store, and use data. Groups are sets of multiple individuals, which are liable to have some social context. Consider a participatory system that is designed to detect concentrations of volatile organic compounds (VOCs). VOCs from sources such as paints are believed to pose a significant health risk to humans, particularly in heavy concentrations. If a system was deployed, the resulting data could be used to identify locations that could be linked back to individuals, as in the IED example from before. In addition to individual privacy being at risk, the sensing campaign might reveal high concentrations that can be traced to neighborhoods or facilities belonging to a particular company. This can result in social loss, such as a negative opinion of the group. Unlike individuals, since groups are comprised of many entities, decision making can become more complicated. This also supports our belief that privacy is a social, as well as ethical concern.

A binary "share or do not share" system is not sufficient to meet the goals of participatory privacy regulation. Instead, regulation is a process. Users decide what information about themselves can be accessed based on the context of requests. This context has "specific, variable, and highly individual meaning in specific circumstances and settings." Throughout the entire sensing campaign, privacy is an issue which is put at risk in different ways. The first place to consider privacy is in deciding which measurements are taken, and how much can be controlled about the measurements. Examples of sampling control are deciding constraints about frequency of samples, the resolution of measurements taken, and metadata about them. Once the data is submitted, privacy decisions must be made based on who can access the data, to what degree, and who can access which results. How long a system keeps data that has been submitted is another detail that must be decided. This retention decision affects the balance between privacy and verifiability of results.

Another point the authors make is that involvement in the process provides an understanding of data policies and information flow. This can help the user make decisions about valuation or willingness to participate in a given campaign. A design philosophy that can be taken away from this view is that context is given through transparency, and providing information about data management is important to participation when designing participatory sensing systems.

The authors present five principles to drive design under participatory privacy regulation. We list those principles along with abridged explanations.

1. "Participant primacy": As previously mentioned, participants should be involved with the system design to avoid participation being invasive. Estrin et al. express this by stating that all participants should also have the role of "researchers." This gives participants an understanding of the entire system from data collection to use in applications, which leaves them better equipped to make decisions about their privacy.

2. "Minimal and auditable information": The sensor platform may allow collecting far more data than necessary for the goals of the campaign. Minimizing the amount of data collected decreases the risk to privacy, and makes both the system and information flow easier to understand. "Coarse control" may allow the user to enable and disable collection, while "fine control" may allow management of retention or data submission on a case-by-case basis. Implementing fine control requires auditing mechanisms that are easy to use, which is a significant challenge.

3. "Participatory design": Participant input into the system's design should be done as a group process. Individuals have a hard time anticipating future risks of a decision based on present privacy decisions [3]. Estrin et al. suggest that communication with participants can highlight spatial or temporal regions of concern or excitement.

4. "Participant autonomy": The system's design should provide a way to avoid "the pitfall of relying entirely on configuration" by making privacy decisions part of the normal participation workflow.

5. "Synergy between policy and technology": Software and hardware alone are insufficient to solve the problems of ethics. "Institutional policies" are required, with the system serving as a tool to help facilitate policies and the enforcement thereof. Participatory policy making should include all parties involved in the system. Whether responsibility for a given policy lies with the policymakers or system is an issue that must be addressed separately for each issue during system design.

We transition to a paper that discusses k-anonymity and l-diversity [17], properties that quantify the degree of anonymity a user has in a dataset. k-anonymity is a term often found in discussions of privacy, however we selected this paper both to show the strengths and weaknesses of k-anonymity, and to study an approach that goes further by introducing l-diversity.

Huang et al. state that in "typical" participatory sensing applications data is "invariably tagged with the location ... and time." This is required since in such applications, time and location are necessary contexts to extract information relevant to the sensing task. However, this is a privacy risk since it can reveal information about users, particularly if multiple submissions can be linked together. The authors then indicate that a priori knowledge of a user's locations can be used to defeat pseudonyms [6], or user identity suppression [40], and describe the effort required as "fairly trivial." Consistent with our beliefs about participatory sensing, the authors recognize that participation is "altruistic," meaning that it is a voluntary act. The violated privacy would introduce a significant cost to participants, and the risk of such a loss may deter participation. Thus, it is important that privacy is considering in designing a participatory sensing system. The emphasis of the paper is on spatial and temporal privacy. There can be other kinds of privacy, for example in a campaign related to health, it may be useful to cluster based on types of ailments. To compare to tiles in 2D space for traffic, the health example might correspond to something like blocks of International Classification of Diseases (ICD) codes.

Tessellation, a technique used in AnonySense [9], takes a real spatial location and reports an artificial region ("tile") with at least k users, instead of the actual location. This type of modification is called generalization. k-anonymity [39] on an attribute, means that at least k distinct users share a value for the given attribute. Generalization is a natural way to achieve k-anonymity by reducing the resolution of an attribute until each class shares at least k members. Because generalization results in a decrease in resolution, the authors are motivated to show that tessellation may not be suitable for applications where higher precision in location information is required. As an example, they mention traffic analysis, which may require knowing which road the measurement comes from. In order to achieve k-anonymity with an acceptable k value using tessellation, the size of tiles may encompass several roads. The system then has no way to determine which road the measurement is used for, and thus cannot effectively perform the task of traffic analysis. The authors modify tessellation to use the coordinates of the tile's center, instead of just a tile ID, when reporting. This alternate method, TwTCR, provides additional context which can help the application determine which station the report is for. For example, simply using the distance between the center and known stations may indicate that only one station is near the center of the tile. Despite the inclusion of tile centers, if the tiles cover a large area, the application may still be unable to determine which station a report is describing. The authors introduce VMDAV, which is described below, for these cases.

The authors also use microaggregation, which is another way to achieve k-anonymity. This decision is influenced by the ability of microaggregation to be used for continuous numeric attributes. Microaggregation creates equivalence classes (ECs). Within an EC, members have a common value for sensitive attributes. The authors observe that the common value is usually an average for that attribute. Clustering is done to try and maximize similarity between members of the EC, where the similarity metric for numerical attributes is often simply the L^2 norm. Unlike tessellation, which is a generalization method, microaggregation is considered a perturbation method since attributes are not generalized, but otherwise altered. The implementation used is Variable-size Maximum Distance to Average Vector (VMDAV) [10].

Having introduced both TwTCR and VMDAV, the authors evaluate if there is a reason to one over the other. Through examples, the authors show that in some situations TwTCR is better, while in other situations VMDAV is better. When the users are distributed nearly evenly across regions, VMDAV performs better. If the user distribution is dense, then TwTCR becomes the better choice. Based on both findings a third method, Hybrid-VMDAV, is proposed. This method works by using TwTCR if a cell has more than k users, and VMDAV otherwise.

To consider privacy vulnerabilities, there must be an adversary who attempts to gain knowledge through the data available in the system. The authors assume that there is an adversary with knowledge about their victims' behaviors (spatially or temporally), but do not have knowledge about the true location or time in reported information. A simple example that the authors provide is the adversary overhearing that their victim will have a medical treatment during a particular part of a specific day. If the adversary has access to the reports, whether this is by being an administrator of the sensing system or by a security exploit (such as eavesdropping), they may be able to use the information from before to determine which group the victim is in. From there, information about the group may reveal specifics, such as a cancer treatment facility being located in that tile. Finally, the adversary can conclude that their victim was treated for cancer, despite the k-anonymity, a failure known as "attribute disclosure." This shows that k-anonymity is not sufficient to protect against what are called "background knowledge attacks" in literature [27].

Two types of disclosure can happen as a result of the information stored in the system:

1. Identity disclosure—linking a specific record to an individual. (k-anonymity protects against this.)

2. Attribute disclosure—private information revealed from the "semantic meaning" of an attribute.

Again using the cancer patient example previously mentioned, the adversary cannot know which of the k records in the tile with the treatment facility belong to their victim, so there is no identity disclosure. However, since the tile is known to be in the region of a treatment facility, which is semantic information, the adversary learns that their victim has a specific medical condition. This is an example of attribute disclosure. In addition to the background information attack, there can be homogeneity attacks. These use "monotony" of attributes to gather information. In the cancer example, background information is used (the time at which the treatment is done) to narrow down possible records, and homogeneity is used to determine that all remaining records share a tile ID which contains the cancer treatment facility.

To rectify the above problem and further protect against disclosure, the authors suggest use of l-diversity. The concept of l-diversity is that within any group, there are at least l distinct values for a sensitive attribute. This prevents the conclusion in the above example, since with higher l-diversity, the adversary cannot conclude that his victim is specifically the report in the cancer treatment facility. In more general terms, monotony corresponds to a low l value, so by increasing diversity, homogeneity attacks are increasingly difficult to perform, if at all possible. The authors note that l-diversity can be applied to a k-anonymity algorithm, and refer to an existing l-diverse implementation of VMDAV, LD-VMDAV [19].

The structure of the system used consists of several components:

1. Mobile nodes (MN)—the actual mobile sensing devices (usually on humans, but could be in cars).

2. Registration authority (RA)—Validates other components; gives security elements necessary for authentication (such as certificates).

3. Task server (TS)—Allows communication from applications to MNs, and does not allow tasks that would violate the MN's required privacy (k,l).

4. Report server (RS)—Combines reports, then submits them to the application.

5. Mix Network (MIX)—Responsible for facilitating anonymous communication; de-couples the identity of the sender from the report so any component/application cannot figure out which reports come from a particular component/application.

6. Anonymization server (AS)—Generates tiles for TwTCR and equivalence classes for VMDAV.

The AS is responsible for facilitating privacy by taking requests from users regarding their required privacy (k and l values), and returning an anonymized value that the MN can then report to the system. The authors assume that "the AS is owned by a third-party operator and is isolated from attacks," and that the AS does not collude with other components to compromise user privacy. The authors also require that the AS has periodic updates about the locations of MNs to effectively provide privacy, and assume that MNs will trust the AS. However, the authors recognize that in actual deployments blind trust in a third party entity also constitutes a single point of failure and thus is not reasonable. To fix this, they suggest using Gaussian perturbation with a normalization factor p. The formal details are not covered in this chapter. The authors recognize that perturbation on its own can be defeated, but suggest it as an added layer of security and use Gaussian perturbation for simplicity.

Authors measure the performance of algorithms in simulations using the Dartmouth campus traces [21]. Performance is measured based on information loss (IL) and positive identification percentage (PIP), which are defined in the original paper. PIP is application specific, since a "correct association" may refer to the system identifying a single attribute or a tuple comprised of several attributes. A notable result is that the performance is not affected by the percentage of users that participate, meaning the algorithms can scale arbitrarily without performance loss. Hybrid-VMDAV achieves higher PIP and lower IL than either VMDAV or TwTCR, however, even Hybrid-VMDAV's PIP is only around 35%. The authors explain that the metric used in experiments was a simple Euclidean distance, and that choosing a more advanced metric could improve performance. Choosing a metric that is suited to the data depends on the environment, application, and resulting attributes, and thus is something that must be considered during system design. Alterations during algorithm execution would result in a new set of tiles or equivalence classes being computed, which would then invalidate older measurements or leave confusion as to which set of anonymized attributes a given report referred.

The authors examined the effect of the Gaussian perturbation on location and found that by increasing p, IL increased. This is expected since perturbation adds noise to the data, and added noise results in higher loss of information. Proper selection of p can keep IL low and balance privacy and system performance. How to tune this parameter is not discussed in the paper, but, as with other designs, finding suitable values for parameters should be part of system design and is likely dependent on the application and current state of the network and environment.

As established in the previously discussed paper by Huang et. al [17], *k*-anonymity is not a sufficient solution for preventing attribute disclosure. PoolView, a participatory sensing application designed with privacy as a goal, develops further privacy measures to overcome the shortcomings of k-anonymity [12]. In PoolView, data collected from users is viewed as a time-series of values called "streams."

PoolView is designed with the belief that there is no trust in the system outside the nodes, and places the responsibility of data protection upon the nodes before data submission. The authors observe that anonymity does not work if there is location information since the resolution of data may have to be quite low in order to provide anonymity through approximate location. Preventing location information in a particular region may still indicate identity, and may create large areas with no measurements by overlapping with areas where position is not measured.

Perturbation cannot protect because correlation between data and correlation between data and context. This correlation can be exploited by statistical tools such as Principal Component Analysis (PCA). Tools can be used to make application-specific noise so reconstruction cannot happen for an individual stream, but for information about the community it can. Since the individual is not at risk using such tools, the authors indicate participants have no need for anonymity. Nodes run by participants send information to pools after agreeing on a noise model. The pools can then be accessed by applications to gain information with little error about aggregate statistics, but cannot obtain information about individual measurements without high error. As in several other systems discussed in this chapter, the application is designed to be simple with the belief that this leads to usability.

The authors consider the application of collecting traffic statistics (which are computed after measurements have been taken), and an ongoing average weight of participants (which happens as data from streams arrive). These are only two examples of time-series data that could be used by PoolView. However, it is worth noting that another design decision that must be made is when data can be accessed. Designing a system that allows partial data to be used means the system cannot have a reliance on knowing the full data in order to process or present results.

The system works by having the pool send a user the noise model when the user joins. The noise model has a distribution of parameters, and the user generates a particular instance of parameters from this distribution. This noise is then added to the user's measurements prior to their submission to the pool. Having a distribution of parameters means the communal noise can be guessed and removed to get aggregate information. The accuracy of this method is higher when there are more participants, since the theoretical parameter distribution will be more closely matched. Getting the model from an untrusted source is risky since the model could be designed to make the stream vulnerable. For example, if the model is a constant, then the stream is simply a shifted version of the actual time series, which as previously indicated, is not adequate protection. More complicated models, such as noise with a known spectral range, can be used since a filter can then be applied to remove the noise. This is not a problem, because the noise model must fit the phenomenon. In other words, even if the noise model describes a person gaining weight over time, but the participant's time series indicates weight loss, this is still acceptable. The participant's time series must be one that the noise model can generate using the parameters available with a probable set of values. If this cannot be done, the model is a poor description of the phenomenon. The user can test with curve-fitting before deciding to submit a stream to a pool. A tool is provided in the PoolView application based on two user parameters, p_1 (the fitting error threshold), and p_2 (the threshold on probability that data was generated by the noise model).

In order for a pool to estimate the actual distribution of a series, the authors show the system must solve a deconvolution problem. The formal math is omitted from this chapter, but provided in the original paper. The approach used to solve this formulation is the Tikhonov-Miller method [41]. In solving, two variables arise. The authors refer to the first as the "regularization coefficient," λ, which comes from needing to provide an error bound, ε. The second parameter is in the method's formulation and is represented by v. A larger value of ε means a larger upper bound on the reconstruction error. The error decreases as the number of participants increases, but even with low numbers of participants the error is reasonably small.

Since the noise is uncorrelated and the resulting signal-to-noise ratio is relatively low, PCA does not work on PoolView's approach. Through experiments, the authors show that PoolView is resilient against PCA, while perturbation by white noise is not sufficient to protect individual measurements. The authors also recognize that the model is available and explain why this does not pose a problem. One attack is to try to estimate the parameters used. Using MMSE (minimum mean squared error), if the noise model is close to the actual phenomenon, and there are many candidate noise streams, the authors deem the approach robust. This creates two ways for a server to be malicious—send a model that does not match the phenomenon (in which case MMSE may give a low-error result), or a very narrow set of possible parameters for the model. If a user suspects a model of being inadequate, they can opt not to submit to the pool. A participant's parameters should not be at a tail of distribution since either the user is unusual or the distribution is not representative. This poses a risk of participants thinking a valid model is untrustworthy if they are anomalous. The authors suggest that in social settings, a user may know if they are representative of the community or not, and this can assist in their decision to trust or not trust the pool. Another vulnerability is that the pool could change models. To solve this, users can simply make a model a permanent decision and not allow the model on a stream to change.

The authors claim that when making a model, there must be some belief about the phenomenon already. Otherwise sensor calibration and validation would be impossible, and no hypothesis could be formed. This poses a roadblock to using PoolView as it has been described for exploratory campaigns. The authors indicate that, in literature, there are methods for extracting a model as a sensor network learns more [15]. However, since the pool must make the modeling decision, it can only adapt based on aggregate as opposed to individual measurements. The authors also suggest that a low number of updates to the model could be sent out, but this must be balanced against the security risk of receiving an updated poor model, a risk previously mitigated by not allowing the model to change.

The authors indicate other techniques which include allowing clustering but protecting privacy via rotation, randomized response for non-quantitative data, and secure multi-party computation. Citations for these approaches are in the original paper. Secure multi-party computation is not feasible because of the high communication overhead, which is not scalable. Scalability is a key requirement of sensor networks. In addition, the approach does not work with dynamic joining/leaving, such as what might happen if the system allows coarse privacy control as described at the beginning of this section.

12.6 Participatory Sensing Systems

So far we have discussed what participatory sensing is, and gone over many issues that arise when thinking about design for participatory systems. We have also introduced the idea of incentives, auctions, and how they relate to participation. Several approaches to privacy have also been discussed. We now examine two systems that are notable for taking several existing ideas and putting them together.

One often-cited participatory sensing system is CarTel, which measured traffic and vehicle information through participants and devices interacting with their cars [18]. Cars were equipped with a GPS, making them into mobile sensors. Due to the fact that cars were driven by humans, human mobility and patterns are integrated into the system by design. One application of CarTel described in the paper was road traffic analysis. The authors found that users had heuristics which influenced their routes. They also found that travel times indicated routes were "reasonably predictable." As an extension, they observe that this means models of traffic delays should be possible to build. In the context of the chapter, it is important to understand that this means even vehicular mobility has patterns and these are influenced by human decision making.

In addition to humans being drivers for vehicles, they may act as data mules and physically transfer data by having an intermediate device such as a USB flash drive or operating a wireless AP that can be utilized by nodes in the system. We do not include an extensive discussion on delay-tolerant networks (DTNs), however many of the ideas about multihop communication and muling are a common topic in DTNs. This suggests that an avenue of research is studying human involvement in participatory DTNs. The authors mention that "unplanned in situ WiFi networks can in fact be used with a delay-tolerant protocol ... such as CafNet..." with further details from their study in a separate paper [4]. The authors also observe that they are not the first to consider WiFi usage, citing Infostations as an example [37].

CarTel uses existing work in "mobile systems, sensor data management, and delay-tolerant networks," and serves as a synthesis of ideas. This approach is worthwhile since, as mentioned several times in this chapter, traditional networks share many of participatory sensing's challenges and a large body of research already exists on addressing these problems in the context of distributed or wireless sensor networks.

The system is composed of three main components:

1. Portal (configuration, control, data storage)

2. ICEDB (delay-tolerant continuous queries)

3. CafNet (carry-and-forward network, delay-tolerant networking protocol)

The portal is the simple interface through which users are able to run applications which can issue queries to nodes and allow users to see the results. Queries contain information about which kinds of data are needed, at what resolution, frequency, and prioritization. The authors recognize that data may exceed bandwidth, and some data may be more important, or receiving summaries prior to the full data set may be desired. Intra-query and inter-query priorities are supported. Once queries are processed, data is sent back to the portal which stores data from nodes in a database. Applications use traditional SQL queries to extract data from the database as it becomes available. This design is of interest since it recognizes that there may be delivery delays, but applications may need results before all data has arrived at the portal. This allows perishable data to be consumed as it is available instead of potentially expiring while the system waits for more measurements, and removes a point of synchronization while still allowing a client-server architecture.

A web interface through the portal lets users look through the data. The authors emphasize visualizations with geo-coded attributes, which indicates the importance of location in their anticipated sensing campaigns. Since location plays a significant role in CarTel's data, the authors claim that traditional search methods such as only using temporal locality may not be useful. Instead, their interface incorporates location-based searching in the design by having operators and areas of interest that are defined graphically. The data available based on these criteria is then displayed. This sort of data can be viewed as a privacy risk. For example, with sufficiently dense geo-traces to infer identity, a user's driving habits could be examined. Several illegal driving activities such as speeding or trespassing could be in the recorded data, and through this inference be tied to a particular user. To address privacy, the portal only allows users to look at their own data. However, this means someone with access to the portal's back end, or identity spoofing, could still compromise users. An alternate that the authors suggest is anonymously reporting data or reporting aggregates. Aggregated queries pose less of a privacy risk, but do not provide as in-depth data exploration opportunities. The authors note that a limitation of CarTel is that there is not a way to aggregate results across users while maintaining privacy. Furthermore, they acknowledge that the correct queries could allow inference of users' locations through targeted aggregate queries. Adapting CarTel to facilitate such queries is left for future work.

Due to bandwidth and connectivity constraints, it is not always possible to deliver all data in order. Authors recognize that data may have different utility based on the application, so local and global prioritization are implemented. This gives each data tuple a score, with global prioritization done by sending a summary of the data and receiving information from the portal. The advantage of the global prioritization is that the portal can see data from all nodes and may be able to make dynamic decisions that an independent node could not. Global prioritization is justified in much the same way as global price vectors, described in the discussion of SORA in Section 12.4, were. From both these examples, it is apparent that in designing systems, adding a way for the system to influence or control network behavior allows global knowledge about measurements and node states to be utilized.

Local prioritization uses two constructs:

- PRIORITY—numeric priority with higher numbers indicating more important data. The authors suggest smaller data such as detecting an event be used for high priority, using raw GPS as an example of low-priority data. This of course depends on the application.

- DELIVERY ORDER—specifies the ordering of results, this can be done by a field name or a user-defined function that computes a score based on a tuple. Authors suggest the example of bisecting data for the purpose of constructing a curve—the rough shape of the curve is more important to deliver before smoothing the curve with additional data points. Simply using ORDER BY (a SQL construct that orders by a column's values numerically or lexicographically) cannot perform contextual ordering like that used in the bisecting function.

Global prioritization only uses SUMMARIZE AS, which specifies a mapping of tuples to summarized tuples by using grouping and aggregation functions. The portal computes an ordering, then sends the ordering to the node in question. The node then returns data ordered in the same way as the server sends it.

To keep general, the authors describe the use of software-based "adapters" which handle configuration of specific sensors on nodes and package measurements in a standardized format to be inserted into the portal database. Adapters are stored on the portal, and when required by an application they are sent to nodes. Nodes can run multiple adapters to regulate different sets of sensors for different applications. Similarly, CafNet has a layer called the **M**ule **A**daption **L**ayer (MAL) which allows any device to be used to transport data providing the appropriate MAL driver exists. While we do not discuss adapters or most technical implementation details, the authors go into much greater detail in their paper.

CarTel is a valuable system not only because it demonstrates a real-world deployment of a participatory sensing network, but because it also provides insight that can be applied to design. It is not necessary to create a system from scratch, but instead by studying past solutions, new systems can be synthesized based on requirements. In CarTel, the goal was usage of distributed nodes through a simple interface with low latency. The use of cars in a participatory sensing system shows that participation can involve more than just carrying a dedicated node with limited power around on foot. Additionally, the discovery that there are patterns in vehicular mobility is valuable in addressing coverage and limitations of participatory sensing even with vehicles.

Having explored the work of many other groups, we now examine the **P**rivacy, **P**ower, and **P**articipation-aware **A**uction **M**echanism (P3AM) which we developed in previous work [16]. Since this chapter emphasizes participatory sensing system design, we describe our approach first.

The initial decision was to consider the general task of a entity that wants to collect measurements with spatial and temporal information in each report. The information flows to a “data sink” which could be the controller, or a data broker which could sell the data independent of the system. In addition to viewing the system as synchronous, we considered that nodes would be some type of phone and thus the existing cellular networks’ infrastructure would be usable. This meant we did not focus on considering how data delivery happened, eliminating the need to design a protocol like CafNet, which was described in CarTel earlier in this section. Additionally, this allowed us to view cellular towers as “data sinks,” and assume that the service provider takes responsibility for the data upon arrival. This responsibility includes any privacy mechanisms, whether they be policy based or a system such as the ones described in Section 12.5. Either type of privacy mechanism can be done on top of the system we describe, and does not affect using incentive for participation. Using cellular service providers was also advantageous because it meant instead of a single controller, there could be distributed controllers. In our experiments, we assume that towers do not communicate with each other, thus allowing for diverse smaller auctions based on locations and mobility patterns of the participants. Lastly, by using service providers, the system has a pre-established channel for distributing incentives.

The work by Lee and Hoh on RADP-VPC in Section 12.4 guided our general approach to using incentive. Like their work, we use risk-neutral adaptive bidding behavior, ROI to model participant tolerance to loss, and the idea that bids should encapsulate perceived costs of the bidder. However, our approach does not use the idea of virtual participation credits, and instead of a single bid value to express concerns, P3AM takes a user’s valuation and modifies it. The modifications come from system-defined curves describing the impact of battery level (to model node resources) and the time since a measurement was last accepted (to model privacy), and blends these with the participant's valuation to produce a bid price. P3AM also has a parameter $P_{cheapest}$ which allows P3AM to operate differently than a first-price auction by prioritizing a percentage of wins based on ROI. This incorporates participation preservation by the ROI model's definition, while still allowing bidding to affect the probability of winning and the amount won in an auction round by a user. User understanding is something we believe is important, so P3AM is designed to be transparent and easy to understand. The functions of battery level and privacy are supposed to be simple functions to allow users to easily visualize the effects of a particular bidding strategy. System designers may consult directly with users when designing the incentive scheme, leading to participant involvement like that of participatory privacy discussed in Section 12.5.

We also consider a second-price auction, PI-GVA [26]. With slight modifications to account for the reverse auction, PI-GVA is a valuable mechanism to compare to because it is designed for recurring auctions and is designed to be incentive compatible. Incentive compatibility makes the bidding decision simple for users who trust the system, since their true valuation will give them the highest utility. However, understanding the mechanism is difficult, so users must know and trust that bidding their true valuation is the best action. Adding in factors such as battery level may change optimal bidding strategies and further complicate a second-price approach, so P3AM uses a first-price (where price is either the bid or the RoI of the user) approach.

In our previous study [16], experiments were run over a variety simulation parameters to examine how first-price auctions, PI-GVA, RADP-VPC, and P3AM behaved. One of the assumptions we made that is not realistic was the use of random mobility. To define how participants move in a more realistic manner, we examined mobility traces of taxi movement in San Francisco [31, 32]. Taxis were chosen since their routes are indicative of paths taken by many individuals in the population, and thus are a good baseline for human mobility in a populated region. We consider a taxi to represent a participant with a single data source only when the taxi is active (i.e. a ride with a customer is in session). Since the taxi traces were originally spread over 3 weeks, we broke each trace apart into individual days of the week and overlaid them on a new 24-hour trace. In other words, there would be one trace for all Mondays for a given taxi, another for all Tuesdays, and so on. This yielded a set of movements that still kept correlation of days and hour of the day but was less susceptible to unusual trips being characterized as probable. A total of 7606 such trips corresponding to the traces are used in simulation with one participant per trace.

The effects of mobility were clearly seen in the average battery level of contributors, shown in Figure 12.1. The fact that holes are present shows that using real traces is important to understanding availability of data sources with respect to both participation and location. The regions of holes is about 30 rounds in length, which corresponds to 5 hours. This is roughly the 01:00 a.m. to 06:00 a.m. time period in which we would expect fewer participants to be active. Broadcasting a stationary location (likely a participant's home) for a 5–8 hour span provides minimal information

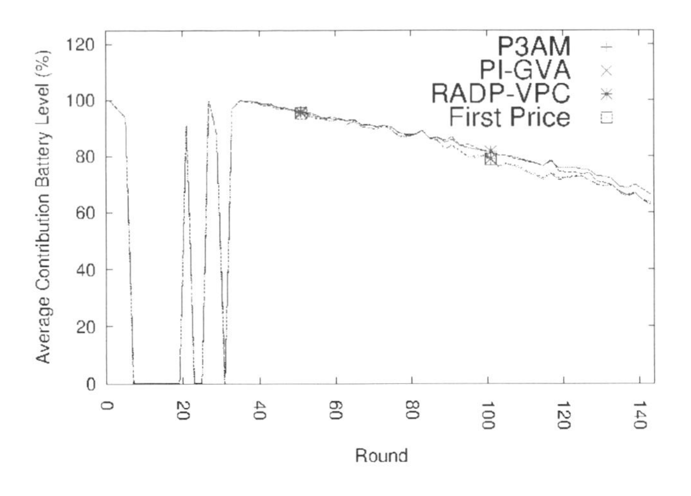

Figure 12.1: Average Contributor Battery Level: Each data point is the average (across all experiments) of the battery levels of all participants winning an auction during a given round. If no data source contributed during a round, there is a "hole" for that round.

to the system while greatly increasing the potential privacy loss for that participant. These observations indicate that the use of taxi traces as a model of human mobility for populous regions is realistic, and highlights the importance of understanding human mobility when designing a participatory sensing system.

Even with extremely low payout settings and very intolerant ROI β, less than 5% of simulated participants stopped participating during the course of the experiments. As a result, the battery level behavior, average price per measurement, and number of samples collected were dependent primarily on data source movement. Unlike the random mobility case where various parameters in mechanisms had effects on behavior, the only effect seen in the trace-based experiments was that the average price per measurement using PI-GVA grew sharply after the "region of holes" described above. By the end of the day, PI-GVA's average price per measurement was still approximately 10 times higher than all other mechanisms, which had very similar average prices to each other. The average price per measurements are shown in Figures 12.2(a) and 12.2(b). Node-based parameters, namely tolerance and true valuation, can cause the average price per measurement under PI-GVA to grow at a much faster rate than any of the other mechanisms we studied.

Due to the difficulty in creating participant dropout from ROI being unsatisfactory, we did not produce a trace-based case of explosion of incentive. However, the fact that parameters needed to be drastically different to produce such an explosion, and that using the same parameters as in the random mobility case we observed very different behavior, indicates that parameters are highly specific to the nodes' behavior. Since this is a participatory sensing application, this translates to needing to understand human behavior's impact on a system. Using testbeds or simulations is an approach that can be used to tune parameters prior to a full-scale deployment [28].

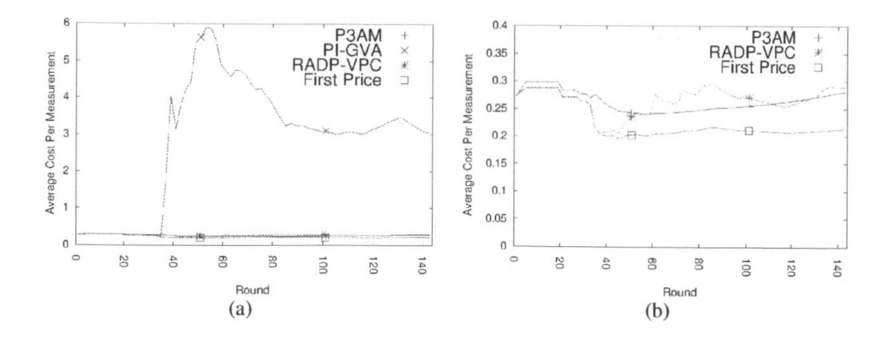

Figure 12.2: Average Price per Measurement: Each point is the average across all experiments of the incentive awarded in exchange for a sample, regardless of if the sample was at a point of interest or not. (a) View of the full curves. (b) Zoomed-in view to compare P3AM, RADP-VPC, and first-price mechanisms.

12.7 Conclusion

Participatory sensing is a type of sensor network applications which uses humans. These uses may include providing mobility to sensing and processing platforms (by carrying hardware), acting as sensors and processors (by deciding when to submit a report or by annotating data), and designing policies for sensor systems. Designing a system that involves human participation requires understanding challenges that arise because of human behavior such as patterns in mobility and difficulties in maintaining involvement.

In this chapter we have discussed only a sampling of the existing literature to explore some of the lessons in designing participatory sensing applications. The increased availability of powerful and versatile mobile hardware, coupled with a wide array of potential applications suggest great potential for growth in the field. Designing systems requires an understanding of the application, and making design decisions about various challenges, such as those described throughout Section 12.3. While our focus has been on incentivizing systems to maintain participation, a real-world deployment requires addressing many issues and combining work done in a variety of independent problems. As the topic of participatory sensing becomes more popular and more mature, we expect to see more effective systems that are more advanced and find innovative ways to address the multitude of design challenges.

[1]Research was sponsored by the Army Research Laboratory and was accomplished under Cooperative Agreement Number W911NF-09-2-0053. The views and conclusions contained in this document are those of the authors and should not be interpreted as representing the official policies, either expressed or implied, of the Army Research Laboratory or the U.S. Government. The U.S. Government is authorized to reproduce and distribute reprints for Government purposes notwithstanding any copyright notation here on.

References

[1] Mobile millennium. `http://traffic.berkeley.edu`.

[2] Urban atmospheres. `http://www.urban-atmospheres.net/`.

[3] Alessandro Acquisti and Ralph Gross. Imagined communities: Awareness, information sharing, and privacy on the Facebook. In George Danezis and Philippe Golle, editors, *Privacy Enhancing Technologies*, volume 4258 of Lecture Notes in Computer Science, pages 36–58. Springer Berlin Heidelberg, 2006.

[4] Vladimir Bychkovsky, Bret Hull, Allen Miu, Hari Balakrishnan, and Samuel Madden. A measurement study of vehicular Internet access using in situ WiFi networks. In *Proceedings of the 12th Annual International Conference on Mobile Computing and Networking*, MobiCom '06, pages 50–61, New York, NY, USA, 2006. ACM.

[5] E. Byrne and P. M. Alexander. Questions of ethics: Participatory information systems research in community settings. In *Proceedings of the 2006 Annual Research Conference of the South African Institute of Computer Scientists and Information Technologists on IT Research in Developing Countries*, SAICSIT '06, pages 117–126, Republic of South Africa, 2006. South African Institute for Computer Scientists and Information Technologists.

[6] Giorgio Calandriello, Panos Papadimitratos, Jean-Pierre Hubaux, and Antonio Lioy. Efficient and robust pseudonymous authentication in VANET. In *Proceedings of the Fourth ACM International Workshop on Vehicular Ad Hoc Networks*, VANET '07, pages 19–28, New York, NY, USA, 2007. ACM.

[7] Maurice Chu, Horst Haussecker, and Feng Zhao. Scalable information-driven sensor querying and routing for ad hoc heterogeneous sensor networks. *International Journal of High Performance Computing Applications*, 16(3):293–313, 2002.

[8] Scott H Clearwater. *Market-Based Control: A Paradigm for Distributed Resource Allocation*. World Scientific Publishing Company, 1996.

[9] Cory Cornelius, Apu Kapadia, David Kotz, Dan Peebles, Minho Shin, and Nikos Triandopoulos. Anonysense: privacy-aware people-centric sensing. In *Proceedings of the 6th International Conference on Mobile Systems, Applications, and Services*, MobiSys '08, pages 211–224, New York, NY, USA, 2008. ACM.

[10] J. Domingo-Ferrer and J.M. Mateo-Sanz. Practical data-oriented microaggregation for statistical disclosure control. *IEEE Transactions on Knowledge and Data Engineering*, 14(1):189–201, 2002.

[11] Shane B. Eisenman, Emiliano Miluzzo, Nicholas D. Lane, Ronald A. Peterson, Gahng-Seop Ahn, and Andrew T. Campbell. Bikenet: A mobile sensing system for cyclist experience mapping. *ACM Trans. Sen. Netw.*, 6(1):6:1–6:39, January 2010.

[12] Raghu K. Ganti, Nam Pham, Yu-En Tsai, and Tarek F. Abdelzaher. PostView: stream privacy for grassroots participatory sensing. In *Proceedings of the 6th ACM Conference on Embedded Network Sensor Systems*, SenSys '08, pages 281–294, New York, NY, USA, 2008. ACM.

[13] Peter Gilbert, Landon P. Cox, Jaeyeon Jung, and David Wetherall. Toward trustworthy mobile sensing. In *Proceedings of the Eleventh Workshop on Mobile Computing Systems: Applications*, HotMobile '10, pages 31–36, New York, NY, USA, 2010. ACM.

[14] Jens Grossklags and Alessandro Acquisti. When 25 cents is too much: An experiment on willingness-to-sell and willingness-to-protect personal information. In *Proceedings of the Sixth Workshop on the Economics of Information Security (WEIS 2007)*, pages 7–8, 2007.

[15] C. Guestrin, P. Bodik, R. Thibaux, M. Paskin, and S. Madden. Distributed regression: an efficient framework for modeling sensor network data. In *Third International Symposium on, Information Processing in Sensor Networks, 2004. IPSN 2004*, pages 1–10, 2004.

[16] Buster O. Holzbauer, Boleslaw K. Szymanski, and Eyuphan Bulut. Socially-aware market mechanism for participatory sensing. In *Proceedings of the First ACM International Workshop on Mission-Oriented Wireless Sensor Networking*, MiSeNet '12, pages 9–14, New York, NY, USA, 2012. ACM.

[17] Kuan Lun Huang, Salil S. Kanhere, and Wen Hu. Preserving privacy in participatory sensing systems. *Computer Communications*, 33(11):1266–1280, 2010.

[18] Bret Hull, Vladimir Bychkovsky, Yang Zhang, Kevin Chen, Michel Goraczko, Allen Miu, Eugene Shih, Hari Balakrishnan, and Samuel Madden. Cartel: A distributed mobile sensor computing system. In *Proceedings of the 4th International Conference on Embedded Networked Sensor Systems*, SenSys '06, pages 125–138, New York, NY, USA, 2006. ACM.

[19] Han Jian-min, Cen Ting-ting, and Yu Hui-qun. An improved v-mdav algorithm for l-diversity. In *2008 International Symposiums on Information Processing (ISIP)*, pages 733–739, 2008.

[20] Philo Juang, Hidekazu Oki, Yong Wang, Margaret Martonosi, Li Shiuan Peh, and Daniel Rubenstein. Energy-efficient computing for wildlife tracking: Design Trade-offs and Early Experiences with Zebra Net. *SIGOPS Oper. Syst. Rev.*, 36(5):96–107, October 2002.

[21] David Kotz, Tristan Henderson, Ilya Abyzov, and Jihwang Yeo. CRAWDAD data set dartmouth/campus (v. 2009-09-09). Downloaded from http://crawdad.cs.dartmouth.edu/dartmouth/campus, September 2009.

[22] V. Krishna. *Auction Theory*. Academic Press/Elsevier, 2009.

[23] Juong-Sik Lee and Baik Hoh. Dynamic pricing incentive for participatory sensing. *Pervasive and Mobile Computing*, 6(6):693–708, 2010. Special Issue PerCom 2010.

[24] Juong-Sik Lee and Boleslaw K. Szymanski. A novel auction mechanism for selling time-sensitive e-services. In *CEC 2005. Seventh IEEE International Conference on E-Commerce Technology, 2005*, pages 75–82, 2005.

[25] Juong-Sik Lee and Boleslaw K. Szymanski. Stabilizing markets via a novel auction based pricing mechanism for short-term contracts for network services. In *2005 9th IFIP/IEEE International Symposium on Integrated Network Management, 2005. IM 2005*, pages 367–380, 2005.

[26] Juong-Sik Lee and Boleslaw K. Szymanski. A participation incentive market mechanism for allocating heterogeneous network services. In *Proceedings of the 28th IEEE Conference on Global Telecommunications*, GLOBECOM'09, pages 2206–2211, Piscataway, NJ, USA, 2009. IEEE Press.

[27] Ashwin Machanavajjhala, Daniel Kifer, Johannes Gehrke, and Muthuramakrishnan Venkitasubramaniam. L-diversity: Privacy beyond k-anonymity. *ACM Trans. Knowl. Discov. Data*, 1(1), March 2007.

[28] Geoffrey Mainland, David C. Parkes, and Matt Welsh. Decentralized, adaptive resource allocation for sensor networks. In *Proceedings of the 2nd Conference on Symposium on Networked Systems Design & Implementation—Volume 2*, NSDI'05, pages 315–328, Berkeley, CA, USA, 2005. USENIX Association.

[29] Leysia Palen and Paul Dourish. Unpacking "privacy" for a networked world. In *Proceedings of the SIGCHI Conference on Human Factors in Computing Systems*, CHI '03, pages 129–136, New York, NY, USA, 2003. ACM.

[30] Galen Pickard, Iyad Rahwan, Wei Pan, Manuel Cebrián, Riley Crane, Anmol Madan, and Alex Pentland. Time critical social mobilization: The DARPA network challenge winning strategy. *CoRR*, abs/1008.3172, 2010.

[31] Michal Piorkowski, Natasa Sarafijanovic-Djukic, and Matthias Grossglauser. CRAWDAD data set epfl/mobility (v. 2009-02-24). Downloaded from http://crawdad.cs.dartmouth.edu/epfl/mobility, February 2009.

[32] Michal Piorkowski, Natasa Sarafijanovoc-Djukic, and Matthias Grossglauser. A parsimonious model of mobile partitioned networks with clustering. In *The First International Conference on COMmunication Systems and NETworkS (COMSNETS)*, January 2009.

[33] Sasank Reddy, Deborah Estrin, Mark Hansen, and Mani Srivastava. Examining micro-payments for participatory sensing data collections. In *Proceedings of the 12th ACM International Conference on Ubiquitous Computing*, Ubicomp '10, pages 33–36, New York, NY, USA, 2010. ACM.

[34] Sasank Reddy, Deborah Estrin, Mark Hansen, and Mani Srivastava. Examining micro-payments for participatory sensing data collections. In *Proceedings of the 12th ACM International Conference on Ubiquitous Computing*, Ubicomp '10, pages 33–36, New York, NY, USA, 2010. ACM.

[35] Sasank Reddy, Deborah Estrin, and Mani Srivastava. Recruitment framework for participatory sensing data collections. In *Proceedings of the 8th International Conference on Pervasive Computing*, Pervasive'10, pages 138–155, Berlin, Heidelberg, 2010. Springer-Verlag.

[36] Katie Shilton, Jeffrey A Burke, D Estrin, Mark Hansen, and Mani Srivastava. Participatory privacy in urban sensing. Retrieved from: http://www.escholarship.org/uc/item/90j149pp, April 2008.

[37] Tara Small and Zygmunt J. Haas. The shared wireless infostation model: A new ad hoc networking paradigm (or where there is a whale, there is a way). In *Proceedings of the 4th ACM International Symposium on Mobile Ad Hoc Networking & Computing*, MobiHoc '03, pages 233–244, New York, NY, USA, 2003. ACM.

[38] Richard S Sutton and Andrew G Barto. *Reinforcement Learning: An Introduction*, volume 1. Cambridge Univ Press, 1998.

[39] Latanya Sweeney. *k*-anonymity: A model for protecting privacy. *International Journal of Uncertainty, Fuzziness and Knowledge-Based Systems*, 10(05):557–570, 2002.

[40] Karen P. Tang, Pedram Keyani, James Fogarty, and Jason I. Hong. Putting people in their place: An anonymous and privacy-sensitive approach to collecting sensed data in location-based applications. In *Proceedings of the SIGCHI Conference on Human Factors in Computing Systems*, CHI '06, pages 93–102, New York, NY, USA, 2006. ACM.

[41] AN Tikhonov and V Ya Arsenin. *Methods for Solving Ill-Posed problems*, volume 15. Nauka, Moscow, 1979.

[42] MP Wellman. Market-oriented programming: Some early lessons. In *Market-Based Control: A Paradigm for Distributed Resource Allocation*, SH Clear water (ed.), pages 74–95, World Scientific Publishing, 1996.

[43] Kamin Whitehouse, Cory Sharp, Eric Brewer, and David Culler. Hood: A neighborhood abstraction for sensor networks. In *Proceedings of the 2nd International Conference on Mobile Systems, applications, and services*, MobiSys '04, pages 99–110, New York, NY, USA, 2004. ACM.

Chapter 13

A P2P Search Framework for Intelligent Mobile Crowdsourcing

Andreas Konstantinidis
Department of Computer Science
University of Cyprus
Nicosia, Cyprus

Demetrios Zeinalipour-Yazti
Department of Computer Science
University of Cyprus
Nicosia, Cyprus

CONTENTS

13.1 Introduction

The advent of *social networks* and the widespread deployment of *smartphone* devices have brought a revolution in location-aware social-oriented applications and services on mobile phones. A *smartphone social network* is a structure made up of individuals carrying smartphones, which are used for sharing and collaboration [1] (i.e., content, interests, comments and places.) For example, Google Latitude, Foursquare, and Facebook Places enable users to check-in to favorite places, provide their location history, as well as numerous other functions. Smartphones can also unfold the full potential of *crowdsourcing*, allowing users to transparently contribute to complex and novel problem solving. A crowd of smartphone users that is constantly moving and sensing providing large amounts of *opportunistic/participatory data* [4, 9, 3] can offer optimality to location-aware search and similarity services [5]. There is already a proliferation of innovative applications founded on opportunistic/participatory crowdsourcing that span from assigning tasks to mobile nodes in a given region to providing information about their vicinity using their sensing capabilities (e.g., noise-maps [38]) to estimating road traffic delay [41] using WiFi beams collected by smartphones rather than invoking expensive GPS acquisition and road conditions (e.g., PotHole [17].)

Currently, the bulk of social networking services, designed for smartphone communities, rely on centralized or cloud-like architectures. In particular, in order to enable content sharing and community search over crowdsourced data, the smartphone clients upload their captured objects (e.g., images uploaded to Twitter, video traces uploaded to YouTube, etc.) to a central entity that subsequently takes care of the content organization and dissemination tasks. Although certain types of objects, such as text-based micro-blogs, will behave reasonably well under this model, significant challenges arise for captured multimedia and sensor data (e.g., data captured by the camera, microphone, WiFi RSS readings, etc.) We claim that the centralization of these object types will be severely hampered in the future due to several constraints including (i) data-disclosure: continuously disclosing user-captured objects to a central entity might compromise user privacy in very serious ways and (ii) energy consumption: smartphones have asymmetric communication mediums with a slow up-link, thus continuously transferring massive amounts of data to a query processor, through WiFi/3G/4G connections, can deplete the precious smartphone battery faster, increase query response time, and quickly degrade the network health.

In this book chapter, we present a P2P search framework for intelligent crowdsourcing in mobile social networks. Our framework uses mobile crowdsourcing primitives, which are expected to unveil the full potential of this novel computation model [5]. Additionally, it uses a P2P search approach, called SmartOpt [25], where captured objects remain local on smartphones and searches take place over an intelligent multi-objective lookup structure we compute dynamically, since the objectives to be optimized conflict with each other. In particular, the multi-objective optimization (MOO) approach utilizes the information contributed by the registered crowd to obtain a diverse and high-quality set of near-optimal solutions (i.e., Pareto Front) to benefit the decision-making process. The book chapter also demonstrates our implemented work over *SmartLab*,[1] which is a programming cloud of 40+ smartphones [26, 30].

The remainder of this book chapter is organized as follows: Section 13.2 provides the background and overviews the related work. Section 13.3 provides our system model and defines the problem in a rigorous manner. Section 13.4 introduces the SmartOpt framework and its internal modules composed of various operators. Section 13.5 details our SmartP2P prototype system and protocol as well as introduces SmartLab, our programming cloud of 40+ smartphones. Our experimental methodology and results are presented in Section 13.6, while Section 15.4 concludes the paper and introduces potential applications of SmartOpt.

13.2 Background and Related Work

In this section, we provide background information and related research work that lies at the foundation of the SmartOpt framework with crowdsourcing. Initially, we introduce several taxonomies and applications of crowdsourcing, followed by research studies on Mobile P2P search and query routing trees. Finally, this sections revisits the area of MOO and introduces multi-objective evolutionary algorithms (MOEAs).

13.2.1 Mobile Crowdsourcing

Crowdsourcing refers to a distributed problem-solving model in which a crowd of undefined size is engaged to solve a complex problem through an open call. Crowdsourcing has still not fully penetrated the mobile workforce, which will eventually unfold the full potential of this new problem-solving model. This is true due to the smartphones' usage characteristics and unique features. Smartphones are in widespread, everyday use and are always connected. Therefore, they offer a great platform for *extending existing web-based crowdsourcing applications* to a larger contributing crowd, making contribution easier and omnipresent. Furthermore, the

[1] Available at: http://smartlab.cs.ucy.ac.cy/

multi-sensing capabilities (geo-location, light, movement, audio, and visual sensors, among others) of smartphones, provide a new variety of efficient means for opportunistic data collection *enabling new crowdsourcing applications*.

Crowdsourcing applications on smartphones can be classified into extensions of web-based applications or as new applications. The former class expands to users that do not have access to a conventional workstation and adds the dimension of real-time location-based information to the service. Instances of such applications are Gigwalk,[2] Jana,[3] and the work of Ledlie et al. [31]. The latter class includes applications for crowdsourced traffic monitoring (e.g., Waze[4]) and road traffic delay estimation (*VTrack* [41]); constructing fine-grained noise maps by letting users upload data captured by their smartphone microphone (*Ear-Phone* [38], *NoiseTube* [40]); identifying holes in streets by allowing users to share vibration and location data captured by their smartphone (*PotHole* [17]); location-based games that collect geospatial data (*CityExplorer* [33]); and real-time fine-grained indoor localization services exploiting the radio signal strength (RSS) of WiFi access points (*Airplace* [29]).

Another key characteristic of mobile crowdsourcing is whether the crowd's contribution is *participatory* or *opportunistic*. Generally speaking, computations performed by users and user-generated data are the inputs for *participatory* crowdsourcing, while the input for *opportunistic* crowdsourcing is data generated from sensors and computations performed by the crowd's devices automatically—i.e., trajectory matching, positional triangulation. The classical crowdsourcing services on the web are participatory, since they require the active participation of the users. The crowdsourcing tasks of the second category are transparent to the user as they usually run in the background using the sensors to collect readings from the environment.

Further crowdsourcing taxonomies are proposed by Geiger et al. [19] and Quinn et al. [35]. Both studies recognize that the value of the input can lie either in the *individual* or the *collective* contribution, where "*the crowdsourcing system strives to benefit from each contribution in isolation or from an emerging property resulting from the system of stimuli*," respectively. Furthermore, [19] divides applications regarding the contribution quality, which can be *homogeneous* or *heterogeneous*. In the former, each contribution has the same weight, whereas in the latter, each contribution is evaluated and can be compared to, compete against, or complete other contributions. In [35], the incentives used for the crowd are also studied, which can be one or more from: *pay, altruism, enjoyment, reputation*, among others. Finally, [35] further classifies applications according to the *human skill* that is exploited including visual recognition, language understanding, and communication. Notice that human skill is only required in applications with *participatory* contribution.

Smartphones feature different Internet connection modalities that provide intermittent connectivity (e.g., WiFi, 2G/3G/4G), as well as peer-to-peer connection capabilities that provide connectivity to nodes in spatial proximity (e.g., Bluetooth, portable WiFi, or the new generation NFC). Notice that each of these connection modalities comes at different energy and data transfer rate characteristics. In par-

[2]Gigwalk Inc., May 2012, http://www.gigwalk.com/

[3]Jana, May 2012, http://www.jana.com/

[4]Waze Ltd., April 2012, http://www.waze.com/

ticular, smartphones have typically energy-expensive communication mediums with asymmetric upload/download links, both in terms of bandwidth and energy consumption, with the upload link being the weaker link.

Classical crowdsourcing applications are developed in a centralized or a decentralized manner. Centralized methods would ship the data generated and collected from the crowd to a server where the answer would be computed. Centralized methods are currently utilized by all social networking sites (such as Twitter, YouTube, Facebook, etc.), Continuously transferring data from the smartphone to the query processor can deplete the smartphone battery, increase user-perceived delays, and quickly degrade the network health. In addition, it demands users to disclose their personal data with a central authority. On the other hand, decentralized methods would send the query to the smartphones, where all computations and communications would be performed locally. This approach might also perform poorly in terms of energy consumption if it invokes expensive computation tasks on all participants.

For location-dependent crowdsourcing applications, localization is usually either i) GPS-only (fine-grained positioning, i.e., a few meters), ii) WiFi-only (semi-fine-grained, i.e., tens of meters), iii) cellular-only (coarse-grained, i.e., tens to hundreds of meters). The latter two methods, which can be combined, require transmitting cellular tower and/or WiFi received signal strength values over the Internet (via WiFi or 3G connection) to the localization server. GPS does not need to communicate any information over the Internet to a localization server.

13.2.2 Mobile P2P Search

Mobile Peer-to-Peer/MANET search can be roughly classified into i) *blind* search [20, 32, 44], where mobile peers propagate the query using an unsophisticated (e.g., random, TTL property) approach to as many nodes in the network as possible, and ii) *informed* search [6, 23, 24, 34, 39], where mobile peers use semantic or location information to forward queries to specific nodes in the network. The proposed search approach presented in this chapter belongs to the latter class with the difference that we utilize a centralized approach where mobile peers (i.e., smartphone devices) subscribe to a centralized registry. Similar to [39], we utilize a content summary mechanism (i.e., profile) for discovering mobile peers that will participate in a query Q by the centralized node. However, in our setting, the content summary of each mobile peer is stored at the centralized node upon its registration thus allowing multiple query users to use this information without requiring the retransmission of the content summary to each mobile peer. In *PeerDB* [34], the authors propose an agent-assisted query processing approach that has the ability to reconfigure the network based on optimization criteria (e.g., channel bandwidth). Although, this can increase the performance of the system (e.g., minimize energy cost, increase time performance), it imposes a high cost for maintaining the agents at each mobile node. In *location-aided routing (LAR)* [24], the authors take into account the physical location of a destination mobile node, reaching in this way only a set of nodes close to the query user, which maximizes the performance of a query (i.e., time, energy). In *SmartOpt*, we additionally augment each mobile node with a social profile, which further decreases the number of participating nodes as only nodes that support a given query will contribute to the results.

Query routing trees (QRTs) in smartphone networks have recently received attention in the context of people-centric applications [4]. Such applications feature continuous sharing of data that can be utilized to create a number of collaborative scenarios (e.g., BikeNet [16]). A central component to realize such scenarios is the availability of some high-level communication structure, such as QRTs. In [42], the authors present a technique that profiles the activities of the user in order to minimize the number of communication packets transmitted in the smartphone network. In contrast to [42], which focuses on a single objective of energy, our proposed technique focuses on two additional objectives: time overhead and recall. In [18], the authors form QRTs using flooding in order to continuously track mobile events and relay data to the query user. Similar to the BFS algorithm, presented earlier, this approach suffers from significant energy waste as all nodes continuously and actively participate in the smartphone network. QRTs have also been extensively studied in the context of an unstructured P2P system (e.g., IS, >RES, RBFS, random walkers, APS, etc. [45]), yet none of these was task into account the resource-constrained nature of smartphone networks. Similarly, query routing structures proposed for sensor networks, such as TAG, ETC, and MHS [2], focus on building routing trees that are near-optimal (with respect to a single objective) but expose good aggregation and data synchronization properties during continuous data percolation to a centralized sink. On the other hand, our setting deals with snapshot query cases and multi-objective query optimization for smartphone social networks.

13.2.3 Multi-Objective Optimization

Multi-objective optimization (MOO) (a.k.a. multi-criteria or multi-attribute optimization) [8, 13] is the process of simultaneously optimizing two or more conflicting objectives subject to certain constraints and can be mathematically formulated as

$$\begin{aligned} &\text{minimize } F(x) = (f_1(x), \ldots, f_m(x))^T \\ &\text{subject to } x \in \Omega, \end{aligned} \tag{13.1}$$

where Ω is the decision space and $x \in \Omega$ is a decision vector. $F(x)$ consists of m objective functions $f_i : \Omega \rightarrow \Re, i = 1, \ldots, m$, where $\Re^m$ is the objective space.

The objectives in (13.1) often conflict with each other and an improvement on one objective may lead to deterioration of another. Thus, no single solution exists that can optimize all objectives simultaneously. In that case, the best trade-off solutions, called the set of Pareto optimal (or non-dominated) solutions, is often required by a decision maker. The Pareto optimality concept is formally defined as follows:

Definition 1. A vector $u = (u_1, \ldots, u_m)^T$ is said to dominate another vector $v = (v_1, \ldots, v_m)^T$, denoted as $u \prec v$, iff $\forall i \in \{1, \ldots, m\}$, $u_i \leq v_i$ and $u \neq v$.

Definition 2. A feasible solution $x^* \in \Omega$ of problem (13.1) is called the *Pareto op-*

timal solution, iff $\nexists y \in \Omega$ such that $F(y) \prec F(x^*)$. The set of all Pareto optimal solutions is called the Pareto Set (PS), denoted as,

$$PS = \{x \in \Omega | \nexists y \in \Omega, F(y) \prec F(x)\}.$$

The image of the PS in the objective space is called the Pareto Front (PF),

$$PF = \{F(x) | x \in PS\}.$$

MOO has numerous applications in virtually all domains of sciences, engineering, and economics. MOO is a relatively new area in mobile/wireless networks, in general, and in smartphone networks in particular. In MOO, it is difficult to apply an existing linear/single objective or systematic method to effectively tackle a multi-objective optimization problem (MOP), giving a set of non-dominated solutions. This is mainly due to the increased complexity and high dimensionality of the search (or decision) space. Our optimizer borrows ideas from *multi-objective evolutionary algorithms (MOEAs)*, which have been shown effective in obtaining a set of non-dominated solutions in a single run. In the literature, several MOPs were proposed in the context of wireless sensor networks and mobile networks [22, 27, 28, 37], tackled in most cases by Pareto-dominance-based MOEAs (e.g., the state-of-the-art Non-Dominated Sorting Genetic Algorithm-II (NSGA-II) [14]) and in few cases by decompositional MOEAs (e.g., multi-objective evolutionary algorithms based on Decomposition (MOEA/D) [46]).

13.3 System Model and Problem Formulation

In this section, we outline our system model and formulate the MOP SmartOpt aims to solve. A table of respective symbols is shown in Table 13.1.

Table 13.1 Table of Symbols

Symbol	Description of Symbols
C	(Centralized) Social Networking Service
U	Users of the Social Network (i.e., $\{u_1, u_2, ..., u_M\}$)
P	User Profiles stored by C for Us (i.e., $\{p_1, p_2, ..., p_M\}$)
o_{ik}	Object k (images, videos, etc.) recorded by user i.
G	Conceptual Graph connecting the users in U.
G'	Social Neighborhood of some arbitrary user.
Q	Query conducted in social neighborhood G' ($G' \subseteq G$).
U'	Users that are connected to C during the execution of Q.
X	Query Routing Tree constructed to answer Q.

13.3.1 System Model

Overview: Let C denote a social networking service that centrally maintains the social profiles $P = \{p_1, p_2, ..., p_M\}$, for each of its M subscribed mobile users (i.e., a crowd $U = \{u_1, u_2, ..., u_M\}$). The profiles record basic user details, authentication credentials, the user interests (e.g., traveling, sports, music, etc.), and friendship relations that define the conceptual social network graph G among the M users. In our setting, a user u_i ($i \leq M$) uses a smartphone (or tablet) device to both perform its day-to-day activities but also to opportunistically capture objects of interest at arbitrary moments (e.g., "take a picture of the Liberty Statue"). Each object o_{ik} might be tentatively *"tagged"* with GPS information and other user tags (e.g., *lat: 40.689201355, long: -74.0447998047, tags: "Statue Liberty Ellis Island"*).

Connection modalities: Each u_i features different Internet connection modalities that provide intermittent connectivity to C (e.g., WiFi, 2G/3G/4G). Each u_i also features peer-to-peer connection modalities that provide connectivity to nodes in spatial proximity (e.g., Bluetooth, portable WiFi, or upcoming NFC available in Android)[5]. We assume that when u_i is connected to C, then C is aware of u_i's absolute location (e.g., GPS) or u_i's relative location (e.g., the cell IDS within u_i's range, WiFi RSSI indicators within u_i's range, or other means utilized for geo-location). Notice that each of the connection modalities comes at different energy and data transfer rate characteristics. For example, we've profiled an Android-based HTC Hero and found that WiFi consumes 39mW/byte, 3G consumes 24mW/byte, and Bluetooth consumes 14mW/byte. Additionally, Bluetooth had a symmetric data rate of 864 kbps, while WiFi an asymmetric data rate of 123 Kbps (up) and 2 Mbps (down) and 3G an asymmetric data rate of 2.7 Mbps (up) and 7.2 Mbps (down). The nominal data rates for the aforementioned modalities might differ significantly, as this is also validated in [21], mainly due to the deployment environment. Moreover, while the power consumption on the different kinds of radios can be comparable, the energy usage for transmitting a fixed amount of data can differ an order of magnitude because the achievable data rates on these interfaces differ significantly [36]. Finally, the availability characteristics of these kinds of modalities can vary significantly. The penetration of some form of cellular availability (e.g., WiFi or 3G) is significantly higher than Bluetooth, on average. Thus, uploading or downloading large data items using Bluetooth can be more energy efficient than using a radio network, but Bluetooth may not always be available and it is often slower.

Search techniques: Let an arbitrary user u_j ($j \leq M$), be interested in answering a query[5] Q over its social neighborhood (i.e., nodes connected to u_j either directly or through intermediate nodes) G' ($G' \subseteq G$). For instance, let Q be a depth-bounded breadth-first search query over u_j's neighbors in the G graph (i.e., in G'). This kind of conceptual query can be realized in the following manners:

1. *Centralized search (CS):* This algorithm assumes that the multimedia objects and tags are all uploaded to C prior query execution. Once Q is posted, C

[5] Without loss of generality, we assume simple Boolean keyword queries over tags.

can locally derive the answers (using its local tag database) and return the answers to u_j. This model, which is currently utilized by all social networking sites (such as Twitter, YouTube, Loopt, etc.), performs well in terms of query response time but performs poor both in terms of *data disclosure* (i.e., o_{ik} objects and tags need to be continuously disclosed to C) and performance (i.e., data transmission of large objects over radio links is both energy demanding and time consuming.)

2. *Distributed breadth-first search (BFS):* This algorithm assumes that the objects and tags are all stored in-situ (on their owner's smartphones.) In order to realize the search task, a querying node u_j downloads from the query processor the addresses (e.g., IP:PORT address) of its first-line neighboring nodes (i.e., $G'' \subseteq G'$). u_j then contacts the nodes in G'' in order to conduct a depth-bounded breadth-first search in a P2P fashion (i.e., using a pre-specified Query Time-to-Live $Q_{TTL} > 0$). Once some arbitrary node $u_x \in G'$ receives Q, it both looks at its local tags, in order to identify an answer, and also forwards the request further until Q_{TTL} becomes zero.

Although the BFS approach improves the data-disclosure drawback of the *CS* approach, it is quite inefficient during search. In particular, Q has to go over a random neighborhood rather than a neighborhood that is contextually related to the query. For instance, in our Liberty Statue query example, we would have preferred querying a friend living in lower Manhattan rather than a person living in California (as the former would have a higher probability of capturing the statue). Also, if u_j had two friends, u_x and u_y, both living in lower Manhattan, with u_x being in spatial proximity to u_j during the query (i.e., within a few meters), while u_y being far away, would have made u_x a better choice for posting the query (as u_x could have been queried through a local link such as Bluetooth).

13.3.2 Optimization Problem Formulation

The *Multi-objective query routing tree (MO-QRT)* structure in this chapter, improves the search operation of the BFS algorithm by optimizing the neighbor selection process. In particular, a node downloads from C a QRT X that is optimized according to the following formulation: *Given a social network of users U, a list of active users U' and their coordinates, the profiles P of these users and a query Q, posted by an arbitrary user u_j, the query processor aims to optimize an X structure using the following* **objectives**:

Objective 1: *Minimize the total* Energy *consumption of X*

$$Energy(X) = min \sum_{\forall (u_a,u_b) \in X \, (X \subseteq U')} e(u_a, u_b), \qquad (13.2)$$

where, $e(u_a, u_b)$ denotes the energy consumption for transmitting one bit of data over the respective edge (WiFi, Bluetooth, and 3G).

Objective 2: *Minimize the* Time *overhead of X*

$$Time(X) = min(max_{(u_a,u_b)\in X} t(u_a,u_b)), \tag{13.3}$$

where, $t(u_a,u_b)$ denotes the delay in transmitting one bit of data over the respective edge.

Objective 3: *Maximize the* Recall *rate of X*

$$Recall(X,Q) = max(\frac{Relevant(Q) \cap Retrieved(X,Q)}{Relevant(Q)}), \tag{13.4}$$

where *Relevant(Q)* denotes the set of all objects in U' that are relevant to Q, formally as:

$$Relevant(Q) = \bigcup_{\forall u_a \forall k (u_a \in U')} (o_{ak})),$$

given that u_a's profile (denoted as p_a) contains terms found in Q. On the other hand, *Retrieved(X,Q)* denotes the set of objects that have been retrieved in response to Q over structure X, formally as:

$$Retrieved(X,Q) = \bigcup_{\forall u_a \forall k (u_a \in X)} (o_{ak})),$$

again given that p_a contains terms found in Q.

In a MOP, there is no single solution X that optimizes all objectives simultaneously, but a set of trade-off candidates. The set of trade-off solutions, commonly known as the Pareto front (PF), is often defined in terms of Pareto optimality. That is, considering a maximization MOP with n objectives: a solution X^* is considered non-dominated or Pareto optimal with respect to another solution Y, iff $\forall i \in \{1,...,n\}, X_i \geq Y_i \wedge \exists i \in \{1,...,n\} : X_i > Y_i$; this is denoted as $X \succ Y$.

13.4 The SmartOpt Framework

In this section, we present the SmartOpt framework (see Figure 13.1), which is composed of three modules: i) *the optimizer module*, which identifies a set of non-dominated QRTs (i.e., Pareto Front) based on geolocated information and social data contributed by the crowd (i.e., a social community of smartphone users); ii) *the decision maker module*, which selects a non-dominated QRT X based on some user-preference criteria from the Pareto Front; and iii) *the search module*, which propagates the QRT X to the P2P network to retrieve the objects of interest.

13.4.1 The Optimizer Module

The SmartOpt optimizer is founded on a MOEA, during which a population of candidate solutions (a.k.a. chromosomes), evolve into better solutions (w.r.t. the objective

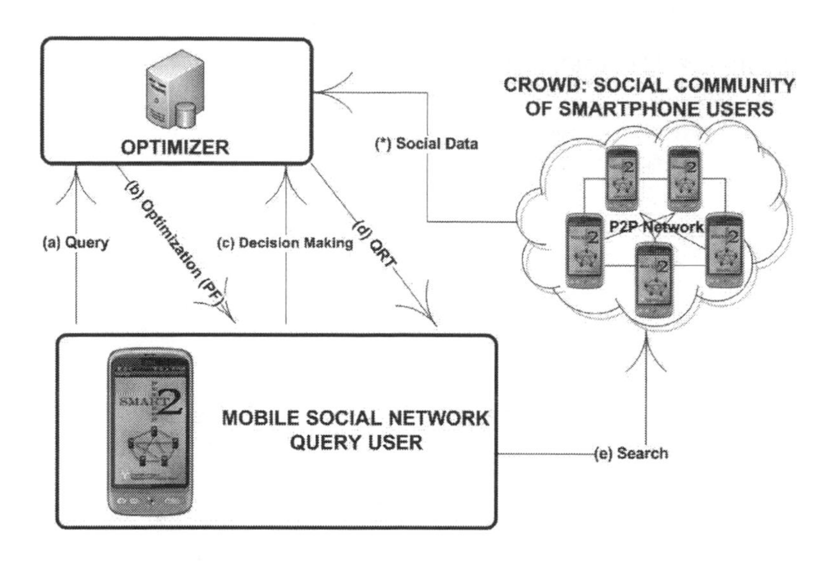

Figure 13.1: The SmartOpt framework: (*) The crowd continuously contributes social data (i.e., text, images, videos, etc.) to the optimizer. (a) A mobile social network user posts a query to the optimizer. (b) The optimizer obtains a set of non-dominated solutions (PF) and sends it back to the user. (c) The user (decision maker) chooses a Pareto-optimal solution based on instant requirements and preferences. (d) The optimizer forwards the selected Pareto-optimal QRT to the user. (e) The user searches the P2P social network for objects of interest.

functions), by utilizing a set of operators (e.g., selection, crossover, and mutation) that are inspired by natural evolution. The given application of operators is inherently stochastic, but applications to numerous domains such as bioinformatics, computational science, engineering, economics, and other fields, have shown that MOEAs can be more effective in difficult multi-objective optimization problems when domain knowledge is incorporated to the operators [27]. In the context of SmartOpt, we introduce both domain expertise into our operators as well as utilize well-known operators that have been proven accurate over the years.

Specifically, we have implemented and specialized the MOEA/D framework [46], which is the state-of-the-art of the decompositional MOEAs and the winner of the Unconstrained Multi-Objective Evolutionary Algorithm competition in the Congress of Evolutionary Computation, 2009. We initially proposed a tree-based encoding representation suitable for the MO-QRT problem and we then designed a MOEA/D composed of our M-tournament selection approach and the two-point crossover and random mutation genetic operators as originally proposed by Zhang and Li in [46]. Furthermore, we hybridized MOEA/D with a problem-specific repair heuristic for identifying infeasible solutions generated by the genetic operators and converting them to feasible. Note that the SmartOpt framework can adopt any MOEA as its Optimizer module (such as NSGA-II [14]) with minor modifications.

Algorithm 20: The SmartOpt Optimizer

Input:

- network parameters (e.g., Q, P, U, G);
- m : population size and number of subproblems;
- T: neighborhood size;
- weight vectors $(w_j^1, ..., w_j^m)$, $j = 1,2,3$;
- the maximum number of generations, gen_{max};

Output: set of non-dominated QRTs, known as the Pareto Front (PF).

Step 0 (Setup): Set $PF := \emptyset$; $gen := 0$; $IP_{gen} := \emptyset$;

Step 1 (Initialization): Uniformly randomly generate an initial set of QRTs $IP_0 = \{X^1, \cdots, X^m\}$, known as the initial internal population;

Step 2:For $i = 1, \ldots m$ **do**

> **Step 2.1 (Genetic Operators):** Generate a new solution (i.e., QRT) Y using the genetic operators.
>
> **Step 2.2 (Local heuristic):** Apply a problem-specific repair heuristic on Y to produce Z.
>
> **Step 2.3 (Update Populations):** Use Z to update IP_{gen}, PF and the T closest neighbor solutions of Z.

Step 3 (Stopping criterion): If stopping criterion is satisfied, i.e., $gen = gen_{max}$, **then** stop and output PF, **otherwise** $gen = gen + 1$, go to Step 2.

MOEA/D requires some pre-processing steps, which consists of representing a QRT and decomposing the problem into a set of scalar sub-problems, before executing its main part, which is outlined in Algorithm 20.

Representation: In our approach, a solution[6] X is a QRT with $|G'|$ active smartphone users that can participate in the resolution of Q. Without loss of generality, let X be represented as a vector in which each index i corresponds to a user u_i and the value of that position corresponds to u_i's parent. The root of the tree is the query user (for simplicity noted as u_1). A negative value -1 in any position indicates that the given user is not currently selected in the query routing tree X.

Decomposition: Initially, the MOP should be decomposed into m subproblems by adopting any technique for aggregating functions, e.g., the Tchebycheff approach used here. In this case, the ith subproblem is in the form

$$maximize \quad g^i(X|w_j^i, z^*) = max\{w_j^i|f_j(X) - z_j^*|,\} \tag{13.5}$$

where f_j, $j = 1,2,3$, are the objectives of our MOP formulated earlier in Subsection 13.3.2, $z^* = (z_1^*, z_2^*, z_3^*)$ is the reference point, i.e., the maximum objective value

[6]The terms *"solution," "vector,"* and *"QRT"* are utilized interchangeably.

$z_j^* = max\{f_j(X) \in \Omega\}$ of each objective $f_j, j = 1,2,3$, and Ω is the decision space. For each Pareto-optimal solution X^* there exists a weight vector w such that X^* is the optimal solution of (13.5) and each solution is a Pareto-optimal solution of the MOP in Subsection 13.3.2. For the remainder of this chapter, we consider a uniform spread of the weights w_j^i, which remain fixed for each subproblem i for the whole evolution and $\sum_{j=1}^{3} w_j^i = 1$.

Initialization Step 1: In Step 1 of Algorithm 20, we adopt a random method to generate m QRT solutions for the initial internal population (i.e., IP_0). Namely, a QRT solution X is initiated by setting each smartphone user $u_i, i = 1 \ldots M$ as a parent. Then, mobile users $u_j, j = 1 \ldots M$ are uniformly randomly selected, and u_i is set as u_j's parent iff $i \neq j$ and u_i is either the root or already has a parent. If u_j already has a parent then we stop and we set as parent the user u_{i+1}. This continues until all users u_i are set as parents once. Then, the IP_{gen} is used to store the best QRT X^i found for each subproblem g^i during the search, i.e., at each generation *gen*.

Genetic Operator Step 2.1: The genetic operators (i.e., selection, crossover, and mutation) are then invoked on *IP* for offspring reproduction, i.e., generate a new QRT solution Y^i for each subproblem $g^i, i = 1 \ldots m$. The following steps summarize the details of each operator:

- **Selection:** We utilize our M-Tournament tree selection [28] for selecting the M closest individual QRTs from the neighborhood of each subproblem g^i, which are then added to a tournament and the two QRTs with the best fitness are selected as parents for crossover. The given selection operator allows us to easily adjust the selection pressure, is simple to implement and works in constant time.

- **Crossover (a.k.a. reproduction or recombination):** This operator allows our algorithm to generate new solutions that share many of the characteristics found in parents, yet are different QRTs. In particular, the *2x-point tree crossover* operator takes as an input two parent QRTs, Pr_1 and Pr_2, and subsequently generates two new QRTs, O_1 and O_2, the offspring. The best offspring O is finally selected as follows:

  - Two crossover points x_1 and x_2 are uniformly randomly selected from numbers 1 to M-1, where $x_1 < x_2$.
  - The pieces of the parents Pr_1 and Pr_2 falling within x_1 and x_2 are exchanged to produce two offspring, e.g., O_1, O_2.
  - The best offspring O is then forwarded to the mutation operator, where $O = O_1$ if $g^i(O_1, w_j^i) < g^i(O_2, w_j^i)$ and $O = O_2$ otherwise.

- **Swap Mutation:** This operator modifies an offspring O to a solution Y with a probability r_m by uniformly randomly swapping the values of two indexes j, z in O.

Repair Step 2.2: In Step 2.2 of Algorithm 20, a problem-specific local search heuristic checks a QRT solution Y and calculates a QRT Z iff:

- **Case #1:** there is a disconnected user u_i in QRT Y (i.e., u_i with or without children that do not have a parent);
- **Case #2:** two or more user IDs i of user u_i are the same in QRT Y;
- **Case #3:** there is an infinite loop in QRT Y.

In all cases, the solution Y is considered infeasible. An infeasible solution can be generated during reproduction (i.e., genetic operation). A local heuristic repairs the QRT solution Y to Z by uniformly randomly generating a parent for the disconnected user u_i in Case #1, replacing the duplicate user u_i with another user u_j in Case #2, and breaking the loop by connecting a random user of the loop with another user out of the loop in Case #3. The repair heuristic continuously repairs solution Y until it does not fall in any of the Cases #1, #2 or #3. In particular, Step 2.2 is repeated after each repair to check for further infeasibility, i.e., whether a new discontinuity, a duplication, or a loop appears in the solution. If this is the case, then the solution is repaired as before, otherwise the feasible Solution Z is used to update the populations of MOEA/D as follows.

Population Update Step 2.3: In Step 2.3, the update phase of Algorithm 20 is processed in three steps. (1) Update IP, $IP/\{X^i\}$ and $IP \cup \{Z^i\}$ if $g_i(Z^i|w^i,z^*) < g^i(X^i|w^i,z^*)$, otherwise X^i remains in IP. (2) Update the neighborhood of Z^i, i.e., the solutions of the T closest subproblems of i in terms of their weights $\{w^1,\cdots,w^m\}$ are updated. If $g^j(Z^i|w^j,z^*) < g^j(X^j|w^j,z^*)$, then $IP/\{X^j\}$ and $IP \cup \{Z^i\}$, otherwise X^j remains in IP, where $j \in \{1,...,T\}$. (3) Update the Pareto front (PF), which stores all the non-dominated solutions found so far during the search. $PF = PF \cup \{Z^i\}$ if Z^i is not dominated by any solution $X^j \in PF$ and $PF = PF/\{X^j\}$, for all X^j dominated by Z^i. The search stops after a pre-defined number of generations, gen_{max}.

13.4.2 The Decision-Maker Module

In the decision-making module, the query user u_j is prompted to decide its preference in terms of *Time* (i.e., Objective 2 calculated by Equation 13.3) and *Recall* (i.e., Objective 3 calculated by Equation 13.4) of the query response to receive from the smartphone P2P network. The decision-maker module of SmartOpt then finds the QRT solution X of the PF that best satisfies the user's decision and it is also the most *energy* efficient (i.e., Objective 1 calculated by Equation 13.2) at the same time. In this way, u_j is responsible to decide the user-oriented objective values (i.e., time and recall) and the decision-maker module decides the system-oriented objective value (i.e., energy), since it is assumed that smartphone users will not be interested in conserving the overall system energy of the network. Here it is important to notice that we proposed a *posterior* approach for giving the opportunity to the user to visually choose a QRT, from the set of Pareto-optimal QRTs obtained by the MOEA/D, based

Algorithm 21: Search Phase

Input: The Query User u_1, A Pareto-optimal Query Routing Tree X, A Query Q

Output: A set of objects $O_j = \{o_{j1} \ldots o_{jk}\}$.

1 **procedure** SEARCH(u_1,X,Q)
2 **if** $(j \neq 1)$ **then Step 1:** Find a set of local objects O_j that satisfy Q
3 $O_j = \bigcup_{\forall i} o_{ji}, satisfy(o_{ji}, Q)$
4 **Step 2:** Send local objects O_j to query user u_1
5 $Send(O_j, u_1)$;
6 **end if**
7 **Step 3:** Forward query u_1 to all children smartphone devices
8 **for** $i = 1$ to $|X|$ **do**
9 **if** j is the parent of i
10 **iIf** $(X[i] == j)$
11 $Search(u_1, X, Q)$;
12 **end if**
13 **end for**
14 **end procedure**

on instant requirements and preferences, instead of choosing a QRT *a priori*, without any knowledge of the obtained Pareto front, or *interactively*, which consumes additional time and energy from the smartphone users.

13.4.3 The P2P Search Module

In the final phase, the query user u_1 receives the Pareto-optimal tree X from the decision maker module of SmartOpt and proceeds with a recursive execution of Algorithm 21 on all smartphone devices participating in the tree X. Recall that X is a vector in which each index i corresponds to a user u_i (IP address and port) and the value of that position corresponds to u_i's parent (IP address and port).

As soon as a smartphone device u_j receives Q it creates a set O_j of all objects o_{ji} that satisfy Q (line 4). Immediately then, u_j transmits these objects to the query user u_1 (line 6) using the most efficient communication technology (i.e., bluetooth, 3G). In the final step, the smartphone device u_j forwards Q to all its child nodes (lines 8-14). This is done by checking each parent entry in X with its own (line 11). If a match u_i is found, u_j transmits Q and X to u_i (line 12). This process executes recursively until all smartphone devices in X receive the query.

13.5 The Smartphone Prototype System

In this section, we provide an overview of our prototype system, called SmartP2P,[7] developed for the ubiquitous Android operating system to demonstrate the

[7]Available at: http://smartp2p.cs.ucy.ac.cy/

applicability of the SmartOpt framework. We particularly evaluated SmartP2P on our programming cloud of smartphones, called the SmartLab testbed.

Our client-side software is developed around the SDK Tools r12 of Android 2.2 and its installation package (i.e., APK) has a size of 327 KB. Our code is written in JAVA and consists of around 7500 lines of code. In particular, our server-code (i.e., optimizer) uses 5000 LOC and runs over JDK 6 and Ubuntu Linux, our smartphone code uses 1600 LOC plus 250 lines of XML elements. The server side also includes a Microsoft SQL server R2 and utilizes the JMATH-PLOT package for drawing the Pareto front images.

The graphical user interface of our system provides a primitive interface for a user to query the active users in the community. Initially, the SmartP2P allows a user to formulate a query in order to find objects of interest. Then the user is provided a group of algorithmic choices including (i) two simple distributed choices, i.e., *random search* and *breadth-first search*, as well as (ii) two MOO choices, i.e., the *MOEA based on Decomposition (MOEA/D)* and the *Non-Dominated Sorting Genetic Algorithm II (NSGA-II)*. The user selects an algorithm and the optimizer calculates a QRT in case (i) or a Pareto front in case (ii). In both cases, the result is returned to the query user. Note that alternative approaches can be easily utilized by our framework with minor modifications. For example, the results can be aggregated at each node and returned back to the user following a reverse path of the same QRT, which is called an aggregation tree [15].

The decision maker is only enabled when the query user selects an algorithm from case (ii) to perform the search. In this case, the Pareto front is forwarded and displayed to the query user. Then the query user makes use of a slide bar below the Pareto front image to set a desired level of time and recall of the search to be initiated. Note that if the user does not choose a desired level of those two objectives, the solution with the minimum energy consumption is automatically chosen. In any case, the decision maker finds the QRT that is closer to the user's choice and downloads it from the optimizer to the user's smartphone. Finally, the query user initiates the search using a P2P protocol and the results of the search as well as the selected QRT are both displayed on the user's smartphone.

The peer-to-peer protocol that lies at the foundation of our prototype system is a text-based protocol, as opposed to a binary protocol, for portability (i.e., endianness) reasons. We also did not choose an XML-based protocol implementation for performance reasons (i.e., minimize annotations.) At a high level, a smartphone user, denoted as *QP*, starts out by registering its obfuscated location (e.g., vector of intercepted cell tower IDs or MAC addresses of WiFi access points) to a well-known host cache (i.e., the SmartOpt SERVER in our case.) After this exchange, the client is considered to be "connected" to the service for a pre-specified amount of time (i.e., k seconds in our setting, after which the lease can be renewed). Now assume that a "connected" client *QP* wants to query the active nodes in the network. *QP* first issues a GET command to the SERVER in order to obtain a tree T that captures its optimization criterions (with respect to time, energy, and recall.) Notice that the SERVER is already aware of the social graph and other statistics used in the optimization process. The issued command is supplemented by a token returned during the registration. The returned tree is serialized in the following format ``NodeIP:NodePort(ParentIP:ParentPort)'', with −1 denoting no parent but is shortened below for ease of exposition. Once T is obtained by QP, QP connects to P0=root(T) and submits its query Q (i.e., {k1,k2,...,kn}), its HOME_ADDR address (i.e., IP:PORT), as well as a hop count parameter. P0 then forwards these parameters to its own children (e.g., P10 and P15), in a recursive manner for N levels (using a predetermined Time-To-Live (TTL) value enforced by the hop count.) Any peer receiving Q, conducts a local search and informs QP directly on HOME_ADDR about possible answers. If a peer in T is not responding for whatever reasons the given branch of the tree is disregarded. The fact that the query tree is optimized for minimum delay, minimum energy, and maximum recall provides an advantage of our approach compared to other approaches for unstructured P2P search, like breadth-first search, random walks [32], as this is presented in our experimental evaluation. In particular, we found that the MO-QRT structure can greatly reduce the number of search nodes, by exploiting meta-relations captured in the social networking graph and the user interests matrix.

13.5.1 The SmartLab Programming Cloud

Experimenting with a large number of devices can be a tedious process as each device needs to be connected to the programming station, the application needs to be installed separately, and the operator needs to manually launch the instances on each device and collect the results. In order to overcome the inherent problems of this setup, we have implemented *SmartLab* [26, 30], an innovative programming cloud of approximately 40+ Android smartphones and tablets, which is deployed at the University of Cyprus (see Figure 13.2). SmartLab is inspired by both PlanetLab [7] and MoteLab [43]. Its intuitive web-based interface is easy to use and provides the ability to reserve and use Android devices for a desired amount of time. Users are able to reboot, list, transfer and remove files, and change Android device settings by using the interactive Android Debugging (ADB) shell session. Additionally, registered users can upload and install executable APK files on their reserved Android devices simultaneously. The SmartLab users are also able to extract application data, output and results automatically from all reserved devices, take screenshots, as well as watch the display of all reserved devices during runtime. Users are also granted access to log files for error and exception handling.

SmartLab implements several modes of user interaction with connected devices using either *Websocket-based interactions* (for high-rate utilities) or *AJAX-based interactions* (for low-rate utilities). In particular, SmartLab supports: i) *Remote Control Terminals (RCT)*, a Websocket-based remote screen terminal that mimics touch-screen clicks and gestures among other functionalities; ii) *Remote File Management (RFM)*, an AJAX-based terminal that allows users to push and pull files to the devices; iii) *Remote Shells (RS)*, a Websocket-based shell developed in house enabling a wide variety of UNIX commands issued to the Android Linux kernels of allocated

[7] Available at: http://smartlab.cs.ucy.ac.cy/

Figure 13.2: Subset of the SmartLab programming cloud: Smartphone fleet connected locally to our datacenter. There are additional devices connected both remotely and wirelessly to our datacenter.[9]

devices; iv) *Remote Scripting Environment (RSE)*, a Websocket-based RCT recording utility that translates user clicks into automation scripts for repetitive tasks; v) *Remote Debug Tools (RDT)*, a Websocket-based debugging extension to the debugging information available through the Android Debug Bridge (ADB); and vi) *Remote Mockups (RM)*, a Websocket-based mockup subsystem for feeding ARDs and AVDs with GPS or sensor data traces encoded in XML, in cases we want to carry out trace-driven experiments with those measurements. A more detailed description of SmartLab can be found in [30].

13.6 SmartP2P Prototype Evaluation on SmartLab

13.6.1 Experimental Setup

Our experimental methodology relies on a *trace-driven real deployment*, during which we deploy our SmartP2P real prototype system implemented in Android over up to 138 users using SmartLab and the traces described next.

Data sets and queries: In our experimental studies, we have constructed two mobile social scenarios from the following three real data sets:

GeoLife [47] (*mobility*): This real dataset by Microsoft Research Asia includes 1,100 trajectories of a human moving in the city of Beijing over a life span of two years (2007–2009). The average length of each trajectory is $190,110 \pm 126,590$ points, while the maximum trajectory length is 699,600 points. Notice that 95% of the GeoLife dataset refers to a granularity of 1 sample every 2–5 seconds or every 5–10 meters.

DBLP [10] (*social*): This real dataset by the DBLP Computer Science Bibliography website, includes over 1.4 million publications in XML format. In particular, the dataset records the paper titles, paper URLs, co-authors, links between papers and authors, and other useful semantics. In order to map this dataset to our problem, we assume that each object is an author's paper. We also assume that each object is "tagged" by the keywords found in the paper title.

Pics 'n' Trails [12, 11] (*mobility and social*): This is a real dataset composed of around 75 GPS traces of a user moving in Tokyo, Japan, during 2007 and a collection of geotagged photos taken along with a short description. In particular, the dataset is comprised of 4179 photos in Japan as well as trajectories with a granularity of 1 sample every 10–15 seconds.

In order to link the above data sets, we have constructed two mobile social scenarios:

Mobile-Social Scenario 1 (MSS-1): This scenario uses the DBLP social dataset and GeoLife mobility dataset. The DBLP dataset is used to construct a social graph G of authors that are related based on their research interests (i.e., keywords of their articles' titles) as well as their co-authorships, which are attributes of the DBLP dataset. Then we have mapped each DBLP author to a trajectory of the GeoLife dataset. Particularly, we have extracted 1,100 authors from the DBLP dataset and we have mapped them to the 1,100 trajectories of the GeoLife dataset using a 1:1 correspondence. This resulted in a social graph with 1,100 mobile DBLP authors moving in the city of Beijing, China.

Mobile-Social Scenario 2 (MSS-2): This scenario uses the Pics 'n' Trails social and mobility dataset. The Pics 'n' Trails dataset is initially used to construct a social graph G of 75 users that are connected based on their interest in taking photos of sightseeing in Japan (i.e., similar tags on their photos taken). Each user is, therefore, carrying a random number of photos tagged with a short description that describes a particular sightseeing in Japan and is associated with a GPS trajectory from the Pics 'n' Trails dataset. This resulted in a social graph with mobile users that carry photos with tags and move in the city of Tokyo, Japan.

In our experiments, we utilize the following three queries:

```
-- Query 1
SELECT S.title, S.url
FROM SmartphoneUsers S, Query Q
WHERE (distance(S.x,S.y,Q.x,Q.y) < 10 KM)
      AND S.Title LIKE '%optimization%';

-- Query 2
```

Table 13.2 Experimental Execution Scenarios and Test Instances

Scenario	Test#	Q	Time	G'	# of Objects	Relevant Objs.
MSS-1	T1	Query1	Morning	49	3877	82
	T2	Query1	Noon	58	5504	73
	T3	Query1	Night	95	8884	121
	T4	Query2	Morning	49	3877	319
	T5	Query2	Noon	58	5504	477
	T6	Query2	Night	95	8884	695
MSS-2	T7	Query3	Morning	26	744	43
	T8	Query3	Noon	66	1877	115
	T9	Query3	Night	47	1456	92

```
SELECT S.title, S.url
FROM SmartphoneUsers S, Query Q
WHERE (distance(S.x,S.y,Q.x,Q.y) < 10 KM)
      AND S.Title LIKE '%networks%';

-- Query 3
SELECT S.title, S.url
FROM SmartphoneUsers S, Query Q
WHERE (distance(S.x,S.y,Q.x,Q.y) < 10 KM)
      AND S.Title LIKE '%Kyoto%';
```

where "S.x,S.y" represent the (x,y) coordinates of a smartphone user in S and "Q.x,Q.y" represent the (x,y) coordinates of the query user.

We execute nine different test instances using the two Mobile-Social Scenarios and the three queries, Query 1, Query 2, and Query 3 as shown in Table 13.2. Our scenarios are executed for various time periods (i.e., during the morning, during noon and during night), in order to capture different mobility patterns that are inherent in the GeoLife and Pics 'n' Trails data sets.

Search algorithms: We have implemented and evaluated the following search algorithms: i) the *centralized search* algorithm (*CS*), presented in Section 13.3.1, which collects all data and metadata tags at the centralized query processor prior query execution; ii) the *distributed breadth-first search* (*BFS*), which conducts a distributed search using a random tree that is generated with a BFS process which visits all nodes in the network, as presented in Section 13.3.1; iii) the *random walker (RW) search* [32], which conducts a distributed search using a list structure that captures a randomly chosen neighbor on each step but that eventually visits all nodes in the network, and iv) the SmartOpt search, which conducts a distributed search using an optimized QRT obtained from the application of ideas presented in this chapter. Smart-trees are inherently smaller in size than their other alternatives, as this structure visits the nodes with a higher probability having more relevant objects (i.e., based on the social graph and the metadata stored for each node.) We evaluate the search algorithms using the following metrics: *Time*, *Energy*, and *Recall*, as these were defined in Section 13.3.2 by using wall clock time along with the PowerTutor power (energy) measuring tool by the University of Michigan, USA. In particular, PowerTutor is a component power management and activity state introspection-based tool that uses an automated power model construction technique for accurate online power estimation in Android.

13.6.2 Experimental Results

Initially, we evaluate the performance of SmartOpt against the CS, BFS, and RW on 100 consecutive timestamps in Mobile-Social Scenario 1 (GeoLife+DBLP) using our model-driven simulator. At each *timestamp (ts)*, we compare the energy consumption, time overhead, and recall of all algorithms.

Figure 13.3 illustrates the results of our experiment for all performance metrics. In Figure 13.3 (top/left) we observe that the energy consumption of SmartOpt is one to two orders of magnitude smaller than its competitors CS, BFS, and RW in all timestamps. BFS seems more efficient than CS as it does not communicate all metadata to the centralized query node. On the other hand, RW is worse than all approaches as the sequential visit to all nodes in the network drains considerable energy (i.e., in each communication only 1 message is sent, as opposed to the rest of the techniques that communicate with several nodes in a single round.) Similar observations apply for Figure 13.3 (top/right) where we demonstrate the time overhead for all algorithms. This happens as the energy is proportional to the time interval the communication transceiver is in active mode. Moreover, in Figure 13.3 (bottom/left), we show that the recall rate for the SmartOpt framework is close to 95%, consistently. The slight decrease of recall with respect to the increase of timestamp is due to the variations in the network. In particular, we noticed an increase of the number of users in the last timestamps, which increases the complexity of the search space, making it harder to find the solutions with maximum recall without increasing the computational effort of the algorithm. In conclusion, the SmartOpt framework consumes less time and less energy and it is able to identify most expected answers at the same time. In Figure 13.3 (bottom, right), we demonstrate the results for a single timestamp (*ts*=70) for all algorithms. The various solutions generated by the SmartOpt optimizer are represented by open squares. The single solutions supplied by the CS, RW, and BFS algorithms are represented by a solid triangle, a solid square, and a solid circle, respectively. We observe that the solution provided by the CS algorithm is the worst w.r.t. BFS and RW in all three performance metrics. However, the CS algorithm demonstrates higher recall (10%) than all solutions provided by the SmartOpt framework. This occurs because CS dictates global participation by all smart objects in the network (i.e., all smart objects forward their results to the query user). However, this has a significantly negative impact on both energy and performance. Specifically, compared to the SmartOpt best solutions, CS, BFS, and RW feature an increase of two orders of magnitude in energy and one order of magnitude in time.

Finally, we evaluate our SmartP2P prototype Android implementation, presented in Section 13.5, over our distributed SmartLab infrastructure as illustrated in Figure 13.4. For the evaluation, we focus only on the distributed search algorithms: *BFS, RW* and *SmartP2P*. We present the query response time, measured in seconds and en-

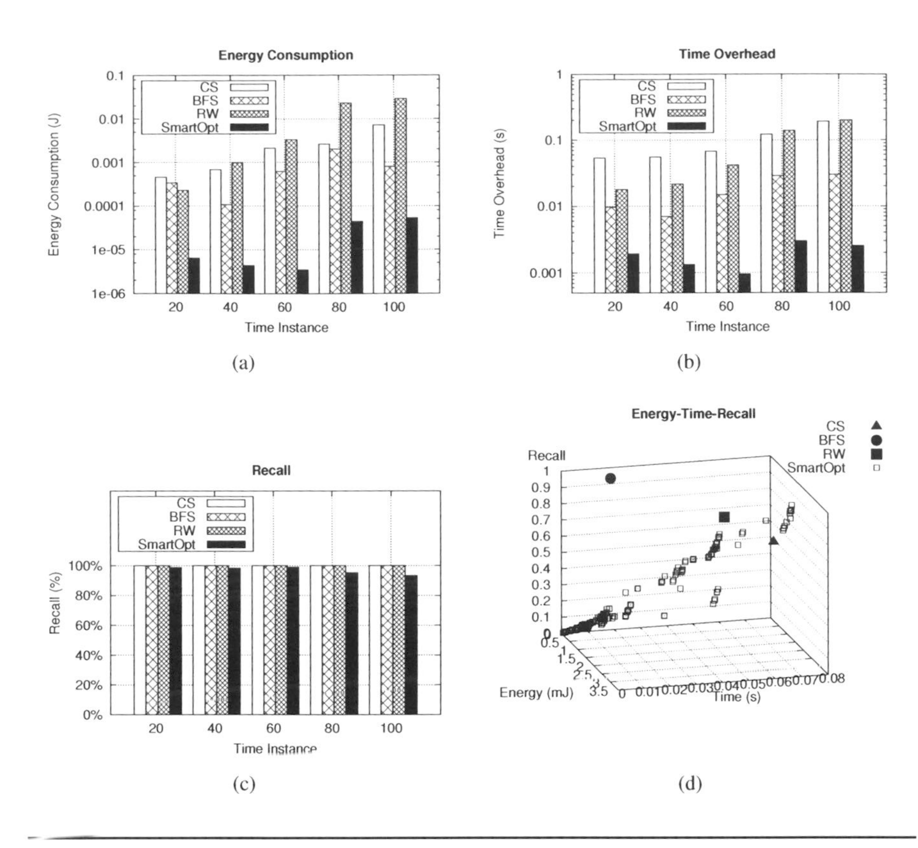

Figure 13.3: Evaluation of the CS, BFS, RW, and SmartOpt search algorithms using the energy, time, and recall performance. The bottom/right figure shows SmartOpt compared to the solutions of CS and BFS in the objective space at timestamp *ts*=70 of Mobile-Social Scenario-1 (MSS-1).

ergy consumption, measured with PowerTutor in Watts and presented in Joules. We utilize five different network sizes in Mobile Social Scenario 1 (MSS-1): 20, 49, 58, 95, and 138 and five different network sizes in Mobile Social Scenario 2 (MSS-2): 20, 35, 50, 61, and 75 to show the scalability aspects of the different search algorithms. In order to accommodate these instances over a physical infrastructure, which was considerably smaller (i.e., 40+ smartphones), we had to run several instances on each of the available physical smartphones (using separate socket servers). For example, an HTC Desire smartphone could easily host tens of instances without any particular performance penalty (recall that these are 1-GHz smartphones with 512 MB of RAM) while the lower-end HTC Hero devices (with a 512-MHz processor and 288 MB of RAM) were excluded from our experiments as they were considerably slower and could not host tens of instances. For practical reasons, we did not utilize the Blue-tooth connection between instances and considered as a local link the socket communication of instances on the same physical smartphone host.

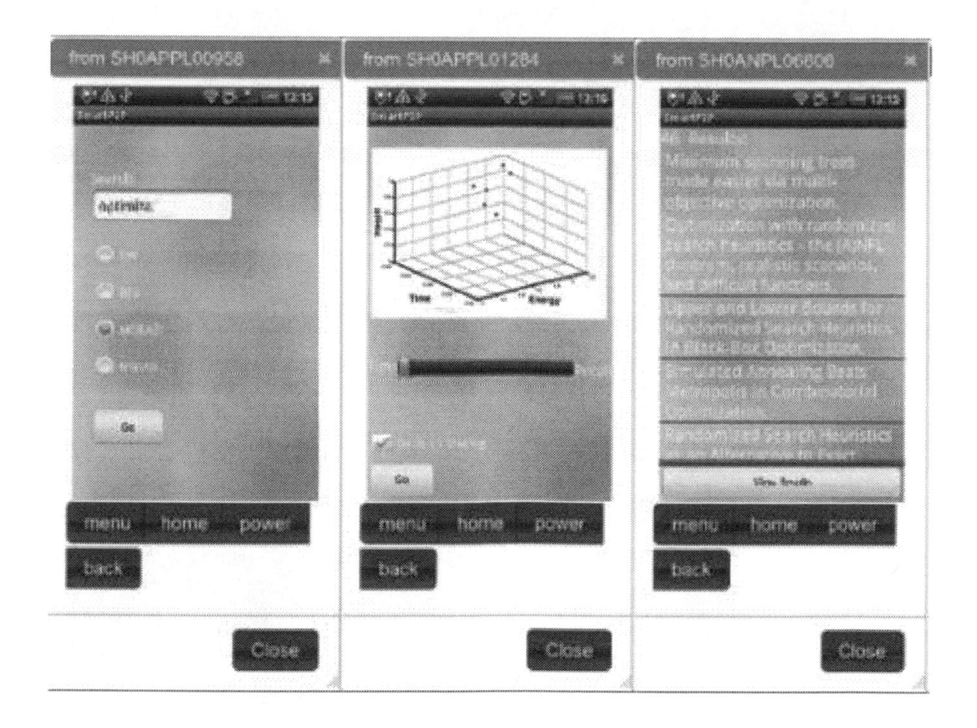

Figure 13.4: A screenshot of SmartP2P on SmartLab using real-time screen capture.

Figure 13.5 (a), presents the response time for the different executions given that all algorithms obtain the complete result set (i.e., maximum recall) in Mobile-Social Scenario 1. We observe that SmartP2P obtains the expected answer in little anywhere between 1.5 seconds and 6 seconds, while both BFS and RW require in many cases as much as 10 seconds. The competitive advantage of SmartP2P over both BFS and RW is considerably better for larger network sizes. This is very encouraging as smartphone networks might consist of thousands of nodes in an area of interest (i.e., within the spatial boundary of a query.) Figure 13.5 (b), presents the energy consumption in Mobile-Social Scenario 1 as this was measured by PowerTutor (i.e., only the energy related to CPU and networking without taking into account costs related to LCD utilization.) The given figure shows that SmartP2P manages to locate the complete answer set utilizing 25% and 33% less energy than RW and BFS, respectively. We also noticed that by bringing down the recall expectation to $\approx$80%, would allow us to obtain great energy savings considerably faster ($\approx$50%). Similarly, Figures 13.5 (c) and (d) show that the SmartP2P search approach is more efficient than the BFS and the RW in MSS-2 as well. In particular, SmartP2P conserves up to 30% time and 25% energy for max recall. Here it is important to notice that the relative performance of the algorithms in terms of energy is the same in both the simulations and the testbed experiments (i.e., SmartP2P consumes less energy than both BWS and RW), although the actual values in the real setting are lower than that of the

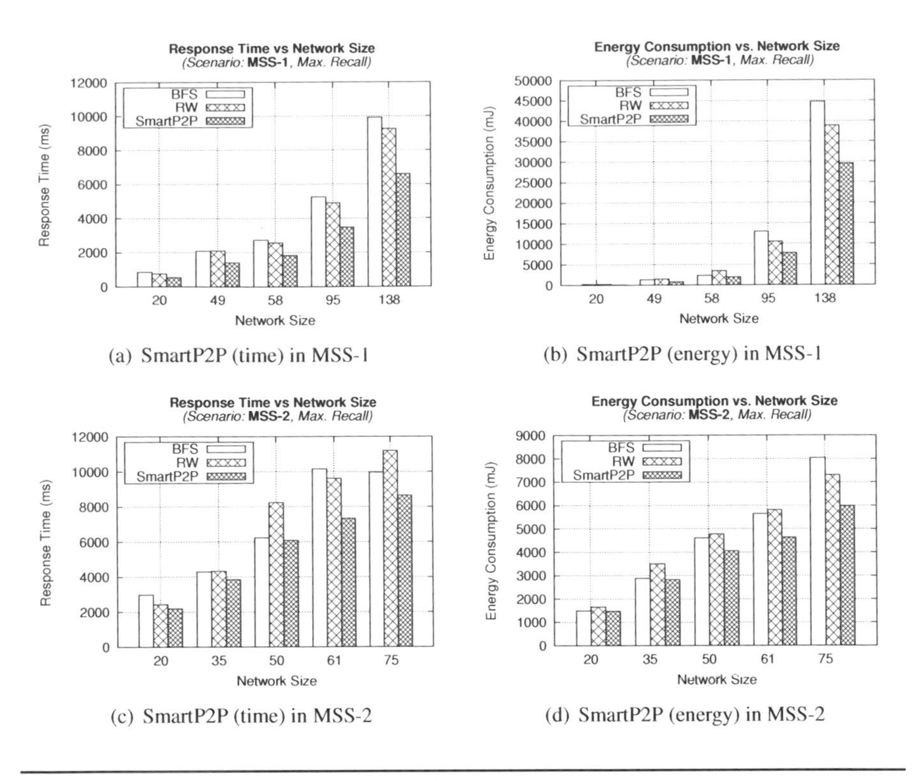

(a) SmartP2P (time) in MSS-1

(b) SmartP2P (energy) in MSS-1

(c) SmartP2P (time) in MSS-2

(d) SmartP2P (energy) in MSS-2

Figure 13.5: Evaluating our SmartP2P prototype system in Android using the SmartLab testbed for different network sizes in both Mobile Social Scenarios 1 (GeoLife+DBLP) and Mobile Social Scenarios 2 (Pics 'n' Trails) with respect to time and energy.

simulations. This is mainly due the fact that the algorithms are implemented in different computational platforms using different network parameter settings (e.g., the network sizes) with additional constraints of the physical infrastructure.

13.7 Conclusions and Potential Applications

In this book chapter, we present the SmartOpt framework for searching objects captured by the users in a mobile social community. Our framework is founded on an *in situ* data storage model and searches then take place over the MO-QRT structure we present in this chapter. Our structure concurrently optimizes several conflicting objectives (i.e., energy, time, and recall). Our experimental assessment uses a trace-driven experimental methodology with mobility and social patterns derived by Microsoft's GeoLife project, DBLP, and Pics 'n' Trails using our real SmartP2P system developed in Android and deployed over our SmartLab testbed of 40+ smartphone devices. Our study reveals that our framework yields high query recall rates and consumes less energy than its competitors. Additionally, our study reveals that the MO-QRT structure is highly appropriate for content search and retrieval in mobile social networks.

The *SmartOpt* framework can be easily adapted and used for several real-life crowdsourcing applications. For example, it can be used as a recommender system where the social crowd of smartphones generates instant information for certain events/places. A querying user can then use *SmartOpt* to generate a query routing tree to retrieve accurate and spatially proximate information about an event/place of interest (e.g. concert, football match, hospital, police station). Furthermore, *SmartOpt* can be utilized for crowdsourcing call assignment tasks, in which the crowd shares its area of expertise as well as technical skills and a query user may retrieve an optimal tree of close-by users that can perform the task more quickly and efficiently. Finally, it can be used as an instant emergency/news system. For example, considering recent disasters such as the Sandy hurricane in New York or the earthquake (and consequently the tsunami) in Japan, the users would have preferred querying and retrieving instant information from the crowd near the areas of disaster than reading arbitrary (and maybe out-dated) information on the web.

References

[1] S. M. Allen, G. Colombo, and R. M. Whitaker. Cooperation through self-similar social networks. *ACM Transactions on Autonomous and Adaptive Systems (TAAS)*, 5(1), 2010.

[2] P. Andreou, D. Zeinalipour-Yazti, A. Pamboris, P.K. Chrysanthis, and G. Samaras. Optimized query routing trees for wireless sensor networks. *Information Systems (InfoSys)*, 36(2):267–291, 2011.

[3] M. Azizyan, I. Constandache, and R.-R. Choudhury. Surroundsense: mobile phone localization via ambience fingerprinting. In *MobiCom*, 2009.

[4] A.T. Campbell, S.B. Eisenman, N.D. Lane, E. Miluzzo, R.A. Peterson, X. Zheng H. Lu, M. Musolesi, K. Fodor, and G.S. Ahn. The rise of people-centric sensing. In *IEEE Internet Computing*, 12(4):12–21, July–August 2008.

[5] G. Chatzimiloudis, A. Konstantinidis, C. Laoudias, and D. Zeinalipour-Yazti. Crowdsourcing with smartphones. In *IEEE Internet Computing*, 16(5): 36–44, September–October 2012.

[6] S.-K. Chen and P.-C. Wang. Design and implementation of an anycast services discovery in mobile ad hoc networks. *ACM Transactions on Autonomous and Adaptive Systems (TAAS)*, 6(1):2, 2011.

[7] B. N. Chun, D. E. Culler, T. Roscoe, A. C. Bavier, L. L. Peterson, M. Wawrzoniak, and M. Bowman. Planetlab: An overlay testbed for broad-coverage services. *Computer Communication Review*, 33(3):3–12, 2003.

[8] C. A. Coello Coello, D. A. Van Veldhuizen, and G. B. Lamont. *Evolutionary Algorithms for Solving Multi-Objective Problems*, volume 5. Kluwer Academic Publishers, 2002.

[9] T. Das, P. Mohan, V.N. Padmanabhan, R. Ramjee, and A. Sharma. Prism: Platform for remote sensing using smartphones. In *MobiSys*, 2010.

[10] DBLP. DBLP Computer Science Bibliography, http://dblp.uni-trier.de/xml/, 2010.

[11] G. C. de Silva and K. Aizawa. Retrieving multimedia travel stories using location data and spatial queries. In *The 17th ACM International Conference on Multimedia*, pages 785–788. ACM, 2009.

[12] G. C. de Silva, T. Yamasaki, and K. Aizawa. Sketch-based spatial queries for retrieving human locomotion patterns from continuously archived gps data. *IEEE Transactions on Multimedia*, 11(7):156–166, 2009.

[13] K. Deb. *Multi-Objective Optimization Using Evolutionary Algorithms*. Wiley and Sons, 2002.

[14] K. Deb, A. Pratap, S. Agarwal, and T. Meyarivan. A fast and elitist multi-objective genetic algorithm: NSGA II. *IEEE Transactions on Evolutionary Computation*, 6(2):182–197, 2002.

[15] A. Deligiannakis, Y. Kotidis, V. Stoumpos, and A. Delis. Building efficient aggregation trees for sensor network event-monitoring queries. In *GeoSensor Networks*, volume 5659 of Lecture Notes in Computer Science, pages 63–76. Springer Berlin Heidelberg, 2009.

[16] S. Eisenman, E. Miluzzo, N. Lane, R. Peterson, G. Seop-Ahn, and A.T. Campbell. Bikenet: A mobile sensing system for cyclist experience mapping. *ACM Transactions on Sensor Networks (TOSN'09)*, 6(1), December 2009.

[17] J. Eriksson, L. Girod, B. Hull, R. Newton, S. Madden, and H. Balakrishnan. The pothole patrol: Using a mobile sensor network for road surface monitoring. In *MobiSys*, pages 29–39, 2008.

[18] A. Gahng-Seop, M. Musolesi, H. Lu, R. Olfati-Saber, and A.T. Campbell. Metrotrack: Predictive tracking of mobile events using mobile phones. In *DCOSS*, pages 230–243, 2010.

[19] D. Geiger, M. Rosemann, and E. Fielt. Crowdsourcing information systems: A systems theory perspective. In *22nd Australasian Conference on Information Systems (ACIS'11)*.

[20] Gnutella. Gnutella peer-to-peer network, 14 March 2000. http://gnutella.wego.com.

[21] H. Inamura, G. Montenegro, R. Ludwig, A. Gurtov, and F. Khafizov. *TCP over Second (2.5G) and Third (3G) Generation Wireless Networks*. RFC 3481 (Best Current Practice), February 2003.

[22] J. Jia, J. Chen, G. Chang, Y. Wen, and J. Song. Multi-objective optimization for coverage control in wireless sensor network with adjustable sensing radius. *Computers and Mathematics with Applications*, 57(11–12):1767–1775, 2009.

[23] V. Kalogeraki, D. Gunopulos, and D. Zeinalipour-Yazti. A local search mechanism for peer-to-peer networks. In *11th International Conference on Information and Knowledge Management (CIKM'02)*, pages 300–307, McLean, Virginia, USA, 2002.

[24] Y. Ko and N. H. Vaidya. Location-aided routing (LAR) in mobile ad hoc networks. *Wirel. Netw.*, 6(4):307–321, 2000.

[25] A. Konstantinidis, C. Aplitsiotis, and D. Zeinalipour-Yazti. Multi-objective query optimization in smartphone social networks. In *12th International Conference on Mobile Data Management (MDM'11)*, pages 27–32.

[26] A. Konstantinidis, C. Costa, G. Larkou, and D. Zeinalipour-Yazti. Demo: A programming cloud of smartphones. In *10th International Conference on Mobile Systems, Applications, and Services (MobiSys'12)*, pages 465–466.

[27] A. Konstantinidis and K. Yang. Multi-objective energy-efficient dense deployment in wireless sensor networks using a hybrid problem-specific MOEA/D. *Applied Soft Computing*, 11(6):4117–4134, 2011.

[28] A. Konstantinidis, K. Yang, Q. Zhang, and D. Zeinalipour-Yazti. A multi-objective evolutionary algorithm for the deployment and power assignment problem in wireless sensor networks. *New Network Paradigms, Elsevier Computer Networks*, 54:960–976, 2010.

[29] C. Laoudias, G. Constantinou, M. Constantinides, S. Nicolaou, D. Zeinalipour-Yazti, and C. G. Panayiotou. The Airplace indoor positioning platform for android smartphones. In *13th International Conference on Mobile Data Management (MDM'12)*, pages 312–315.

[30] G. Larkou, C. Costa, P. Andreou, A. Konstantinidis, and D. Zeinalipour-Yazti. Managing smartphone testbeds with smartLab. In *Proceedings of the 27th USENIX Large Installation System Administration Conference*, LISA'13, 2013, pages 115–132.

[31] J. Ledlie, B. Odero, E. Minkov, I. Kiss, and J. Polifroni. Crowd translator: On building localized speech recognizers through micropayments. *ACM SIGOPS'10 Operating Systems Review*, 43(4): 84–89.

[32] Q. Lv, P. Cao, E. Cohen, K. Li, and S. Shenker. Search and replication in unstructured peer-to-peer networks. In *16th International Conference on Supercomputing (ICS'02)*, pages 84–95, New York, USA, 2002.

[33] S. Matyas, C. Matyas, C. Schlieder, P. Kiefer, H. Mitarai, and M. Kamata. Designing location-based mobile games with a purpose: collecting geospatial data with city Explorer. In *International Conference on Advances in Computer Entertainment Technology*, pages 244–247, 2008.

[34] W. S. Ng, B. C. Ooi, K.-L. Tan, and A. Zhou. Peerdb: A p2p-based system for distributed data sharing. *International Conference on Data Engineering*, 0:633, 2003.

[35] A. Quinn and B. B. Bederson. Human computation: A survey and taxonomy of a growing field. In *Annual Conference on Human Factors in Computing Systems (CHI'11)*, pages 244–247.

[36] M. Ra, J. Paek, A. Sharma, R. Govindan, M. H. Krieger, and M. J. Neely. Energy-delay tradeoffs in smartphone applications. In *MobiSys*, pages 255–270, 2010.

[37] R. Rajagopalan, C. K. Mohan, P. K. Varshney, and K. Mehrotra. Multi-objective mobile agent routing in wireless sensor networks. In *Proc. IEEE CEC'05*, pages 1730–1737, Edinburgh, Scotland, September 2005.

[38] R. K. Rana, C. T. Chou, S. S. Kanhere, N. Bulusu, and W. Hu. Ear-phone: An end-to-end participatory urban noise mapping system. In *IPSN*, pages 105–116, 2010.

[39] T. Repantis and V. Kalogeraki. Data dissemination in mobile peer-to-peer networks. In *6th International Conference on Mobile Data Management (MDM'05)*, pages 211–219, Ayia Napa, Cyprus, 2005.

[40] M. Stevens and E. D. Hondt. Crowdsourcing of pollution data using smartphones. *Ubiquitous Computing (UbiComp'10)*.

[41] A. Thiagarajan, L. Ravindranath, K. LaCurts, S. Madden, H. Balakrishnan, S. Toledo, and J. Eriksson. Vtrack: Accurate, energy-aware road traffic delay estimation using mobile phones. In *SenSys '09: Proceedings of the 7th ACM Conference on Embedded Networked Sensor Systems*, pages 85–98, New York, NY, USA, 2009. ACM.

[42] H. Tomiyasu, T. Maekawa, T. Hara, and S. Nishio. Profile-based query routing in a mobile social network. In *7th International Conference on Mobile Data Management, 2006. MDM 2006*, pages 105 – 105, May 2006.

[43] G. Werner-Allen, P. Swieskowski, and M. Welsh. Motelab: A wireless sensor network testbed. In *Fourth International Symposium on Information Processing in Sensor Networks, 2005. IPSN 2005*, pages 483 – 488, April 2005.

[44] B. Xu, O. Wolfson, and C. Naiman. Machine learning in disruption-tolerant MANETs. *ACM Transactions on Autonomous and Adaptive Systems (TAAS)*, 4(4), 2009.

[45] D. Zeinalipour-Yazti, V. Kalogeraki, and D. Gunopulos. Exploiting locality for scalable information retrieval in peer-to-peer systems. *Information Systems (InfoSys), Elsevier*, 30(4):277–298, 2005.

[46] Q. Zhang and H. Li. MOEA/D: A multi-objective evolutionary algorithm based on decomposition. *IEEE Transactions on Evolutionary Computation*, 11(6):712–731, 2007.

[47] Y. Zheng, L. Liu, L. Wang, and X. Xie. Learning transportation mode from raw GPS data for geographic applications on the web. In *WWW*, pages 247–256, 2008.

Chapter 14

Encounter-Based Opportunistic Social Discovery in Mobile Networks

Udayan Kumar

Department of Computer and Information Science and Engineering
University of Florida
Gainesville, Florida

Ahmed Helmy

Department of Computer and Information Science and Engineering
University of Florida
Gainesville, Florida

CONTENTS

14.1 Introduction

Neighbor selection is a non-trivial problem faced in the design of opportunistic networks. Consider any opportunistic networking applications ranging from social discovery to P2P mobile games to DTNs and MANETs, almost all of the applications require neighbor discovery followed by some kind of neighbor selection. A DTN application may select a neighbor that has higher chances of relaying the message to the destination or a P2P mobile gaming application may select a neighbor based on duration of past encounters. However, due to diversity of requirements, we argue that there is no one optimal method of selecting neighbors for all applications.

Selecting a subset of neighbors from all the available neighbors becomes challenging in the case of opportunistic network because (1) neighbor selection is application dependent, (2) identities of all the neighbors may not be known, (3) users may not be comfortable communicating with unknown neighbors, (4) without incentives, neighbors may not be willing to participate (mobile device are constrained for power & processing), (5) not all neighbors may meet the requirement of the application, and 6. some neighbors may have malicious intent.

At the same time, there are several unique characteristics of opportunistic networks that provide an arsenal to tackle many of the above challenges such as 1. physical proximity that enables easier verification of identity (one can have face-to-face meetings and also exchange out-of-band cryptographic keys), 2. tight coupling of mobile devices and users can allow customization of selection based on user needs, and 3. availability of location and other contextual information (e.g. mode of transport, importance of location) can give valuable insights when making neighbor selection. Face-to-face meeting, verification of user profile, and setup of out-of-band keys are comparatively low-cost operations for neighbors in radio range as they are in physical proximity (for e.g. Bluetooth 4.0 range is < 50 m) when compared to wired networks.

Utilizing the above-mentioned characteristics, we provide a brief overview of studies in the areas of encounter-based neighbor and social discovery, context awareness and recommendation (and reputation) systems for opportunistic network establishment. Furthermore, we present detailed discussion of the design and analysis of a new encounter-based framework, ConnectEnc (Connections based on Encounters) as a solution to the problem of neighbor selection. The framework is fully distributed, self-bootstrapping, privacy preserving, and integrates attack resilience mechanisms. This framework utilizes mobile encounters as a primitive to address the problem of neighbor selection. A mobile encounter signifies the detection of radio signals (WiFi, Bluetooth, etc.) from another device (current neighbor). The use of short-range radios (e.g., Bluetooth, WiFi) enables detection and utilization of proximity and encounters. Furthermore, the tight coupling between users and mobile devices enables new and accurate ways to establish behavioral profiles that can be used to fine-tune the neighbor selections based on application requirements; e.g., by selecting, say, only the users encountered at multiple locations. Along with this encounter framework, we also promote face-to-face interaction between peer-to-peer users that allows authentication peer identity and establishment of out-of-band encryption keys [14] that can be later used to establish secure a P2P/opportunistic communication channel.

At the core of ConnectEnc, we use encounter rating metrics called encounter filters. These encounter filters analyze mobile encounters, proximity, location, and context data in novel ways, to augment the user's (and application's) network/neighbor view and awareness. Its goal is to rate opportunities in terms of neighbor selections based on weighted filter scores that are coupled with the user's input and application requirements. We investigate and detail five different algorithms applied to filter the encountered devices.

It is the fusion and integration of these multi-dimensional data, that provides the promise of selecting better neighbors in opportunistic networking in ways we could not before, and in ways that are not possible in wired networks due to lack of connectivity proximity. An opportunistic network application can now state its neighbor/peer selection criteria to the ConnectEnc framework (such as neighbor with highest probability of meeting again or a neighbor who met at a particular location before, etc). The ConnectEnc framework, based on encounter filters can provide the most suitable candidate(s) out of all the current neighbors.

This study introduces a systematic framework and new protocol for gathering and processing the encounter information to build encounter-based profiles of the neighbors. Evaluation of the 'ConnectEnc' framework and mobile application is a three-phase process: 1. real-world mobile networks trace statistical analysis, 2. extensive trace-driven simulation of the framework components, and 3. prototype implementation and participatory testing on smartphones. First, we use wireless network traces from 3 different major university campuses spanning 9 months with over 70K users and 150 million encounters. We find that several filters possess desirable stability characteristics, and that selecting neighbors with high encounter scores in general forms a small world. Resilience to attacks (neighbors attempting to inflate encounter statistics), using anomaly detection, achieves less than 10% false positives and 7% false negatives. Second, we measure the effectiveness of ConnectEnc on epidemic routing in DTN, with selfishness using neighbor recommendations by ConnectEnc and obtain higher network performance reducing the effects of selfishness. Third, we conduct a series of surveys and participatory experiments using ConnectEnc's mobile application to evaluate the performance of the framework against the ground truth. We find users' selection of trustworthy peers/neighbors (for opportunistic communication) has a statistically strong correlation with ConnectEnc's recommendations. Further, ConnectEnc filters can capture 80% of the already known users within the top 25% of the encountered users.

Key contributions of this work include: 1. introducing a framework to augment the mobile user's perception and awareness of the network neighborhood by fusing multi-dimensional encounter and contextual data for better neighbor selection, 2. analyzing various trust adviser filters with extensive network traces, 3. proposing a model for anomaly or attacker detection, 4. developing a mobile app 'ConnectEnc' that integrates the filters and contextual information to aid the user in neighbor classification and selection, and 4. deploying ConnectEnc as a proof-of-concept mobile application to evaluate the framework based on ground truth via participatory testing.

14.2 Overview

This overview is organized into 4 different subsections; each corresponds to a step involved in establishing short-range-radio based mobile P2P networks. These 4 steps are i. neighbor discovery: here information for all the available devices is obtained, ii. neighbor selection: here a subset of all the available devices is selected, iii. connection establishment: after selection, peers exchange/negotiate connection parameters based on security and authentication, and iv. applications: here we list some of the popular P2P applications.

14.2.1 Neighbor Discovery

In any P2P scenario, if the peers have unpredictable behavior (availability) in either space or time, there will be a need to discover peers that are currently available for interactions. Most of the popular P2P applications, whether communicating over the Internet or via an ad hoc radio network employ some kind of P2P discovery mechanism. There are primarily two ways to discover i. using a central infrastructure (torrents) and ii. ad hoc (sensor networks, DTNs). For opportunistic networking, the latter is more commonly utilized. To discover other peers, generally peers send out a discovery radio beacon to solicit response from all the neighboring peers. Several popular radio protocols, such as Bluetooth and WiFi-Direct natively support this kind of discovery. Since a peer may be continuously moving (surrounding peers may also move), searching for available peers can be an expensive process in terms of energy consumption. Several energy-efficient methods have been proposed (including one by the authors of this chapter) [35, 20]. There is also a research direction where based on the previous discovery patterns of a peer, predictions are made about future discovery of that peer [30]. Researchers find that human movement patterns are predictable at a coarse granularity based on the fact that peer discoveries can also be predicted.

14.2.2 Neighbor Selection

Once a set of neighbors is discovered, there may be a need to select a subset of neighbors based on the requirements of the application. This step may be necessary when an application does not want to interact with all the available peers. A DTN routing application such as epidemic routing [34] may interact with all the available peers, however, several other DTN routing protocols such as [21, 26, 16, 36] may require selecting peers based on their encounter history. The main focus is on optimum end-to-end routing and less on one hop (next-hop) node selection. They may not be privacy preserving or may not provide stable recommendations and are not easily configurable to the needs of an application. Similarly, a gaming application may want to select peers who may be encountered again to finish off the game. Unlike neighbor discovery, there is no one way to select a peer. Different applications have different criteria.

The idea of neighbor selection in P2P networks has been well explored in wired networks where neighbors are selected based on geographic proximity, latency, bandwidth available, etc. [27]. These ideas, however, do not hold when a mobile application wants to leverage a P2P based direct radio connection. Mobile P2P networks face greater set of challenges since the peers are mobile and there is a high possibility of peers moving out of radio range. There exist several DTN routing protocols that employ node selection algorithms [21, 35, 16] but focus mainly on optimum end-to-end routing and thus focus is less on one hop-node selection. They may not be privacy preserving or may not provide stable recommendations and are not easily configurable to the needs of an application. The lack of any optimized one-hop P2P neighbor selection is also a challenge for the P2P mobile application development community [6]. There are several P2P applications available [5, 33, 4] but without any automatic strategy for peer selection, it is left to the user to decide. But how will a user decide? ConnectEnc attempts to solve this problem by providing background information about the peers to make an informed selection. To the best of our knowledge, there is no existing solution to this problem. Later in this chapter, we propose, as a solution to this neighbor selection problem, a multi-criteria neighboring peer selection framework, ConnectEnc.

Several researchers have proposed novel approaches in peer selection using reputation-based schemes, incentive-based schemes, and game theory. The reputation-based schemes target better peer selection based on previous interaction records by rating interactions with each peer. In [11], a node detects misbehavior locally by observation and use of second-hand information. In [10], a fully distributed reputation system is proposed that can cope with false information, where each node maintains a reputation rating for peers. In [31, 9, 15, 8], analysis of reward provisions and punishment is conducted based on game theoretic approaches to provide incentives for message delivery. In [13], authors propose a game-theoretic model to discourage selfish behavior and stimulate cooperation by leveraging Nash equilibria with socially optimal behavior. In [38], authors propose a pricing mechanism to give credits to nodes that participate in the message forwarding mechanism. The cooperation is developed based on the number of messages transferred by the users.

A common theme in these works is the reliance on peer *interaction* to evolve the reputation/credit scores. Inherently, this creates an undesirable *circular dependence*, where interaction requires technology adoption (reputation/credit system), which in turn requires trust in specific instances of these systems. Hence, there is a compelling need for a *bootstrap* mechanism, which we directly address in our proposed design. Further, unlike other studies, we do utilize encounter context in this paper. Our work contributes toward solving this challenge by providing inputs from a user's location preferences and contextual (e.g., social) behavior.

14.2.3 Connection Establishment

Once a neighbor is selected, the next thing to do is to establish a connection with the neighbor to enable data/information exchange. The connection establishment involves several challenges, mostly from the security and privacy standpoint. Challenges such as establishing secure connections between the nodes and identifying a peer when meeting again are some of the bigger challenges faced by mobile P2P connection establishment. Authors of [25, 14, 28] propose an explicit authentication mechanism to generate trust and cooperation in network. These approaches are better modeled for small groups [25] and require exchange of public keys and the installation of the private key on the user's device [14]. Another step to secure a connection can be to meet the peer/neighbor face to face (since radio range of mobile devices is limited, peers must be co-located physically), verify the peer, and can also set up out-of-band encryption keys. The out-of-band encryptions keys can be set up by the peers simply by exchanging a secret code word when meeting face to face or can also achieve cryptographic strength by using [14].

14.2.4 Applications

Whether it is a DTN application or a mobile P2P application, once a connection is established, these applications can start leveraging the established connection. A few examples of existing P2P applications are P2P multiplayer mobile gaming [5], cooperative sensing [23], mobile proxy [4], social discovery [3], personal safety [33], and cellular offloading [19]. This generation of applications do not employ a neighbor selection method, hence they are mainly human driven in terms of neighbor selection; with very minimal or no automated sensing and selection of neighbors.

In the following sections we present our proposed ConnectEnc framework, which can automate the neighbor selection process by providing automatic requirement-specific neighbor selection. The framework provides novel ways to rate the neighbors and integrates within itself existing peer-rating systems such as recommendation and reputation systems. We begin with the rationale for the design and then proceed toward design principles. Following the design, we present details of encounter filters, comprehensive analysis of the framework, and a section on validation, along with a summary of the ConnectEnc user study.

14.3 Rationale and Architecture

In this section, we present the rationale, design goals, and high-level design of the ConnectEnc framework.

14.3.1 Rationale and Approach

P2P mobile applications can be only used with peers who are in the radio range. In cases where several peers may be around to participate in a P2P activity, which out of those should be picked? For some application, any set of peers would work but many applications as mentioned earlier would require an informed selection based on the requirements of the application. For e.g. which peer has a higher chance of encountering a peer again or for a longer encounter session. Also users may not want to interact with randomly selected peers (Section 14.6.1).

Looking at this problem from a user's perspective, who will be the peers this user may meet again (or for a longer duration, etc.). These will be the peers who are similar to the user in their behavior (being at the same place and the same time). In social science this is known as the principle of homophily [29]. Homophily can be measured is several different ways and using encounter history we can measure spatio-temporal homophily. We propose several encounter filters to measure spatio-temporal homophily. For greater trust and reliability (optional) on a peer, a user can meet this peer face to face. This is easier in mobile P2P networks as the peers are in physical proximity and some peers may already be socially known (although we do not make any such assumptions). Face-to-face meetings can be utilized to verify the authenticity of a peer's claim and can be used to set up out-of-band encryption keys [14, 25]. If these keys are stored with ConnectEnc, then other applications can use this key to securely communicate with the specific peer.

14.3.2 Design Goals

The main design goals for ConnectEnc include:

1. Balanced Discovery: In our peer selection (and discovery) framework, identifying peers known to the user (i.e., a perfect matches) is not always our goal. Instead, we aim to provide the user with a balance between acquaintances and new matches as a more useful and realistic measure. We achieve this by generating encounter scores over several filters and allowing application-specific peer selection.

2. Stability: The peer recommendation should be stable over time and insensi-

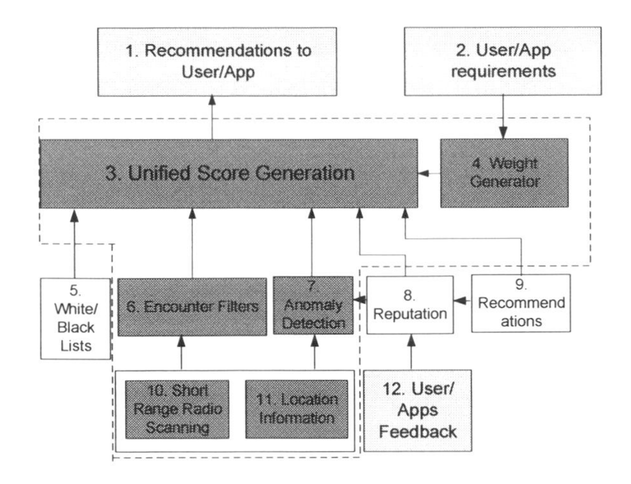

Figure 14.1: Block diagram overview of the ConnectEnc architecture. Dotted lines enclose the modules of ConnectEnc (3, 4, 6, 7, 10, and 11). The blocks 1, 2 and 12 illustrate the components that interact with other applications and user of the device. Blocks 5, 8 and 9 illustrate the integration of external systems with ConnectEnc.

tive to minor, temporary changes and noise in user behavior. Outliers and anomalies should be detected and removed.

3. Distributed Operation: ConnectEnc should be able to provide all the functionalities in a distributed fashion without the need for a centralized infrastructure or trusted third party. All operations should be performed locally on the users device. No sharing of user information should be required by the system for privacy preservation.

Other goals include resilience (against attacks), power efficiency, and flexibility to utilize external sources (reputation and recommendation).

14.3.3 Overall Design

An architectural overview of the ConnectEnc framework and its related subsystems is provided in Figure 14.1. Overall there are 3 categories of blocks: 1. Blocks 3, 4, 6, 7, 10, and 11 indicating the core components of ConnectEnc, 2. Blocks 1, 2, and 12 indicating the modules that interact with the applications and users, and 3. Blocks 5, 8 and 9 indicate examples of external systems that can be integrated with ConnectEnc.

All the core components of ConnectEnc are fundamental to the design of the framework. These modules are required to meet the design goals. The basic functionality of each of the modules is as follows: the *Short Range Radio Scanning* module provides basic encounter information (for e.g., Bluetooth, WiFi AP discovery). The *Location Information* module provides the device's positioning data. This data is now received by encounter filters and *Anomaly Detection* modules. The encounter filters are the block that generates encounter scores using a family of filters (described in the next section). *Anomaly detection* provides a recommendation regarding suspicious encounter activities. The *Unified Score Generation* module combines the output of encounter filters with the output from *anomaly detection*, *recommendation system*, *reputation system*, and *black and white lists* using the weights provided by the *weight generator*. The *weight generator* provides weights that decide how much importance is to be given to the different inputs to *Unified Score Generation*. The selection of weights is done based on the requirements given by the application.

Blocks 1, 2, and 12 in Figure 14.1 indicate how applications and user can interact with the ConnectEnc framework. We perceive the applications will interact with ConnectEnc framework by first setting up the requirements by specifying either relatively or absolutely the importance (weight) of each input considered by the *Unified Score Generation*. Once the weights are selected, ConnectEnc will generate a rank-ordered list of the encountered peers (Block 1) in the neighborhood. Once the application finishes the transactions with the neighboring devices, it provides (optional) feedback about the experience with the users. This feedback is feed into the *Reputation* block.

Blocks 5, 8, and 9 are optional and external components of the ConnectEnc framework. These modules can enrich the peer selection process but are not required. Any existing systems providing necessary functionality can easily be integrated with this framework. The Reputation block receives peer feedback from applications based on application's experience with this peer device. The Recommendation block runs an external recommendation service and provides input to the framework. The White/Black List allows users to explicitly give a score to a device. This can empower the user to add peers without even encountering them.

With this conceptual understanding of the system, we now describe the heart of ConnectEnc framework, encounter filters.

14.4 Encounter Filters

Encounter Filters rate encounters in multiple dimensions so that applications and users can make selections based on a rich set of choices. Due to lack of space we are going to discuss and analyze 5 major filters, however, the design of ConnectEnc is modular and can easily integrate more filters (if needed). The filters we propose and investigate are based on: *i.* simple encounter (frequency and duration) ranking and *ii.* spatial correspondence.

14.4.1 Simple Encounter Ranking

These filters rate encounters by aggregating the encounter data using simple statistics. They are:

Frequency of encounters (*FE*): This filter ranks encountered devices based on total number of encounters over a window of history, regardless of the duration. So if a peer *A* is encountered more times than peer *B*, peer *A* will get a higher rank than *B*. For an encounter session (continuous uninterrupted encounters) the FE score for the peer is increased only once by one. This filter score can be useful for applications when they have to decide between peers based on the chances of meeting again. Simply put a higher FE score means higher chances of meeting.

Duration of encounters (*DE*): This filter ranks encountered devices based on the duration of encounters. Encounter duration can be measured in two ways: i. total duration of encounters, ii. average session duration per encounter. An application using this metric to decide between peers will find that a higher DE score would mean that the peer may have a higher chance of having a longer-duration encounter session. Through our trace-based analysis we find that both measures of DE have a statistically strong correlation with each other. Since, the first measure requires less storage and computation, we choose its score to represent the DE score.

14.4.2 Spatial Correspondence

Spatial correspondence-based measures rate encounters on similar location visitation patterns. Higher spatial correspondence means that the peer is very similar in visiting locations as the user herself. Selecting a user with higher spatial correspondence means selecting a user who may be encountered more at locations preferred by the user. Spatial correspondence can be measured in multiple ways, we present some of those techniques below.

Profile vector (*PV*): To capture spatial correspondence, we have designed a PV filter that stores location visitations of a user in a single dimensional vector. It is assumed for this filter that a device has some localization capability, which is quite common for today's devices. Each device maintains a vector. The columns of the vectors represent the different locations visited by a user and the values stored in each cell indicate either duration or count of the sessions at that particular location. At each location visit, the vector is updated with respect to the location.

To get an encounter score, this vector is exchanged with other users and the inner product of the two vectors is computed. This score is higher if the two *PV*s are similar and can be zero, if the users do not have any visited location in common. Here, implicit weight is given to locations based on the count/duration spend. We can also provide an option to the user, where locations can have explicit weights.

However, this filter is not privacy preserving and can introduce attacks in the system, where a peer can tamper with its vector; also there are communication costs involved in exchanging the vectors. This problem is solved by *LV* filters at the cost of having lesser information to compute similarity scores with.

Location vector (*LV*): The *LV* filter is very similar to the *PV*, except that a user not only maintains a vector for itself but also for each encountered peer. The columns of the vectors represent the different locations visited by a user and the values stored in each cell indicate either duration (*LV-D*) or count (*LV-C*) of the sessions at that particular location. For every encounter, the vector for the encountering peer is updated with respect to the encounter location. See the illustration in Figure 14.2.

Since vectors for all the encountering peers are maintained locally on the device, LV requires no exchange of vectors among users for calculating similarity. This is

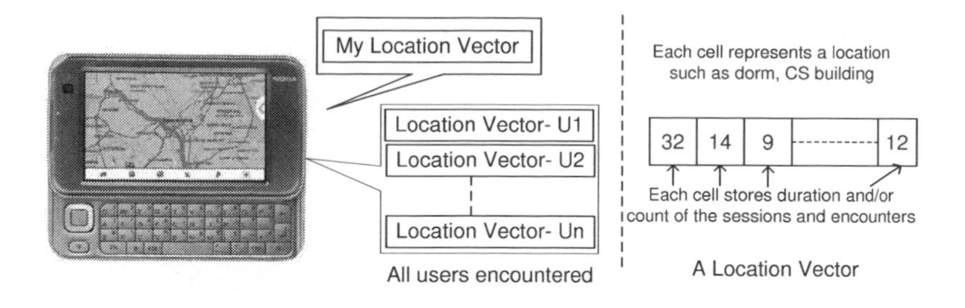

Figure 14.2: Location vector *LV* for a user.

more privacy-preserving and more resilient to attacks since only first hand information is used (equivalent to what the user might have observed). This privacy comes at the cost of requiring extra storage space for storing vectors for each user. Considerable storage optimization is achieved by storing (for each encountering user) only the locations where encounters happened. LV similarity calculations are similar to PV.

Behavior matrix (*BM*): The behavior matrix captures a spatio-temporal representation of user behavior. Columns of the behavior matrix denote locations and rows represent time units (days in our case). The value stored at each cell is a fraction of the online time spent by the user at a particular location on a particular day (see Figure 14.3). Each user maintains their own matrix. To get the correspondence score, users can exchange and compare the two matrices.

To make the behavior similarity check efficient (in terms of space and computation complexity) and privacy-preserving (as only the summary of the matrix is exchanged), we use the eigenvalues of the behavior matrix for exchange between the two users. The eigenvalues are generated using SVD (singular value decomposition). SVD is applied to a behavior matrix *M*, such that:

$$M = U \cdot \Sigma \cdot V^T, \tag{14.1}$$

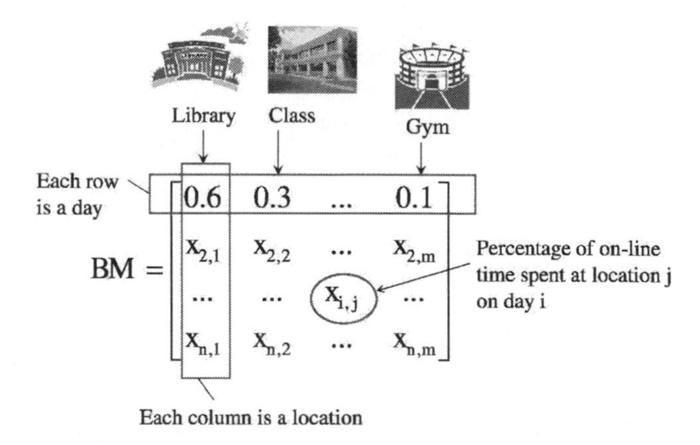

Figure 14.3: Behavior matrix for a user.

where a set of *eigen-behavior* vectors, $v_1, v_2, \ldots, v_{rank(M)}$ that summarize the important trends in the original matrix M can be obtained from matrix V, with their corresponding weights, $w_{v_1}, w_{v_2}, \ldots, w_{v_{rank(V)}}$ calculated from the eigen values in the matrix Σ. This set of vectors is referred to as the *behavioral profile* of the particular user, denoted as $BP(M)$, as they summarize the important trends in user M's behavioral pattern. The *behavioral similarity* metric between two users' association matrices A and B is defined based on their *behavioral profiles*, the vectors of a_i and b_j and the corresponding weights, as follows:

$$Sim(BP(A), BP(B)) = \sum_{i=1}^{rank(A)} \sum_{j=1}^{rank(B)} w_{a_i} w_{b_j} |a_i \cdot b_j|, \quad (14.2)$$

which is essentially the weighted cosine inner product between the two sets of *eigen-behavior* vectors.

BM, like *PV*, is not privacy-preserving, but can provide better spatio-temporal similarity calculations. Due to its privacy preservation, in the following sections, we have only used the LV filter for spatial correspondence.

14.4.3 Hybrid Filter (HF)

Each filter provides a different perspective on an encounter or behavioral aspect. The hybrid filter provides a systematic and flexible mechanism to combine the scores from all filters and present a unified score to the users. The selection of weights for various filters would depend on several factors including the user's preference and feedback (check Section 14.6.1) and application requirements. A generic hybrid filter score (H) for a user U_j can be generated by using the following:

$$H(U_j) = \sum_{i}^{n} \alpha_i F_i(U j), \quad (14.3)$$

where $F_i(U_j)$ is the normalized score for user U_j according to filter i. The α_i is the weight given to filter score F_i and n is the total number of filters used. We select α_i such that $\sum \alpha_i = 1$, and $0 \leq \alpha_i \leq 1$.

This linear combination is chosen for its simplicity.[1] Our implementation allows users to customize these weights. From the analysis of user feedback (Section 14.6.2), we find that not all the users prefer the same weights.

The processing and storage overheads for all the filters are shown in Table 14.1.

14.4.4 Decay of Filter Scores

Users may have a change in lifestyle (e.g. move to a different city, switch jobs) and may not very often encounter some of the previously highly rated peers. So, there may be a need to decay the score of peers, if they have not been encountered in

[1] Other non-linear combinations shall be investigated in future work.

Table 14.1 Overhead of Filters in Terms of Processing and Storage

Filter	Processing Overhead	Storage Overhead
FE	$O(m)$	$O(n)$
DE	$O(m)$	$O(n)$
PV	$O(m)$	$O(l)$
LV	$O(m)$	$O(nl)$
BM	$O(m)$	$O(ld^2)$ for SVD
HF	$O(n)$	$O(n)$

Note: m is the total number of records in the encounter file, n is the number of unique encountered users, l is number of locations visited, d represents the number of days used for BM calculations. We also assume that $m >> n$.

Table 14.2 Facts about Studied Traces

Trace Source	U1	USC [22]	Dartmouth [1]
Time/duration of trace	Fall 2007	Spring 2007	Fall 2005
Start/End time	09/01/07-11/30/07	01/01/07-03/30/07	09/01/05-11/30/05
Unique Locations	845 APs	137 buildings	133 APs
Unique MACs analyzed	34694	32084	4906

a while. To design the decay of encounter scores, we borrow from social science studies that have shown that social relationships are dynamic and require frequent interactions to prevent decay. The strength of a relationship wanes with an increase in the time between interactions. This decay follows a exponential decay pattern with half time dependent on the relationship type [12] (3.5 years for family, 6 months for colleagues). We use a similar function to decay the filter scores with a user configurable half-time with 6 months set as the default.

14.5 Trace-Based Analysis

To evaluate our design of encounter filters, we consider anonymizied trace sets from three universities (see Table 14.2; the information provided in the traces is anonymized; the name of University U1 is also anonymized). The advantage of using WLAN traces is that they are much closer to reality in terms of user mobility (also representative of a larger population) than the existing synthetic mobility models. However, due to lack of ground truth in WLAN traces, we also collected traces with ground truth by deployment of ConnectEnc.[2] The results from the deployment are discussed after this section. The WLAN traces, much like other real traces, have a small percentage of noise and error. We assume that users associating to the same wireless access point (AP) encounter each other as AP range is generally less than 50 meters indoors and most of the traces are from indoor usage. It is assumed that each unique device (identified by MAC address) represents a user.

[2]MIT Reality Mining [17] traces have ground truth in terms of survey data. However, the average number of friends per person is close to 1 (including several users who have listed themselves as their friends). Therefore, this trace set cannot be meaningfully used for evaluations.

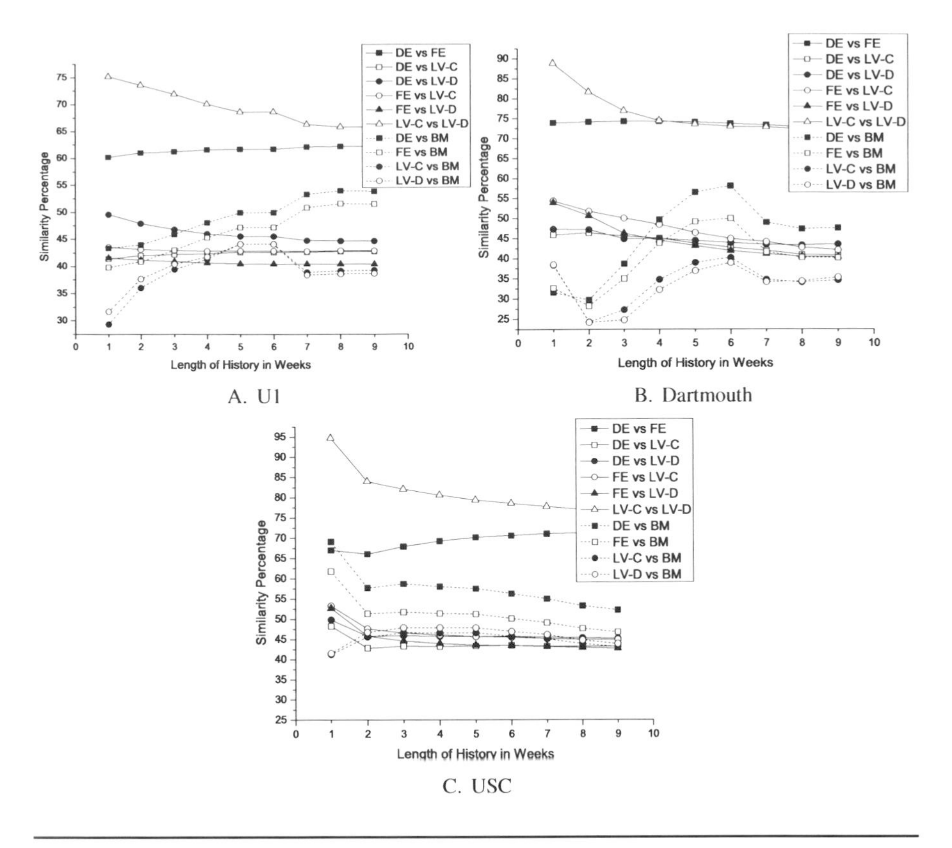

Figure 14.4: Correlation between the encounter lists produced by various filters at threshold, T=40%.

We use the WLAN traces to generate an Encounter Filter score for each user found in the trace. The WLAN trace is converted to encounter traces for each user by determining all the other users who had overlapping sessions with this user at the same AP (location). Encounter Filters take this encounter trace as an input and produce a ranked list by encounter score. For analysis, we pick the top $T\%$ of peers of a user from the ranked list. We investigate three properties of the filters: 1. correlation among filters, 2. stability, and 3. small-world characteristics.

14.5.1 Filter Correlation

We examine the degree of similarity (correlation) among scores from different filters. While high similarity indicates redundancy of the filters, low similarity implies orthogonality of the recommendations. For this investigation, we have considered 9 week-long traces and threshold the score list at $T = 40\%$ for varying length (at 1-week intervals) of encounter history (results for other T values show a similar trend).

As Figure 14.4 shows, the trends are similar across the traces. $LV-D$ and $LV-C$ filter results show $\sim$70% similarity as the list stabilizes around 9 weeks of history.

FE vs. DE stabilize around 60% to 70%. The rest of the filters stabilize between 55% to 30%, meaning they produce different sets of lists. The low similarity indicates that filters are not redundant and can be used to generate a rich set of recommendations.

14.5.2 Filter Stability

When an application requests a node recommendation, giving the criterion for selection, it may want to know if this recommendation will hold true in the future. For example, will a peer who had frequent encounters in the past maintais a similar trend in the future (user is assumed to maintain the same lifestyle). Basically, are ConnectEnc's recommendations stable in time? Moreover, instability can confuse users and reduce the effectiveness of in-application cache. Therefore, it is imperative to examine stability of the peer recommendation over time. We investigate the stability of the peer lists at $T = 40\%$ using 9 weeks of U1 traces (other T values and traces show a similar trend). Peer lists from multiple trace lengths are used to examine stability.

More than 90% similarity is found between 1-and 9-week traces for DE, FE, and LV-C filters (see Figure 14.5), implying that users selected in 1st week of encounter continued to be in the peer recommendation list of the 9-week-long encounter history. The BM filter shows high stability when the difference in history is less than 2 weeks (80%) and falls to 55% for 1 week and 9 weeks. The LV-D filter shows similarity of about 40% between any lists, implying that every week the list changes by 60%. This indicates that users may encounter regularly (by stability in LV-C) but may spend different amounts of time encountering over the weeks. Overall, we note that some filters (DE, FE, and LV-C) stabilize in just 1 week of history, which makes them suitable for recommendations when encounter history is short. The time interval between the recommendation list regeneration can also be long (reducing processing requirements).

14.5.3 Graph Analysis

We analyzed the effect of peer recommendations on the network graph and compared it with the regular and random graphs while increasing selection threshold (T) (using the DE filter; other filters show similar results). An edge is added between a pair of nodes only when at least one of them is peer recommended by each other (un-directed graph). We note that the clustering coefficient (CC) [7] of the network increases with $T\%$ and the path length (PL) decreases with an increase in $T\%$. For example, using a 9-week U1 trace, CC is 0.171 at $T = 10\%$ and becomes 0.201 at $T = 100\%$. However, in the same scenario, path length decreases from 3.64 to 2.59. More than 99% of the nodes were connected even at $T = 10\%$.

A small-world analysis is performed as described in [7]. We find that normalized CC (NCC) is close to the CC of the regular graph and the normalized PL (NPL) is close to the PL of the random graph (Figure 14.7 shows NCC and NPL for different lengths of traces and values of T). It appears that the network created by peer recommendations is a small-world network (results for other traces are similar).

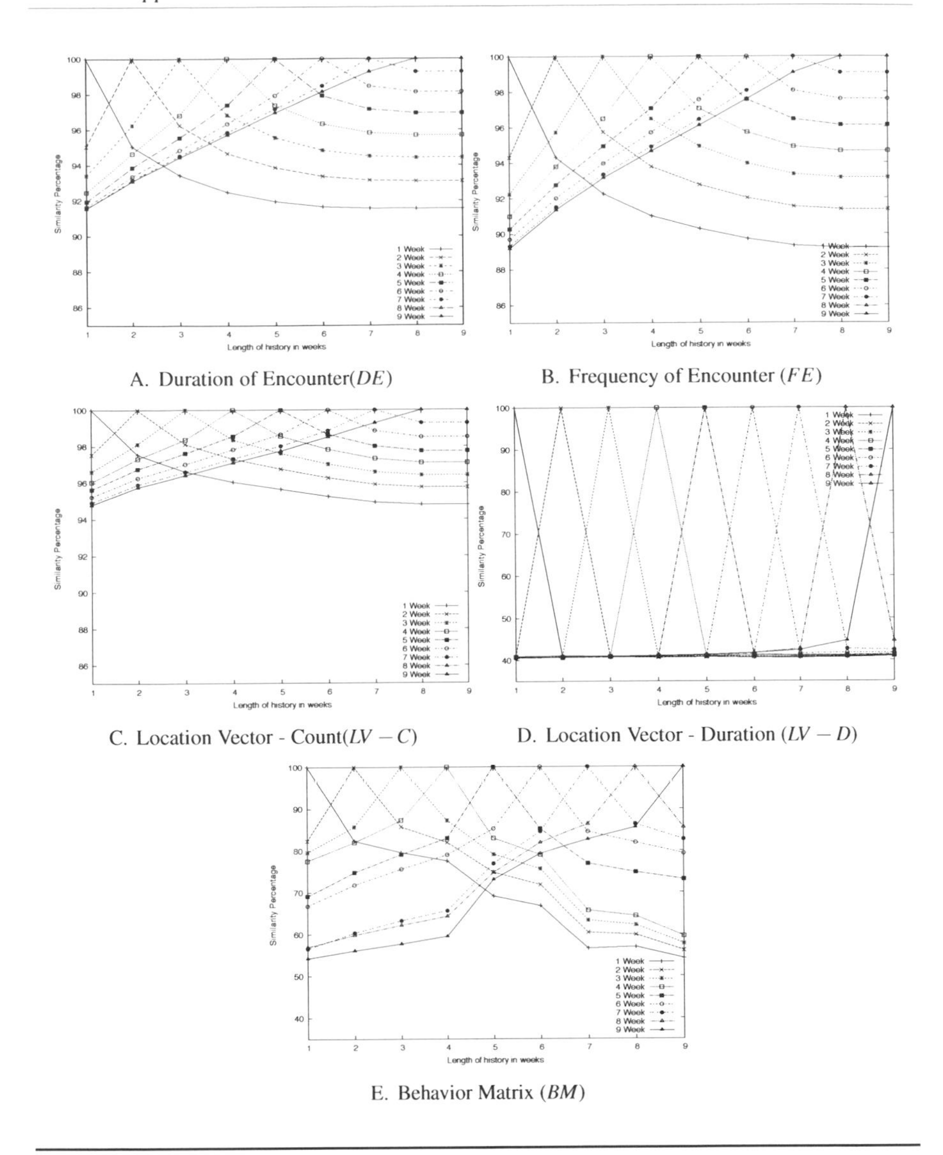

Figure 14.5: Comparison of encounter score lists belonging to different history for various filters at T=40% (note that the y-axis scale for *DE* starts at 85% and for *LV − D* and *BM* the scale starts at 35%).

14.6 Implementation and Simulation

In this section we show our validation of three major questions regarding the design of ConnectEnc: i. Do people prefer connecting with peers they already have some information on, ii. Is ConnectEnc able to discover peers that users may want to con-

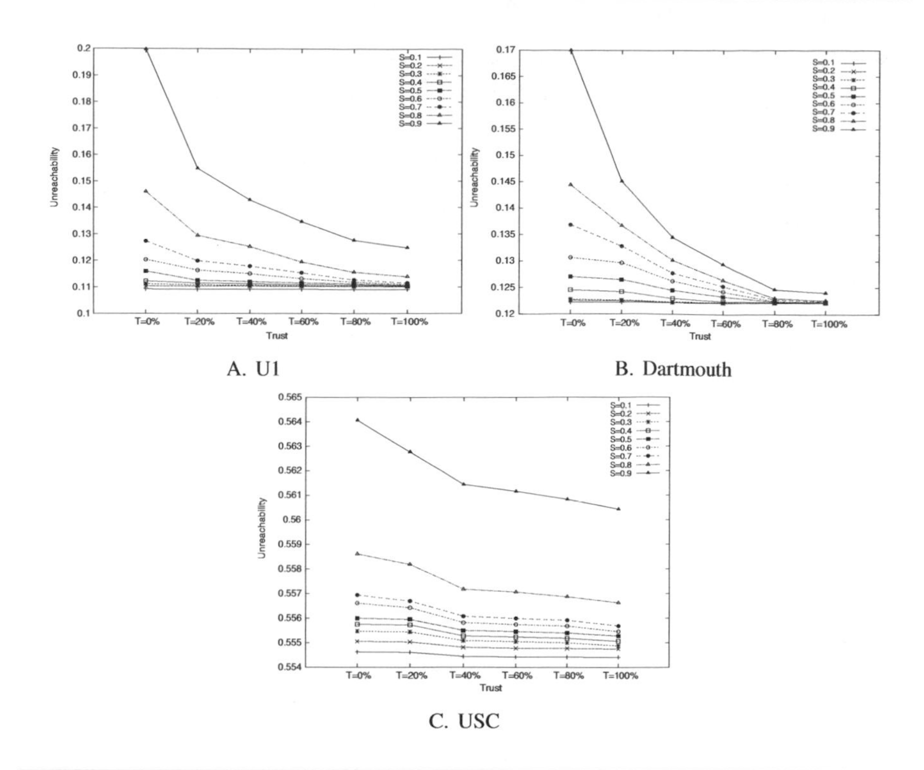

Figure 14.6: Average unreachability with varying encounter score threshold, T, and selfishness, S, using the DE filter.

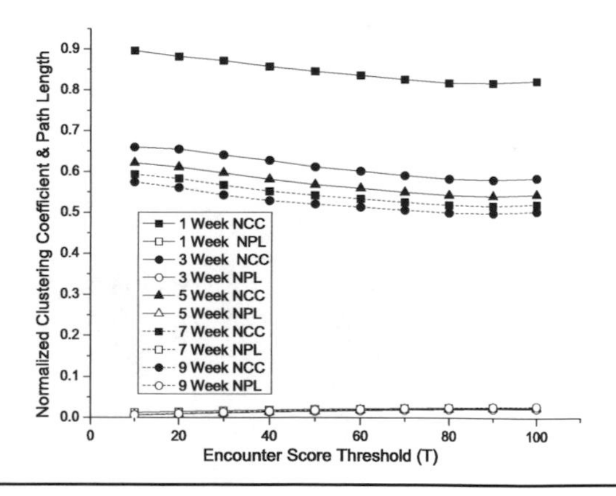

Figure 14.7: Normalized clustering coefficient and path length.

nect to? iii. Can ConnectEnc recommendations be useful in a P2P communication scenario? The first point is to check the premise of our assumption that a user may have preferences in selecting a peer that can affect how applications select neighbors (user may add constraints such as "I only want to play this game when there are higher chances of finishing this game later"). We tackle this question with a survey. The second point is to validate that if users prefer selecting peers who have a higher encounter score, is ConnectEnc able to discover them? We perform a user study using the ConnectEnc mobile application to address this question. For the third point, we take DTN routing as our P2P application. We show, with the help of large-scale trace-driven simulations, that ConnectEnc recommendations can lead to better routing in DTN networks having selfish nodes.

14.6.1 Survey

To investigate whether people prefer connecting with peers they already have some information on, we conducted a survey at a major computer network conference; this population has a good understanding of computer networks. Participants were asked to indicate their willingness to communicate (using P2P applications) under different scenarios on a scale of 1 to 10. We received 32 usable responses. As Figure 14.8 shows, willingness of the users to cooperate with an unknown user/device is low (mean is 2.31). However, willingness increases when users have knowledge about the encounter history. This reinforces the approach of ConnectEnc of using encounters to make peer selections. We also observe that users give more importance to combined scores (FE and DE scores are high) than individual scores (FE is high or DE is high). This justifies ConnectEnc's use of the hybrid filter for combining encounter scores. Standard deviations in results suggest that although most users want information about encountered users before cooperating, the individual importance of the filters may vary. This flexibility is made available in ConnectEnc's hybrid filter

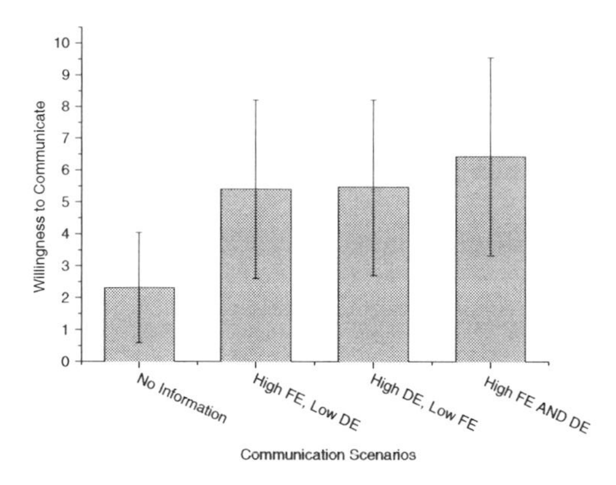

Figure 14.8: Survey results showing the user's propensity to communicate with other users in various communication scenarios.

(more generically by Unified Score Generation) by assigning weights according to the user's preference.

14.6.2 ConnectEnc Application

To investigate whether ConnectEnc is able to discover peers that users may want to connect to, we developed a ConnectEnc mobile application and conducted a user study. The application measures the mobile encounters (over Bluetooth radio) and rates the peer devices based on the scores of encounter filters. The application allows the user to mark a device as trusted if they would like to have any P2P communication with that device in the future. We collect this selection data and correlate the user selections with ConnectEnc recommendations (based on encounter score) to validate our approach.

Currently, ConnectEnc is available for the Android platform and the Linux-based Nokia Tablet N810 [2]. It provides the ability to rate encounter users based on FE, DE, LV, and hybrid filters. Encountered users can be sorted by any filter and weights because the Hybrid filters are user configurable. If some of the encountered users are currently discoverable, their listing would have a green circular mark as shown in Figure 14.9A. The application provides inbuilt facilities for scanning Bluetooth devices and wireless access points (for localization as GPS is energy-wise expensive; the user can select GPS, if needed). On selecting a particular user, encounter details (Figure 14.9B are presented and clicking on the map option, one can see encounter locations on a map (Figure 14.6.2). Apart from the filter scores, other statistics such as distribution of encounters with a peer over time are also available. Encountering devices can be rated for trust (P2P communication oriented) by the user on scale from -2 (no trust) to 2 (high trust). This application is also capable of providing peer selection information to other applications. This application can also be used a social discovery application, where it can alert the user about neighboring peer devices and give context by showing history and location of past encounters. We note that use of ConnectEnc does not affect privacy of the users. ConnectEnc only stores information on discoverable Bluetooth devices. Any Bluetooth-capable device can capture the same information that ConnectEnc captures.

Application evaluation: Twenty-two students (grad and undergrad) from the CS major ran the ConnectEnc app for at least a month. Users were asked to mark devices they trust (for P2P communication) in the application. On average, the number of trusted peers marked by each user is 15 and the number of unique devices encountered per user is 175. We use this data to investigate if recommendation by encounter filters correlates with trusted user identification. We note that not all encountered users who may be trusted/non-trusted may have been marked and not all trusted users may have discoverable Bluetooth. This issue will be of lesser concern as the adoption of ConnectEnc increases.

We rated the performance of ConnectEnc for each of the 5 filters (including the hybrid filter, referred as the combined filter (CF) in the app, with equal weights) on 2 metrics: 1. number of trusted peers in range top 1 to 10, 11 to 20, etc. of ConnectEnc recommendations (also known as the precision metric in information retrieval liter-

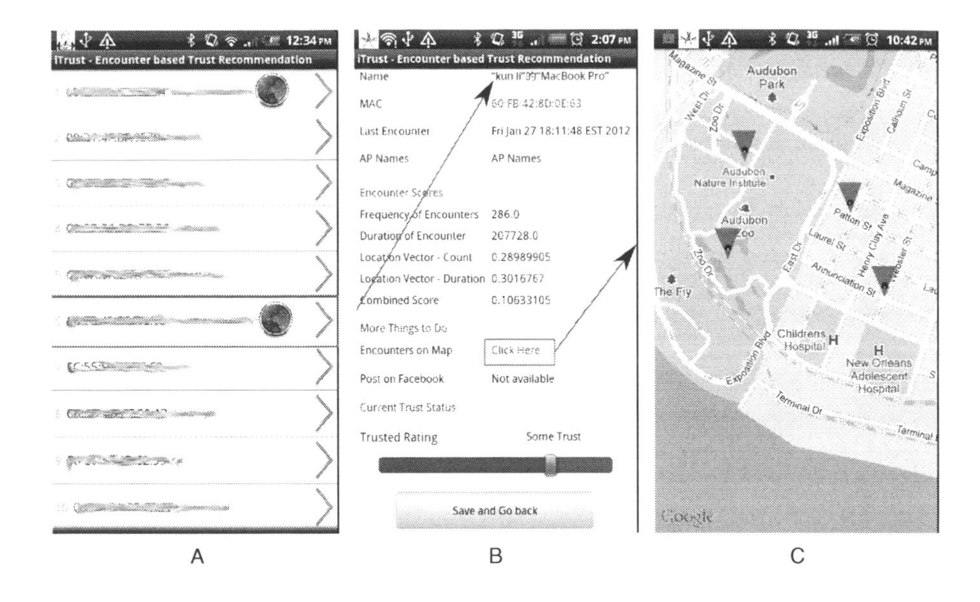

Figure 14.9: Selected screenshots of the ConnectEnc application (earlier it was named iTrust). A shows the main screen where encounter users are sorted by the filter score (Names and MAC are blurred intentionally). Current encounters are marked with circles. Marked known users are shown in lines 1, 2, 3, 5, 6, and 10. B shows details for an encountered user. C shows user encounters on the map. Annotations (arrows) are added to show the application flow.

ature) and 2. fraction of encounter peers needed (from top) to capture $x\%$ of trusted peers for each filter. The above metrics are chosen to measure how well the filters perform when compared to the user's selection. Here, ranking is based on the filter score.

For metric 1, we note that ConnectEnc is able give high ranks to trusted peers (Figure 14.10A.). On average, out of the top 10 ranked peers recommended by FE, DE, and CF, 5 (50%) or more peers are marked as trusted. We see that the LV filter's top 10 ranks have 3 to 4 peers on average, however, if we consider top 20 peers, all filters capture 6–8 trusted peers (more than 50% of the total trusted peers). The number of trusted peers in the rest of the ranges continue to fall except in the last range as it contains all the peers ranked beyond 80. For all the filters, there is a strong statistically significant correlation between the score and the rank of trusted peers (e.g., for LVC, $r=0.84$, $p <0.01$). Evaluations using metrics 2 shows that 80% of the trusted peers are captured by the top 25% of the encountering peers as ranked by the filters and there is a strong statically significant correlation (Figure 14.10B.). This shows that users willingness to trust others (for P2P communication) in a mobile network statistically correlates with recommendation given by ConnectEnc. We also note that there are peers who have high rank, yet they are not trusted. We believe these can be the encountered peers, who are very similar to the user and can provide new

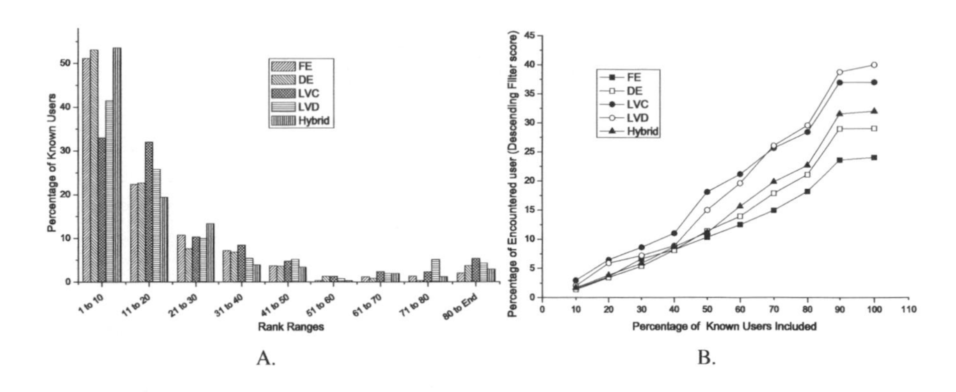

Figure 14.10: ConnectEnc evaluations based on application usage. Figure A shows the percentage of trusted users in 1 to 10 top users, 11 to 20 top users for each filter. Figure B shows a fraction of encounter users needed (from top) to capture $x\%$ of trusted users for each filter.

interaction opportunities to the user and can be utilized by other mobile applications (including social networks).

Another finding from the deployment is that the average storage requirement for ConnectEnc to store one month of data is 6.2 MB including raw and processed data (75 MB per year). This implies that with the current availability of mobile devices with multi-GB storage capacity, ConnectEnc's storage requirements can easily be met. We have also used this deployment data to create an energy-efficient encounter scanner as explained below.

Energy efficiency: Scanning of Bluetooth and WiFi devices consumes considerable power (since the scanning process is periodic). After receiving the traces (which were scanned at 1-min intervals), we noted that due to spatial locality in the traces, we can skip the scanning rounds if we find the same devices again in the next round, assuming that the user remains in the same location. The number of rounds we skip is $(2^n - 1)$, where n is the number of times the same devices are found consecutively, with an upper threshold (MaxThres). If after a scan round, the devices change, we make $n = 0$. We note that reducing the scanning period increases the loss of encounter information. Since we have the ground truth (traces scanned at 1 min), we can find out the information lost using the L1 norm on the distribution of AP (WiFi trace) and Bluetooth devices in both cases. We note that $n = 2$ gives us 64% savings in scanning, yet the loss is of 6.5% (more in Table 14.3). The current version of the ConnectEnc application incorporates this energy-efficient scan mode. We also foresee that the ConnectEnc framework can save considerable energy when multiple P2P applications are running by providing encounter information to all the applications and thus preventing each of those applications from running its own scanning process.

Table 14.3 Trade-off between Saving in Terms of Scans and Loss of Information, W and B, indicates WiFi and Bluetooth Trace, Respectively

MaxThres	Loss(W)%	Saving(W)%	Loss(B)%	Saving(B)%
3	6.52	64.21	6.79	66.31
7	10.52	75.27	11.40	76.61
15	15.11	81.53	15.02	82.29

14.6.3 Simulation Evaluation

To test the utility of ConnectEnc recommendations on a larger scale, we use trace-based simulation. The goal of this simulation is to investigate if ConnectEnc recommendations can make a difference in routing messages over a delay-tolerant Network (DTN) with selfish nodes. DTNs are infrastructure-less networks that work on cooperation among the nodes. Since nodes spend their resources in routing messages, the nodes may only route messages for nodes they know or when they have some incentive (and thus become selfish). Here we use the ConnectEnc framework to help a node decide from which of its peers to accept packets and route then further while being selfishness to other peers. Since ConnectEnc selects nodes that are similar in terms of spatio-temporal similarity, several nodes having high encounter scores may be already known to the user (homophily [29]). Therefore routing messages for nodes that have high encounter scores may give the user social incentive [24].

Setup: To examine the effectiveness of ConnectEnc, we use the epidemic routing protocol [34]. Epidemic routing performs a controlled flooding and has been proved to provide lower bound in performance in terms of hops, delay, and unreachability. These properties make it an appropriate tool for the purpose of our evaluations. We use WLAN traces (converted into encounter trace) from 3 campuses for this simulation.

Figure 14.11 shows the flow chart for ConnectEnc routing used by each node. When a node receives a message from a peer with an encounter score above a threshold (T), it accepts the packet and attempts to route it. Otherwise, the node accepts the packet based on factors such as user-configured selfishness. Selfishness is defined as the probability (S) that a node will not accept and route packets for a peer who is below a set encounter score threshold.

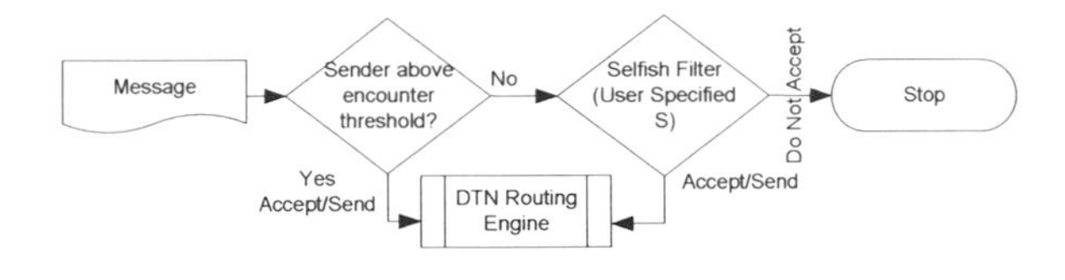

Figure 14.11: Flow chart for DTN routing using ConnectEnc's peer selection.

The performance of epidemic routing is measured using three metrics: *i.* Unreachability—the number of nodes out of all receivers that could not be reached by a given source, *ii.* Delay—the ratio of average time taken by a message to reach all the possible receivers over the max possible delay, and *iii.* Overhead—average number of hops a message took to reach all the possible receivers using the shortest path. Since overhead and delay were seen to vary directly with unreachability, we have skipped their results.

For the simulations, we use first 60 days of traces to create preliminary encounter scores and run epidemic routing on traces for the next 30 days. Encounter scores are updated weekly during the run of epidemic routing (to mimic a mobile device as computing encounter scores after every encounter or daily would be resource intensive for the device). Around 800 nodes are randomly selected as sources for the epidemic routing.

Results: Intuitively, selfishness should cripple the connectivity in the network. Figure 14.6 shows that the network unreachability increases as S increases (and $T = 0$). To the benefit of our scheme, we find that as social incentive is introduced based on the encounter scores in the network, the effect of selfishness is reduced. Here we use encounter scores from the DE filter (other filters show a similar trend). For U1, when $T = 0\%$ and $S = 0.9$, unreachability increases by 83% from the case when $S = 0$. However, when increasing the threshold to $T = 40\%$ ($S = 0.8$), unreachability remains only 31% from the case when $S = 0$. Likewise, for Dartmouth, when $T = 0$ and $S = 0.9$, unreachability increases by 40% from the case when $S = 0$. However, when increasing the threshold to $T = 40\%$ ($S = 0.9$), unreachability remains only 10% from the case when $S = 0$. For USC, $T = 0$ and $S = 0.9$ increases unreachability by 1.7% of the case when $S = 0$. However, increasing threshold to $T = 40\%$ ($S = 0.9$) brings unreachability to only 0.48% from the case when $S = 0$. The effect of ConnectEnc peer recommendations is higher when selfishness is high, which makes ConnectEnc more suitable in networks with high selfishness. The effect of peer selection by ConnectEnc (or selfishness) is not significant in USC traces, which could be a result of high unreachability in the network even at $S = 0$ (5 times U1 or Dartmouth).

We now compare the performance of hybrid filters (using 5 different weight combinations). The highest unreachability (worst performance) is produced by using only the *BM* filter score and the lowest by using the *FE* filter (Figure 14.12). The combination of filters at equal weights has unreachability close to the *FE* filter. This analysis shows that and combination of filter scores can produce better results (a also avoids user confusion) than using individual filters. Better performance of *FE* over *BM* does not imply that we should not use *BM*, but it implies that for this particular application, *FE* is a better Encounter Filter.

14.7 Other Modules

This section discusses the remaining modules as mentioned in the architecture diagram (Figure 14.1). These modules are not needed for basic functionality of ConnectEnc, but can enhance its capabilities. These modules include Anomaly Detection,

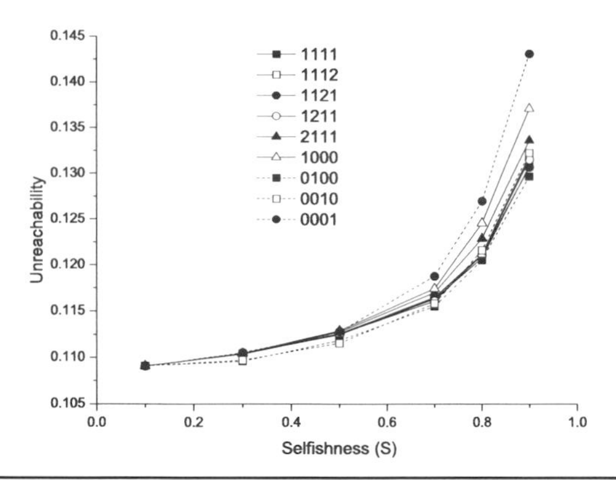

Figure 14.12: Hybrid filter results when T=40%. Number on the legend indicated the ratio of score from each filter. For example, 1211 implies $\alpha_{DE} = 0.2$, $\alpha_{FE} = 0.4$, $\alpha_{LV-D} = 0.2$, and $\alpha_{BM} = 0.2$ and 0100 implies $\alpha_{DE} = 0$, $\alpha_{FE} = 1$, $\alpha_{LV-D} = 0$, and $\alpha_{BM} = 0$ (Section 14.7.3).

External Inputs, and Unified Score Generation. Due to unavailability of any suitable existing anomaly detection system, we have designed our own. External Inputs and Unified Score Generation are provided to give a high-level idea about the framework, however, more research in the future is needed.

14.7.1 Anomaly Detection

Incorporating resilience to attacks is a primary requirement for our design. Here, the attack on the ConnectEnc system includes an attempt by a peer to gain an encounter score in a relatively short time by injecting many encounter events (e.g., via stalking). A growth of encounter scores in this fashion can be considered an anomaly (or an attack), and a specialized anomaly detection system is needed to combat such attacks. Since ConnectEnc scores individual encountered peers, at present we consider single attacker scenarios.

An attacker would want to get a high encounter score as soon as possible to have high returns for limited effort. The goal of the anomaly detection design would then be to considerably raise the level of effort needed for a successful attack, to be no less than genuine trusted nodes and friends, which may entail weeks of consistent encounters at trusted locations by the attacker. The spatio-temporal granularity of the filters determines such attack effort and provides us with the anomaly we aim to detect. Note that in our implementation, *Unified Score Generation* takes input from the anomaly detection unit. The role of anomaly detection would then be to raise a red flag (and also lower the unified encounter score of a peer) on suspicion of attack.

Anomaly detection, theoretically, can be achieved using supervised or unsupervised learning techniques. However, due to the present lack of learning data (from real attacks), we only consider the unsupervised technique. Our anomaly detection investigates the evolution of encounter patterns over time (without information exchange between nodes). The anomaly detection mechanism considers the growth slope of encounter statistics (including scores generated by the encounter filters). The detection system learns normal behavior over time, and incorporates deviations from the normal to detect suspected nodes and trigger user alerts.

Based on the approach mentioned above, we have created and tested this anomaly detection system with the help of trace-driven simulations (and by creating an attacker's model). The anomaly detection we designed is able to detect attackers with less than 8% false positives and 6% false negatives. However, to due to lack of space, we are skipping the details.

14.7.2 External Inputs

i. Recommendation and reputation systems: ConnectEnc is designed to take inputs from existing recommendations [32, 18] and reputation systems [10]. ConnectEnc can also *bootstrap* a recommendation system, since recommendation system scores start to evolve only after initial direct interaction. Recommendation systems can receive peer recommendations from other peers. Reputation systems can receive feedback on peers from applications and utilize it to raise the overall score of a peer who has a low encounter score but high reputation (or reduce the score for a peer with bad reputation).
ii. Black list and White list: Users can use these lists to explicitly add and rate (including not encountered) users. This functionality allows addition of infrequently encountered yet known peers.

14.7.3 Unified Score Generation

ConnectEnc needs to provide easily understandable information to the application or the user. Providing scores from independent modules separately may confuse the user or complicate an application design. As a first step to simplify the output, we earlier created a hybrid filter (HF), combining the Encounter Filter scores. A similar idea can be used to combine the scores from all the modules discussed above and generate a single encounter score for an encountered peer. The scores can be combined using the following:

$$U(P_j,\alpha,\beta,\delta) = \delta H(P_j,\alpha) + (1-\delta)(\sum_{i=1}^{m} \beta_i R_i(P_j))), \tag{14.4}$$

where $U(P_j,\alpha,\beta,\delta)$ represents the unified encounter score for an encountered peer P_j; it is always between 0 (lowest) and 1 (highest). $H(P_j)$ is the score from the hybrid filter. β_i represent the weights for other normalized external inputs (R_i) such as anomaly detection, recommendation system, reputation systems among others. Here $\sum_{i=1}^{m} \beta_i = 1$ and $0 \leq \beta_i \leq 1$. The factor δ decides the combination ratio of Hybrid Filter and other external inputs. δ varies between 0 and 1, so the combined score is also between 0 and 1. If the peer (P_j) is included in the *white list*, then this peer automatically gets the highest encounter score. However, if a peer exists in the *black list*, she will be always be removed before sending the list to an application or the user.

The modules discussed in this section are presented for the sake of completion and would require further research in the future (out of scope for this work). For example, a challenge now lies in finding out the correct weights $(\alpha, \beta, \& \delta)$ to combine different inputs. These weights depend on the user and application preferences.

14.8 Conclusion and Future Work

This work introduces ConnectEnc, an effective encounter-based framework for making informed peer selection choices in mobile P2P applications in an efficient, privacy-preserving, and resilient manner. ConnectEnc is driven by encounter filters that leverage increased sensing capabilities of the mobile devices and their close association with users, which enables them to capture peer similarity with encountered devices at multiple levels.

We use four novel encounter filters, based on encounter frequency, duration, location behavior-vector and behavior-matrix. The score reflects the level of similarity to aid the user or application to select peers in coordination with personal preferences, location priorities, contextual information, and/or encounter-based keys. The calculations are fully distributed, eliminating the need for any server or trusted third party.

Three-phase evaluation reveals that most filters possess high stability and form a small world among the users. A series of surveys and participatory experiments shows that statistically strong correlation exists between the filter scores and the selection of peers. This validates the encounter filter-based approach used by ConnectEnc. Selfishness analysis using social incentive based epidemic routing shows that it is possible to efficiently use peer recommendations by ConnectEnc without sacrificing network performance in DTNs. Further, resilience to attack using anomaly detection achieves less than 10% false positives and 7% false negatives.

ConnectEnc has been designed to inspire several potential applications that can be enabled in the future. However, there are a few avenues that require further research. In the future, we plan to address some of these questions, such as handling multiple devices belonging to a user or MAC address spoofing (several techniques exist [37]), as part of future research. Future work will include analysis of other filters for measuring behavioral similarities. We also want to develop and deploy ConnectEnc for popular mobile platforms and study the effect of its usage on a larger scale. There is a need to conduct more research in order to understand how to effectively leverage P2P connections in mobile societies. We hope that this research contributes to that effort.

References

[1] CRAWDAD, August 2008.

[2] iTrust/ConnectEnc mobile app, Dec. 2012. https://code.google.com/p/itrust-uf/.

[3] Bizzabo, Nov. 2012.

[4] Open Garden, Nov. 2012.

[5] P2P games, Nov. 2012.

[6] Query by a game developer, Nov. 2012.

[7] R. Albert and A. L. Barabsi. Statistical mechanics of complex networks. *Rev. Mod. Phys., Vol. 74*, pp. 47–97, 2002.

[8] Eitan Altman. Competition and cooperation between nodes in delay tolerant networks with two hop routing. In *NET-COOP*, 2009.

[9] Eitan Altman, Arzad A. Kherani, Pietro Michiardi, Refik Molva, Pietro Michiardi, and Refik Molva. *Non-cooperative Forwarding in Ad-Hoc Networks*. Technical report, PIMRC, 2004.

[10] Sonja Buchegger and Jean-Yves Le Boudec. A robust reputation system for mobile ad hoc networks. In *P2PEcon*, 2003.

[11] Sonja Buchegger and Jean-Yves Le Boudec. Self-policing mobile ad hoc networks by reputation. *IEEE Comm. Mag.*, 43(7):101, 2005.

[12] Ronald S. Burt. Decay functions. *Social Networks*, 22(1):1–28, 2000.

[13] Levente Buttyan et al. Barter-based cooperation in delay-tolerant personal wireless networks. In *WoWMoM*, 2007.

[14] Chia-Hsin Owen Chen et al. GAnGS: Gather, authenticate 'n group securely. In *MobiCom '08*, pages 92–103, 2008.

[15] Jon Crowcroft, Richard Gibbens, Frank Kelly, and Sven Östring. Modelling incentives for collaboration in mobile ad hoc networks. *Performance Evaluation*, 57(4):427–439, 2004.

[16] E.C.R. de Oliveira and C.V.N. de Albuquerque. Nectar: A DTN routing protocol based on neighborhood contact history. In *Proceedings of the 2009 ACM Symposium on Applied Computing*, pages 40–46. ACM, 2009.

[17] N. Eagle, A. Pentland, and D. Lazer. Inferring social network structure using mobile phone data. *PNAS*, 2007.

[18] Elizabeth Gray, Jean-Marc Seigneur, Yong Chen, and Christian Jensen. Trust propagation in small worlds. In *Trust Management, Lecture Notes in Computer Sciences*, 2692:239–254, 2003.

[19] B. Han, P. Hui, VS Kumar, M.V. Marathe, G. Pei, and A. Srinivasan. Cellular traffic offloading through opportunistic communications: A case study. In *ACM Chants Workshop*, 2010.

[20] Bo Han and A. Srinivasan. ediscovery: Energy efficient device discovery for mobile opportunistic communications. In *(ICNP), 2012.*

[21] Weijen Hsu, Debojyoti Dutta, and Ahmed Helmy. Profile-Cast: Behavior-aware mobile networking. In *IEEE WCNC*, 2008.

[22] Weijen Hsu and Ahmed Helmy. MobiLib, June 2008. http://www.cise.ufl.edu/helmy/MobiLib.htm

[23] Youngki Lee, Younghyun Ju, Chulhong Min, Seungwoo Kang, Inseok Hwang, and Junehwa Song. Comon: cooperative ambience monitoring platform with continuity and benefit awareness. In *MobiSys*, 2012.

[24] Qinghua Li, Sencun Zhu, and Guohong Cao. Routing in socially selfish delay tolerant networks. In *Infocom*, 2010.

[25] Yue-Hsun Lin and et al. Spate: small-group pki-less authenticated trust establishment. In *MobiSys*, 2009.

[26] Anders Lindgren, Avri Doria, and Olov Schelén. Probabilistic routing in intermittently connected networks. *LNC*, pages 239–254, 2004.

[27] E.K. Lua, J. Crowcroft, M. Pias, R. Sharma, and S. Lim. A survey and comparison of peer-to-peer overlay network schemes. *IEEE Communications Surveys and Tutorials*, 7(2):72–93, 2005.

[28] Jonathan M. McCune, Adrian Perrig, and Michael K. Reiter. Seeing is believing: Using camera phones for human authentication. *Int. J. Secur. Netw.*, 4(1/2):43–56, 2009.

[29] Miller McPherson, Lynn S. Lovin, and James M. Cook. Birds of a feather: Homophily in social networks. *Annual Review of Sociology*, 27(1):415–444, 2001.

[30] Sungwook Moon and Ahmed Helmy. Understanding periodicity and regularity of nodal encounters in mobile networks: A spectral analysis. In *GLOBECOM 2010.*

[31] Vikram Srinivasan Pavan, Vikram Srinivasan, Pavan Nuggehalli, Carla F. Chiasserini, and Ramesh R. Rao. Cooperation in wireless ad hoc networks. In *IEEE Infocom*, 2003.

[32] Glenn Shafer. Perspectives on the theory and practice of belief functions. *Int. Journal of Approximate Reasoning*, 4(5-6): 323–362, 1990.

[33] G.S. Thakur, M. Sharma, and A. Helmy. Shield: social sensing and help in emergency using mobile devices. In *GLOBECOM*, pages 1–5, 2010.

[34] Amin Vahdat and David Becker. Epidemic routing for partially-connected ad hoc networks. Technical report, Duke University, 2000.

[35] Wei Wang, Vikram Srinivasan, and Mehul Motani. Adaptive contact probing mechanisms for delay tolerant applications. In *Proceedings of the 13th Annual ACM International Conference on Mobile Computing and Networking*, MobiCom, 2007.

[36] Jie Wu and Yunsheng Wang. Social feature-based multi-path routing in delay tolerant networks. In *INFOCOM, 2012*.

[37] Kai Zeng, Kannan Govindan, and Prasant Mohapatra. Non-cryptographic authentication and identification in wireless networks. *Wireless Communications*, 2010.

[38] Sheng Zhong, Jiang Chen, and Richard Yang. Sprite: a simple, cheat-proof, credit-based system for mobile ad hoc networks. In *INFOCOM*, 2002.

Chapter 15

VANETs as an Opportunistic Mobile Social Network

Anna Maria Vegni
Department of Engineering
University of Roma Tre
Rome, Italy

Thomas D.C. Little
Department of Electrical and Computer Engineering
Boston University
Boston, Massachusetts

CONTENTS

15.1 Introduction

The delay-tolerant network (DTN) routing mechanism in mobile scenarios provides capabilities for nodes to opportunistically communicate with each other depending on the current environment. Basically, this mechanism requires nodes to cooperate on the level of packet forwarding: when a node wants to transmit a message to another node, the message can be opportunistically routed through relay nodes, under the assumption that each node is willing to participate in the forwarding process.

In Vehicular Ad hoc NETworks (VANETs), opportunistic routing is well exploited through the concept of *bridging* between mobile nodes (i.e., vehicles). In VANETs, packets are exchanged between vehicles traveling on constrained paths at different speeds, and forming clusters. Partitioning may exist between vehicles traveling in the same direction of the roadway. Several works have investigated the use of the DTN in VANETs [52], and a few of them have also included social aspects that can occur among vehicles. In [28], the authors investigate the use of DTNs together with social ranking between members of a social group. As a vehicle encounters a neighbor, it checks the social ranking of the owner, and if the ranking is higher the message is passed on to that vehicle. The aim is to "elect" users with high sociability as the carrier of the message, in order to increase the probability that a message will reach its destination.

Novel routing algorithms for message forwarding can exploit *cooperative behavior* among multiple communities of vehicles in VANETs. Vehicles belonging to the same communities (i.e., vehicles driving in clusters along the same route direction, or in the same time windows) may share common interests and information relevant to the whole community. As an instance, a group of people all driving (or walking or cycling) to a football game can experience traffic on the route to the stadium, and they are highly expected to encounter others with like interests (i.e., supporters of the same team) or will otherwise be enjoying the same shared experience. Based on such considerations, in [60] the Multi-Community Evolutionary Game Routing (MCEGR) algorithm is presented. It aims to reach a possible equilibrium of vehicular ad hoc multiple communities by forwarding packets within a multi-community evolutionary game theoretical framework.

Opportunistic networking applications are naturally about social networking (e.g., introduction services, friend finders, job recommendations, content sharing, gaming, and so on), as well as human mobility (i.e., exploited for forwarding in opportunistic networks) is directly related to the social behavior of people. For these reasons, in VANETs vehicles can benefit from user social networks, and many forwarding schemes can rely on social information for efficient packet forwarding decisions. It follows that several open issues of social networking in VANETs still need to be investigated, such as (*i*) to define specific social needs for vehicles or groups of people (i.e., traffic/event information messages, advertisements, etc.), (*ii*) to measure if the system has supported the needs of the users (i.e., metrics), and (*iii*) to define what is opportunistic (i.e., users' social ranking).

An increasing number of mobile applications aim to enable "smart cities," exploiting different contributions from individuals equipped with mobile devices with sensing abilities. Such applications are crowdsourcing-based solutions, with a participative online activity of individuals in order to solve a given task. As a consequence, "smart roads" can be designed by means of crowdsourcing-based applications, like smart parking systems [17]. In [23], the authors introduce MobiliNet, a system for vehicular environments based on an "extended" social network, i.e., not limited to human participants, but also including vehicles, corporations, parking spaces, and other objects. MobiliNet allows traveling to be more comfortable and less stressful, and then more efficient for the travelers (e.g., allowing high-level trip planning or self-determined mobility for user groups with special needs).

Finally, still in the context of smart cities, it is important to mention applications for pollution reduction, like aggregated carbon emission-aware traffic control, as well as for vehicular fuel consumption, by enabling drivers to change their driving behavior and be more eco-friendly. The Eco-Driving Coach [4] is a mobile-social application suite, enabling individuals to track their daily vehicular carbon emissions, and share on social networks (i.e., Facebook, Twitter, Google+). Eco-Driving Coach is intended primarily to raise social awareness of vehicular carbon emission, and then to encourage a more correct driving behavior, and also serves as a research platform for data collection for traffic management, carbon emissions, and user behavior analysis in social network-based applications.

This chapter is organized as follows. In Section 15.2, we introduce the concept of opportunistic social networks with basic characteristics, and briefly give an overview of main types of such networks. Particularly, we address social networks, and discuss routing protocols based on social networking. Then, in Section 15.3, we investigate the case of VANETs, as a particular class of opportunistic networks in vehicular environments. VANETs are described by means of their envisioned set of applications and main features. In Subsections 15.3.1–15.3.3, we present how social networking can be exploited for innovative applications in VANETs. We show that several safety and entertainment applications rely on social networking solutions through the cooperation of vehicle communities. We provide an overview of the main state-of-the-art representative techniques, particularly approaches based on crowdsourcing, eco-friendly applications, and social-based routing protocols for data dissemination based on relevant (*interesting*) information. Finally, conclusive remarks and future works will be summarized at the end of the chapter.

15.2 Opportunistic Social Networks

Opportunistic mobile ad hoc networks consist of mobile devices that communicate with each other in a *store-carry-and-forward* fashion, without any infrastructure [46]. They present distinct challenges compared to classical networks, such as the Internet, which assumes the availability of a contemporaneous, reasonably low propagation delay, low packet loss rate path between the two end points that communicate. In opportunistic networks, disconnections and highly variable delays caused by human mobility are the norm. The underlying topology of opportunistic networks is highly dynamic. Therefore, establishing and maintaining end-to-end paths between nodes is generally infeasible. In addition, node mobility that is leveraged because forwarding is often unpredictable.

Opportunistic networking techniques allow mobile nodes to exchange packets on the basis of mobility patterns and node storage capacity. The "store-carry-and-forward" approach allows messages to be stored in a mobile node and then forwarded toward the next hop, as soon as there is an available link. This represents a kind of opportunity exploited in order to retransmit the message without additional delay. Opportunistic networks are also considered as a special kind of *delay-tolerant network* (DTN) [25], which can provide high connectivity despite long delays and frequent disconnections due to mobility (e.g., nodes moving out of transmission range). DTNs belong to the class of dynamic networks since they are partitioned wireless ad hoc networks with intermittent connectivity. Indeed, DTNs are never fully connected at any point in time: nodes may not know the availability of future encounters but can predict points of disconnection over time.

There are several types of opportunistic networks, such as mobile sensor networks, pocket-switched networks, social networks, and vehicular ad hoc networks. In the case of Pocket Switched Networks (PSNs) i.e., a type of DTN for pervasive environments, a multitude of devices carried by people are dynamically networked. A PSN uses contact opportunities to allow humans to communicate without a centralized network infrastructure [47]. As a consequence, an efficient data forwarding mechanism within the PSN is required to cope with dynamic network topology caused by human mobility.

The traditional approach used in Mobile Ad hoc NETworks (MANETs) of building and updating routing tables is not cost effective for a PSN, since mobility patterns are often unpredictable and topology changes can be rapid. A PSN is formed by people, and the social relationships among people may prove to be a more stable network structure. In this system, unicast and multicast forwarding approaches are emergent properties of the community structure and community interests, respectively. Thus, a *social backbone* can be used for better forwarding decisions.

Recent years have seen the emerging of numerous online social networks (OSNs). Well-known cyber communities such as Facebook, MySpace, Google+, LinkedIn, and Orkut, where users can interact on the Internet, have emerged as top-ten sites globally in terms of traffic. Today, everyone on the Internet knows the concept of *social networking*. Social web communities (e.g., Facebook or LinkedIn), as well as content-sharing sites that also offer social networking functionality (e.g., YouTube), have captured the attention of millions of users and made millions of dollars by venture capitalists [13, 54]. Social networking over the Internet offers many interesting functionalities including a network of friends lists, person surfing, private messaging, discussion forums or communities, and media uploading. Although social networking is a buzzword popular all over the Internet, online social networks exist everywhere (e.g., at school and workplace, as well as within families and social groups); as a result, social networking allows people to work together over common activities or interests.

The study of social interactions is the main pillar for the analysis of social networks. Social networks are formed when individuals socialize through interaction and communication. Basically, humans socialize with other individuals with whom they are connected in some way, such as family ties or friendship. There are several ways to socialize, through casual interactions at the bar, in the store, on the street, or more formally in associations, clubs, and communities.

OSNs are represented as a graph, which contains a huge amount of *sparsely connected nodes* (i.e., persons), and *bidirectional edges*, since a mutual agreement is required before friendship links are established. In OSNs, one of the most notable phenomenons is the resemblance to the so-called 6 degrees of separation [42]. This phenomenon states that on average, every person is "connected" to any another random person via 5 other people in the real world. More in detail, online social networks consist of a series of *hierarchically embedded tiers of layers* (i.e., the circles of acquaintanceship) that vary in composition, frequency of contact, and emotional closeness. The strength of connections among members of a social network distinguishes in three layers of the network i.e., the two innermost layers of (*i*) the support clique and (*ii*) the sympathy group, and (*iii*) the outermost layer of the active network.

Social networks belong to the class of *egocentric* (or *bottom up*) networks, where a social world of an individual has her/himself at the center, and is surrounded by a series of layers of friends and acquaintances whose emotional valency declines as one proceeds outward through the successive layers. A person node in OSN, also called *individual ego*, can be envisaged as sitting in the center of a series of concentric circles of acquaintanceship, which increase in size with a scaling ratio of 3 [27, 65]. As the number of alters in each layer of the personal network increases, the level of emotional intimacy and level of interaction between ego and alter decreases. In contrast, conventional (*top-down*) networks represent the social world as seen by an outside observer, as consisting of a series of modular units (*strong links*) linked together by weak links. The innermost layer of the personal network is the *support clique*, which can be defined as all those individuals from whom one would seek advice, support, or help in times of severe emotional or financial distress [24], and averages about only five members [42]. The next layer out is the *sympathy group*, which can be defined as those whom an individual contacts at least monthly, and averages 12–15 members [14, 24].

Many *ego characteristics* have been shown to affect network size and composition. For example, individuals who are single and without children tend to have larger networks than egos who are married [24]. Other examples have shown that social networks have strong homophyly by gender, such that female networks are dominated by females, and male networks by males [41, 48]. Also, socio-economic status, as measured by education level, occupation, and income, is positively correlated with network size and also with network diversity [41]. Finally, social network size tends to decline for individuals older than 65.

In OSN settings, the standard definition of a network of devices evolves into a *network of people*, where information about users and their surrounding context is fundamental to provide efficient communications and data exchange among mobile users. In [43], the authors conducted a study in order to analyze the main features of very large OSNs, such as Orkut, YouTube, MySpace, LinkedIn, and LiveJournal. The study highlighted an important point, that while web pages are based on content,

Table 15.1 Main Features of OSN Topology, with a Short Description

Feature	Description
Network of people	Nodes are identified by individuals, and edges by human interactions.
Egocentric	Individual at the center of the social network.
Ego characteristics based	Gender, age, socio-economic status, etc., affect network size and composition.
Sparsely-connected nodes	6 degrees of separation phenomenon supports OSNs.
Bidirectional edges	A mutual agreement is required to connect nodes.
Hierarchically tiers of layers	Three layers of OSNs, from the innermost to the outermost layers.

online social networks are based on users. Moreover, the researchers concluded that the most trustworthy nodes (i.e., members of a social network) are those users who established the largest number of friends within the online network. As a matter of fact, those users establish themselves as close to the core of that social network as possible. This means that the closer to the core of a social network that a node is, the faster the node is able to propagate information out to a wider segment of the network.

In Table 15.1, we have collected the main features of OSN topology. Moreover, to summarize, we have enlisted a few basic characteristics as typical of a social network:

1. **User-based**: OSNs are built by users themselves, which populate the network with conversations and content.

2. **Interactive**: social networks are interactive. This means that a social network is not just a collection of chatrooms and forums, but also provides network-based gaming applications.

3. **Community-driven**: OSNs are built and driven by community concepts. As an analogy with communities or social groups, where members hold common interests or hobbies, social networks allow users to find sub-communities of people sharing commonalities (e.g., alumni of a particular high school, or people belonging to the same political group).

4. **Relationships**: The more relationships a user has within the network, the more an established user is toward the center of that network. User content can then proliferate out across their own network of contacts and sub-contacts.

From the introduction of OSNs, the Social Networking Service (SNS) has evolved rapidly and has carried a new trend in social interaction. SNS has emerged as an inter-connectivity forum, which facilitates and encourages individuals to stay in touch with their own network of friends/acquaintances, and also provide means to widen this network.

In SNS, the concept of information sharing is considered one of the basic cornerstones in the formation and expansion of SNS. For example, questions like "*where are you now?*" and "*what's happening?*" are often replied to by users in an OSN. Moreover, the prevalence of mobile devices, intended as an easy tool for information sharing, helps to answer such questions among a network of interested friends and hence maintains the concept of inter-connectivity as brought by SNS.

Opportunistic social networking [19, 46] has emerged as a new communication paradigm, which exploits user mobility to establish communications and content exchange between mobile devices in pervasive and mobile computing environments. As an instance, content sharing through YouTube or Flickr currently represents one of the most popular services.

Mobile users are becoming the principal actors of the network. Efficient development of this kind of service in opportunistic networks imposes mobility support, requiring the knowledge of user context and social behavior. Information about users and their habits, interests, and social interactions, plays a fundamental role, since it allows the system to generate routes *on the fly* to correctly deliver messages to the intended recipients. Several projects have investigated routing solutions for message delivery in social networks, based on social relationships and human behaviors [5, 10, 47].

In [18] Conti et al. present a social and context-aware content-sharing service, developed in the framework of the European Commission's Information Society Technologies/Future and Emerging Technologies (FET) Haggle project [5]. The proposed service exploits a context definition designed for opportunistic networks. This is composed of the (*i*) user context (i.e., name, address, habits, timetables, and identifiers of the social community affiliation), (*ii*) service context (i.e., list of shared files and interests for content-sharing services), and (*iii*) device context (i.e., battery lifetime, capacity, and available embedded technologies). In the Haggle project, users are grouped into social communities with common interests (e.g., group of coworkers, flatmates, family, or friends). User social-behavior patterns allowed identification of content that might be relevant to the communities, and mobile users who want to participate in the service should declare information about the contents to share, as well as personal information that enables the system to trace their own social interactions and mobility patterns.

In the framework of the Haggle project, an opportunistic mobile social networking middleware, named MobiClique [47] was designed and implemented. MobiClique exploits user mobility and social relationships to forward messages in an opportunistic manner, using a *store-carry-forward* technique. MobiClique has been assessed in three example applications including (*i*) social networking, (*ii*) asynchronous messaging among friends and interest groups, and (*iii*) distributed voting. Finally, another project aiming to address opportunistic networking in social networks is SOCIALNETS [10], which exploits the concept of the human relationship and behavior to design a pervasive and mobile computing environment.

Since the advent of SNS and, more recently, the introduction of smartphones and new wireless technologies, there have been several solutions aiming to improve the social interaction between people and their physical world. In [49], the authors present the SenseShare project, which allows users to share sensor data with their friends. Also, the ShopLovers solution [15] offers consumers in shopping malls a new shopping experience; based on RFID (Radio Frequency Identification) / NFC (Near Field Communication) technology, ShopLovers provides shopping items connected to the Internet.

It is easy to understand that routing in opportunistic social networks should be different from traditional routing protocols in MANETs, since such protocols are not suitable in this field due to intermittent and quick disconnections, and human behavior and mobility patterns. As an instance, nodes (i.e., persons) may be drawn to particular geographic regions or influenced by the behavior of other nodes (i.e., neighbors, friends, co-workers, and so on). In MANETs, a basic assumption for routing protocols is that the mobility process is ergodic and stationary, so that it is possible to predict the future from past behavior. However in opportunistic social networks, this assumption may not be valid. Typical features of opportunistic social networks should be taken into account and exploited to design novel routing mechanisms. As an instance, slower changing attributes, like social connections, may be leveraged to enable efficient message delivery, since social relationships are expected to vary more slowly with respect to the transmission links among mobile nodes in a mobile ad hoc network [28].

15.2.1 Social-Based Routing Protocols

The application of social networks' theory has been largely exploited to model delay-tolerant networks and provide the design of a new class of routing solutions.

In opportunistic networks, the communication and social layers are not disjoint, since social relationships can impact human mobility, and then the available connections. Leveraging this aspect, social-based routing approaches are particularly relevant solutions for opportunistic networks with social components, like pocket-switched and mobile peer-to-peer networks.

Social-based protocols exploit knowledge of the social network structure to forward information among nodes. Furthermore, social-based protocols may identify socially similar nodes, and utilize context information, like shared interests or community affiliations [21], for message delivery. As an instance, users belonging to the same "group" (e.g., co-workers, friends, fans of a sport team) are very likely to meet in common places. This fact can be leveraged to opportunistically deliver messages related to the group's main interest.

In Figure 15.1 we show typical use cases, where students with a common study interest will visit the same classrooms to attend the same courses, customers with the same shopping interests often visit the same shops, and so on. Based on these social characteristics, the authors in [61] propose a Community-Aware Opportunistic Routing (CAOR) algorithm, assuming that mobile users with common interest, autonomously form a community.

In [20], Costa et al. present SocialCast, a routing framework for publish-subscribe that exploits predictions based on metrics of social interaction like patterns of movements among communities. When a node subscribes to a service, it identifies its in-

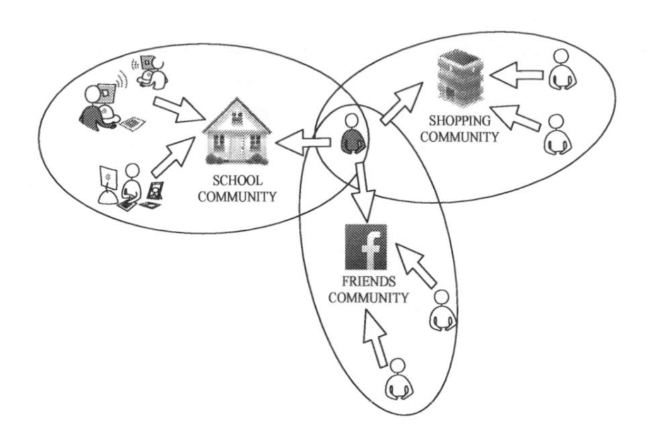

Figure 15.1: Home-aware community model [61]. Mobile users with common interests belong to the same community, and the frequently visited location is their common "home."

terest (e.g., "rugby" or "baseball"). When a message is published (e.g., "Six Nations Results" or "Red Sox game"), it is tagged with the related interest, and SocialCast will deliver the message to the nodes with at least one interest matching the one in the message.

Many forwarding algorithms like SimBetTS [21] and BUBBLE [28] consider the importance of the role of a node in the social structure of the network, in order to make efficient routing decisions. Compared to traditional routing protocols, where the existence of available wireless links is exploited in order to forward packets, in disconnected networks, routing protocols aim to utilize pairwise contacts to enable opportunistic communication.

In [51], Schurgot et al. classify the routing protocols on the basis of the different network graphs such as (*i*) the dynamic *wireless graph* composed of every available link in time, (*ii*) the *contact graph* calculated from the aggregation of past wireless links, and (*iii*) the *social graph* formed by interpersonal relationships. The wireless graph can be depicted as a three-dimensional graph, where dynamic connections among nodes are monitored over time i.e., for each time epoch (TE). It is a dynamic undirected graph with an edge between nodes, indicating the presence of a wireless link in both directions.

In a real scenario, the complete global knowledge of the wireless graph is an issue, and routing protocols do not have deterministic information on future connectivity in a wireless network. As a solution, the *contact graph* (*CG*) aims to predict all these future encounters, by means of the concept of statistical aggregation taken from a wireless graph. The contact graph estimates future connectivity links from statistics of the wireless graph, assuming the mobility process is ergodic and stationary. In this way, a node does not need to store a snapshot of the network at each past time epoch, and this results in a reduction of the amount of information stored and processed by nodes. Then, in a contact graph, connectivity links are weighted between 0 and 1, which are values calculated during an aggregation window composed of a series of time epochs.

Several solutions dealing with routing protocols exist, and among them, three benchmark protocols are used for performance comparisons by almost all more recently proposed protocols, such as Epidemic [56], Spray and Wait [53], and PRoPHET [35]. The Epidemic protocol [56] is based on general broadcasting of messages. Nodes replicate messages on each encounter until a message has reached a predefined maximum hop count. Messages are not exchanged if a copy is already present in the node buffer. Epidemic was shown to have a good packet delivery ratio, but it suffers from very high overhead given the large number of packet copies flooding the network.

The Spray and Wait protocol [53] outperforms Epidemic for a large range of network connectivity scenarios. It consists of two phases i.e., (*i*) *spray* and (*ii*) *wait*. During the *spray* phase, copies of a packet are "sprayed" to relay nodes in the network. These nodes enter the *wait* phase, until they meet the destination and the message is delivered. Finally, the PRoPHET algorithm [35] makes routing decisions on the basis of node delivery predictability for a specific destination. Although PRoPHET does not explicitly define a contact graph, the delivery predictability is a metric calculated from the aggregation of the wireless graph over time. In ProPHET, nodes with higher message delivery predictability will be selected as relay nodes to forward messages.

All previous techniques are based on the concept of the random mobility model. However, in a real scenario, this assumption is not properly correct, since mobile users tend to have mobility patterns influenced by their social relationships, life style, and social behavior. The introduction of social aspects in routing approaches can enhance performance of message delivery in opportunistic networks.

A social graph represents a virtual social network, where information about the interpersonal relationships of users can be used in social-based routing protocols to make forwarding decisions. Basically, a social graph can be viewed as an extension of the contact graph, where the links may be known *a priori* or inferred from the frequency of observed contacts. With the knowledge of link weights, different rules may be considered to extract the social graph connections (*SG*). It is known that in a social network, nodes become popular on the basis of the number of connections and their ability to bridge the partitioned network; nodes belonging to two separate clusters can never communicate, while well-connected nodes have high popularity.

The characterization of the social network influences routing protocol performance. Social-based routing protocols use information extracted from a social layer of the social graph. A social layer could be inferred from shared context, identified by the application of social network analysis on the contact graph, or constructed from interpersonal relationships available to the network designer.

Among the main social-based routing protocols, the most widely referenced techniques are HiBOp [11], SimBetTS [21], and BUBBLE [28]. HiBOp (History-Based Opportunistic Routing) considers mobility aspects together with past and current context information from the virtual social layer, in order to calculate message delivery probabilities [11]. The context information (e.g., shared attributes and history of encounters) describe the users environment and provide information about social relationships in the network. Thus, a message is transferred if a potential forwarder a the delivery probability for a given destination that is greater than the current node. Compared to Epidemic and PRoPHET in community-based mobility simulations, HiBOp reduces the consumption of resources and message loss rate for limited buffer scenarios.

In the SimBetTS algorithm [21], the betweenness metric measures the bridging capability of nodes, and similarity identifies nodes socially similar to the destination. Basically, SimBetTS utilizes the bridging capability of weak ties and the strong relationships that bind clusters. Tie strengths are also exploited as an indicator of link availability, measured by the frequency of encounters, the duration of encounters, and how recently the contact occurred. Performance analysis has shown that message delivery in SimBetTS outperforms PRoPHET and is close to Epidemic with less overhead.

Finally, in [28] Hui et al. evaluate the impact of *community* and *centrality* on message forwarding in social networks. Indeed, community is an important attribute of PSNs; cooperation binds, but also divides human society into communities, and within a community, some people are more popular i.e., have high centrality, and interact with more people than others. Hui et al. [28] have proposed a hybrid algorithm, called BUBBLE, whose name implies a bubble as a metaphor for a community. BUBBLE selects high-centrality nodes and community members of destination as node relays. In BUBBLE, for each node two centrality measures are calculated, based on its (*i*) *global* popularity in the whole network, and (*ii*) *local* popularity within its set of communities. BUBBLE forwards messages to those nodes with higher global rankings (*high global centrality*), until the carrier node encounters a node with the same community label as the destination node. The message is then forwarded to nodes with higher local rankings (*high local centrality*), until the successful delivery. It is worth noticing that this method does not require knowledge of the ranking of all other nodes in the network, but only the node's neighbors.

A detailed overview of the main DTN routing protocol, including the social-based protocols, is provided in [50], where each solution has been classified based on the network graph considered (i.e., the wireless graph, contact graph, and social graph).

15.3 Vehicular Ad Hoc Networks

After describing the features of opportunistic social networks, and the main routing protocols adopted in such networks, in this section we will give attention to a particular class of opportunistic networks i.e., the Vehicular Ad hoc NETworks (VANETs), and then we will investigate how social networking can be applied in VANETs. As a result, the concept of vehicular social networks (VSNs) will be presented through main features and applications. Finally, a comparison of VSNs with traditional VANETs, and social networks will be provided.

As previously introduced, VANETs belong to the family of MANETs, with the particular feature that mobile nodes are vehicles, able to communicate with each oth-

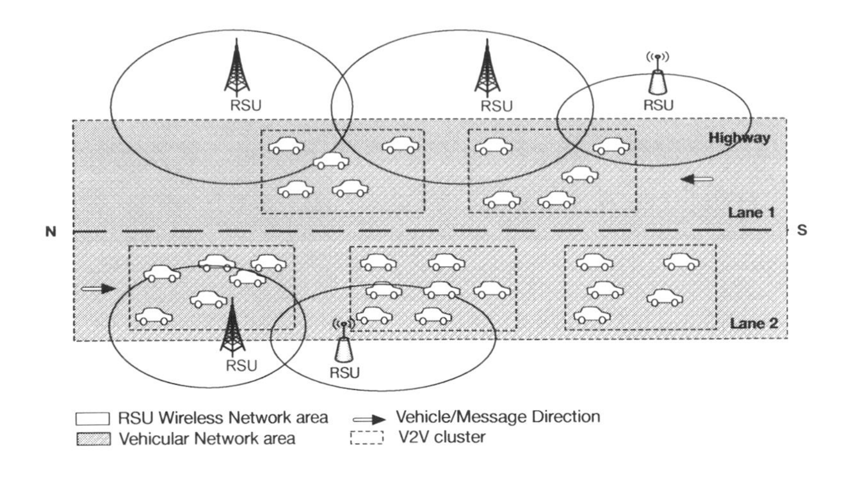

Figure 15.2: Vehicular grid with an overlapping heterogeneous wireless network infrastructure. Vehicles can communicate via V2V, as well as V2I [44].

ers via opportunistic wireless links [26]. Vehicles travel on constrained paths (i.e., roads, highways) and exchange safety and entertainment messages among neighboring vehicles. Different communication modes are allowed in VANETs, depending on available connectivity links. For example, a vehicle can transmit traffic information messages to its neighbors via vehicle-to-vehicle (V2V) mode, while it can receive data from a traffic light i.e., a roadside unit (RSU), via the vehicle-to-infrastructure (V2I) approach. Figure 15.2 depicts a vehicular grid with an overlapping heterogeneous wireless network infrastructure, comprised of RSUs. Notice that vehicles move along different lanes i.e., Lane 1 (2) is from south (north) to north (south), forming clusters.

There are several wireless access technologies used for vehicular communications. On-board devices are equipped with IEEE 802.11 and Wireless Wide Area Network interface cards, like Long-Term Evolution (LTE) and Worldwide Interoperability for Microwave Access (WiMax), as well as Global Navigation Satellite System (GNSS) receivers for vehicle positioning and tracking [57]. Particularly, the IEEE 802.11*p* standard is intended to operate with the IEEE 1609 protocol suite, which provides the Wireless Access in Vehicular Environments (WAVE) protocol stacks [55]. Finally, short-range communications are also provided within Personal Area Networks, through Bluetooth technology.

VANETs and MANETs are different in their dedicated applications. VANETs arise from the need to enhance traffic and reduce the number of road victims and accidents. Safety and traffic management applications are typical for VANETs, requiring timely and reliable message delivery, as compared to MANETs. As a consequence, VANETs well fit into the class of opportunistic networks, since messages are forwarded according to the store-carry-and-forward approach: messages are stored in

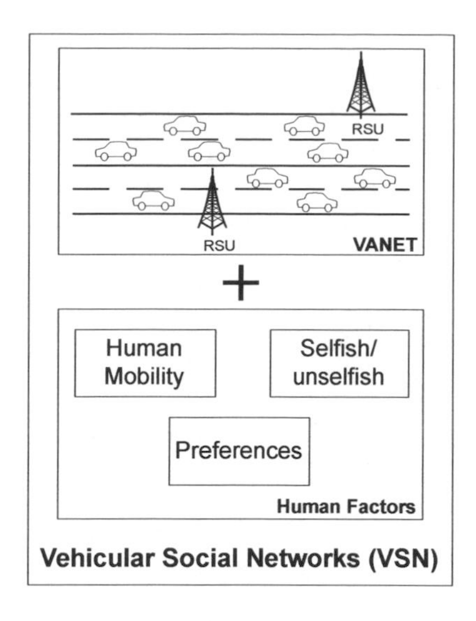

Figure 15.3: Concept of a Vehicular Social Network (VSN). VSNs arise from VANET plus human factors (i.e., human mobility, selfishness, and preferences).

a vehicle and quickly forwarded over an available wireless link. Connectivity links are then *opportunistically* exploited to forward messages within the network.

Vehicular communications can be considered as the "first social network for automobiles," since each driver can share data with other neighbors e.g., Ford's concept car Evos can directly form a social network with the car driver's friends [1]. In VANETs, vehicles are usually driven by citizens in an urban environment, and thus vehicular mobility is directly affected by people's behavior and intentions. Human factors are involved in VANETs, not only due to the safety-related applications, but also for non-safety-related applications i.e., entertainment.

Social characteristics and human behavior largely impact VANETs, and this affects the VSNs [38]. A VSN is a VANET, that includes the traditional V2V and V2I communication protocols, but also highlights the influence of human factors i.e., mostly human mobility, selfishness, and user preferences, on vehicular connectivity (see Figure 15.3). Since people are involved in vehicular DTNs, human factors will affect the network characteristics. Therefore, vehicular DTNs can be regarded as one kind of VSN.

From the nature of vehicular ad hoc networks, traffic patterns can provide social interactions. For instance, in heavy traffic scenarios (e.g., during morning rush hours), the vehicular density is very high and traffic patterns are relatively static. Thus, such scenario becomes a popular social place for vehicles to connect to each other, and share information (e.g., traffic information, weather news, and so on). Obviously, if a roadside infrastructure is deployed in these high-traffic areas and then used to assist in forwarding data packets in vehicular DTNs, reliability in vehicular communications can be dramatically improved with incurred costs under control.

Differently from VANETs, VSNs highlight human factors like (*i*) the human mobility model, (*ii*) the human selfishness, status, and (*iii*) human preferences. In VSNs, vehicles are driven by people with their own decision ability and driving style. Then, the mobility model is no longer a random waypoint and some realistic human mobility models in a city environment should be adopted. For example, drivers select the shortest path toward a destination instead of traveling along the longest path. This is easily computed by means of an on-board GPS (Global Positioning Service) navigator.

Other mobility models follow collective human behavior (i.e., community). In the community-based mobility model, it is assumed that there exist several points of interest with high social attractivity (e.g., restaurants, malls, theaters, etc.). The attractivity is a dynamic process given by the number of vehicles that are currently stopping in a point of interest. The higher the attractivity level of a point of interest, the higher the probability that a vehicle, moving in the area of that place, will stop there. Finally, mobility in vehicular ad hoc networks is also affected by a time-variant model, such as a vehicle moving toward a given spot at a given time of a day e.g., people goes to the office in the morning, and back home in the evening, while on Sunday people prefer to relax at home, and then traffic is very low in urban areas.

In VSNs, the design of novel non-safety applications should consider not only these realistic mobility models, but also human behavior i.e., the *human selfishness status* and *preferences*. For the first factor, not all drivers are nonselfish, but some people will behave selfishly and not participate in some non-safety applications. For instance, for some reason (e.g., the need to conserve buffer and computing resources), a selfish vehicle may be reluctant to cooperate with other neighboring vehicles, if this is not directly beneficial to it. Therefore, selfishness is a very challenging issue for non-safety-related applications in VSNs.

Based on human preferences, it is possible to create novel non-safety applications. In VSNs, especially in an urban scenario, a great number of vehicles move between home and office every day, so their mobility pattern is predictable spatially and temporally. Groups of vehicles moving along the same road and at the same time can form some virtual communities, to discuss some interesting topics. For example, in [63] the authors have developed a social Ubiquitous-Help-System (UHS) for vehicular networks, based on context awareness. Through social relations, like Friend-of-a-Friend (FOAF), only relevant and reliable information has to be shared between nodes.

In a vehicular social network scenario, a mobile node is a vehicle equipped with advanced technology (i.e., multiple wireless network interface cards and GPS receivers), and is allowed to belong to one or more dynamic vehicular social networks. Indeed, vehicular social networks can form *on the fly*, through available connectivity links, so that a vehicle can access a social network and share data with other members, and then leave the network when the journey has terminated. This is a quite different from the concept of OSNs, where members of a social network are people with social interactions. In VSNs, social networks are more *dynamic* than traditional

OSNs, because members (i.e., vehicle drivers) are intended to access communities only when they are in mobile. For this reason, connectivity in a VSN is affected by mobility, causing limited access to members.

Due to specific features of VANETs (i.e., limited connectivity with short-life wireless links), we observe that in a vehicular environment, social networks are formed *on the fly*, and represent a *dynamic* process, where vehicles can connect with each other for short time periods (e.g., during travel time). Interactions and data sharing with neighbors occur only in given scenarios i.e., for a given *position*, *content*, and *social relationships*. As an instance, when a vehicle is approaching an area of interest, she can check for available social networks. Through the exchange of *query* and *reply* messages about a given topic (e.g., multimedia file sharing, traffic information, shopping experience, etc.), a vehicle can access a social network and stay for a limited time depending on vehicle journey duration. Moreover, a vehicle can take part of a known i.e., previously visited, social network (e.g., a co-worker's social network), whenever approaching a specific area of interest (e.g., driving on the road near the office).

To summarize, we can state the following characteristics of VSNs, which are different from traditional online social networks: (*i*) VSNs are mostly dynamic, and social connections among members occur even if they do not know each other, and (*ii*) members are not strong friends, but only contacts that can become acquaintances, and eventually friends (e.g., members of a vehicular social network are mostly people with common interests, not friends or family members).

In the following, we will discuss several techniques and applications used in VSNs, which exploit social aspects. In Subsection 15.3.1, we will address the main crowdsourcing-based applications for VSNs, while in Subsection 15.3.2 we will present different eco-friendly applications. Finally, the social-based protocols suitable for VSNs are discussed in Subsection 15.3.3.

15.3.1 Crowdsourcing-Based Applications

The concept of social networks, designed as systems for vehicles, has been largely exploited. For example, SignalGuru [32] uses windshield-mounted phones for detecting current traffic signal states with their cameras. By combining multiple measurements, SignalGuru creates a schedule and gives its users a Green Light Optimal Speed Advisory (GLOSA) warning or suggests an alternative route that efficiently bypasses the predicted stops. As a result, test drives have shown a potential savings 20.3% vehicle fuel consumption on average.

Many techniques for VSNs are also used for *smart city* applications. Indeed, an increasing number of mobile applications are emerging with the aim to enable smart cities [6]. Nowadays, cities are simultaneously addressing the challenge of combining competitiveness and sustainable urban development. This challenge reflects the impact on issues of urban quality, such as housing, economy, culture, social, and environmental conditions. A smart city is a city that performs well in six characteristics, built on the smart combination of activities of self-decisive, independent, and aware citizens [6]. These characteristics are smart economy, smart mobility, smart environment, smart people, smart living, and smart governance.

As a practical example of smart city applications, we can consider the parking problem in big cities. Studies show that an average of 30% of the traffic in busy areas is caused by vehicles cruising for vacant parking spots [59]. The situation is getting worse in developing countries, like China, where the number of private cars has soared recently, while the investment in parking facilities has lagged. The additional traffic causes significant problems from traffic congestion to air pollution and energy waste.

The huge demand for transportation-related services to simplify daily life is the basis for mobile *crowdsourcing* applications. By assuming each citizen is equipped with a mobile device with sensing ability, she can share information with her own neighbors. This represents the concept of crowdsourcing, which considers a variety of online activities that exploit collective contribution and intelligence to solve complex problems [2]. With the popularity of mobile social networking and the emergence of participatory sensing, mobile crowdsourcing can fix issues related to real-time data collection, and coordination among a large number of participants. The desired effect is to save the time and the fuel spent in cruising, reduce unnecessary walking, and reduce traffic congestion and fuel waste, as well as improve the quality of information necessary for a given request (e.g., what restaurants to go to, what low price fuel station, etc.).

In [17] Chen et al. describe a real scenario for smart parking that is a system employing information and communication technologies to collect and distribute real-time data about parking availability. More specifically, information collected through coordinated crowdsourcing for assisted parking guidance is integrated into traditional road navigation system. The proposed system is based on a client-server architecture, consisting of three components such as (*i*) smart parkers, (*ii*) client servers, and (*iii*) central devices. Smart parkers are the drivers who have access to the smart parking service, through their client devices. Through the use of a GPS navigator, a smart parker will receive recommendations from a central server about potential free parking slots, whenever approaching a given destination.

Central servers collect real-time data from drivers, who report their current location and destination, car speed, and parking availability on a certain street. The central servers check the dynamic map for potential parking vacancies, according to the smart parker's current location and destination. Then they inform the client device of the search result, which might be either a specific location of a parking spot or the direction of the next turn to the parking spot. Finally, the client devices e.g., smartphones, tablet PCs, and versatile GPS navigators, are installed on-board, able to communicate with the server, and have GPS capability and Internet connection. They upload geo-tagged data and can download the result of queries regarding parking slot availability.

The flow of data transfer is depicted in Figure 15.4. Notice that the participation rate is more important than the volume of information each individual provides. If the participation rate is low, a sophisticated data collection mechanism should be adopted to compensate for the lack of data sources. Note that crowdsourcing information-

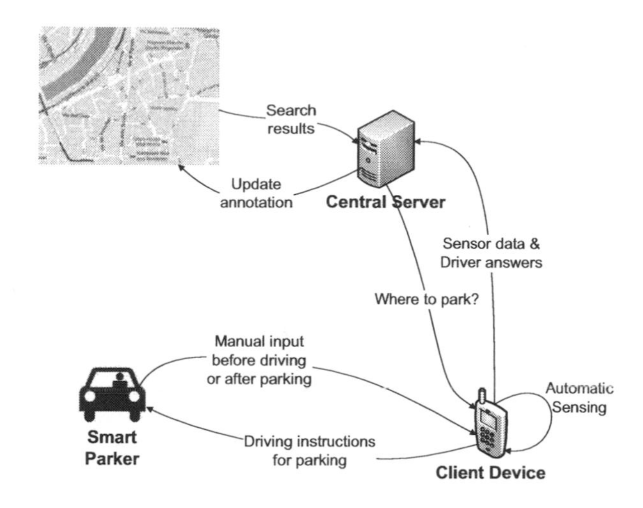

Figure 15.4: Data flow in the crowd-based smart parking approach [17].

based approaches can have confidentiality issues, since a driver may not want to send personal information (i.e., positioning), to be stored by an untrusted peer. It follows that these kinds of applications are sustainable if they can attract a sufficient number of involved users.

Another example of crowdsourcing service is Waze [9], a social mobile application available on smartphones allowing users to publish and consume real-time maps and traffic information. Waze "*outsmarts traffic*," since it exploits crowdsourcing information to provide vehicles with updated traffic information. Thanks to data crowdsourced through thousands of mobile devices, drivers are able to pick a better route to avoid a road segment that was detected as congested by Waze. This approach targets more efficient transportation, also employing ramifications to encourage participants to contribute and compare themselves against other contributors.

More in general, crowdsourcing applications in vehicular environments are related to many fields. For example, drivers can refill at a gas station with a lower price using GasBuddy application [7], and also find a parking place using applications like Open Spot [30]. Similarly, taxi drivers can select routes on the basis of colleagues' trajectory in order to improve their route [34]. A dedicated application for commuters is Roadify, which provides real-time transit information and updates from other commuters [8]. Finally, CrowdPark [62] assumes a seller-buyer relationship between drivers, to help other users find parking spots. Through a combination of simulation and real-world experiments, CrowdPark has shown to achieve over 90% successful parking reservation with a few minutes of waiting time, and has been proven to effectively detect malicious users with accuracy of over 95%.

MobiliNet [23] is a user-oriented approach for optimizing mobility chains, providing innovative mobility across different types of mobility providers, from public transports (e.g. buses, short-distance train networks) to personal mobility means (e.g.

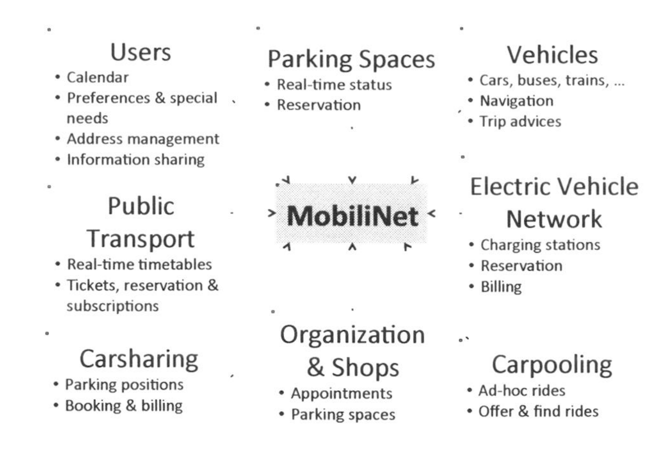

Figure 15.5: Overview of possible MobiliNet participants. MobiliNet platform connects all mobility-related things [23].

car sharing). Basically, it is a platform for connecting mobility-related institutions, systems, and services, as depicted in Figure 15.5, which shows all the objects involved. Notice that MobiliNet allows high-level trip planning, but also takes care of users with special needs, by providing self-determined mobility.

MobiliNet is based on the concept of social networks, not limited to human participants, but it extends to objects (i.e., vehicles, corporations, parking spaces, public transport stations, and other mobility-related systems and services). MobiliNet users can configure personal information like mobility preferences (e.g., cost limits, requirements, restrictions, etc.), and travel-related information (e.g., private car to potentially share), in order to fully experience MobiliNet service. As an instance, people with a mobility handicap can get a closer parking space, and people with babies or toddlers could get assigned a broader parking space so that the getting in and out would be more comfortable.

Other preferences, such as the preferred mode of transportation (e.g. with a private vehicle), can be used to refine the system's behavior. In this way, users can also connect to friends and other known people, and based on the degree of confidence, they can share their profile information. Notice that by default, the profile is private, but parts of it can also be made publicly available so that friends can find each other. However, cars can also participate in MobiliNet without being linked to a user; they could, for example, detect and report free parking spaces, or report traffic data to MobiliNet's central route calculation system. Passengers in public transportation bus are also part of MobiliNet, and they could provide information about available places, ticket cost, lateness of the bus, and so on.

Finally, Clique Trip [31] connects drivers and passengers in different cars, when traveling as a group to a common destination (i.e., multicar traveling group). As an instance, a group of people all driving to a football game can experience traffic on the route to the stadium. In order to establish the feeling of connectedness, the system automatically switches to an alternative navigation system when the cars tend to lose each other. Whenever the cars are within a defined distance, the system further establishes a voice communication channel between the vehicles.

15.3.2 Eco-friendly Applications

Nowadays, the ongoing discussion of climate change and the ever-increasing demand for fossil fuel have led to the need to reduce the vehicular carbon footprint and still represents an issue for both environmental and economical reasons. Although other factors such as the vehicle's model, engine type, age of the vehicle, etc. also have impact on vehicular carbon emissions, estimating emissions from trip information is a good approach to measure how individual driver's behavior can affect carbon emissions.

Applications oriented to reducing the carbon footprint are of great importance. As an instance, the Carbon Footprint/Fuel Consumption Aware Variable Speed Limit (FC-VSL) [37] traffic control scheme aims to minimize average vehicular fuel consumption, and also outperforms standard variable speed limit schemes that are designed to smooth traffic flow without considering fuel consumption.

Aside from aggregated carbon emission-aware traffic control, vehicular fuel consumption could also be reduced by enabling drivers to change their driving behavior and become more eco-friendly. Eco-Driving is a style of driving that reduces fuel consumption and GreenHouse Gas (GHG) emissions through smart and smooth driving techniques [4]. The Eco-Driving Coach [36] is a mobile-social application suite, designed to enable individuals to track their daily vehicular carbon emissions, and share this information on social networks. It has the dual purpose of (*i*) raising social awareness of vehicular carbon emissions by providing detailed feedback as measurement data for carbon emissions, and (*ii*) serving as a research platform for data collection in vehicular traffic management, and carbon emissions.

Eco-Driving Coach is achieved through a suite of applications including a mobile application for the Android platform for tracking and estimating carbon emissions, and a Facebook application for past driving record presentation and analysis. Eco-Driving Coach is supported by eco-feedback technologies (i.e., through smartphones and social networks), which provide feedback on individual or group behaviors with a goal of reducing environmental impact.

Figure 15.6 shows the main components of Eco-Driving architecture and how they are connected to each other. The mobile client is responsible for recording and visualizing trip and carbon emission information. The Facebook application is mainly responsible for visualizing and analyzing historic records, reporting travel and carbon emission information published by the user, and providing social gaming functionalities. Finally, the back-end server stores all the records submitted by the mobile clients, and provides data to the Facebook application. Notice that as different driving styles result in different vehicular carbon emissions, one functionality provided by the Eco-Driving Coach is to classify driver's driving behavior into

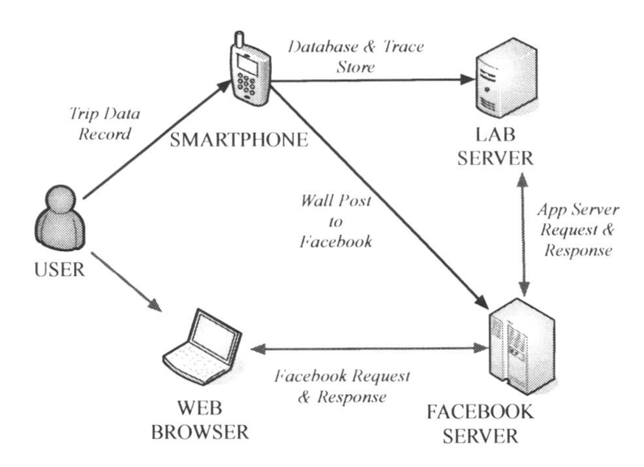

Figure 15.6: The Eco-Driving Coach system architecture mainly contains three components i.e., (*i*) a mobile client, (*ii*) the Facebook application, and (*iii*) a back-end server [36].

different driving styles i.e., normal, aggressive, and eco-driving. The driving style classification can help drivers to improve their style to a more eco-friendly one.

15.3.3 Social-Based Routing Protocols in VSNs

Traditionally, information gathering in vehicular ad hoc networks has been addressed through *push* models, where potentially useful (*relevant*) data, e.g., emergency braking warning, traffic congestion alarm, highlights of available parking spaces, etc., are pushed toward vehicles. A correct method of data dissemination, that allows vehicles to receive the relevant information efficiently i.e., with low delay and overheads, is still an open issue [16].

With a push model, it is difficult to communicate all to every vehicle, as this would consume too much bandwidth. Information about other (*minor*) events will not be disseminated, with the result that information sharing among a small set of interested vehicles (i.e., a vehicular social network) is not provided. In contrast, the use of *pull* models can allow users to send queries to a set of cars, in order to find the desired information. This provides more flexibility in terms of the types of queries that can be considered, as opposed to the approaches based on a push model, since a query could, in principle, be diffused far away to retrieve remote data.

In the autonomous environment of VANETs, vehicles randomly move with high speed and rely on each other for a successful data transmission process. The routing can be difficult or impossible to predict in such intermittent vehicles connectivity and highly dynamic topology. However, the knowledge that behavior patterns exist in real-time urban vehicular networks can represent a basis to design efficient routing protocols. The social nature of VSNs is then largely exploited, since vehicles are used by humans and their behaviors are based on social networks. Information about human behavior in vehicular environments is considered and represents the main basis for routing in VSNs.

It is worth noticing that in opportunistic social networks, mobility models should be restricted to human behaviors and then form a novel class of *social-aware models* [12]. These models are based on the observation that people move because they are attracted toward other people they have social relationships with, or toward physical places that have special meaning with respect to their social behavior. In short, user movements are driven by their social relationships and behavior.

Social-aware mobility models are suitable for opportunistic networks since they closely reproduce statistical features of real movement traces. Actually, mobility modeling is one of the most active and challenging areas in the opportunistic networking field, since mobility models should be reconsidered and re-designed based on real user mobility traces available to the community. Such traces can be stored in wireless network data archives, like CRAWDAD (Community Resource for Archiving Wireless Data) [3], which provides wireless trace data from many contributing locations. Data captured from live wireless networks can help in understanding how real users use real networks under real conditions. This data helps us to identify and understand the real problems, to evaluate possible solutions, and to evaluate new applications and services.

The concept of sociability-based routing in vehicular networks has been introduced in [58], where the authors present Sociable Routing, which chooses the set of best forwarders for an efficient message delivery among nodes having high sociability indicators (e.g., the frequency and type of node encounters). Simulation results show that sociability-based routing can achieve good performance in terms of delay performance and cost.

Another social-based routing protocol in vehicular environments is SPRING (Social-based PRivacy-preserving packet forwardING) [40]. It is characterized by the deployment of intelligent RSUs at high social intersections (e.g., placed over traffic lights), in order to assist vehicles in packet forwarding. By temporarily storing packets through V2I communications, whenever next-hop vehicles are not available, RSUs act as relay nodes. Forwarding and packet reliability are then guaranteed with a reduced probability of packet drop. Furthermore, the SPRING protocol not only addresses the social-based RSU deployment for enhancing the message delivery ratio, but also discusses the privacy preservation issues and security attacks in vehicular DTNs.

Security issues in VSNs have been largely discussed, specially to protect user identity and location privacy. Most of the existing security solutions are associated with authentication mechanisms, which usually require expensive cryptography and an assumption of a central authority. Furthermore, several existing works lack collaborative effort among nodes to create a trusted vehicular community. As an instance, the Socialspot-based Packet Forwarding (SPF) protocol is very effective in achieving high-level location privacy in vehicular social networks [39]. A socialspot is referred to as the location in a city environment that many vehicles often visit, such as a shopping mall, a restaurant, or a cinema. Socialspots are exploited as the relay nodes for packet forwarding, in order to achieve an improvement of packet delivery. Moreover, since many vehicles visit the same socialspot, the socialspot cannot be used to trace a specific vehicle's other sensitive locations, and the receiver's location privacy is protected.

Several works have addressed query processing in vehicular social networks [22, 33, 45, 64]. PeopleNet [45] is an infrastructure-based proposal for information exchange in a mobile environment. It relies on the existence of a fixed network infrastructure to send a query to an area that may contain relevant information, and then it uses epidemic query dissemination through short-range communications within the area. Then, the answer to the query is communicated back through the fixed network infrastructure.

FleaNet [33] is a "virtual market" organized over a vehicular social network. Basically, it relies on a mobility-assisted query dissemination protocol, where the node that submitted the query periodically advertises only to its one-hop neighbors. As a result, the query spreads based on the motion of the vehicle that submitted the query, in order to avoid network overloads with multiple messages. Similar to FleaNet, Roadcast [64] is a content-sharing scheme, where vehicles send keyword-based queries to neighboring vehicles encountered along the same road. Replies to a query are returned as the most popular—but not necessarily the most relevant—content relevant to that query, since the content of the replies is expected to be useful for more vehicles in the future.

In GeoVanet [22], the authors assume that mobile users need to spread queries in the VSN, for which the answer is expected in a bounded time, with a minimum delay. For instance, let us consider a tourist driving a car in a city, and searching for information about the most interesting places to see. Queries about what to see are broadcasted to neighboring vehicles. Selected vehicles (e.g., vehicles of tourists sharing information about the sites they have visited) send answers to the tourist, and if the shared information matches the user's needs, it has to be delivered to her. As opposed to traditional query processing techniques, whose objective is to deliver the query result as quickly as possible, in GeoVanet the goal is to guarantee that the maximum amount of results will be delivered in a bounded time.

Finally, in [29] the authors propose the Fuzzy-Assisted Social-based rouTing (FAST) protocol, which takes the advantage of social behavior of humans on the road to make optimal and secure routing decisions. FAST uses the global knowledge of real-time vehicular traffic for packet routing from a source to the destination; then, a fuzzy inference system leverages a friendship mechanism to make critical decisions at intersections. In FAST, three types of human relationships are considered, including (*i*) direct friends, (*ii*) indirect friends (friends-of-friends), and (*iii*) non-friends. In direct friendship, the vehicles may establish relations using personal judgment in daily life experiences, while indirect friendship is based on the good reputation of other vehicles. In Figure 15.7, the nodes can start to establish mutual relations in the office and can be later direct friends using Facebook, Twitter, Google+, LinkedIn, etc. The nodes can also establish their relations in other places such as a residential area, playground, shopping mall, etc. Notice that there are some advantages of these types of friendship in terms of security, packet delivery ratio (PDR), and average delay. In-

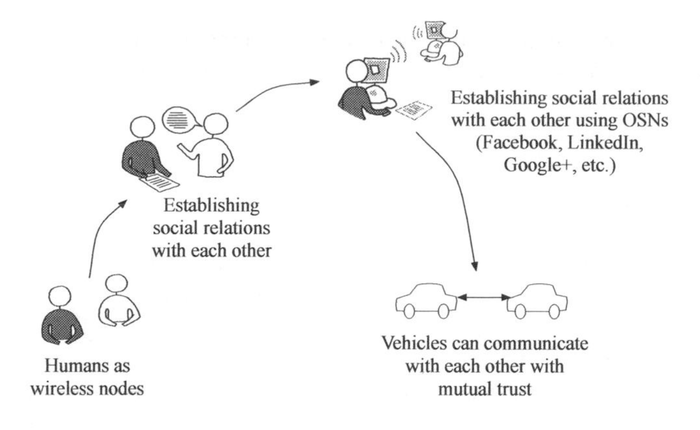

Figure 15.7: Schematic of social relation establishment between vehicles based on personal experiences [29].

deed, simulation results in urban vehicular environments show that FAST performs best in terms of PDR with up to a 32% increase, a decrease of 80% in average delay, and a hops count decrease of 50% compared to the state-of-the-art VANET routing solutions.

15.4 Conclusions

In this chapter, we have investigated the opportunistic mobile social networks as a novel communication paradigm, exploiting opportunistic encounters among human-carried devices and social networks for mobile social networking and collaborative content dissemination.

We first described opportunistic social networks through the main features (see Table 15.1), which distinguish such networks from traditional mobile wireless networks. It has emerged that opportunistic social networking represents a new communication paradigm, which exploits user mobility to establish communications and content exchange between mobile devices in pervasive and mobile computing environments. Thus, we presented an overview of the main projects addressing routing solutions for message delivery in social networks, based on social relationships and human behaviors [5, 10, 47].

We highlighted that opportunistic networking applications are strictly driven by social networking, and the fact that human mobility is directly related to the social behavior of people. Leveraging those considerations, particular attention has been given to the context of vehicular ad hoc networks, since vehicles move according to real-life human mobility and social interactions.

VANETs have been presented as a particular class of opportunistic mobile networks, under the assumption of social networking for several vehicular applications. In VANETs, vehicles can benefit from user social networks, and many forwarding schemes can rely on social information for efficient packet forwarding decisions. The concept of vehicular social networks (VSNs) was then presented, and the main differences between VSNs and VANETs, and online social networks have been discussed.

Furthermore, approaches based on *crowdsourcing* and eco-friendly applications have been addressed as the most significant solutions for vehicular social networking, while the main social-based routing protocols for data dissemination in VANETs have been presented at the end of this chapter. Social-based routing protocols in VANETs were distinguished from traditional data dissemination approaches in mobile ad hoc networks, since packets are routed based on the social behavior of drivers and, most in general, humans in urban scenarios. The friendship mechanism is opportunistically exploited instead of simply *bridging* partitions in the network, and forwarding the messages to the next available node toward a destination.

References

[1] Available online, http://mashable.com/2011/09/01/evos-social-networking-car/.

[2] Available online, http://www.crowdsourcing.org/.

[3] Available online, http://crawdad.cs.dartmouth.edu/index.php.

[4] Ecodriving USA. Available online at http://www.ecodrivingusa.com/.

[5] European Commission's FET Haggle project. Available online, http://www.haggleproject.org.

[6] European Smart Cities. http://www.smart-cities.eu/.

[7] GasBuddy, Find low gas prices in the USA and Canada. http://gasbuddy.com.

[8] Roadify. http://www.roadify.com/.

[9] Waze. http://www.waze.com.

[10] SOCIALNETS project: Social networking for pervasive adaptation. http://www.social-nets.eu, 2008.

[11] C. Boldrini, M. Conti, and A. Passarella. Exploiting users social relations to forward data in opportunistic networks: the HiBOp solution. *Pervasive and Mobile Computing*, 4(5):633–657, 2008.

[12] C. Boldrini, M. Conti, and A. Passarella. Social-based autonomic routing in opportunistic networks, In *Autonomic Communication*, pages 31–67. Springer, 2009.

[13] J. Breslin and S. Decker. The future of social networks on the Internet: the need for semantics. *IEEE Internet Computing*, 11(6):86–90, 2007.

[14] C.J. Buys and K.L. Larson. Human sympathy groups. *Psychological Reports*, 45:547–553, 1979.

[15] U.B. Ceipidor, C.M. Medaglia, V. Volpi, A. Moroni, S. Sposato, and M. Tamburrano. Design and development of a social shopping experience in the ioT domain: The ShopLovers solution. In *Software, Telecommunications and Computer Networks (SoftCOM), 2011 19th International Conference on*, pages 1–5, 2011.

[16] N. Cenerario, T. Delot, and S. Ilarri. Vehicular event sharing with a mobile peer-to-peer architecture. *Transportation Research Part C: Emerging Technologies*, 18(4):584–598, August 2010.

[17] X. Chen, E. Santos-Neto, and M. Ripeanu. Crowd-based smart parking: a case study for mobile crowdsourcing. In *Proc. of 5th International Conference on MOBILe Wireless MiddleWARE, Operating Systems, and Applications (MOBILWARE)*, Berlin, Germany, 2012.

[18] M. Conti, F. Delmastro, and A. Passarella. Social-aware content sharing in opportunistic networks. In *Sensor, Mesh and Ad Hoc Communications and Networks Workshops, 2009. SECON Workshops '09. 6th Annual IEEE Communications Society Conference on*, pages 1–3, 2009.

[19] M. Conti and M. Kumar. Opportunities in opportunistic computing. *IEEE Computer*, 43(1):42–50, 2010.

[20] P. Costa, C. Mascolo, M. Musolesi, and G.P. Picco. Socially-aware routing for publish-subscribe in delay-tolerant mobile ad hoc networks. *IEEE Journal on Selected Areas in Communications*, 26(5):748–760, 2008.

[21] E. Daly and M. Haahr. Social network analysis for information flow in disconnected delay-tolerant MANETs. *IEEE Transactions on Mobile Computing*, 8(5), May 2009.

[22] Thierry Delot, Nathalie Mitton, Sergio Ilarri, and Thomas Hien. Decentralized pull-based information gathering in vehicular networks using GeoVanet. In *Proceedings of the 2011 IEEE 12th International Conference on Mobile Data Management—Volume 01*, MDM '11, pages 174–183, Washington, DC, USA, 2011. IEEE Computer Society.

[23] S. Diewald, A. Muller, L. Roalter, and M. Kranz. MobiliNet: a social network for optimized mobility. In *Proceedings of the Social Car Workshop at the 4th International Conference on Automotive User Social Interaction for Digital Cities, Interfaces and Interactive Vehicular Applications*, Portsmouth, NH, USA, October 2012.

[24] R. Dunbar and M. Spoors. Social networks, support cliques, and kinship. *Human Nature*, 6(3):273–290, 1995.

[25] K. Fall. A delay-tolerant network architecture for challenged internets. In *Proc. of Special Interest Group on Data Communications*, pages 27–34, Karlsruhe, Germany, August 2003.

[26] H. Harteinstein and K.P. Labertaux. *VANET Vehicular Applications and Inter-Networking Technologies*. John Wiley & Sons, Ltd, March 2010.

[27] R. Hill and R. Dunbar. Social network size in humans. *Human Nature*, 14(1):53–72, 2003.

[28] P. Hui, J. Crowcroft, and E. Yoneki. BUBBLE Rap: social-based forwarding in delay tolerant networks. *IEEE Transactions on Mobile Computing*, 10(11), November 2011.

[29] R. H. Khokhar, R. Noor, K. Z. Ghafoor, C.-H. Ke, and A. Ngadi. Fuzzy-assisted social-based routing for urban vehicular environments. *EURASIP Journal on Wireless Communications and Networking*, 2011(178):1–15, 2011.

[30] J. Kincaid. *Google's Open Spot Makes Parking a Breeze, Assuming Everyone Turns into a Good Samaritan*. Technical report, 2010.

[31] M. Knobel, M. Hassenzahl, M. Lamara, T. Sattler, J. Schumann, K. Eckoldt, and A. Butz. Clique trip: feeling related in different cars. In *Proceedings of the ACM Designing Interactive Systems Conference, DIS*, pages 29–37, Newcastle, UK, June 11–15 2012.

[32] E. Koukoumidis, L.-S. Peh, and M. R. Martonosi. Signalguru: Leveraging mobile phones for collaborative traffic signal schedule advisory. In *Proc. of the 9th ACM International Conference on Mobile Systems, Applications, and Services, MobiSys*, pages 127–140, New York, NY, USA, 2011.

[33] U. Lee, J. Lee, J.-S. Park, and M. Gerla. FleaNet: a virtual market place on vehicular networks. *IEEE Transactions on Vehicular Technology*, 59(1):344–355, 2010.

[34] Bin Li, Daqing Zhang, Lin Sun, Chao Chen, Shijian Li, Guande Qi, and Qiang Yang. Hunting or waiting? Discovering passenger-finding strategies from a large-scale real-world taxi dataset. In *2011 IEEE International Conference on Pervasive Computing and Communications Workshops (PERCOM Workshops)*, pages 63–68, 2011.

[35] A. Lindgren, A. Doria, and O. Scheln. Probabilistic routing in intermittently connected networks. *ACM SIGMOBILE Mobile Computing and Communications Review*, 7(3):19–20, July 2003.

[36] B. Liu. Next generation vehicular traffic management enabled by vehicular ad hoc networks and cellular mobile devices. Submitted in partial satisfaction of the requirements for the degree of Doctor of Philosphy in computer science, Office of Graduate Studies of the University of California, June 2011.

[37] Bojin Liu, D. Ghosal, Chen-Nee Chuah, and H.M. Zhang. Reducing greenhouse effects via fuel consumption-aware variable speed limit (FC-VSL). *IEEE Transactions on Vehicular Technology,* 61(1):111–122, 2012.

[38] R. Lu. Security and privacy preservation in vehicular social networks. Submitted in partial satisfaction of the requirements for the degree of doctor of philosophy in electrical and computer engineering, University of Waterloo, 2012.

[39] R. Lu, X. Lin, X. Liang, and X. Shen. Sacrificing the plum tree for the peach tree: a socialspot tactic for protecting receiver-location privacy in VANET. In *Proc. of GLOBECOM*, pages 1–5, 2010.

[40] R. Lu, X. Lin, and X. Shen. Spring: A social-based privacy-preserving packet forwarding protocol for vehicular delay tolerant networks. In *Proc. of INFOCOM'10*, pages 1229–1237, San Diego, CA, USA, March 2010.

[41] M. McPherson, L. Smith-Lovin, and M.E. Brashears. Social isolation in America: changes in core discussion networks over two decades. *American Sociological Review*, 71:353–375, 2006.

[42] R.M. Milardo. Comparative methods for delineating social networks. *Journal of Social and Personal Relationships*, 9:447–461, 1992.

[43] A. Mislove, M. Marcon, K.P. Gummadi, P. Druschel, and B. Bhattacharjee. Measurement and analysis of online social networks. In *Proc. of the 7th ACM SIGCOMM Conference on Internet Measurement (IMC'07)*, pages 29–42, 2007.

[44] A. Mostafa, A.M. Vegni, T. Oliveira, T.D.C. Little, and D.P. Agrawal. Qoshvcp: hybrid vehicular communications protocol with QoS prioritization for safety applications. *ISRN Communications and Networking*, 2012:14, 2012.

[45] M. Motani, V. Srinivasan, and P. S. Nuggehalli. PeopleNet: engineering a wireless virtual social network. In *Proc. of ACM 11th Annual International Conference on Mobile Computing and Networking (MobiCom)*, pages 243–257, 2005.

[46] L. Pelusi, A. Passarella, and M. Conti. Opportunistic networking: data forwarding in Disconnected mobile ad hoc networks. *IEEE Communication Magazine*, 44(11), November 2006.

[47] A.-K. Pietilinen, E. Oliver, J. LeBrun, G. Varghese, and C. Diot. MobiClique: middleware for mobile social networking. In *Proc. of the 2nd ACM Workshop on Online Social Networks (WOSN '09)*, pages 49–54, 2009.

[48] S.G.B. Roberts, S. Wilson, R. Fedurek, and R.I.M. Dunbar. Individual differences and personal social network size and structure. *Personality and Individual Differences*, 44:954–964, 2008.

[49] T. Schmid and M. B. Srivastava. Exploiting social networks for sensor data sharing with SENSESHARE. Available online at `http://www.escholarship.org/uc/item/4919w4vh/`.

[50] Mary R. Schurgot, Cristina Comaniciu, and Katia Jaffrès-Runser. Beyond traditional DTN routing: social networks for opportunistic communication. *CoRR*, abs/1110.2480, 2011.

[51] M.R. Schurgot, Cristina Comaniciu, and K. Jaffres-Runser. Beyond traditional DTN routing: social networks for opportunistic communication. *Communications Magazine, IEEE*, 50(7):155–162, 2012.

[52] Y. Shao, C. Liu, and J. Wu. Delay-Tolerant Networks in VANETs. In *Handbook on Vechicular Networks*, M. Weigle and S. Olariu (eds.), Taylor & Francis, Boca Raton, FL, (accepted for publication).

[53] T. Spyropoulos, K. Psounis, and C.S. Raghavendra. Efficient routing in intermittently connected mobile networks: the single-copy case. *Journal IEEE/ACM Transactions on Networking (TON)*, 16(1):63–76, 2008.

[54] S. Staab, P. Domingos, P. Mika, J. Golbeck, L. Ding, T. Finin, A. Joshi, A. Nowak, and R. R. Vallacher. Social networks applied. *IEEE Intelligent Systems*, 20(1):80–93, 2005.

[55] R. Uzcategui and G. Acosta-Marum. WAVE: a tutorial. *IEEE Comm. Mag.*, 47(5):126–133, 2009.

[56] A. Vahdat and D. Becker. *Epidemic Routing for Partially Connected Ad Hoc Networks*. Technical report, April 2000.

[57] A. M. Vegni, M. Biagi, and R. Cusani. Smart vehicles, technologies and main applications in vehicular ad hoc networks, In *Vechicular Technologies—Department and Applications*, L. Galati Giordano and L. Reggiani (eds), pages 3–20. InTech, 2013.

[58] R. Verdone and F. Fabbri. Sociability based routing for environmental opportunistic networks. *Journal of Advances in Electronics and Telecommunications*, 1(1), April 2010.

[59] P. White. No vacancy: Park Slope's parking problem and how to fix it. Available online, `http://www.Transalt.org/newsroom/releases/126`.

[60] D. Wu, J. Cao, Y. Ling, J. Liu, and L. Sun. Routing algorithm based on multi-community evolutionary game for VANET. *Journal of Networks*, 7(7):1106–1115, July 2012.

[61] Mingjun Xiao, Jie Wu, and Liusheng Huang. Community-aware opportunistic routing in mobile social networks. *IEEE Transactions on Computers*, 99(PrePrints):1, 2013.

[62] T. Yan, B. Hoh, D. Ganesan, K. Tracton, T. Iwuchukwu, and J.-S. Lee. *CrowdPark: A Crowdsourcing-Based Parking Reservation System for Mobile Phones*. Technical report, 2011.

[63] A.-U.-H. Yasar, N. Mahmud, D. Preuveneers, K. Luyten, K. Coninx, and Y. Berbers. Where people and cars meet: social interactions to improve information sharing in large scale vehicular networks. In *Proc. of the ACM Symposium on Applied Computing, SAC*, pages 1188–1194, 2010.

[64] Y. Zhang, J. Zhao, and G. Cao. RoadCast: A popularity aware content sharing scheme in VANETs. In IEEE Computer Society, editor, *Proc. of 29th IEEE International Conference on Distributed Computing Systems (ICDCS)*, pages 223–230, 2009.

[65] W. X. Zhou, D. Sornette, R. A. Hill, and R. I. M. Dunbar. Discrete hierarchical organization of social group sizes. *Proceedings of the Royal Society B: Biological Sciences*, 272(1561):439–444, 2005.

Chapter 16

Network Emulation Testbed for Mobile Opportunistic Networks

Razvan Beuran
Hokuriku StarBED Technology Center
National Institute of Information and Communications Technology
Ishikawa, Japan

Toshiyuki Miyachi
Hokuriku StarBED Technology Center
National Institute of Information and Communications Technology
Ishikawa, Japan

Shinsuke Miwa
Hokuriku StarBED Technology Center
National Institute of Information and Communications Technology
Ishikawa, Japan

Yoichi Shinoda
Research Center for Advanced Computing Infrastructure
Japan Advanced Institute of Science and Technology
Ishikawa, Japan

CONTENTS

16.1 Introduction

Research on opportunistic mobile networks has become more and more active in recent years, and this is partly motivated by the advent of mobile devices—such as smartphones—in our daily life. These widely used devices offer an ideal framework for deploying opportunistic mobile networks, including in the emerging area of social networks.

Mobile devices have a series of restrictions in terms of computing resources, battery life, and so on. Moreover, mobile devices use wireless connectivity for network access, and this makes them prone to communication disruptions. Any protocol or application developed in the context of opportunistic networks has to cope with all these challenges in order to operate correctly and to provide a satisfactory user experience.

We have developed a network emulation testbed—called QOMB—that can be used to validate the correct operation of opportunistic network applications and protocols in scenarios that involve both node mobility and wireless communication. The testbed was developed by integrating the wireless network emulation set of tools QOMET with the large-scale network testbed StarBED of the National Institute of Information and Communications Technology, Japan. We note that most of our work related to opportunistic networks was done on the particular case of delay/disruption-tolerant networks (DTN).

Researchers can deploy implementations of opportunistic network applications and protocols on our testbed, and carry out experiments in realistic mobility scenarios, so that the applications and protocols are thoroughly validated through real-time repeatable and controllable experiments. This makes it possible to determine and optimize the performance characteristics of the applications and protocols under test, and to verify that any improvements and fixes that may be required have the intended effects.

The remainder of this chapter is organized as follows. First we present an overview of our approach that emphasizes the motivation and the key concepts used in our work (Section 16.2). Then we discuss the main design challenges related to the implementation of our network emulation testbed (Section 16.3). In Section 16.4 we provide more practical details regarding the actual implementation of the QOMB testbed, and in Section 16.5 we focus on the necessary extensions for opportunistic network emulation. Following that, in Section 16.6, we illustrate the practicality of our testbed with several experiment results and observations. The chapter ends with related work, a series of conclusions, and references.

16.2 Approach Overview

The goal of our work can be summarized as trying to provide an experiment infrastructure for opportunistic network experiments, so that protocols and applications that are being developed for these networks can be thoroughly tested. This helps researchers make sure that the developed protocols or applications behave as intended when subsequently deployed in a live/production network.

The key element in our approach is the use of the experiment technique called network emulation. In what follows we motivate this choice and we provide an overview of our approach.

16.2.1 Experiment Techniques

When researchers develop protocols or applications for opportunistic networks, a key requirement is to validate the behavior of the developed technologies, so as to ensure that they behave as expected.

Validation at algorithm level can be done partially through analytical methods, and it is often done through network simulation techniques. When simulation is used, the developed algorithms must be implemented in the framework of the network simulator employed, and integrated with the other network protocols with which it is supposed to interact. Moreover, suitable scenarios must be created so that the behavior of the algorithms can be thoroughly validated in various circumstances.

Once validation through simulation is completed, the algorithms are typically implemented in a programming language which may differ from that used by the simulator, and which depends on the deployment target. Changes are usually necessary also at the level of the interface with the network protocol stack of the target system. The software implementation is then deployed in actual networks, where real-world trials take place. Once this phase is completed, the developed protocol or application is finally deployed in production networks.

The validation approach described so far has two points of failure, as follows:

1. Even though an algorithm may have been thoroughly tested through formal methods and network simulation, it is usually difficult to guarantee that the software implementation used in the real-world trials is identical in functionality with the initial algorithm or with its implementation used in simulation.

2. For many kinds of networks, and in particular for opportunistic networks, it is difficult to reproduce the characteristic conditions representative of these networks in an actual network environment; moreover, it is often impractical to conduct repeatable experiments in such environments.

In case the software implementation differs from the simulated model (first point of failure above), then the results of the validation experiments done through simulation are not applicable anymore, as they *do not refer* to the same algorithm. This is an important issue, since it is the actual software implementation that will be subsequently deployed in production networks, and not the simulated model. Therefore, any potential differences and bugs in the software implementation will have undesired effects once the system is deployed, and may even disrupt the network environment into which the technology is deployed.

To counter this potential issue, one possibility is to validate as thoroughly as possible the software implementation itself through real-world trials performed in suitable scenarios. We estimate that most of the time such a validation does not occur in practice for the following reason: a thorough validation through real-world trials is time consuming and costly, hence it is usually skipped or shortened in virtue of the simulation experiments that were already performed (even though they may not be fully equivalent, as discussed above). As time constraints and financial costs become prohibitive, the parameter space covered by real-world trials will typically be significantly smaller than that covered by the simulation experiments.

The task of performing real-world trials is further complicated by the need to have *repeatable and controllable* experiment conditions, so that the same circumstances can be re-created a number of times until each implementation aspect is thoroughly tested and validated (the second point of failure above). For instance, repeatability is mandatory for iterative testing during debugging and/or optimization procedures.

16.2.2 Network Emulation

Network emulation is a hybrid experiment approach that attempts to circumvent the two points of failure described above by combining the essential features of network simulation and real-world trials. The concept of network emulation is thoroughly analyzed in [2] including descriptions of actual tools and emulation mechanisms.

The main methodology differences between network simulation and network emulation are typically as follows. On one hand, network simulation uses software implementations of protocol and application models that are executed in logical time over a modeled network in order to predict system performance. On the other hand, network emulation uses real-world software implementations of applications and protocols that are run in real time over networks that contain both modeled and real components; performance is then assessed by actual measurements at protocol and application levels.

Consequently, the key characteristics of network emulation that address the two issues discussed in Section 16.2.1 are:

1. Through network emulation experiments, one tests actual software implementations of the protocol or application under test, and thus the experiment results are directly applicable to practical scenarios related to their deployment. This property addresses the first point of failure above.

2. Network emulation uses models for several aspects of real-world networks, such as wireless network technologies, mobility, etc. Therefore, experiment conditions can be reproduced in a controllable and repeatable manner, allowing for a thorough validation of the protocol or application under test. This characteristic addresses the second point of failure above.

A graphical comparison of network emulation with simulation and real-world trials is shown in Figure 16.1. Thus, the level of control and scalability characteristics are highest for network simulation, they decrease somewhat for network emulation, and are lowest for real-world trials. On the other hand, costs are highest for real-world trials, they decrease for network emulation, and are lowest for network simulation. These characteristics emphasize the trade-off that network emulation makes between real-world trials and network simulation, thus positioning network emulation as a "golden mean" between the other two experiment techniques that aptly combines their respective advantages.

16.2.3 Network Emulation Testbed

As discussed in [2], there are two ways to do network emulation in practice: in a *centralized* or in a *distributed* manner. For the centralized case, the scalability of the setup is limited by the processing capabilities of the emulator, and by the number of end nodes that can be connected to it. The distributed approach, however, makes it possible to increase the scale of the experiment by sharing the emulation task among multiple hosts.

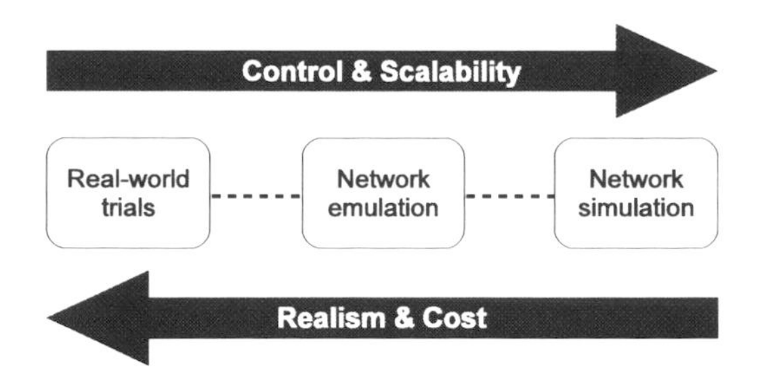

Figure 16.1: Network emulation versus other experiment techniques.

Scalability is an important requirement for experiment setups, in particular in the context of opportunistic networks. Tests with a small number of nodes may be relatively easy to do through real-world trials, but costs become quickly prohibitive as the number of devices increases. In simulation it is easy to increase scale, albeit at the expense of execution time. Similar to simulation, emulation has good scalability characteristics (especially when used in a distributed manner), but the challenges and costs are higher for emulation because of the additional constraints of real-time execution and the use of real equipment for some parts of the experiment setup (see also Section 16.3.4).

A typical instance of distributed emulation methodology that focuses on scalability is the use of a *network emulation testbed.* We have followed the same approach in our work since we believe that it is particularly suited for the case of opportunistic networks. The testbed that we have developed for such experiments will be described in Section 16.4. However, before plunging into technical details, we shall first discuss the main challenges that have driven our design and implementation process.

16.3 Design Challenges

There are several challenges that we encountered when designing the emulation testbed for opportunistic mobile networks, as we tried to strike a balance between user control and experiment realism, complexity and execution speed, and so on. Thus, we distinguish two main categories:

- challenges that are particularly related to mobile opportunistic networks, such as the emulation of node mobility and communication disruptions; and
- challenges that have a more generic nature, such as the emulation of wireless communication, or experiment scalability.

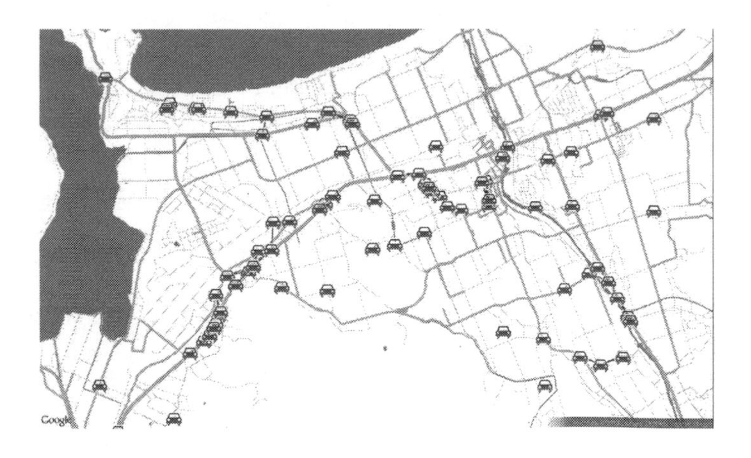

Figure 16.2: Screenshot of an emulation experiment that uses a vehicular network and the DTN protocol for post-disaster network recovery.

16.3.1 Mobility

Opportunistic networks are formed in a dynamic manner, with each node attempting to detect systems in their vicinity and to integrate them into the network if this action has a potential benefit [19]. Mobility is a key characteristic of opportunistic networks, because it makes it possible to find new nodes to be integrated with the network.

Real-world trials which involve mobility are difficult to organize, especially when the number of nodes is not negligible. One other difficulty is *repeating* mobility experiments, because the human factor makes it so that the movement trajectory and timing are typically different between each run. The network emulation technique plays an essential role in this context, as by emulation one is able to repeat the same experiment any number of times, while executing actual software implementations (and not models as in simulation) [2].

Our emulation testbed provides support for various kinds of mobility models by means of the wireless network emulation set of tools called QOMET, which includes several often-used mobility models (for instance, random walk) [7]. Other more realistic models are also supported, such as the *behavioral mobility model* for pedestrian movement, which produces realistic mobility traces for areas with geographic constraints (e.g., the road and building topography of urban locations) [5]. Mobility traces produced by other mobility generators, or even those created from GPS data, can also be imported in QOMET, thus giving the user complete freedom over the mobility models utilized.

In Figure 16.2 we illustrate the mobility features of our testbed with a vehicular motion scenario that was used for network emulation experiments done as a feasibility study for using the DTN protocol for post-disaster network recovery. Visualization was done by Tomoya Inoue of the Japan Advanced Institute of Science and Technology using the Google Maps API.

Table 16.1 Areas in Which Faults Can Occur in Network Environments (based on [21])

Area	Type of faults
Physical	Network equipment faults, communication errors
Specification	Protocol or standard logical errors
Implementation	Application or protocol implementation errors
Configuration	Hardware or software configuration errors

Mobility is typically generated in advance when making emulation experiments, thus making it possible that, when an experiment is executed, all the computation resources are dedicated to communication emulation (in addition, of course, to application and protocol execution). In some cases, offline computation of the mobility traces is not possible, for instance, when the mobility of the nodes depends on the application being executed, such as a node trajectory being determined as a function of the content of the messages being received. This methodology can be employed on our testbed if the application determines node mobility by itself and interfaces with the appropriate QOMET libraries for communication emulation; such a technique was used successfully for robot motion planning experiments [8].

16.3.2 Disruptions

Fault injection is an important feature of network testbeds, as evaluation of applications and protocols must be done in wide variety of network conditions [21]. There are several reasons why one may wish to purposefully inject network faults into experiment environments:

- the need to create realistic situations in which communication disruptions that are similar to those in real networks occur; and
- the desire to perform experiments in critical network conditions, for instance, in order to study the behavior of the application and protocols under test in circumstances similar to those occurring in disaster situations, or in conjunction with emergency response training, etc.

According to [21] there are several areas in which faults can occur in network systems, as shown in Table 16.1. In this work we focus on fault injection in the physical area, and in particular on communication errors, but we wish to emphasize the fact that our testbed can be used to inject, discover, and fix faults in all the other areas as well.

In typical wired networks, communication disruptions only occur when there is a network malfunction. However, in mobile opportunistic networks, disruptions often occur naturally, because of the specific characteristics of these networks: mobility and wireless communication. This type of disruptions takes place naturally on our testbed as well: for instance, communication conditions will vary according to the changing relative position of the nodes and of their communication environment as the nodes move in the virtual world created through emulation.

In addition to the natural disruptions mentioned above, when dealing with opportunistic networks, researchers may find the need to create communication disruptions on demand, so as to test certain aspects of the protocols and applications they develop. This requires specific *additional* fault injection mechanisms similar to the ones discussed in [21]. We distinguish here two main categories:

- communication environment disruptions, such as the injection of electromagnetic noise; and
- network condition disruptions, such as intentional delay and jitter, bandwidth limitations, or packet drop.

Communication environment disruptions are created in our emulation testbed by the artificial injection of faults in the wireless communication environment between the emulated nodes. A typical example in this fault injection category is the introduction of artificial electromagnetic noise in the emulated virtual world. As electromagnetic noise will interfere with the communication in the virtual environment, it will affect, in an indirect manner, the network conditions in the experiment; support for this feature was added to the appropriate QOMET library. Electromagnetic noise can be controlled by the user in terms of position of the noise source, its intensity, as well as the starting time and duration of noise generation, but the user has *no direct control* over the resulting communication conditions.

Network condition disruptions refer to the case when the user defines *explicitly* the network conditions between a certain pair of nodes, and QOMET libraries recreate the specified conditions without performing any model computation. Although this approach is not very realistic, it provides a good quantitative control over the network conditions, such as directly indicating the available bandwidth, the delay and jitter, and the loss rate during a certain time interval, thus making it easy to describe the time-varying conditions that characterize opportunistic networks.

16.3.3 Wireless Communication

Wireless communication is another key characteristic of opportunistic networks. In order to account for this aspect, our approach relies on the QOMET set of tools for wireless network emulation. QOMET provides support for several wireless network technologies, such as IEEE 802.11a/b/g, IEEE 802.15.4, and IEEE 802.16e (the latter being currently under development). As for propagation, we support the log-distance path loss propagation model. We note here that the architecture of QOMET is modular, and its functionality can be easily extended to new wireless network technologies or new propagation models, as needed.

When performing network emulation, wireless communication conditions must be reproduced by QOMET in real time as the experiment is running, so as to provide to the applications and protocols under test a communication environment that is similar to that of a real wireless network. This is achieved by using probabilistic models for the wireless communication, and through the real-time execution of a network condition management module [7]. The two-stage scenario-driven approach

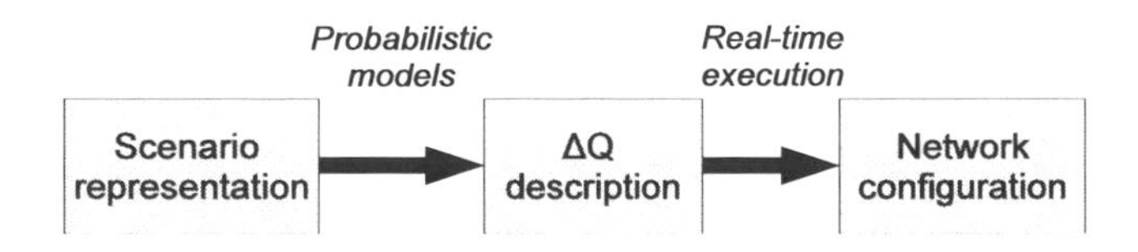

Figure 16.3: QOMET's two-stage scenario-driven network emulation approach.

that makes this possible is shown in Figure 16.3, where ΔQ is used to denote the continuously changing network conditions (more details about QOMET will be provided in Section 16.4.2).

16.3.4 Scalability

In order to increase experiment scale when performing network simulation, most often it is sufficient to change the number of nodes specified in the simulation scenario. Then one has to face the longer simulation time caused by the increase in the number of nodes. This is because simulation often uses logical time and a discrete-event approach; therefore, the time it takes to complete an experiment depends on the number of events to be processed. Typically this number is proportional with N^2, where N is the number of nodes in the simulation scenario. Therefore, the duration of a simulation run increases very quickly with the number of nodes: a one-hour scenario may even take days to complete if the number of nodes is sufficiently large.

In the case of network emulation, in order to increase experiment scale, one has to provide the necessary resources for the actual execution of the additional nodes. However, experiment duration will not increase, as emulation runs in real time (also called "wall-clock time").

We use a *multi-layer emulation* approach for our testbed, so that the most appropriate scaling method is used for any given scenario. Thus, we distinguish three methods of allocating emulated nodes to the physical hosts in our testbed, as illustrated in Figure 16.4:

1. Emulated nodes run directly on physical hosts, and there is a one-to-one correspondence between them.

2. Emulated nodes run as virtual machines (VMs), with several instances being executed on each physical host.

3. Emulated nodes run as processes on physical hosts, with multiple processes being executed in parallel on each host.

The most straightforward method for performing emulation is to employ a physical computer host (a "bare-metal machine") for each emulated node, as per method 1 above. This approach has the advantage of providing exclusive use of all the resources of a host to the emulated node it represents, which results in good performance characteristics. However, high costs are associated with this approach,

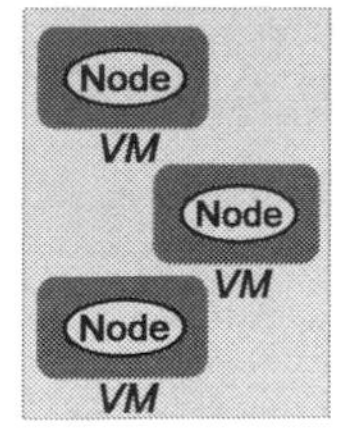

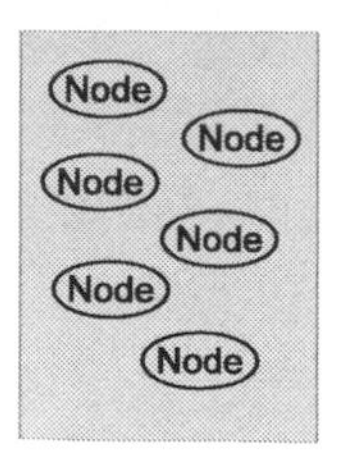

Figure 16.4: Various approaches for allocating emulated nodes per host.

since increasing experiment scale requires physically adding computer hosts to the experiment setup. Hence, this method is most appropriate for relatively low-scale experiments with nodes that require a large amount of computing resources. We note that, with more than 1300 experiment hosts (as of August 2013), our testbed infrastructure—StarBED—provides sufficient physical resources even for multiple experiments that use hundreds of physical hosts each.

A method to further increase experiment scale, and to reduce costs for large experiments, is to use virtualization techniques so as to run multiple emulated nodes on a physical host (method 2 above). Depending on the virtualization technique used, both the software and hardware layers of a target system are emulated, or only software ones, so that the target system is reproduced with the necessary degree of accuracy. In some cases, virtualization may be the only solution for emulation, for instance, when testing a networked sensor, because such a device is very different from a typical computer, hence it cannot be represented directly by a bare-metal machine.

The following remarks regarding virtualization techniques are relevant in the context of our work:

- The more accurate the virtualization needs to be, the larger the required amount of physical resources is. For instance, CPU instruction-level virtualization needs more computing resources than function call level virtualization.

- Multiple virtual machines running on a physical host will share its resources, with potential effects on the performance of each VM instance; usually, by keeping the number of VMs below a certain (application-specific) threshold one can ignore these effects.

- Virtualization techniques can be either generic, such as for general-purpose x86 systems, or very specific, such as emulating a certain type of microcontroller with its particular CPU, amount of memory, sensors, etc.

The relatively low cost of scaling up experiments by using virtualization techniques makes this method suited especially for medium- to large-scale experiments,

Table 16.2 Characteristics of Various Experiment Scale-Up Methods

Method	Experiment scale	Fidelity	Cost	Testing target
Physical	Low to medium-scale	Highest	High	System
Virtualization	Medium to large-scale	Average	Medium	System/protocol
Process	Very large scale	Lowest	Low	Protocol

in which the performance of a target system as a whole is to be assessed. If this method is used, one has to take into account the potential influence of the variable performance of the VMs on the accuracy of the results. As an example, we successfully used KVM (kernel-based virtual machine) technology and OpenWrt VMs to run 100-node mesh network emulation experiments on 5 physical StarBED hosts (hence with 20 emulated nodes per physical host).

Finally, the most light-weight methodology is to represent each emulated node by a process running on a physical host, as per method 3 above. This approach makes it very easy to extend experiment scale, as adding more nodes is equivalent to creating more processes. However, this method has the least amount of fidelity, since most of the components of the system are abstracted. One could actually consider this approach as a particular case of virtualization in which the virtualization layer represents the target system in a very abstract way, typically I/O oriented. Process-level emulation is most appropriate for evaluating the logical behavior of a certain application or protocol implementation, and for very-large-scale experiments. For instance, StarBED was successfully used to emulate a one million peer-to-peer node network by using 100 physical hosts (hence with a total of 10,000 emulated nodes, i.e. processes, per host) by using the PIAX framework [25].

The most important characteristics of the above methodologies for scaling up experiments are summarized in Table 16.2. Thus, the "physical" method (emulated nodes correspond directly to physical hosts) can be used for low- to medium-size experiments, and has a high fidelity, but also a high cost. Virtualization is most suited for medium- to large-scale experiments, and has average fidelity and also average costs. Process-level emulation is the only solution for very large-scale experiments, but has the lowest fidelity; the costs are low as well. The testing targets for each methodology are typically overall systems and application/protocol combinations in the first two cases, and mainly simpler protocol evaluations for the third method.

We conclude this subsection by stressing the fact that the above emulation approaches are not mutually exclusive. Thus, it is possible to combine two or even all of the methods in one experiment, so as to achieve a good overall experiment scale while maintaining sufficient accuracy for some portions of the experiment. For instance, one can imagine running a 100-node experiment by using 10 physical hosts for 10 of the emulated nodes—hence with high fidelity—and by using a small number of physical hosts to run the remaining 90 emulated nodes through virtualization techniques. Related to experiment accuracy, we also mention that it is possible to interface real network devices with emulated ones, so that both complete realism (for the real portion of the setup) and scalability can be achieved simultaneously in the same experiment [3].

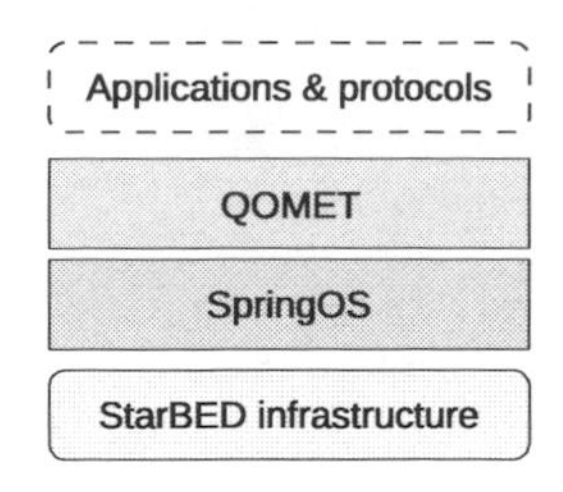

Figure 16.5: Logical hierarchy of QOMB components.

16.4 QOMB Testbed

We created the QOMB wireless network emulation testbed by integrating the QOMET wireless network emulation set of tools with the large-scale network experiment testbed that is StarBED. QOMB was initially developed for IEEE 802.11 network emulation; its general architecture and utilization examples in this context were presented in [10]. The modular architecture of QOMB and of its components makes it possible to extend the testbed to other wireless network technologies and devices in a straightforward manner. For instance, by adding the necessary support, we could employ QOMB for ubiquitous network emulation experiments using the IEEE 802.15.4 technology [9].

The logical hierarchy of the QOMB elements that are used in the context of opportunistic network emulation is shown in Figure 16.5. SpringOS is the experiment-support software used to perform experiments on StarBED. Although not components of QOMB *per se*, for clarity purposes the applications and protocols under test are also indicated at the top level.

In what follows we shall introduce in more detail each of the above components, starting with StarBED and its support software SpringOS, followed by QOMET and its libraries, and ending with the overall architecture used on QOMB for opportunistic network emulation.

16.4.1 StarBED

StarBED is a large-scale wired-network testbed at the Hokuriku StarBED Technology Center of the National Institute of Information and Communications Technology, located in Ishikawa, Japan [22]. With over 1300 interconnected PCs made available for experiments, users can perform a wide range of network experiments on StarBED, which represents the physical infrastructure of QOMB.

An important point about StarBED is that the experiment network, which carries the traffic generated during the execution of the applications and protocols under test, is completely separated from the control network, which is used to manage the experiment hosts, collect logs, etc. This ensures that there is no undesired interaction between the experiment and the control traffic.

SpringOS is the main experiment-support software tool for StarBED, and it allows users to easily perform complex experiments with a large number of hosts. The most important functions of SpringOS are:

1. *Experiment preparation:* Configure the experiment hosts and network so that they are ready for experiment execution.

2. *Experiment execution*: Effectively carry out the experiment by performing as required the necessary commands on the experiment hosts.

SpringOS has a message-based client-server architecture. Thus, a module called *SpringOS master* is in charge of controlling the experiment execution on all the experiment hosts. This is done with the assistance of local modules, called *SpringOS clients*, executed on each experiment host.

16.4.2 QOMET

QOMET (Quality Observation and Mobility Experiment Tools) is a set of tools for wireless network emulation [7]. QOMET allows the definition of various complex scenarios, including experiments with node mobility and urban settings. QOMET provides the necessary mechanisms for performing wireless network emulation in a distributed manner by reproducing the communication conditions between the nodes that are part of an experiment. QOMET relies on the experiment management mechanisms of StarBED for its distributed execution. Thus, the integration of StarBED, SpringOS, and QOMET effectively gives birth to a new entity, the wireless network emulation testbed QOMB.

In the context of this chapter, the most important components of QOMET that make wireless network emulation possible are the libraries called *deltaQ* and *wireconf*, as follows.

The deltaQ library is in charge of computing the communication conditions between wireless nodes given a user-defined scenario. The scenario specifies the properties of the wireless nodes (position, parameters of the network technology, mobility patterns, etc.), and of the environment in which they are placed (attenuation, shadowing, street and building structures, and so on). These properties are used to create a "virtual world" that corresponds to the user-defined scenario, in which the wireless nodes move and communicate with each other. We emphasize the fact that, although the movement is virtual, the communication that takes place is real, since real protocol and application implementations are executed in our experiments.

The communication conditions computed by the deltaQ library are re-created during the real-time experiment by the wireconf library. This library is in charge of controlling the communication conditions between the wireless nodes in the experiment according to the user-defined scenario, by applying the corresponding network degradation (packet loss, delay, bandwidth limitation) to the traffic being generated and received by the nodes. The latter task is done by means of the ipfw/dummynet link emulation tools (see Section 16.5.1 for more details).

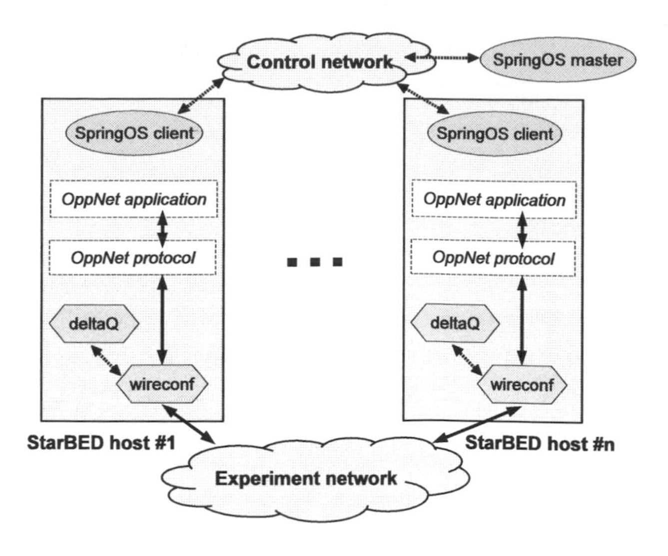

Figure 16.6: Architecture of the opportunistic network emulation framework.

16.4.3 Overall Architecture

In Figure 16.6 we show the architecture of the QOMB-based network emulation framework used in our opportunistic network experiments, created by combining the components mentioned so far. Note how SpringOS manages the experiment executed on StarBED hosts via the global SpringOS master and the local SpringOS client modules. These operations are done via the control network in StarBED. SpringOS also plays a role in experiment preparation, as indicated in Section 16.4.1.

The communication conditions between the emulated nodes are re-created in the experiment network by the wireconf library which interacts with the deltaQ library for condition computation purposes. The instances of the wireconf library on all the experiment hosts exchange information with each other in order to adjust the communication conditions in function of the continuously changing state of the experiment network, for instance, in order to take into account the contention created by the wireless nodes as they use the communication channel.

The traffic through the experiment network is generated by opportunistic network applications that we denoted generically by "OppNet application" in the figure. The applications run on top of their corresponding protocols, denoted by "OppNet protocol." Examples of such opportunistic network applications and protocols in the context of DTNs will be provided in Section 16.6.1.

16.5 Opportunistic Network Emulation

In this section we shall review several new features and components that were required in order to make opportunistic network emulation possible on QOMB.

16.5.1 Linux Support

In the original version of QOMET, the wireconf library was designed on top of the FreeBSD ipfw/dummynet system (in particular, the legacy version FreeBSD 5.4). However, many recent opportunistic network implementations, such as DTN2, are mainly implemented and supported on Linux. As a consequence, it was necessary to port the wireconf library to Linux.

For this purpose we took advantage of the fact that, as of version ipfw3, the ipfw system itself was ported to Linux by its developers [11]. Therefore, we made the necessary changes in the source code of the wireconf library so that it can be compiled and executed on top of ipfw3. The operating systems that we selected as targets for experiment execution are Scientific Linux 6.0 and CentOS 6.0, which are both free variants of RedHat Enterprise Linux (RHEL) 6.0.

The main changes required in wireconf in this context were:

- Modify the interface used by wireconf to enforce the scenario communication conditions, so that it can drive the new ipfw3 module.
- Modify the interface used by wireconf to obtain routing information from the kernel, so as to be able to perform this task on the new target OSs.

These modifications implied adapting the wireconf code so that it takes into account the changes in the ipfw3 source code compared to the previous version that was running on FreeBSD 5.4. As for the interface with the kernel, some modifications were required since kernel headers and structures are different between FreeBSD 5.4 and RHEL 6.0.

We stress in this context the fact that the other main component of QOMET, the deltaQ library, doesn't have similar limitations regarding execution, since it is mainly a computation engine. Hence, from the beginning, deltaQ could be executed both on FreeBSD and Linux, as well as on Windows or Mac OS X.

16.5.2 Multi-Interface Support

In previous QOMET versions, the user was limited to defining a single wireless interface per node in the emulation scenario. However, modern wireless systems, such as smartphones, typically have multiple wireless interfaces, for instance GSM/3G/4G and WLAN. As a consequence, we considered it important to provide the possibility to describe such experiments in QOMET.

This was accomplished by adding the "interface" element to the XML scenario description, which is the input of the deltaQ library. This description was introduced as a sub-element of the "node" element that existed previously and implicitly defined a single network interface. With the new "interface" element, users can define several network interfaces for each node in the scenario, each with its own properties (transmit power, receive sensitivity, etc.).

At this moment, we assume in the wireconf library that the experiment host on which a multi-interface node is emulated has several physical network interfaces available for experiments, and they are assigned in a one-to-one manner to the emulated interfaces. We envisage that in the future it may be possible to assign virtual network interfaces to the emulated interfaces of a certain node, thus allowing us to run experiments on hosts that have fewer physical interfaces than the emulated ones.

16.5.3 Node Mobility

As discussed in Section 16.3.1, mobility is an important characteristic of opportunistic networks. In order to provide more flexibility regarding mobility traces, we added support in QOMET for importing motion trajectories, such as those obtained from GPS devices during real-world experiments.

We implemented such support for importing mobility data by adding the possibility to use the QualNet trace format to specify this data. This format is text based and versatile, hence it can be easily created from any kind of motion data. Many mobility generators, such as BonnMotion [1], have built-in support for exporting mobility data in QualNet format, hence they can be used to create the input mobility trace for QOMET.

In our work we also used the ONE (Opportunistic Network Environment) simulator to generate mobility data [15]. In particular, for a project not discussed in this document, we employed the Shortest Path Map-Based Movement (SPMBM) model to generate realistic vehicular mobility using actual road data for Japanese cities. Therefore, we extended the ONE simulator with support for outputting motion traces in the QualNet format, so that the ONE-generated data can be used on our emulation testbed as well.

16.5.4 Fault Injection

Assessing the behavior of opportunistic network protocols and applications implies being able to create the necessary disruptions of the communication conditions. The challenges related to this aspect were discussed in detail in Section 16.3.2.

As a consequence we added support in QOMET so that such disruptions can be introduced not only through mobility or other similar high-level causes, but directly and on-demand, for basic assessments and debugging of the applications and protocols under test.

Support for both communication environment disruptions and network condition disruptions were added by extending the scenario specification language used in QOMET. In order to introduce communication environment disruptions, one can configure a node as a "noise source," and also specify the time interval in which the noise source is active. To support network condition disruptions, we added an element called "fixed_deltaQ" for each connection. This element has attributes that allow the user to specify the bandwidth, loss rate, and delay conditions, as well as the time interval in which they should be applied. Time-varying disruptions can be introduced by using multiple "fixed_deltaQ" elements.

16.6 Experimental Results

In this section we present a selection of experimental results that illustrate the practicality of our testbed. For this purpose we use two DTN implementations, DTN2 and IBR-DTN. DTN is a typical protocol for opportunistic networks, and our experiments will show how these two DTN implementations can be evaluated by utilizing our testbed.

16.6.1 DTN Implementations

DTN2 is the reference implementation of the *bundle protocol*, which is a general overlay network protocol (RFC 5050) representing the main focus of the implementation effort in the Delay Tolerant Networking Research Group (DTNRG) of the Internet Research Task Force (IRTF) [13].

The other DTN implementation we used is IBR-DTN, which is "a lightweight, modular and portable bundle protocol implementation and DTN daemon" developed by the Technical University, Braunschweig, Germany [24].

In both cases, the DTN protocol used in our experiments is represented by the module "dtnd" that must be running on all the nodes in the DTN. To configure the dtnd module we used the default parameters, with a few exceptions. In order to establish links between nodes we used the built-in discovery protocol. For multi-hop experiments we used two routing protocols:

- *flood*: Performs flood routing by sending a message to all the known neighbors of a node (i.e., through all the active links).
- *dtlsr*: A delay-tolerant link state routing implementation similar to OLSR (*only available in DTN2*).

The two DTN applications used in our experiments were:

- *dtnping*: An equivalent for the typical "ping" command, it measures the round-trip time (RTT) between issuing requests and receiving replies.
- *dtnperf*: An equivalent for the widely used "iperf" command for network performance assessment, that measures the duration and goodput for a data transfer (*only available in DTN2*).

16.6.2 Experiment Summary

In what follows we summarize our observations based on experiments done by running the DTN implementations mentioned above on our network emulation testbed. We performed the following series of experiments:

- *Basic experiments*: Simple 2- to 3-node tests that validate the basic functionality of the two DTN implementations.

- *Large-scale experiments*: Several 26-node tests that revealed certain scalability issues of DTN2, even though IBR-DTN behaved as expected.
- *Mid-scale experiments*: Several 10-node experiments that confirmed that at mid-scale, DTN2 performs according to expectations on our emulation testbed.

The simple 2- to 3-nodes scenarios confirmed that the two DTN implementations function as expected, and we noticed reasonable performance characteristics for dtnperf, especially for large bundle sizes. However, the RTT results shown by dtnping are typically one order of magnitude higher for DTN2 compared to IBR-DTN. Moreover, the performance of an older DTN2 version, dtn-2.7.0, exceeds that of the newer ones, dtn-2.8.0 and dtn-2.9.0. For IBR-DTN we used version 0.6.5.

The more complex experiments were done in urban scenarios in which we modeled human mobility using the behavioral motion model in QOMET. Our extensive evaluation led us to the following general conclusions that demonstrate the need to perform repeatable experiments with DTN applications and protocols.

DTN2 performance degrades quickly with scale, leading to poor results for scenarios with 26 nodes, even if only some of them are mobile. We believe that, at least partially, this low performance is caused by the low performance of dtnping itself, which causes high CPU utilization on all the participating DTN nodes, practically causing bottlenecks. Results are somewhat better for goodput measurements conducted using dtnperf. IBR-DTN always performed according to our expectations.

The implementation of flood routing generally behaved as expected, having good performance in sparsely connected networks, and poorer performance in good connectivity conditions and with many traffic sources. On the other hand, while dtlsr in DTN2 had relatively good performance in good connectivity conditions, the performance was poorer than expected in sparse networks and in the presence of mobility, which leads us to believe that the performance of the dtlsr implementation can be further improved.

16.6.3 DTN2 Performance

Our experimental results were discussed in more detail in [4] and [6]. For illustration purposes, we focus here on several experiments that show how the performance of one of the DTN implementations, namely DTN2, changes with experiment scale. This demonstrates the possible use of our testbed for performance assessment and optimization procedures of opportunistic network applications and protocols.

16.6.3.1 26-Node Experiments

The first series of experiments uses 26 nodes in an urban environment, including 25 mobile nodes as well as a fixed one acting as a gateway (GW0). The mobile nodes all start from the same initial position near the gateway, and move toward individual destinations (D01 to D25) within the 400 x 300 m area. The destinations are indicated in Figure 16.7, which shows a screenshot from this scenario at the end

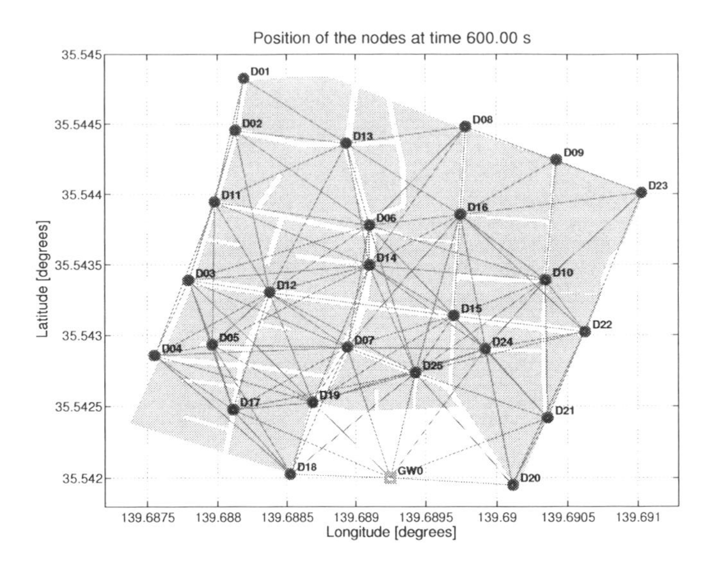

Figure 16.7: Urban mobility scenario for experiments with 26 nodes.

of the experiment; each experiment run lasted 10 minutes. We used either flood or dtlsr as routing protocols. Other conditions were: transmit power 10 dBm (802.11b), and attenuation 3.32. The interval between dtnping requests was 10 s.

We denote the above scenario series by the letter "A." We ran three types of dtnping experiments under these circumstances, as follows:

- *A-all*: All nodes (including the gateway) send dtnping to GW0.
- *A-5m*: A total of 5 mobile nodes (#1, #3, #6, #18, and #22) and the gateway send dtnping to GW0.
- *A-1m*: Only mobile node #1 and the gateway send dtnping to GW0.

We then simplified the above scenario to only include 5 mobile nodes, namely #1, #3, #6, #18, and #22, whereas all the other nodes are placed from the beginning at their corresponding destinations. We denote the simplified scenario by the letter "B." We conducted four types of dtnping experiments under these circumstances:

- *B-5m*: All the 5 mobile nodes and the gateway send dtnping to GW0.
- *B-1m*: Only the mobile node #1 and the gateway send dtnping to GW0.
- *B-5f*: A total of 5 fixed nodes, #8, #11, #15, #17, and #20, and the gateway send dtnping to GW0.
- *B-1f*: Only fixed node #8 and the gateway send dtnping to GW0.

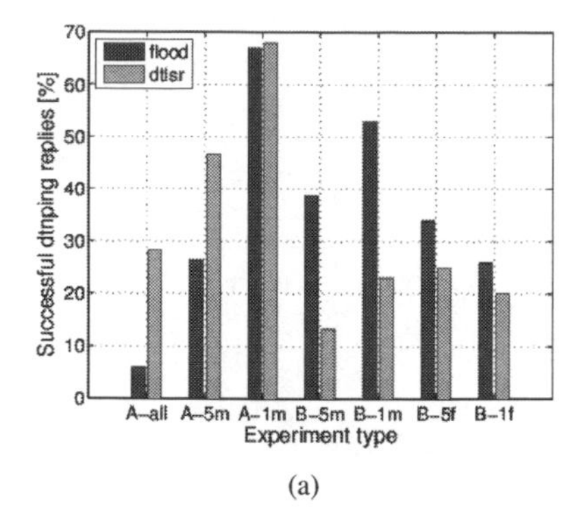

(a)

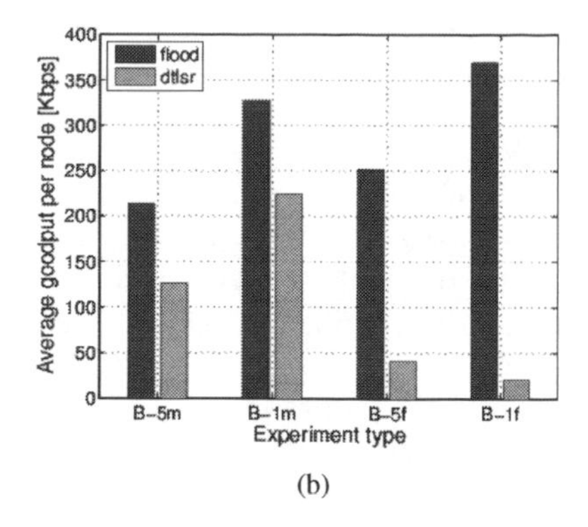

(b)

Figure 16.8: Results for flood and dtlsr routing in 26-node experiments: (a) Successful dtnping replies; (b) average dtnperf goodput per node.

The dtnping results of the A and B series of experiments discussed so far are summarized in Figure 16.8(a) for the two routing protocols, flood and dtlsr. For experiment A-all (25 nodes mobile, all nodes sending), the general performance is rather poor, especially for flood routing, which saturates the network, whereas dtlsr has a somewhat better but still low performance (almost 30% success rate). For experiment A-5m (25 nodes mobile, only 6 nodes sending) performance is more than tripled for "flood" and almost doubled for "dtlsr." In experiment A-1m (25 nodes mobile, only 2 nodes sending), performance is again improved, being more than double for "flood" when compared to A-5m; the differences between the two routing algorithms almost disappear, but the success rate is still under 70%. We stress that the expected success rate was around 100%, given the low request rate (once every 10 s) and the good connectivity of the nodes.

Going to the B series of experiments, we first observe that the differences between flood and dtslr are reversed, with the first one showing better performance in all cases. This difference is caused by the smaller number of mobile nodes compared to series A, and by the remote placement of the fixed nodes, which leads to a decrease of both the network overload caused by flooding and of the "density" of the topology that is constructed by dtlsr. Thus, for experiment B-5m (5 mobile nodes, 6 nodes sending), flooding leads to almost 40% dtnping success rate, whereas dtlsr only produced less than half of this. In experiment B-1m (5 mobile nodes, 2 nodes sending), performance increases, although dtlsr still constructs a sparse network topology. The results from experiments B-5f (5 mobile nodes, 6 fixed nodes sending) and B-1f (5 mobile nodes, 2 fixed nodes sending) are poorer compared to those of their counterparts B-5m and B-1m, respectively. This could be explained by the smaller number of connection opportunities of the fixed nodes compared to the mobile ones. The only exception is the fact that the success rate for dtlsr in experiment B-5m (around 13%) is lower than the corresponding rate in experiment B-5f (around 25%); this could be caused by the fact that it is easier for dtlsr to manage the topology for the fixed nodes compared to the mobile ones.

It is also interesting to compare similar tests, such as A-5m and B-5m, or A-1m and B-1m, in which the senders are the same, and the only difference is that 20 other nodes are moving (for series A) or are placed at their destinations (for series B). The results are significantly better for these particular series A experiments because of the initial period in the experiments in which the mobile nodes all start from the vicinity of the gateway interval during which they achieve direct communication for a period of 2–3 minutes.

For the B series of experiments, we also conducted an evaluation using dtnperf in the same conditions as above, except that the gateway GW0 was not sending any traffic to itself. The results in Figure 16.8(b) show the same trends as before, and in a certain sense mirror the dtnping results, with performance being lower for dtlsr compared to flood. Similarly, performance in B-1m (1 mobile sender) is better than in B-5m (5 mobile senders). However, for flood routing goodput in B-5f and B-1f is better than that in B-5m and B-1m, respectively; moreover, throughput is very low for dtlsr. We interpret these results as showing that while flooding manages to find good routes in static conditions, dtslr fails to do so optimally. We do emphasize that dtnping and dtnperf results cannot be directly compared, since dtnping is a two-directional test whose performance is evaluated as message counts, whereas dtnperf is mainly uni-directional (albeit with bundle receive confirmations traveling in the opposite direction) and is evaluated via throughput.

16.6.3.2 10-Node Experiments

Given the generally poor performance of DTN2 in the 26-node experiments, for a third series of tests we decreased the scale to a total number of 10 nodes. One of them is the gateway GW0; among the other nodes, 3 are mobile (#6, #18, and #22) and the rest are fixed. Note that for clarity sake we use the same indexes as in the experiment series A and B above, even if the node count decreases. These new experiment runs lasted for about 7 minutes each; a snapshot obtained at the end of the experiment is shown in Figure 16.9. All the other conditions are the same as in the experiments above.

The new experiment series was denoted by “C,” as follows:

- *C-all*: All the nodes including the gateway send dtnping to GW0.
- *C-3m*: Only the three mobile nodes and the gateway send dtnping to GW0.
- *C-3f*: Only three fixed nodes (#15, #17, #20) and the gateway send dtnping to GW0.

The dtnping results presented in Figure 16.10(a) show that the performance of the flood routing protocol is relatively low when there are 10 senders, as the success rate is only around 40% (experiment C-all). One may attribute this low performance to the overload caused by the packet flood, but this is not expected given the low rate of requests (one request message per 10 s from each node, hence a total average of 1

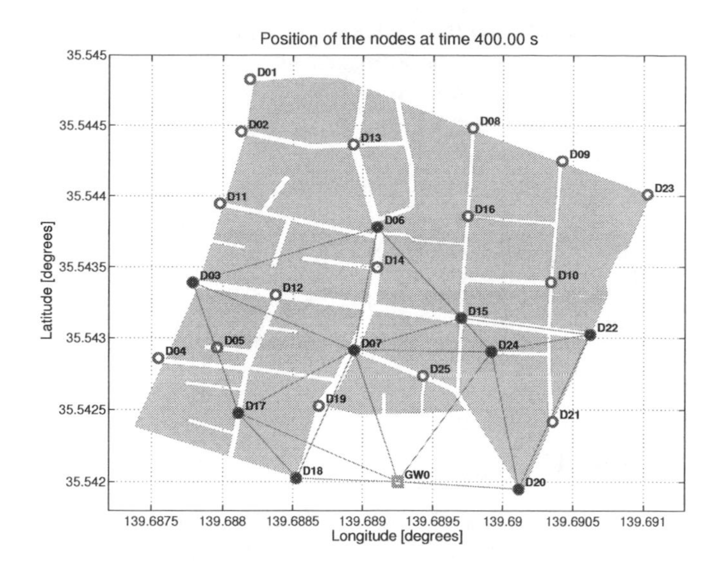

Figure 16.9: Urban mobility scenario for experiments with 10 nodes.

request per second). However, a more important reason for this poor performance is the CPU overload caused by dtping itself in the case with 10 senders. Nevertheless, the behavior for the dtlsr case is much closer to our expectations, with success rates exceeding 90%, presumably because of the lower network load observed in this case. When the number of senders decreases to a total of 4 (experiments C-3m and C-3f),

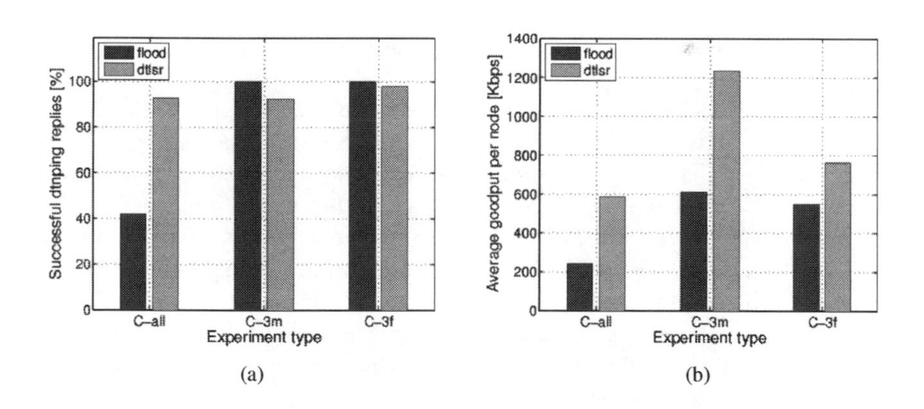

Figure 16.10: Results for flood and dtlsr routing in 10-node experiments: (a) successful dtnping replies; (b) average dtnperf goodput per node.

both for flood and dtlsr routing the results are close to 100%, emphasizing the fact that for low network loads DTN2 performance is according to expectations.

Using the same conditions, we also did experiments in which we replaced dtnping with dtnperf (and node GW0 did not send any traffic to itself). The results shown in Figure 16.10(b) confirm our expectations regarding the superiority of dtlsr over flood for each experiment type, as flood is known to cause network overload, and hence decrease the overall network capacity. Results of experiment C-3m are better than for C-all, as expected given the reduced number of nodes sending traffic. As for C-3f, the fixed placement of the three senders seems to have a negative impact on performance when compared to C-3m, although this impact is strongest for dtlsr, as we have noticed before.

16.7 Related Work

Although to the best of our knowledge no equivalent to our opportunistic network emulation testbed exists, we mention below some of the related projects with focus on the field of DTNs.

DTN implementations have been previously evaluated, but only at a low scale. For instance, in [14] two scenarios with one sender, one receiver, and up to 4 hops are compared through emulation experiments done on Emulab. Other works focus on space communication, such as the results presented in [26] for a total of 3 nodes, by again using emulation. The work in [23] uses 4 real wireless nodes for the evaluation, and includes a low-level performance analysis. One main characteristic of our work is that we address large-scale scenarios, and in this document we have presented experiments with more than 25 nodes.

On the other hand, there is a considerable number of works that do DTN evaluations at large scale through simulation, such as [20], which used a realistic scenario with 600 vehicles. By contrast, our testbed allows one to assess performance of *actual* DTN implementations, and hence leads to results that have direct practical application.

DTN2 and IBR-DTN are not the only current implementations of the DTN bundle protocol, and several other implementations exist. Some groups have started from earlier versions of DTN2, and have rewritten the code, as is the case for SPINDLE, developed at BBN under the DARPA DTN program [17]. Others implementations are completely independent, while still being based on the same RFC 5050. For instance, ION is a project developed by the Jet Propulsion Laboratory and partially maintained by Ohio University [18].

A DTN testbed using real nodes is DTN-Bone, described as an "effort to establish a worldwide collection of nodes running DTN bundle agents and applications" [12]. This testbed currently connects around 9 institutions, but makes available only about a dozen nodes. Hence, we position DTN-Bone more as an inter-operability platform, given that 5 different implementations of DTN are being run on it, rather than a testbed for DTN performance testing *per se*. Another DTN testbed presented in [16] includes 12 geographically spread nodes. Our testbed has the potential resources for experiments with hundred of nodes, and it is not limited to specific DTN applications or protocols.

16.8 Conclusion

In this chapter we presented an emulation testbed intended for mobile opportunistic network application and protocol experiments. We discussed in detail the challenges involved in designing and implementing this testbed, as well as the main changes that were necessary in order to make possible such experiments when basing our work on the QOMB wireless network emulation testbed.

We employed this emulation testbed for a series of experiments that assessed the performance characteristics of the DTN2 and IBR-DTN implementations of the DTN bundle protocol. We have shown that the DTN2 implementation exhibits poor performance in scenarios with 26 nodes, even in cases when only 5 of those nodes are mobile. Nevertheless, for smaller 10-node scenarios—with lower mobility and a reduced number of senders—performance is closer to our expectations. The IBR-DTN implementation performed as expected in all the tested scenarios. These results emphasize the need to perform repeatable large-scale experiments with DTN applications and protocols in order to validate their behavior before deployment in a wide range of conditions.

Our experiments demonstrated how the emulation testbed that we designed and implemented can be used for a thorough assessment of the performance characteristics of opportunistic network applications and protocols. This usage also makes possible performance optimization procedures. Thus, researchers can identify performance bottlenecks through precise controlled experiments, and then test the improved implementation in exactly the same scenarios so as to validate that the problems were fixed. Such performance testing is mandatory should one wish to apply the opportunistic network paradigm to everyday network scenarios, in which node count is large and network conditions are difficult to predict.

References

[1] N. Aschenbruck, R. Ernst, E. Gerhards-Padillar, and M. Schwamborn. BonnMotion: A mobility scenario generation and analysis tool. In *Proceedings of the 3rd International ICST Conference on Simulation Tools and Techniques*, pages 1–10, 2010.

[2] R. Beuran. *Introduction to Network Emulation*. Pan Stanford Publishing, 2012.

[3] R. Beuran, S. Miwa, and Y. Shinoda. Making the best of two worlds: A framework for hybrid experiments. In *Proceedings of the 7th ACM International Workshop on Wireless Network Testbeds, Experimental Evaluation and Characterization (WiNTECH 2012), in Conjunction with MobiCom 2012*, pages 75–81, 2012.

[4] R. Beuran, S. Miwa, and Y. Shinoda. Performance evaluation of DTN implementations on a large-scale network emulation testbed. In *Proceedings of the 7th ACM International Workshop on Challenged Networks (CHANTS 2012), in Conjunction with MobiCom 2012*, pages 39–42, 2012.

[5] R. Beuran, S. Miwa, and Y. Shinoda. Behavioral mobility model with geographic constraints. In *Proceedings of the 9th IEEE International Workshop on Heterogeneous Wireless Networks (HWISE 2013), in Conjunction with AINA 2013*, pages 470–477, 2013.

[6] R. Beuran, S. Miwa, and Y. Shinoda. Network emulation testbed for DTN applications and protocols. In *Proceedings of the 16th IEEE Global Internet Symposium (GI 2013), in Conjunction with INFOCOM 2013*, pages 3607–3612, 2013.

[7] R. Beuran, J. Nakata, T. Okada, L. T. Nguyen, Y. Tan, and Y. Shinoda. A multipurpose wireless network emulator: QOMET. In *Proceedings of the 22nd IEEE International Conference on Advanced Information Networking and Applications (AINA 2008) Workshops, FINA 2008 Symposium*, pages 223–228, 2008.

[8] R. Beuran, J. Nakata, T. Okada, Y. Tan, and Y. Shinoda. Behavioral mobility model with geographic constraints. In *Proceedings of the 1st International Conference on Simulation Tools and Techniques for Communications, Networks and Systems (SIMUTools 2008)*, 2008.

[9] R. Beuran, J. Nakata, Y. Tan, and Y. Shinoda. Emulation testbed for IEEE 802.15.4 Networked Systems. *IEICE Transactions on Communications*, E95-B(9):2892–2905, 2012.

[10] R. Beuran, L. T. Nguyen, T. Miyachi, J. Nakata, K. Chinen, Y. Tan, and Y. Shinoda. QOMB: A wireless network emulation testbed. In *Proceedings of the IEEE Global Communications Conference (GLOBECOM 2009)*, 2009.

[11] M. Carbone and L. Rizzo. Dummynet revisited. *ACM SIGCOMM Computer Communication Review*, 40(2):12–20, 2010.

[12] Internet Research Task Force Delay-Tolerant Network Research Group. DTN-Bone website: http://dtnrg.org/wiki/DtnBone.

[13] Delay-Tolerant Network Research Group, Internet Research Task Force. DT-NRG website: http://www.dtnrg.org/wiki.

[14] M. Demmer, E. Brewer, K. Fall, S. Jain, M. Ho, and R. Patra. *Implementing Delay Tolerant Networking*. Intel Research Berkley Technical Report, IRB-TR-04-020, 2004.

[15] Ari Keränen, Jörg Ott, and Teemu Kärkkäinen. The ONE simulator for DTN protocol evaluation. In *Proceedings of the 2nd International Conference on Simulation Tools and Techniques (SIMUTools '09)*, 2009.

[16] E. Koutsogiannis, S. Diamantopoulos, G. Papastergiou, I. Komnios, A. Aggelis, and N. Peccia. Experiences from architecting a DTN testbed. *Journal of Internet Engineering*, 3(1):219–229, 2009.

[17] R. Krishnan. The SPINDLE disruption-tolerant networking system. In *Proceedings of MILCOM 2007*, 2007.

[18] H. Kruse. Interplanetary overlay, Website: https://ion.ocp.ohiou.edu/.

[19] L. Lilien, Z. H. Kamal, V. Bhuse, and A. Gupta. Opportunistic networks: The concept and research challenges in privacy and security. In *Proceedings of NSF Intl. Workshop on Research Challenges in Security and Privacy for Mobile and Wireless Networks (WSPWN 2006)*, 2006.

[20] P. Luo, H. Huang, W. Shu, M. Li, and M. Wu. Performance evaluation of vehicular DTN routing under realistic mobility models. In *Proceedings of IEEE Wireless Communications and Networking Conference (WCNC 2008)*, pages 2206–2211, 2008.

[21] T. Miyachi, R. Beuran, S. Miwa, Y. Makino, S. Uda, Y. Tan, and Y. Shinoda. Fault injection on a large-scale network testbed. In *Proceedings of the 7th Asian Internet Engineering Conference (AINTEC 2011)*, pages 4–11, 2011.

[22] T. Miyachi, K. Chinen, and Y. Shinoda. StarBED and SpringOS: Large-scale general purpose network testbed and supporting software. In *Proceedings of the Intl. Conf. on Performance Evaluation Methodologies and Tools (Valuetools 2006)*, 2006.

[23] E. Oliver and H. Falaki. Performance evaluation and analysis of delay tolerant networking. In *Proceedings of ACM/USENIX Conference on Mobile Systems, Applications, and Services (MobiSys 2007), Workshop on System Evaluation for Mobile Platforms (MobiEval)*, 2007.

[24] S. Schildt, J. Morgenroth, W.-B. Pottner, and L. Wolf. IBR-DTN: A lightweight, modular and highly portable Bundle Protocol implementation. *Electronic Communications of EASST*, 37:1–10, 2011.

[25] Y. Teranishi. PIAX: Toward a framework for sensor overlay network. In *Proceedings of the 6th IEEE Conference on Consumer Communications and Networking Conference*, pages 1212–1216, 2009.

[26] R. Wang, X. Wu, Q. Zhang, T. Taleb, Z. Zhang, and J. Hou. Experimental evaluation of TCP-based DTN for cislunar communications in presence of long link disruption. *EURASIP Journal on Wireless Communications and Networking*, 2011.

Index

D